CYTOPLASMIC GENES AND ORGANELLES

CYTOPLASMIC GENES AND ORGANELLES

Ruth Sager

Department of Biological Sciences
Hunter College
New York, New York

ACADEMIC PRESS New York and London

ACADEMIC PRESS, INC.
111 Fifth Avenue, New York, New York 10003

United Kingdom Edition published by
ACADEMIC PRESS, INC. (LONDON) LTD.
24/28 Oval Road, London NW1 7DD

LIBRARY OF CONGRESS CATALOG CARD NUMBER: 71-182609

PRINTED IN THE UNITED STATES OF AMERICA

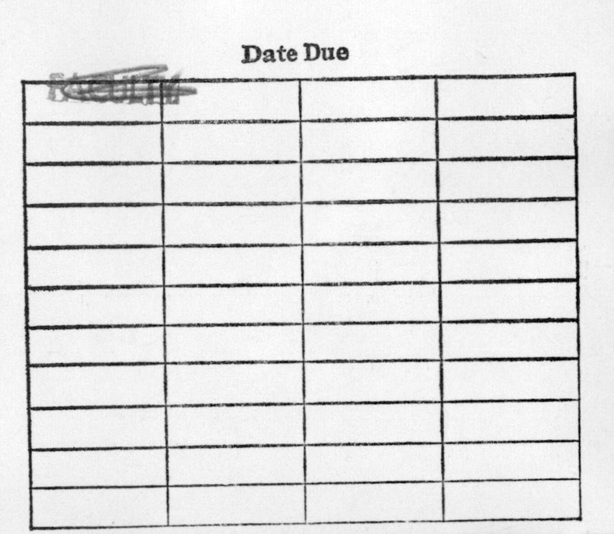
Date Due

NEW MEXICO
EASTERN
UNIVERSITY
E
N
M
U
LIBRARY

To Adam

Contents

GENETIC ANALYSIS OF CYTOPLASMIC SYSTEMS

3. Cytoplasmic Genes in Chlamydomonas

4. Mitochondrial Genetics of Yeast

5. Cytoplasmic Genes in Neurospora and Other Fungi

6. Cytoplasmic Genes in Higher Plants

9. Cytoplasmic Genes and Cell Heredity

Appendix

Preface

Cytoplasmic genes are an essential part of the total genetic endowment of all eukaryotic cells. Their existence was first described in the early 1900's, and their association with cytoplasmic organelles was clearly perceived at that time. It is only now, some sixty years later, that experimental systems have been developed with which to pursue the relationships between cytoplasmic genes, cytoplasmic DNA's, and cytoplasmic organelles.

We are still on the threshold of knowledge about cytoplasmic genes. This very lack of knowledge presages a fruitful future, for it indicates a crucial area of ignorance in our understanding of cellular and molecular biology. Present evidence suggests that cytoplasmic genes are centrally concerned with organelle function and membrane specificity, two areas of outstanding importance for the understanding of normality and of disease.

The principal aim of this book is to apply the concepts and methods of cytoplasmic genetics to the problems of cell and molecular biology to which they can uniquely contribute. I have tried to show geneticists the many attractive problems in this area awaiting their attention; I have tried to show cell biologists and biochemists the usefulness of cytoplasmic genetic analysis in their endeavors; but above all, I have tried to delineate for the student the potential power of an integrated experimental approach using cytoplasmic genes together with the more conventional tools of biochemistry and electron microscopy in the investigation of organelle biogenesis.

It is now clearer than ever that in the analysis of organelle biogenesis and regulation investigators need the incisive analytical tool that mutations provide. The genetic control of organelle formation and function is turning out to involve a tightly intermeshed network of nuclear and organelle gene products and signals, the complexity of which is just beginning to be revealed. In this circumstance, the availability of genetically mapped and biochemically defined cytoplasmic mutants may be expected to contribute enormously to the dissection of the network of biogenesis, even more than mutants are contributing, for example, to the analysis of phage morphogenesis.

Furthermore, the possibility remains open that further classes of cytoplasmic genes may exist, beyond those in chloroplasts and mitochondria, on DNA's associated with other cytoplasmic organelles, structures, or membrane systems. Indeed, if such DNA's do exist, mutations may provide a more powerful means for their identification than the direct physical isolation of the DNA's themselves.

Ruth Sager

Acknowledgments

This book was written largely during summers spent in the superlative library of the Marine Biological Laboratory, Woods Hole, Massachusetts. I wish to express my appreciation to the library staff and to the Trustees who have over the years cherished and maintained this library.

Our recent research efforts, discussed in this book, have been supported by the National Institutes of Health, the American Cancer Society, and the Research Foundation of the City University of New York, to all of whom I feel deeply indebted. I am above all thankful to the National Institutes of Health for their continuous support since 1955, support which made these studies of cytoplasmic inheritance possible.

Institutions, libraries, and financial support provide the essential framework, but research is a human effort, and its success depends upon the abilities and above all upon the spirit of the participants. I wish to express my gratitude to my associate and friend over the years, Mrs. Zenta Ramanis, whose unflagging enthusiasm for cytoplasmic genetics, great competence, and calm persistence have been such crucial factors in our progress; and to Dorothy Lane for her willingness to tackle a very difficult and very important problem, the molecular basis of maternal inheritance. I am grateful to everyone who has contributed at one time or another to the work of our laboratory.

Many colleagues helped generously with unpublished data, interpretations, and criticism. I wish especially to thank W. Bertsch, L. Bogorad, H. O. Halvorsen, H. Mahler, J. Marmur, A. B. Pardee, T. Pittenger, M. M. Rhoades, G. Schatz, G. Schlanger, and R. A. E. Tilney-Bassett for the critical reading of particular chapters; and G. Attardi, A. L. Colwin, R. H. Kirschner, E. Racker, D. L. Ringo, H. Swift, E. F. J. Van Bruggen, J. Vinograd, and D. von Wettstein for the illustrations they provided. My special thanks are due to G. E. Palade for the beautiful electron micrographs of *Chlamydomonas,* including the one on the book jacket. I would also like to thank Maureen Jones for the rendition of many of the original linecuts. I am indebted to the staff of Academic Press for their efficiency and cooperation.

1

What Are Cytoplasmic Genes

This book is about cytoplasmic genes: what they are and what they do. The principal cytoplasmic genes we know about at this time are located in cytoplasmic organelles: chloroplasts and mitochondria. These organelles, which are the basic energy-generating elements of the cell, contain not only their own unique DNA's, but also unique RNA's, enzymes, and ribosomes for transcription and protein synthesis.

The existence of cytoplasmic genes was suggested as long ago as 1909 when the first examples of non-Mendelian inheritance were described by Correns and by Baur (cf. Chapter 6). Nonetheless, most of our knowledge about these genes comes from recent investigations. Direct evidence for the existence of cytoplasmic genes comes from the genetic analysis of non-Mendelian mutations; and this evidence is supported by the presence in organelles of DNA and the machinery for its transcription and translation.

Historically, the idea that chloroplasts and mitochondria might be autonomous organelles goes back to the early 1900's. However, genetic evidence of non-Mendelian genes accumulated slowly between 1910 and 1960, a period when the great advances in Mendelian genetics and then in molecular genetics were being made and consolidated. During this period, the study of non-Mendelian genes was fervently pursued by a few groups of investigators, but ignored or discounted with equal fervor by most geneticists. Indeed, the literature of cytoplasmic genetics was viewed more as a blot on the escutcheon of the science than as part of a more comprehensive genetic theory.

Slowly, the tide of opinion began to shift. In general, European geneticists were more receptive to the concept of cytoplasmic genes than were the Americans. However, in the 1940's several studies of cytoplasmic inheritance, especially those by M. M. Rhoades in the corn plant, *Zea mays,* influenced American geneticists to admit the possibility of some genetic autonomy in the cytoplasm. Plant breeders began to use cytoplasmically inherited pollen sterility extensively in hybrid seed production without much concern about the basic significance of the phenomenon. Then, with the growing popularity of microbial systems for genetic analysis, cytoplasmic genes were discovered in the algae, fungi, and yeast.

The first systematic investigation of cytoplasmic genetics in a microbial system, beginning with the development of a suitable mutagen and the collection of a stockpile of cytoplasmic mutations, was carried out with the green alga *Chlamydomonas.* This work led to the demonstration of recombination and linkage of cytoplasmic genes, and, subsequently, to the recognition of the first cytoplasmic linkage group or "chromosome." It should be noted that these genetic studies were well underway prior to the identification of cytoplasmic DNA's. As with genetic analysis in other systems, studies of cytoplasmic genes can proceed by methods which are operationally independent of the existence of DNA.

Thus, it is particularly interesting, from the viewpoint of the development of scientific ideas, that the discovery of cytoplasmic DNA's in chloroplasts and mitochondria played such a central role in the acceptance of cytoplasmic genes as fact and not artifact. The pendulum of opinion had swung from one extreme—cytoplasmic genes do not exist because we do not see cytoplasmic chromosomes—to the other extreme—cytoplasmic DNA's exist, and therefore there must be cytoplasmic genes.

Clearly both extremes are formally invalid because genetic evidence depends upon genetic methods and is in that sense a self-contained science, whereas the presence of DNA does not itself demonstrate its informational content. Nonetheless, both extreme positions have validity in a looser and more intuitive sense. The evidence for the existence of cytoplasmic genes was purely formal, unsupported by any independent evidence of a physical carrier of these genes, whereas the mere presence of DNA urged its genetic role.

All that is history. At present, investigators of cytoplasmic genes are striving to collate evidence from genetic, biophysical, and biochemical studies of mutant strains and their DNA's in an effort to use whatever knowledge can be brought to bear in the analysis of these systems.

The existence of unique genetic systems in organelles means that all organisms, from the green algae, yeast, and fungi to the higher plants and animals, contain cytoplasmic genes. Indeed organelle genetic systems are a fundamental feature of the organization of all eukaryotic cells, i.e., cells with true nuclei.

Thus we must face a more complex situation than had previously been appreciated: the interaction in the cell of multiple genetic systems, at least two in the animals and three in the plants. Before we can begin to analyze the interactions among these systems, obviously we need some understanding of the properties of each one alone. This book will be concerned primarily with bringing together the available knowledge about cytoplasmic genes themselves, emphasizing methods for studying their properties and functions. This knowledge is a prelude to the research that lies ahead: investigating the mechanisms and consequences of interactions between nuclear and cytoplasmic genetic systems in the cell. It should also be kept in mind that cells may contain cytoplasmic genes in locations other than those already identified. Speculation on this possibility will be discussed in Chapter 9.

Cytoplasmic genes are being investigated by the same methods that have been developed for the analysis of nuclear and bacterial systems: genetic analysis utilizing mutants for the detection of recombination and linkage and for mapping; biophysical studies of the corresponding DNA's; and biochemical studies of transcription, translation, and the macromolecular consequences of individual mutations.

In this book, the following aspects of cytoplasmic genetic systems will be treated: (1) the properties of cytoplasmic DNA's, (2) the genetic analysis of cytoplasmic systems, and (3) the functions of cytoplasmic genes in organelle biogenesis. In this first chapter the principal findings will be summarized to provide the reader with a bird's eye view of the subject. We will begin with DNA because its properties and its size set the boundary conditions for its genetic role.

CYTOPLASMIC DNA'S

Cytoplasmic DNA's of chloroplasts and mitochondria were first characterized as double-stranded DNA's of high molecular weight and distinctive base composition by extracting them from isolated and purified organelles. Differences in nucleotide composition between organelle and nuclear DNA's from the same cells (in some organisms) made their separation relatively easy and also provided the first line of evidence that the genetic identity of the two classes might be different.

All organelle DNA's so far examined are double-stranded, and most of them are circular or include circular forms. The prevalence of circular DNA's suggests that this form is of fundamental biological importance, in agreement with the evidence from bacterial and viral systems. Circles are also extremely useful to the investigator, permitting special techniques for their isolation and making possible accurate measurements of their size.

Mitochondrial DNA's from animal cells are found primarily in the circular form, averaging about 1×10^7 daltons in molecular weight. This size corresponds to a coding potential for about 15,000 amino acids, or about 100 proteins. When allowance is made for the transcription of mitochondrial ribosomal RNA's and tRNA's, enough coding potential remains for some twenty to thirty proteins. This value represents the minimum amount of genetic information in any of the known cytoplasmic DNA's. The mitochondrial DNA's of eukaryotic microorganisms like yeast, *Neurospora,* and *Tetrahymena* are some three to five times the size of those from animal cells; the size of plant mitochondrial DNA's is not known.

Chloroplast DNA's are much larger than the known mitochondrial DNA's. Individual chloroplasts of algae like *Chlamydomonas* and *Euglena* contain about as much DNA as does a bacterial genome: some $1–4 \times 10^9$ daltons. However, recent evidence from measurements of reannealing kinetics indicates considerable redundancy in these DNA's, so their informational content may be only about $1–2 \times 10^8$ daltons. This amount is still considerably larger than that of mitochondrial DNA, providing a coding potential for several hundred proteins.

These findings suggest that chloroplasts contain a substantially larger genome than do mitochondria, especially mitochondria of animal cells. Indeed, the mitochondrial DNA of animal cells seems to have reached an irreducible minimal value.

If, as it appears, the chloroplast genome is larger than that of mitochondria, the chloroplast may be the organelle of choice for a detailed analysis of organelle genetics. Already, the methods developed with *Chlamydomonas* have been useful in studies of mitochondrial genetics in yeast. In both systems, the dire predictions that organelle genetics would be impossible to study because of the presence of many copies of the genome have not been borne out experimentally. For this reason alone it is fortunate that genetic analysis preceded studies of organelle DNA's and proceeded on its own methodology.

FORMAL GENETICS OF CYTOPLASMIC SYSTEMS

Genetic analysis of cytoplasmic systems began with studies of higher plants. Following the initial discoveries by Correns and by Baur in 1909, extensive studies in the succeeding decades were carried out with a few plants, in particular, *Oenothera* and *Epilobium*. The results provided ample evidence of non-Mendelian genes influencing chloroplast development, pollen sterility, and a host of other morphogenetic properties. However, in no instance was any evidence adduced of linkage or linked recombination, and consequently no further genetic analysis was achieved, beyond recognition of many phenotypes under non-Mendelian genetic influence.

Further progress seemed to require a simple organism with a fast sexual life cycle, biochemically distinct mutant phenotypes, and, above all, a chloroplast, so that the most widely known class of cytoplasmic mutations, those affecting chloroplasts, could be studied. These criteria led to the choice of *Chlamydomonas* as a model organism for the investigation of cytoplasmic genetics in chloroplasts and wherever else in the cell non-Mendelian mutations might lead us.

The studies with *Chlamydomonas* led to the collection of a large number of mutations, each exhibiting the same pattern of non-Mendelian uniparental transmission. The phenotypes of these mutant strains included loss of ability to grow photosynthetically, poor growth on all media, temperature sensitivity (i.e., ability to grow at 25°C but not at 35°C), and resistance to a number of different antibiotics, each known to inhibit protein synthesis by bacterial (and chloroplast) ribosomes. All of the mutations so far studied have been found to lie within a single cytoplasmic linkage group or "chromosome," which on the basis of extensive indirect evidence, is located in chloroplast DNA.

Vegetative cells of *Chlamydomonas* are genetically diploid for this linkage group, and genetic analysis with multiply marked crosses has revealed regular distribution at cell division, extensive exchanges between homologs leading to recombination (frequently during vegetative multiplication but rarely in meiosis), and circularity as judged from genetic evidence. These genetic studies, discussed in Chapter 3, have provided methods to identify cytoplasmic genes, to distinguish mutations in different genes, to locate genes on particular DNA's (e.g., chloroplast or mitochondrial), and to correlate the behavior of the organelle DNA in biophysical studies with its inferred behavior from genetic analysis.

Parallel with these studies, Ephrussi and his students, especially

Slonimski, initiated an intensive investigation of the *petite* mutation in yeast, a cytoplasmically inherited change responsible for loss of mitochondrial function. The system was investigated with great thoroughness, yielding knowledge of basic importance on mitochondrial biochemistry and biogenesis, but no other cytoplasmic mutations were found. Then, with the availability of methods to examine mitochondrial DNA, Slonimski and his students found that the mitochondrial DNA of *petites* was substantially altered from that of wild-type cells, leading to large losses in genetic coding potential. Subsequently, it was found that some *petite* strains have no mitochondrial DNA whatsoever. No wonder no further mitochondrial mutations were found in *petites.*

A new era in mitochondrial genetics dawned with the isolation by Wilkie and Linnane of a new class of non-Mendelian mutations from wild-type yeast. These mutations to antibiotic resistance, analogous to similar mutations in *Chlamydomonas,* have been located in mitochondrial DNA, as judged by the fact that they are wiped out when wild-type drug-resistant strains become *petite.* The mapping of these mutant strains, now in full swing in several laboratories, is providing us with a new cytoplasmic genetics of mitochondrial systems, as discussed in Chapter 4.

These studies of mitochondrial genetics in yeast provide a conceptual framework for evaluating the related but much less extensive studies of cytoplasmic genes in fungi.

Studies with *Neurospora* have given further evidence of the activity of mitochondrial genes, as well as new methods for their investigation, i.e., mycelial fusion, hyphal tip isolation, and microinjection. In addition to providing further evidence of mitochondrial genes, the studies with fungi presented in Chapter 5 pose problems in cytoplasmic heredity that were not encountered in the *Chlamydomonas* or yeast systems. For instance, the humanly intriguing phenomenon of senescence or *aging* was shown to be regulated by a cytoplasmically transmitted genetic factor in the ascomycete *Podospora,* a relative of *Neurospora.* Other properties such as differentiation of fruiting bodies, growth rate, and incompatibility were also shown to be influenced by cytoplasmically transmitted factors. However, the intercellular location of these factors remains obscure, and the question of whether they represent DNA-based cytoplasmic genes or not remains unanswered.

Thus, in Chapter 5 we touch for the first time in the book upon examples of cytoplasmic heredity that have not yet been sufficiently analyzed to permit a decision or even a strong judgment on their identity. Many more examples are known of what we may call cytoplasmic *heredity,* not sufficiently understood to be part of cytoplasmic *genetics.* These phenomena include many characteristics of *Paramecium* and related protozoa, such as the inheritance of serotypes, of mating types, and of other properties showing many generations of clonal inheritance but no direct basis in a permanent cytoplasmic DNA. These phenomena, while outside of the main theme of this book, may be of great importance as model systems in the analysis of differentiation, and as such will be discussed under the heading of "epigenetic phenomena" in Chapter 9.

Our discussion in Chapter 6 of cytoplasmic genes in higher plants ranges from the early studies of Correns, Baur, Renner, Michaelis, and others, who together laid the foundation of this field, to the current studies of biparental non-Mendelian inheritance in *Pelargonium* and *Oenothera.*

Of special importance are the impeccable studies of cytoplasmic inheritance in maize by M. M. Rhoades who established the cytoplasmic genetic basis not only of a chloroplast abnormality, but also of a mutation causing pollen sterility in maize. Both studies influenced geneticists' ideas about cytoplasmic heredity, and the work on cytoplasmic pollen sterility also had a far-reaching effect on agriculture, leading to the incorporation of cytoplasmic genes for pollen sterility into the inbred lines used in hybrid corn seed production. The genetics of hybrid seed production and some unexpected consequences of using cytoplasmic pollen sterility will be discussed in Chapter 6.

Two patterns of transmission of cytoplasmic genes have been recognized in the higher plants: one is strictly maternal, and is found in maize and in most plants so far examined; the other is biparental, with aberrant non-Mendelian ratios but some transmission of cytoplasmic genes from both parents, seen typically in *Pelargonium* and *Oenothera.* With the recognition of recombination of chloroplast and mitochondrial genes in *Chlamydomonas* and yeast, it would seem reasonable to look for recombination of cytoplasmic genes in a higher plant.

In terms of evolution, the relation between cytoplasmic genes and maternal inheritance, or more generally stated, the preferential transmission of cytoplasmic genes from one parent, poses a fascinating puzzle. Preferential transmission occurs in every cytoplasmic genetic system that has been described. Thus it appears to be a fundamental property of these systems. A genetic consequence of preferential transmission is the infrequency of any opportunities for recombination.

Thus, preferential transmission inhibits the formation of cytoplasmic heterozygotes or *cytohets,* and when they do manage to appear, they are quickly dispersed by means of somatic segregation. It seems clear, then, that elaborate mechanisms have appeared in the course of evolution to minimize the occurrence of recombination of cytoplasmic genes.

In summary, the genetic analysis of cytoplasmic systems has revealed the presence of elaborate, well-integrated sets of cytoplasmic genes present in organelle DNA's that show great stability, permanence, and importance in survival and evolution of eukaryotic microbes and higher organisms.

Given the presence of these well-established cytoplasmic genomes, what do they do? The latter part of the book, Chapters 7 and 8, considers the role of cytoplasmic genes in the biogenesis of mitochondria and chloroplasts. As the reader will quickly discover, very little is known about the functions of organelle genes, and most of the evidence presented in these two chapters is very indirect.

The most direct approach to this problem would be to identify the products of transcription and translation of organelle DNA's. A beginning has been made in this direction with the hybridization of organelle RNA's and the corresponding organelle DNA's. Ribosomal and transfer RNA's present in chloroplasts and mitochondria have been identified as transcripts of organelle DNA's. However, the method is not powerful enough to use in the identification of individual messenger RNA's. Consequently, to find out which proteins are the products of organelle genes, one must examine the proteins themselves.

The principal experimental approaches that have been tried in an effort to identify specific proteins synthesized in organelles have not been notably successful. Studies with isolated organelles have shown low rates of incorporation of labeled amino acids. In such experiments, radioisotope label has been found only in the membrane fraction in mitochondria whereas in chloroplasts, both membrane and soluble fractions were labeled. However, the incorporation rates were so low that the results were not very meaningful. Essential cofactors may have been lost during organelle isolation.

Turning to intact cells, investigators have devised various techniques to distinguish protein synthesis in the cytoplasm from that occurring within the organelle system. These techniques, such as the use of antibiotics and shifts in growing conditions, each have their pitfalls, and the conclusions from these lines of experimentation have not been very persuasive. In general, the picture that emerges with respect to mitochondria, whether from yeast or from animal cells, is that as much as 95% of the proteins of the organelle are of cytoplasmic origin. The situation is somewhat different in chloroplasts, where one of the major soluble proteins, RuDP-carboxylase, may be synthesized in the chloroplast.

Synthesis within the organelle is of course not conclusive evidence that the protein was coded by organelle DNA. The possibility must be considered that messenger RNA's may move across organelle membranes, just as proteins do. Thus the presence of a protein within an organelle does not mean it was made there, and its synthesis within the organelle does not mean its messenger RNA was transcribed there.

Proof that a given protein is coded by an organelle DNA depends upon correlating the protein directly with the gene. Classically, this proof has depended upon recovering altered forms of the protein resulting from mutations in the corresponding gene. Identification of the genes which code for particular organelle proteins is of fundamental importance, not only for the understanding of cytoplasmic gene function, but also as a step in the further investigation of regulatory mechanisms at the genetic level.

At this point the problem of cytoplasmic gene function begins to merge with the more general problem of organelle biogenesis, that is, the genetic and biochemical processes that regulate organelle formation. Chloroplasts and mitochondria are, of course, not mere bags of sequestered soluble enzymes. Most of the essential organelle functions of electron-transport and coupled ATP production, as well as of cation and substrate transport, occur in and on membranes. Thus the membrane-bound proteins are of particular importance in organelle formation, and the control of membrane growth and specificity is an intrinsic part of the puzzle of biogenesis.

Both nuclear and organelle genes are known to influence organelle formation, and it seems likely that each genetic system codes for particular proteins. Identifying the proteins coded by nuclear genes has, thus far, been a frustrating problem since large numbers of nuclear genes seem to be involved indirectly in the regulation of organelle development and function. As yet, the only nuclear gene clearly demonstrated to code for an organelle protein is the gene for mitochondrial cytochrome c in yeast.

It is here one may anticipate that cytoplasmic genetics can make a direct contribution to the biogenesis problem. Organelle DNA's contain a selected set of genes, each of which presumably carries the information for an essential organelle function. Thus, the investigation of strains carrying cytoplasmic gene mutations should provide direct information of the functions of cytoplasmic genes and their role, which is undoubtedly a central role, in organelle biogenesis.

Opportunities for investigations of this sort are only now becoming available, with the development of cytoplasmic genetics in *Chlamydomonas* and in yeast. Proper utilization of mutants for biochemical studies will require genetic manipulation of mutations, just as has been essential in bacterial and viral systems. Thus it is necessary not only to isolate mutant strains and sort out the nuclear from the cytoplasmic mutants, but also to map them. Mapping is essential in order to distinguish mutations of different genes from those within the same cistron, to recognize operons if they exist, and to relate the genetic map and behavior of the linkage group with that of the corresponding DNA. Thus the power of cytoplasmic genetic analysis as a tool in the study of cellular biogenesis will depend upon the development and utilization of cytoplasmic genetics per se.

Having presented a thumbnail sketch of what this book is about, I would like to note some topics that were not discussed. Cytoplasmic genetics is a part of the science of genetics, i.e., the analysis of hereditary mechanisms. Cytoplasmic heredity is a larger subject, including phenomena that are poorly understood. Except for a short section in Chapter 5, we have not considered instances of cytoplasmic heredity other than those involving established or semiestablished DNA-based genetic systems.

Three problem areas other than organelle genetics have emerged from studies of cytoplasmic heredity in various orgamisms, and they have been dramatically illustrated in *Paramecium.* These problem areas representing major puzzles in cytoplasmic heredity, not related in any obvious way to the cytoplasmic genetics of organelles, are listed below.

1. *Relation of viruses and other symbionts to the established cytoplasmic genomes of cells.* Beginning with studies of lysogenic viruses in bacteria, investigators have kept in mind the possibility that foreign DNA's from viruses and other symbionts may become integrated into the genomes of eukaryotic cells. Indeed, organelles with their bacteria-like organization of DNA and protein-synthesizing apparatus may provide easier access by foreign DNA's to eukaryotic cells than do the nuclei. Thus, whether or not organelle DNA's were originally of symbiotic origin, they may now provide a haven for DNA of exogenous origin. The test of whether a particular DNA is foreign or native depends upon its current function in the organism: Is it essential for survival? Is it always present? Is the genetic information unique within the organism? The answers to these questions may depend upon when in the evolution of the organism the questions are being asked.

2. *The molecular basis of long-lasting phenotypes showing persistence of clonal differences on a common genetic background.* Examples such as the inheritance of mating type and serotypes in *Paramecium,* while ultimately under control of the micronucleus, also depend upon macronuclear inheritance. Do the cells of higher organisms contain metabolic systems which function in a fashion analogous to the ciliate macronu-

cleus? This possibility has been raised with respect to some aspects of cellular differentiation under the title of "epigenetics."

3. *The inheritance of pattern.* Recent studies of the inheritance of cortical arrangements of organelles on the surface of *Paramecium* have provided descriptive evidence of a kind of heredity involving the spatial relations of various structures, not obviously related to specificity at the DNA level. In a more general sense, this phenomenon has been called "patterned growth," a term embracing not only the arrangement of the ciliate cortex, but also the growth of structures, subcellular as well as multicellular, in a unique pattern.

These classes of phenomena will be discussed in Chapter 9 in relation to some speculations about the future directions of research in cytoplasmic heredity. In Chapter 9 we will also consider some of the new developments in cytoplasmic genetics per se, in terms of their future potential impact on basic science and its applications to the fields of human genetics, medical research, and agriculture.

2

Cytoplasmic DNA's

The demonstration that chloroplasts and mitochondria contain specific DNA's of their own came as a great surprise to most biologists, although partial genetic autonomy for these organelles had been postulated intermittently by a few cytologists and geneticists for 50 years. Before 1910, Meves (*50*) had proposed that mitochondria originate from preexisting structures of the same kind and carry their own heredity. And in the following decades, the origin of mitochondria—whether they arise from preexisting mitochondria or *de novo*—was the subject of many speculations. But methods were not available for a fruitful experimental approach.

Although chloroplasts are larger than mitochondria, the investigation of their origin proved no more tractable. Following the discovery in 1909 (*2, 15*) of the first cytoplasmic genes and their effect on chloroplast formation, extensive research efforts were expended, particularly in Germany in the 1920's and 1930's, to investigate cytoplasmic heredity (*16*). Considerable evidence was amassed for the existence of cytoplasmic genes influencing chloroplast development, pollen formation, and other aspects of morphogenesis in higher plants, but most geneticists balked at the pluralistic concept of multiple genetic systems. Efforts were made by cytologists and biochemists in the 1950's to look for nucleic acids in chloroplasts, but their experiments met with technical difficulties and their results suffered a cool reception.

The climate of opinion changed dramatically in the early 1960's. Excellent new cytochemical electron microscopic and biochemical lines of evidence for the presence of DNA in chloroplasts (*66*) and in mitochondria (*53, 54, 67a*) were reported, but still did not find wide acceptance on their own. The hard evidence for the existence of chloroplast DNA (*11, 67*) and subsequently of mitochondrial DNA (*45*) came from the use of a powerful new experimental method: the separation of different DNA's in cesium chloride density gradients in the ultracentrifuge (*49*). This method, which has played and continues to play a decisive role in the solution of research problems involving DNA, is described in the Appendix. Instances of its application to organelle biogenesis will be discussed in several sections of this book.

This chapter summarizes the present state of our knowledge about the classes of organelle DNA's in eukaryotic cells—chemical and physical properties of molecules, replication patterns, and informational content. Considerations of transcription, translation, and interactions of these DNA's with the corresponding nuclear genomes will be discussed in subsequent chapters.

IDENTIFICATION OF ORGANELLE DNA'S

Organelle DNA's are present in small amounts; chloroplast and mitochondrial DNA's typically comprise 1–10% of the total cellular DNA. The identification of an organelle DNA requires both its detection within the organelle and its characterization as a unique component. Cytochemical methods are valuable for the detection of organelle DNA but do not provide material for its characterization. Extraction procedures per se do not distinguish organelle DNA from nuclear contamination. The unambiguous demonstration that an extracted DNA is associated with a particular organelle requires a means of recognizing the DNA as a distinct molecular species.

As we now know, organelle DNA's are distinct entities, comprising a unique set of genes. This uniqueness has provided the basis for their identification. In experimentally favorable organisms, the average nucleotide composition of chloroplast and mitochondrial DNA's has been sufficiently different from the nuclear DNA to permit their separation in cesium chloride density gradients (see Appendix).

Typically the DNA extracted from whole cells has been compared with DNA from isolated organelles. An example of this procedure, the identification of chloroplast DNA in *Chlamydomonas* is shown in Fig. 2.1 (*67*). The result is definitive because: (*a*) chloroplast DNA is present in high enough concentration to be seen in whole cell extracts, (*b*) the

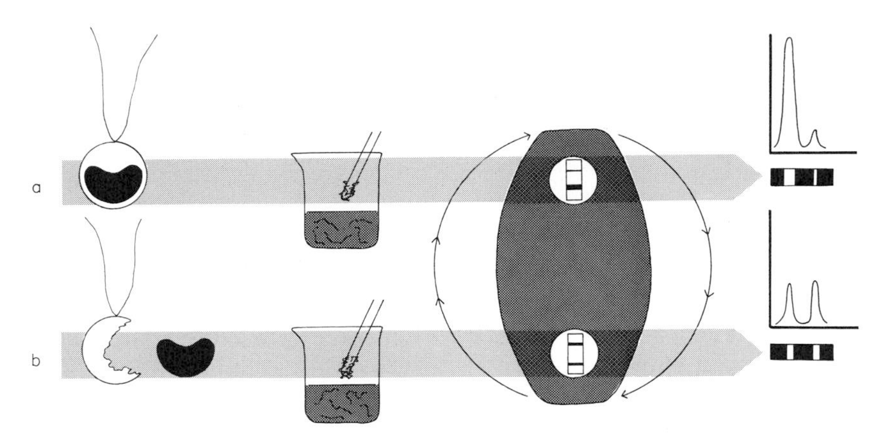

Fig. 2.1. Identification of chloroplast DNA in *Chlamydomonas* by CsCl density gradient centrifugation. DNA is extracted from whole cells (a) or from isolated chloroplasts (b) and centrifuged to equilibrium in cesium chloride (see Appendix for method) in the analytical ultracentrifuge. The position of the banded DNA's is photographed with UV light and the film traced with a microdensitometer to produce the tracings diagrammed in the figure. (a) Major peak is nuclear DNA and small peak is chloroplast DNA. This identification is based on the observation that in the DNA from the chloroplast fraction (b), the major peak is reduced and the small peak is enriched eightfold.

buoyant densities of nuclear and chloroplast DNA's are different enough to provide excellent separation in the gradients, and (*c*) chloroplast DNA can be extracted from isolated chloroplasts and compared with that of whole cell extracts.

Thus, the results depend on compositional differences between organelle and nuclear DNA's, on the presence of enough organelle DNA to be seen in whole cell extracts, and upon recovery of the organelle DNA from isolated organelles. Fortunately these conditions have been met in many organisms. However, identification becomes much more difficult when organelle DNA's closely resemble the nuclear complement and are present in very small amount. Techniques have been devised to improve the resolution of minor DNA components in gradients, including the use of mercuric ions in cesium sulfate gradients to magnify small compositional differences (*18*), and ethidium bromide to separate covalently closed circles from linear DNA molecules (*59*). Mercuric ions bind preferentially to adenine, greatly increasing the bouyant density of AT-rich DNA's over that of DNA's with high GC content (*18*). Mercuric-treated DNA's must be centrifuged in dense solutions provided by cesium sulfate. The differential binding effects of ethidium bromide will be discussed below (p. 25).

Density gradient centrifugation is the method of choice not only for identifying different DNA components, but also for recovering them for further study. However, other methods of detecting and examining DNA's have been used successfully under particular conditions. These methods include: (*a*) cytochemical identification of DNA in the light microscope by the Feulgen reaction or by fluorescence microscopy after reaction with dyes such as acridine orange; (*b*) radioautographic localization of radioisotopes taken up into DNA; (*c*) identification of DNA fibers in electron micrographs of sectioned cells; and (*d*) chemical extraction of radioisotope-labeled DNA from cell fractions. Chemical methods include fractionation on hydroxyapatite (*3, 50a*) and on MAK (methylated albumen and Kieselguhr) (*46a*) columns.

Each of these methods has advantages and pitfalls. Cytochemical staining methods are invaluable tools in the localization of DNA in particular structures, but they require careful comparison with DNase-treated controls. The same is true for identifying DNA in electron micrographs of sectioned cells, in which one simply notes the presence or absence of rather nondescript fibers. The uptake of radioisotopic precursors into DNA as visualized with autoradiographs represents a sensitive detection method, limited by the specific activity of the label and by the nonspecific background activity. In general, however, even with excellent controls, cytochemical and isotope incorporation methods require independent confirmation by a different technique, since they do not identify DNA unequivocally.

Chemical extraction of the characteristic DNA from its cell fraction depends for success on the purity of the isolated cell component, and the same is true for studies of radioisotope incorporation into cell fractions. An excellent way to purify organelle DNA is by DNase treatment of the isolated intact organelle. The success of this method depends on the intactness of the outer organelle membranes and their impermeability to the enzyme. This method has worked well with mitochondria from *Neurospora.*

These methodological considerations are of especial interest because of the possibility that some cytoplasmic DNA's remain undetected, owing to the inadequacy of existing methods to find them. DNA has been demonstrated unequivocally in chloroplasts, mitochondria, and the kinetoplasts of some protozoa, as will be discussed below, but we do not know whether there are additional classes of cytoplasmic DNA's with essential cell functions.

The technical difficulties in this problem have been highlighted by reports, based on cytochemical observations, of DNA in basal bodies of cilia in *Tetrahymena* (*60*) and *Paramecium* (*72*) (cf. p. 367). These cytochemical observations have not been confirmed by chemical characterization. The underlying question remains unresolved: Do basal bodies contain DNA and, if so, is it genetically unique? New methods of detection and isolation of cytoplasmic DNA's are urgently needed for cytoplasmic DNA's present in very small amounts.

ORGANELLE DNA'S OF PLANT CELLS

Chloroplast DNA was the first cytoplasmic DNA to be identified in extracts of isolated organelles and characterized in terms of base composition and physical properties as a double-stranded molecule of high molecular weight (*11, 67*). As already described, in *Chlamydomonas,* whole cell DNA was found to have a small satellite, banding at a different density from the predominant nuclear fraction; and in isolated chloroplasts, the satellite component was greatly enriched (Fig. 2.1). The nuclear and chloroplast DNA's of *Chlamydomonas* are shown in Fig. 2.2. Another component as yet unidentified is seen as a light shoulder on the nuclear DNA peak.

Chun *et al.* (*11*) described DNA's from two higher plants, spinach and beet, and from two algae, *Chlamydomonas* and *Chlorella.* In all their preparations, the "chloroplast fraction" contained a major component with buoyant density similar to nuclear DNA which they attributed to nuclear contamination, and a single satellite component which they called chloroplast DNA. Their inferences were correct for the algal preparations, but wrong for the higher plants. The component they identified as chloroplast DNA in beet and spinach was actually mitochondrial, and the so-called nuclear contaminant was in fact the chloroplast DNA (*38*).

After a prolonged period of confusion, recent investigations have clarified the identification of higher plant DNA's as shown in Table 2.1. The confusion resulted from two unexpected circumstances: (*a*) the close similarity of composition of nuclear and chloroplast DNA's of higher plants, and (*b*) the presence of mitochondrial DNA in the isolated chloroplast fractions.

In higher plants, mitochondrial DNA has been identified unambiguously from roots and tubers in which the ratio of mitochondrial to chloroplast DNA is much more favorable than in green tissues. All DNA's from isolated plant mitochondria so far examined have very similar buoyant densities of 1.706–1.707 gm/cm^3. Values around 1.705–1.707 gm/cm^3 previously ascribed to chloroplast DNA's have

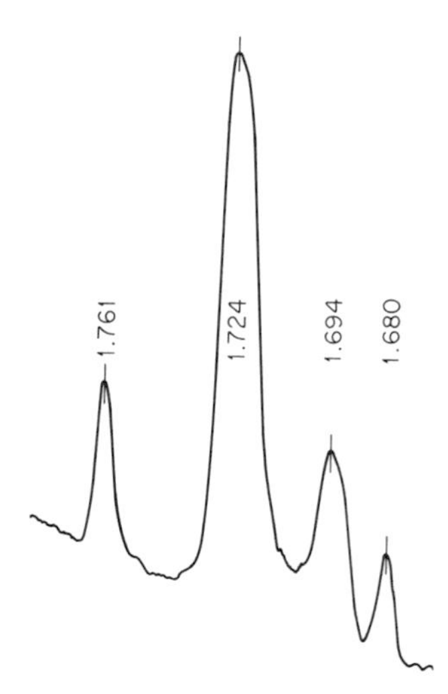

Fig. 2.2. Principal DNA's of *Chlamydomonas*. Microdensitometer tracing of DNA's from gametes (mating type plus) centrifuged to equilibrium in CsCl density gradient. Bands seen are: nuclear DNA at 1.724 gm/cm^3 chloroplast DNA at 1.694 gm/cm^3 as computed from markers at 1.761 gm/cm^3 (SP-15 phage DNA from Dr. Marmur) and 1.680 gm/cm^3 (crab poly dAT from Dr. Sueoka). In gametes, chloroplast DNA is 7% of total DNA, based on calibration with known amount of SP-15 DNA. From (*42*).

been shown to represent mitochondrial contamination in the chloroplast fraction (*38*).

The distinction between chloroplast and mitochondrial DNA has been unambiguously established in *Euglena* (*22, 62*). The chloroplast DNA of *Euglena* was identified as a band of density 1.686 gm/cm^3 on the basis of its enrichment in partially purified cell fractions of chloroplasts (*7*). This DNA is present in dark-grown *Euglena* in which chloroplasts do not develop beyond the proplastid stage (Chapter 8), but it is not found in some of the colorless mutants which have permanently lost the ability to form normal green chloroplasts. The use of mutants which lack chloroplast DNA helped in the identification of mitochondrial DNA, a small fraction with a buoyant density of 1.691 gm/cm^3 which had been overlooked in earlier studies. This DNA fraction was subsequently extracted from isolated mitochondria of colorless cells lacking chloroplast DNA. Both mitochondrial and chloroplast components can be seen in cesium chloride density gradients heavily overloaded with unfractionated DNA from normal green cells, as shown in Fig. 2.3.

TABLE 2.1

Organelle DNA's of Algae and Higher Plants[a]

Organism	DNA (gm/cm^3)				Reference
	Nuclear	Chloroplast	Mitochondria	Other	
Chlamydomonas	1.724	1.695	–	(1.715)	*11, 67*
Chlorella	1.716	1.695	–	–	*11*
	1.717	{ 1.692 (1.717) }	1.712	– –	*36*
Euglena	1.707	1.685	1.690	–	*7, 21, 61, 62*
Acetabularia	(1.702)	(1.704)	1.714	1.722	*29*
	(1.702)	(1.695)	–	1.724	*26*
Porphyra tenera	1.720	1.696	–	–	*34*
Tobacco	1.698	1.698	–	–	*46*
	1.690	[1.703]	–	–	*69*
	1.696	[1.706]	–	1.715	
	1.695	1.697	–	–	*88*
	1.698	[1.703]	–	–	*83*
	1.697	1.697	–	–	*92*
Spinach	1.695	[1.719]	–	–	*11*
	1.694	1.696	–	–	*92*
	1.694	1.696	–	–	*88*
Swiss chard	1.689	1.700	–	–	*41*
	1.694	1.696	1.705	–	*92*
Mung bean	1.691	–	1.706	–	*80*
(*Phaseolus aureus*)	1.695	1.697	–	–	*88*
Turnip (*Brassica rapa*)	1.692	1.695	1.706	1.700	*80*
Sweet potato (*Ipomoea batatas*)	1.692	–	1.706	–	*80*
Onion	1.689	–	1.706	1.718	*80*
	1.691	1.696	1.706	–	*88*
Beet	1.695	[1.719]	–	–	*11*
Wheat	1.702	1.698	–	–	*88*
Sweet pea	1.695	1.697	–	–	*87*
Lettuce	1.694	1.697	–	–	*87*

[a] Figures in parentheses not certain; figures in brackets later shown to be incorrect (*38*). All values standardized to *E. coli* DNA = 1.710 gm/cm^3.

IDENTIFICATION OF MITOCHONDRIAL DNA'S

In 1963, the presence of DNA in the mitochondria of mammalian cells was described on the basis of extensive cytochemical and electron microscopic studies (*53, 54*). Subsequently, the unequivocal identification of mitochondrial DNA from *Neurospora* by cesium chloride density

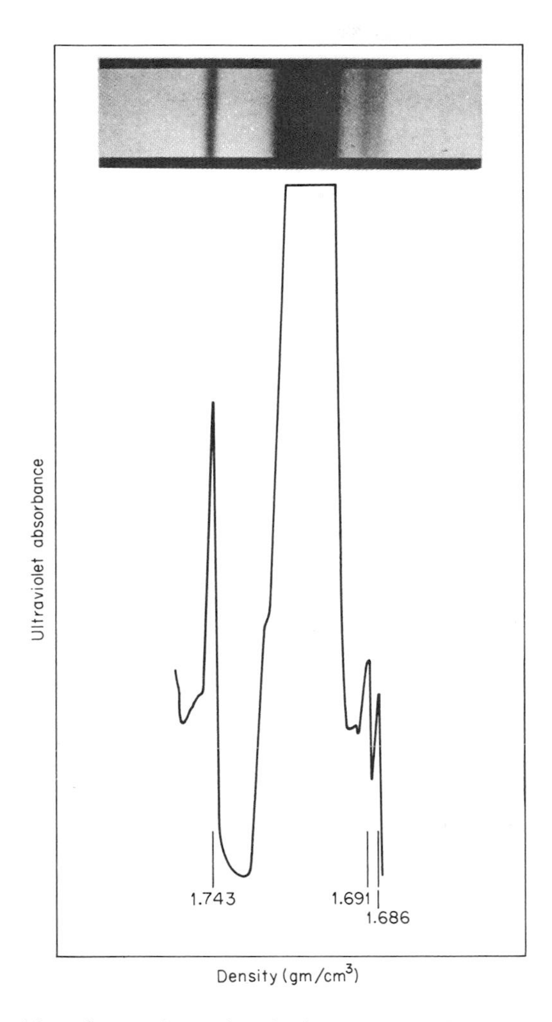

Fig. 2.3. Nuclear, chloroplast, and mitochondrial DNA's of *Euglena*. CsCl density gradient of extracted DNA (40 μg) from wild-type cells. Bands appearing are density standard (1.743 gm/cm^3), greatly overloaded main band, mitochondrial DNA (1.691 gm/cm^3), and chloroplast DNA (1.686 gm/cm^3). From (*21*).

gradient centrifugation was reported (*45*). Mitochondrial DNA with a buoyant density of 1.701 gm/cm^3 was seen as a shoulder on the nuclear DNA banding at 1.712 gm/cm^3, and compared with DNA extracted from isolated DNase-treated mitochondria of *Neurospora*.

Table 2.2 lists mitochondrial DNA's extracted from various nongreen microorganisms and from animal cells. The buoyant densities show

TABLE 2.2

DNA's of Microbial and Animal Cells[a]

Organism	DNA (gm/cm³) Nuclear	Mitochondria	Other	Reference
Yeast (*Saccharomyces cerevisiae*)	1.698	av. 1.684	1.704	*14, 51, 52, 82, 84*
Neurospora crassa	1.713		1.698	*63*
N. crassa abn-1 mutant		{1.702, 1.726}	–	*63*
N. sitophila	1.713	{1.702, 1.692}	1.698	*63*
Tetrahymena	1.688	1.684	1.693	*79*
	1.692	1.686	1.700	*79*
	1.685	1.685	1.698	*79*
Paramecium	1.689	(1.702)	–	*81*
Physarum polycephalum	1.700	1.686	–	*23*
Leishmania henrietti	1.721	1.699	–	*20*
Trypanosoma cruzi	1.710	1.699	1.686	*65*
Sea urchin	1.694	1.704	–	*56*
Xenopus	1.700	1.702	–	*19*
Rana pipiens	1.702	1.702	–	*19*
Siredon mexicanum	1.704	1.695	–	*93*
Necturus maculosus	1.707	1.695	–	*93*
Carp	1.697	1.703	–	*84*
Chick embryo	1.701	1.707	1.718	*58*
Pigeon	1.700	1.707	–	*6*
Duck	1.700	1.711	1.708	*6*
Guinea pig	1.700	1.702	1.704	*14*
Mouse liver	1.701	1.701	1.690	*6*
Beef liver	1.703	1.703	1.715	*6*
Beef heart	1.704	1.702	1.715	*6*
Sheep	1.703	1.703	1.714	*6*
Ox	1.704	1.702	1.715	*6*
Man (leukemic leukocytes)	1.695	1.705	–	*13*

[a] Figures in parentheses not certain. All values standardized to *E. coli* DNA = 1.710 gm/cm³.

some regularity: the lowest densities are found in yeast, *Tetrahymena*, *Euglena*, and the slime mold *Physarum*. Mitochondrial DNA's from higher plants so far described have a buoyant density of 1.706 gm/cm³, whereas in the algae *Acetabularia* and *Chlorella* the values are 1.714 gm/cm³ and 1.712 gm/cm³, respectively. Among the animals, as pointed out by several investigators, mammalian mitochondrial DNA is in the range of 1.701–1.704 gm/cm³, while that of birds is 1.707–1.711 gm/cm³. From an evolutionary viewpoint, *Euglena* seems well isolated from the algae and much closer to the fungi and higher protozoa; this relationship is in keeping with other indications that *Euglena* is indeed an "animal," a protozoan carrying functional chloroplasts (*86*).

In summary, organelle DNA's have been identified primarily on the basis of their distinctive buoyant densities. Further characterization requires isolation and purification of individual organelle DNA's. One may then determine nucleotide composition, and examine the DNA with respect to other parameters of interest. Data on size, shape, uniqueness, replication, and mutational alterations of organelle DNA's to be discussed below are based primarily on studies with purified components. The relationship between base composition and buoyant density in cesium chloride which is so useful in comparing different DNA's does not hold well with DNA's of very high AT (adenine–thymine) content (*4*).

THE UNIQUENESS OF ORGANELLE DNA

The striking differences in average nucleotide composition of organelle and nuclear DNA's, as reflected in buoyant density differences, strongly suggest differences in specificity, but do not prove it unequivocally. More direct evidence requires a method which examines sequence specificity. In the absence of direct methods to determine nucleotide sequences in DNA, a powerful method presently available is nearest-neighbor frequency analysis (see Appendix).

This method examines the frequencies of the sixteen possible pairs of adjacent dinucleotides present along single strands of DNA. The differences in these frequencies have been found to be sufficient to distinguish between DNA's coming from different species of bacteria and viruses. Thus the method can be used to evaluate relatedness of nucleotide sequence of different DNA's.

Mitochondrial and nuclear DNA's from the slime mold *Physarum polycephalum* were compared by this method (*17*). The four doublets ending in G were examined, and definite differences were observed between nuclear and mitochondrial DNA. In a study of chloroplast and nuclear DNA's of *Chlamydomonas,* all sixteen doublets were compared and the results are shown diagrammatically in Fig. 2.4 (*77*). The wide divergence between the two DNA's is clearly evident in these data.

In algae and in higher plants, a consistent difference in the percentage of methylated bases between nuclear and chloroplast DNA's has been reported. Nuclear DNA's contain as much as 6% 5-methyl cytosine, whereas no methylated bases have been detected (at the 1% level) in any chloroplast DNA's (*38*).

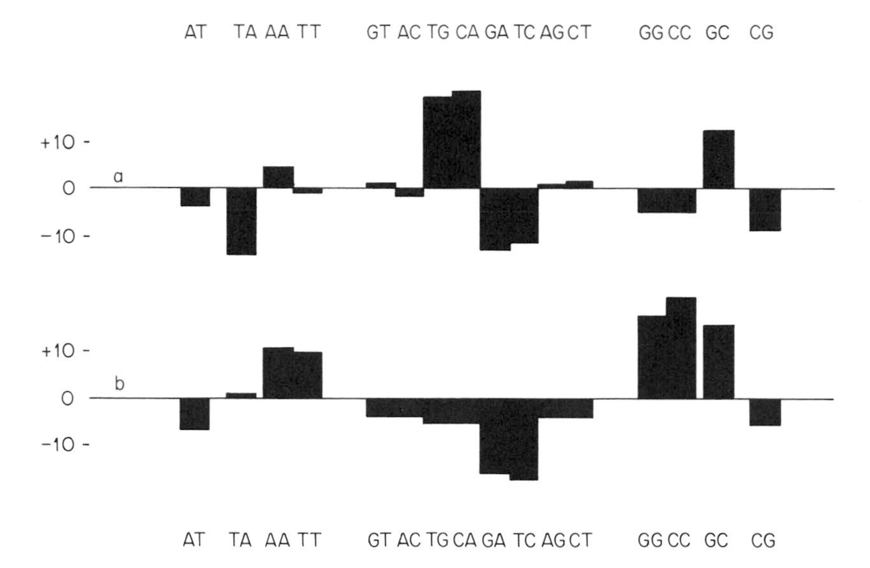

Fig. 2.4. Nearest-neighbor frequencies of nuclear (a) and chloroplast (b) DNA's of *Chlamydomonas.* The increase (+) or decrease (−) in frequency of each doublet compared with random frequencies (adjusted for the average nucleotide composition of each DNA) is shown as bars above and below the base line. From (*77*).

Methylation of organelle DNA's is of particular interest because it provides a possible mechanism for maternal inheritance (cf. Chapters 3 and 6). In bacteria, host restriction and modification of DNA is regulated by methylation of special sites sensitive to a restriction enzyme (exonuclease) in the absence of protective methyl groups (*1*). An analogous process could distinguish organelle DNA's from male and female parents. The number of methylated sites could be so low that special methods would be required to identify them. Thus, the published evidence that chloroplast DNA's are not methylated does not exclude the possibility of low level methylation.

SIZE AND CIRCULARITY OF ORGANELLE DNA

The chloroplasts of most algae and higher plants contain as much DNA as a bacterial genome. *Chlamydomonas,* for example, which has only one chloroplast per cell, contains a minimum of about 4×10^9 daltons of chloroplast DNA in gametes (*78, 42*) and two to four times as much in growing cells. In *Euglena,* each of its ten to twelve chloroplasts contains about $4\text{–}6 \times 10^9$ daltons of DNA, and in the cells of higher plants, the estimated average values range from $0.6–6 \times 10^9$ daltons per chloroplast (*39*). For comparison, *E. coli* contains $2–3 \times 10^9$ daltons per genome.

The extraordinary unicellular alga *Acetabularia*, which is 2 cm or more long, contains more than 10^6 chloroplasts, with an average DNA content of about 6×10^7 daltons (*26*), distinctly less than that in other organisms. However, recent evidence indicates that only about 20% of the chloroplasts of *Acetabularia* contain any DNA at all (*95*). If this report is correct, the average value for DNA per chloroplast should be revised upward into the range of other organisms.

The biological significance of chloroplasts without DNA is not clear at this time. The chloroplasts of *Acetabularia* do fuse and come apart, and therefore could be considered a network, rather than a suspension of distinct and separate units. DNA might then be localized at definite attachment sites on chloroplast membranes, and the number of attachment sites may be fewer than the number of chloroplasts seen at any one time. There are several indirect lines of evidence that a similar situation may exist for mitochondria (cf. p. 120).

How is chloroplast DNA organized? Is it one enormous molecule or many separate ones? Unpublished studies in several laboratories have indicated that chloroplast DNA's of algae and higher plants are indeed very long molecules and very difficult to recover intact without breakage.

In a recent electron microscope study of chloroplast DNA from *Acetabularia*, lengths up to 400 μ (8×10^8 daltons) were seen attached to fragments of chloroplast lamellar membranes (*27*). If such lengths represent single molecules, they approach the size of bacterial DNA's. The recovery of circular DNA's from *Euglena* chloroplasts has now been reported by Manning *et al.* (*46b*). The circles average 43 μ in contour length, corresponding to about 9×10^7 daltons and correlating well with the genomic size of 1.8×10^8 daltons (*76*) adjusted to 9×10^7 daltons by allowing for low GC content (*91*).

In contrast to the paucity of information about the size and organization of chloroplast DNA's, a considerable body of evidence is available about mitochondrial DNA's. Those which have been studied most intensively are small and covalently closed circular molecules. Before discussing them, it is necessary to consider the methods which have been applied to their examination.

The existence of a double-stranded circular DNA was first demonstrated with polyoma virus (*86a*). Then, using polyoma DNA as a model system, Vinograd and his associates developed general methods for detecting and investigating circular DNA molecules (*85*). They found that covalently closed circular duplexes have a higher buoyant density in cesium chloride than do open circles of the same size and composition. Covalently closed circles are supercoiled, a configuration which increases their buoyant density. A single-stranded nick is sufficient to release the supercoiled configuration, allowing the twisted molecule to open up. Covalently closed circles are more resistant to alkali and heat denaturation than are open circles or linear molecules. The relations among linear, open, and twisted circles are shown in Fig. 2.5.

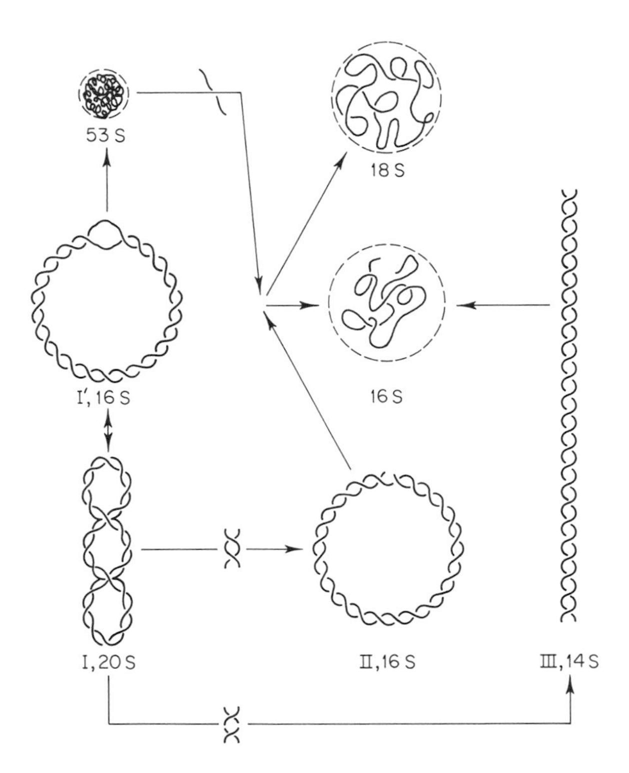

Fig. 2.5. Sedimentation values and configurations of several forms of polyoma virus DNA. Component I is covalently closed and supercoiled; Component II has a single-stranded nick which produced the open configuration and lower sedimentation coefficient. Component III is the linear form, produced by a double-stranded break in I. I′ is a covalently closed circle in open configuration resulting from a single strand break followed by repair after unwinding. The 53 S component is alkali denatured form of I; the 16 S and 18 S components are, respectively, single-stranded linear and circular forms derived from II. The dashed circles around the denatured forms indicate the relative hydrodynamic diameters. The sedimentation coefficients were measured in neutral and in alkaline NaCl solutions. The twist in I should be right-handed. From Vinograd and Lebowitz (*85*).

The difference in buoyant density between open and closed circles can be increased by ethidium bromide and, even more effectively, by propidium iodide, two acridine dyes that bind to DNA (*32a*). Because of their supercoiled configuration, closed circles bind less dye than do open circles or linear molecules at dye saturation. Consequently the closed circles are heavier in buoyant density than the other components. The various dye–DNA complexes can be separated by banding them in preparative cesium chloride density gradients (Fig. 2.6) and the separated components can then be identified in the electron microscope.

With these methods, investigators have been able to extract DNA from cells and from organelles and assess the frequency distribution of the various molecular configurations. These techniques are particularly useful for examining small circles in the range of $1–20 \times 10^6$ daltons. Larger DNA molecules are more prone to breakage. However, Hickson *et al.* have isolated episomes of about 1×10^8 daltons (*31a*).

All the mitochondrial DNA's of animal cells that have been examined so far consist of covalently closed circles of uniform size, approximately 5 μ in circumference, corresponding to a molecular weight of about 1×10^7 daltons. Circularity has been a great aid in establishing length.

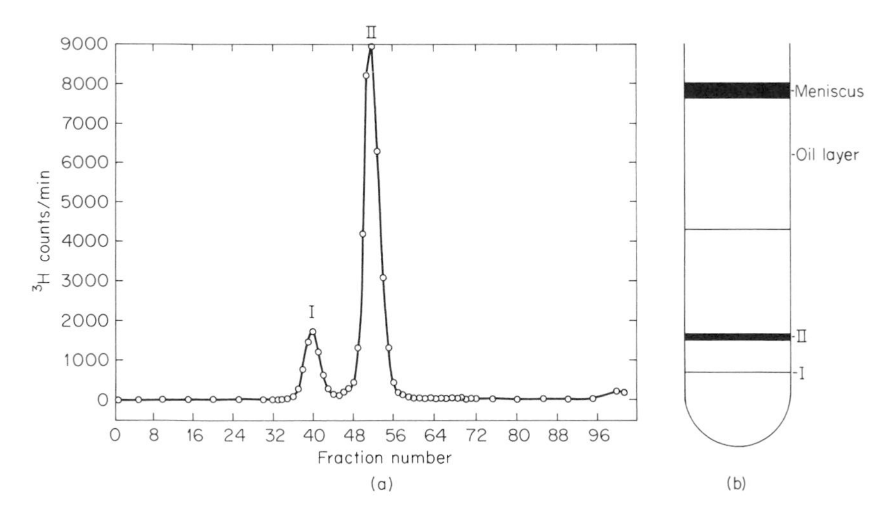

Fig. 2.6. Separation of twisted (I) and open (II) circles of polyoma DNA in an ethidium bromide–CsCl density gradient in the preparative ultracentrifuge. DNA was prelabeled with ^{3}H-thymidine to measure the relative amounts of the two configurations. (a) The band maxima are 12 fractions apart, permitting excellent separation of the two components as shown later in electron micrographs of DNA from the two peaks. (b) A diagram of the centrifuge tube prior to drop collection. From (*59*).

In general, uniformity of contour length has been seen within species, and some variation has been noted from one species to another, as shown in Table 2.3. Whether linear molecules are also present in animal cell mitochondrial DNA's is not known. Some are always found but they have been attributed to breakage.

TABLE 2.3

Size and Circularity of Mitochondrial DNA's[a]

Source	Structure	Contour length (μ)
Mammalia		
Man	Circular	4.8–5.3
Monkey	Circular	5.5
Ox	Circular	5.1–5.3
Sheep	Circular	5.4
Rat	Circular	4.9–5.4
Mouse (liver)	Circular	5.0–5.1
Mouse (L-cells)	Circular	4.7
Guinea pig	Circular	5.6
Hamster	Circular	5.1
Aves		
Chick	Circular	5.1–5.4
Duck	Circular	5.1
Amphibia		
Frog (*R. pipiens*)	Circular	5.9
Toad (*X. laevis*)	Circular	5.7
Axolotl (*S. mexicanum*)	Circular	4.9
Mud puppy (*N. maculosus*)	Circular	4.9
Osteichthyes		
Carp	Circular	5.4
Echinoidea		
Sea urchin (*L. pictus*)	Circular	4.6–4.9
Insecta		
Fly (*M. domestica*)	Circular	5.2
Protozoa		
Tetrahymena pyriformis	Linear	17.6
Fungi		
Saccharomyces	Circular	25.0

[a] Based upon P. Borst (5), Tables 1 and 2.

In plants and in eukaryotic microorganisms, the mitochondrial DNA's are much longer than in animal cells, and may not be circular. It is very difficult to determine the intracellular length of linear molecules, unless the extracted molecules are very uniform. On the assumption that short molecules are the result of breaks, the longest molecules may be considered to approach native length. However, artificially long molecules can be formed by the action of the enzyme ligase during DNA extraction. Indeed, the possibility exists that some circles found in extracted DNA were formed during extraction. Recently, investigators have been heating cells before extracting DNA to inhibit ligase action.

An example of the difficulties in determining size and circularity is provided by yeast mitochondrial DNA. Both linear molecules and circles of various lengths have been reported (*31, 32, 68*). Recently, large circular molecules, 25 to 30 μ in circumference ($5-6 \times 10^7$ daltons), have been recovered (*32*). However, these large circles were rare, with the majority of molecules being linear. It seems likely that the 25–30 μ circumference circles represent a native state of mitochondrial DNA, but the question of whether linear molecules may also be present at some stages of the cell cycle remains unsolved.

The longest noncircular mitochondrial DNA yet reported is from *Tetrahymena,* in which uniform linear molecules with a mean length of 17.6 μ (about 35×10^6 daltons) were observed; no circles were seen (*79*).

The presence of DNA in the kinetoplasts of parasitic flagellates has been known for many years on the basis of Feulgen staining. More recently, electron microscope studies showed that the organelle contains DNase-sensitive fibers as well as typical mitochondrial cristae within the same double-membraned organelle (*64, 44*). Biochemical studies have shown that the organelle is mitochondria-like in both its structure and function (*11a, 83a*).

Kinetoplast DNA has been identified and characterized by density gradient centrifugation. In *Trypanosoma cruzi,* the kinetoplast DNA represents 15–20% of the total DNA of the cell and has a buoyant density of 1.699 gm/cm^3, whereas the nuclear DNA has a bouyant density of 1.721 gm/cm^3. The first studies of kinetoplast DNA were carried out with the related trypanosome, *Leishmania henrietti* (cf. Table 2.2).

Further characterization of the DNA from *Trypanosoma cruzi* by electron microscopy was undertaken to investigate the size and shape of the molecules. Riou and Delain (*64*) isolated total DNA and then recovered the presumed kinetoplast fraction in preparative gradients in the ultracentrifuge. They found primarily very small circular DNA molecules with an average contour length of 0.45 μ, as well as a few linear molecules representing only about 1% of the total DNA, with lengths of 2–12 μ.

Laurent and Steinert (*44*) examined kinetoplast DNA from the related organism *Trypanosoma mega*. They found that long linear molecules are frequent and that small circles about 0.74 μ in contour length are also seen in the same samples. In this study, the DNA was extracted from isolated kinetoplasts directly onto grids for the electron microscope, rather than from whole cells. The investigators suggest that their preparation may represent a more valid sample of the *in vivo* distribution than appears in preparative gradients.

What is the biological significance of circularity? Circularity was first seen in viral DNA's. Examples are known among the viruses of covalently closed circles, of hydrogen-bonded circles, and of rods with terminal redundancy making possible transient circularization associated with recombination. Bacterial DNA's are circular. Some investigators have proposed an essential role for circularity in the regulation of replication. Perhaps all organelle DNA's are effectively circular at some critical stage in their replication cycle.

A remarkable application of physical studies of mitochondrial DNA to a medical problem was discovered and is being investigated by Vinograd and colleagues (*12, 13*). Mitochondrial DNA's from the leukocytes of patients with certain types of leukemia were found to differ greatly from the normal. The abnormal DNA's contained substantial numbers of circles which were multimers of the normal size, that is dimers and larger multiples present as open circles and as interlocked circles, which are the catonated forms. Most remarkable of all, in a few patients undergoing chemotherapy, the frequency of these abnormal forms was sharply reduced.

The origin of these abnormal configurations has not been established. However, they resemble the products of recombination between closed circles in which a single exchange event produces either a double-sized open circle or two interlocking circles, depending on the topology of the event. Preliminary genetic evidence has indicated that recombination of mitochondrial DNA molecules does occur in yeast zygotes (cf. Chapter 4), but whether recombination normally occurs at other stages and in other organisms is not known. In the chloroplast DNA of *Chlamydomonas* (cf. Chapter 3) recombination does occur during normal vegetative growth.

If recombination between circular mitochondrial DNA's is a common event, then a mechanism would be required to insure the absence of abnormal forms, perhaps by a special repair enzyme associated with recombination. The high frequency of abnormal forms in the leukemic cells might then result from the absence of such an enzyme or conversely from an elevated frequency of exchange events. Whatever the etiology, the correlation between aberrant mitochondrial DNA and leukemia raises numerous questions and opens new lines of experimentation.

GENOMIC SIZE OF ORGANELLE DNA'S

Measurement of the total amount of an organelle DNA present per cell does not indicate whether the DNA is present in one or many copies. Historically, that question has been answered by means of genetic analysis. A powerful new method for the overall evaluation of the genomic size of DNA was developed recently (*8, 92*). This method, described in the Appendix, is based on the rate of reannealing of a denatured sample of DNA under standardized conditions. In a given amount of DNA, complementary strands will find each other and reanneal faster if many copies are present than if only a few are to be found. Since reannealing follows strict second-order kinetics, the rate of reannealing provides a quantitative measure of the number of copies present.

Several organelle DNA's have been examined by this method and compared with the DNA's of phage T4 and *Escherichia coli,* which are in the organelle size range. The results given in Table 2.4 represent the best estimate of genomic size as proposed by each investigator. The values should be considered as order of magnitude estimates rather than as precise values. The number of copies is estimated by dividing the calculated genomic size into the measured analytical amount of DNA in the organelle.

The four estimates given for chloroplast DNA of *Chlamydomonas, Euglena,* and two higher plants are in remarkably good agreement. These data indicate that chloroplasts contain on the average about twenty copies of the major component. A fast renaturing component representing a short sequence present in many copies has also been seen in chloroplast DNA of *Chlamydomonas* and higher plants (*87, 89*).

In *Chlamydomonas,* genetic analysis (cf. Chapter 3) has demonstrated that the chloroplast linkage group behaves as if it were present in two copies, not in twenty copies. How can the apparent discrepancy in the number of copies estimated by genetic analysis and renaturation kinetics be reconciled? One possibility is that a slowly renaturing fraction is present which corresponds to the genetic markers but is undetected by the renaturation procedure. This problem is clearly in need of further investigation.

In contrast to the similarity in size of chloroplast DNA's from algae

TABLE 2.4

Estimates of Genomic Size of Organelle DNA's by Renaturation Kinetics[a]

Organelle	Estimated genomic size	Estimated analytic size	Reference
Chloroplast DNA			
Chlamydomonas	2.0×10^8	4×10^9	*89*
Euglena	1.8×10^8	6×10^9	*76*
Lettuce	1.4×10^8	2×10^9	*87*
Tobacco	1.1×10^8	3×10^9	*83*
Mitochondrial DNA			
Neurospora	0.66×10^7	–	*94*
Guinea pig	1.1×10^7	1×10^7	*5*
Yeast	5.0×10^7	5×10^7	*5*
Lettuce	$>10.0 \times 10^7$	–	*87*
Standards			
E. coli	2.5×10^9	2.5×10^9	*9*
T4	1.8×10^8	1.3×10^8	*91*

[a] All data in daltons.

and higher plants, there is great diversity in the genomic and analytic sizes of mitochondrial DNA's from different organisms, as shown in Table 2.3. The genomic size of animal cell mitochondrial DNA corresponds closely to the analytic size of approximately 10^7 daltons. The genomic size for yeast mitochondrial DNA corresponds to the estimated analytic size based upon the contour length of closed circles. The mitochondrial DNA's of other eukaryotic microorganisms are also much larger than those of animal cell preparations. No analytic size estimates are yet available for the mitochondrial DNA's of algae or higher plants.

HERITABLE ALTERATION IN YEAST MITOCHONDRIAL DNA

DNA isolated from wild-type strains of yeast (*Saccharomyces cereviseae*) contains three components distinguishable in cesium chloride density gradients: a major band at 1.699 gm/cm^3 and two satellite bands, one at 1.683 and the other at 1.705 gm/cm^3. DNA isolated from the mitochondrial fraction contains primarily the 1.683 gm/cm^3 component. The 1.705 gm/cm^3 component is not present in mitochondria and its identity is as yet unknown.

Non-Mendelian mutations in yeast called *petites,* which lead to the loss of mitochondrial function, have long been known and have been investigated intensively (cf. Chapter 4). Although mitochondrial DNA was postulated as the site of these mutations, the evidence until recently was indirect. A major breakthrough in our understanding of the *petite* mutation came with the discovery (*51*) that mitochondrial DNA's of different cytoplasmic *petite* strains differ from the wild type and from one another in buoyant density. Some examples are given in Table 2.5.

Recently three such strains have been intensively examined in comparison to their wild-type parents, with respect to several physical parameters (*4*). The results shown in Table 2.6 demonstrate clearly that significant changes have occurred in the mutant DNA's. Furthermore it is noteworthy that the nucleotide composition values, when determined chemically, differ considerably from the values calculated from empirical equations relating composition to buoyant density and to the melting temperature (see Appendix).

Some understanding of these discrepancies comes from a careful study of the melting curves of these DNA's. When bacterial or viral DNA's are heated, the optical density of the solution undergoes a sudden hyperchromic shift resulting from the disruption of hydrogen bonds that occurs at a temperature determined by the average base composition of the DNA. The sharpness of the transition results from the cooperative nature of the forces that stabilize the helical configuration of native DNA. The first derivative of the melting profile will

TABLE 2.5

Buoyant Densities of Yeast Mitochondrial DNA's from Different Petite Strains

Strain	Buoyant density (gm/cm^3)[a]	Reference
D243-2B-R_1 (ρ^+)	1.683	*51*
D243-2B (ρ^+)	1.683	*51*
D243-2B-g (ρ^-)	1.679	*51*
D243-2B-13 (ρ^-)	1.683	*4*
DM_1 (ρ^-)	1.675	*3*
D243-2B-106 (ρ^-)	1.683	*4*
(982-19 d p 6/6-M 15-1B-2/1b) (ρ^-)	1.678	*4*
D310-4D (ρ^+)	1.683	*47*
D310-4D-21 (ρ^-)	1.673	*47*
D310-2A-184 (ρ^-)	1.676	*47*
D310-4D-76 (ρ^-)	1.681	*47*

[a] Standardized to *E. coli* DNA = 1.710 gm/cm^3.

TABLE 2.6

Comparison of Mitochondrial DNA's from Wild-Type and Petite Yeast Strains[a,b]

Strain	Chemical analysis	Computed values		
		From buoyant density	From T_m	From Y^c
A (wild type)	17.4	23.5	13.2	17.6
a_1 (*petite*)	15.5	23.5	10.0	14.7
a_2 (*petite*)	15.6	23.5	10.2	16.3
B (wild type)	16.8	23.5	10.7	14.7
b (*petite*)	12.6	19.0	9.8	12.3

[a] Based upon G. Bernardi *et al.* (*4*), Table 4.
[b] All data in mole percent guanine + cytosine.
[c] Y is a function relating the hyperchromic shift per °C to the G + C content; it is particularly useful for non-Gaussian melting curves and computed as shown in Fig. 2.7. T_m is the melting temperature.

show a symmetrical Gaussian curve if the DNA is homogeneous. The mitochondrial DNA's from the cytoplasmic *petites* under investigation did not show symmetrical Gaussian curves, but rather showed considerable heterogeneity, as indicated in Fig. 2.7. This heterogeneity is thought to reflect the presence of sizeable stretches of DNA localized along the molecule with quite different average nucleotide composition. If this interpretation is correct, it is not surprising that equations derived from homogeneous DNA's do not apply well to these molecules. A similar kind of heterogeneity has been seen in the chloroplast DNA of wild-type *Chlamydomonas* (Fig. 2.8).

REPLICATION AND REPAIR OF ORGANELLE DNA'S

The first studies of the synthesis of any organelle DNA were those of Iwamura (*35*) who found a DNA fraction in *Chlorella* which took up ^{32}P label much faster than did the nuclear fraction. In subsequent studies Iwamura and Kuwashima (*36*) identified the fast-labeling fraction as chloroplast DNA on the basis of its buoyant density. Since that time, several investigators have observed incorporation of radioisotope label into chloroplast DNA (*69, 73*). Whether this incorporation occurs during replication or whether it represents repair was not considered in these studies.

Radioisotope incorporation is an excellent method for studying DNA synthesis but is not a dependable method for examining DNA replication because the occurrence of DNA repair leads to insertion of radio-

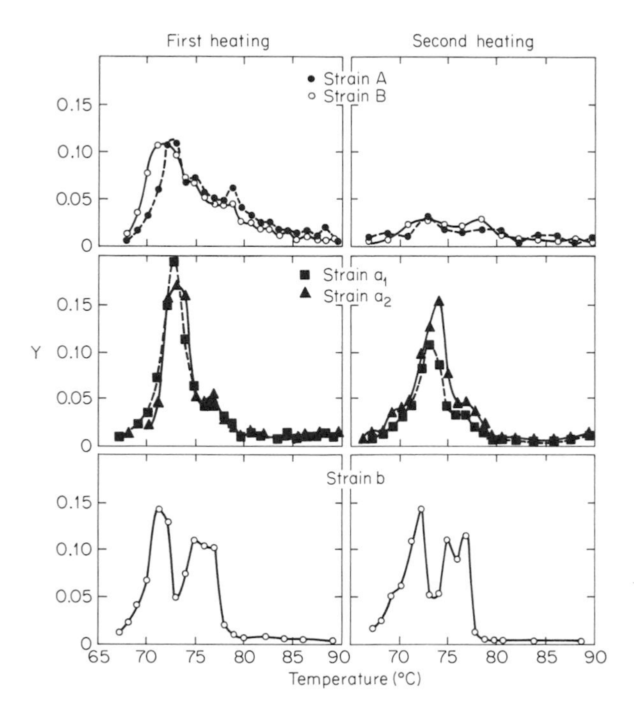

Fig. 2.7. Differential melting curves obtained with the mitochondrial DNA's listed in Table 2.6. The ordinate, *Y*, indicates the increment in relative absorbance per degree:

$$Y = \frac{(At_1 - At_2)/A100 - A25}{(t_1 - t_2)}$$

where At_1, At_2, A100, and A25 are absorbances measured at temperatures t_1, t_2, 100°C, and 25°C, respectively. Abscissa values are equal to $(t_1 + t_2)/2$. The same computation was used for the values of *Y* in Table 2.6. Similarities between first and second heating curves indicate a return to native configuration during renaturation between the two melts. From (*4*).

isotope label into nonreplicating molecules (*55b*). Experiments in which density changes are examined over an entire molecule are far more reliable than radioisotope uptake data for demonstrating the pattern of replication in cells or organelles.

The replication of DNA can best be examined with isotopic transfer experiments using the methodology introduced by Meselson and Stahl to examine the pattern of replication of DNA in *E. coli* (*48*) (see Appendix). The power of this method lies in its precision, but there are circumstances in which the results are not easily interpretable, e.g., when large precursor pools are present or when incomplete or nonsynchronous replication, extensive recombination, or repair occur.

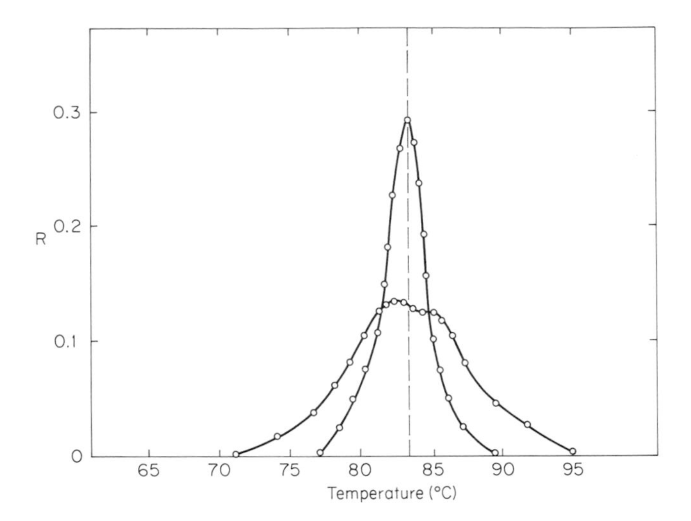

Fig. 2.8. Differential melting curves of native *Chlamydomonas* chloroplast (○–○–○) and T4 (●–●–●) DNA's. Each point is the mean of values from three separate experiments. The ordinate represents the rate of melting of the DNA.

$$R = \frac{(At_2 - At_1)/(A100 - A20)}{(t_2 - t_1)}$$

At_2, At_1, A100, A20 are the 260 mμ absorbances (corrected for water expansion) at temperatures t_2, t_1, 100°C, and 20°C, respectively. The abscissa represents the mean temperature of the intervals and is equal to $(t_2 + t_1)/2$. The denaturation conditions were $1 \times$ SSC at a temperature rise 10 minutes per °C. The DNA's were not sonicated and the single-stranded sizes were 1.2×10^6 daltons and 1.8×10^6 daltons for chloroplast and T4 DNA, respectively. From (*89*).

Evidence of semiconservative replication of chloroplast DNA comes from the work of Chiang and Sueoka, shown in Fig. 2.9 (*10*). They examined synchronous cultures of *Chlamydomonas* by the ^{15}N–^{14}N transfer method. Using a 12-hour light, 12-hour dark cycle to synchronize growth, they found that chloroplast DNA was replicated during the light period and that nuclear DNA was replicated some 12 hours later during the dark period. In their experiments, cells divided synchronously at one time in the cell cycle, each cell giving rise to four or eight daughter cells. By sampling at short intervals during the light period, they found a time at which all the chloroplast (β) DNA was hybrid, ^{14}N–^{15}N, and somewhat later they found that it was now half hybrid and half ^{14}N, corresponding to two cycles of semiconservative replication. By the end of the light period, the ^{14}N peak was somewhat larger than the hybrid peak indicating that a third doubling of some but not

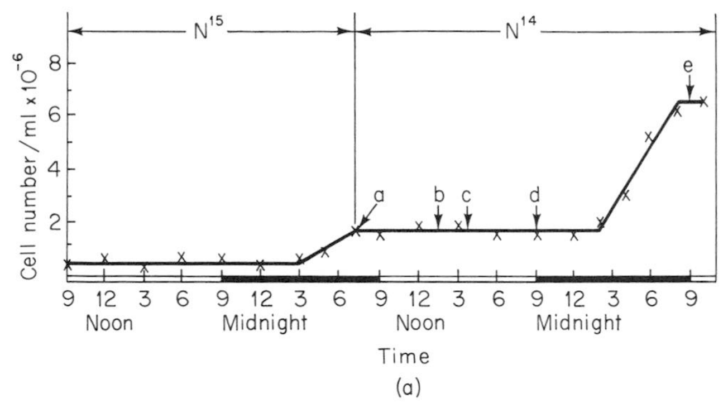

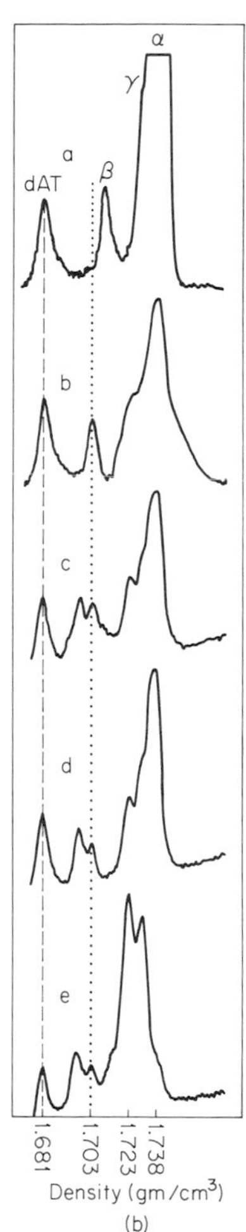

Fig. 2.9. Semiconservative replication of chloroplast DNA in *Chlamydomonas,* shown by an ^{15}N to ^{14}N density shift experiment. Cells were grown synchronously in an ^{15}N medium from an initial concentration of 10^4 cells/ml. When the culture reached 1.5×10^6 cells/ml, it was washed and transferred to an ^{14}N medium, as shown in (a). Aliquots were taken for DNA preparation at the times indicated as *a, b, c, d,* and *e.* Analyses of these samples in CsCl density gradients are shown in (b), where each profile corresponds to a time point shown in (a). Chloroplast DNA (β) shows fully heavy density of about 1.710 gm/cm^3 in *a* and hybrid density of 1.703 gm/cm^3 in *b* after 7 hours in the light cycle. At later times, both hybrid and fully light peaks are seen, representing a second round in *c,* and part of a third round of replication in *d* and *e.* From *(10).*

all DNA molecules had occurred. This inference was corroborated by the cell count.

The replication of mitochondrial DNA in yeast has recently been investigated in synchronized cultures of *Saccharomyces lactis* (*71*). The time of replication of mitochondrial DNA was related to that of nuclear DNA by extracting total DNA at a number of specific times over the cell cycle and examining it in the analytic ultracentrifuge, using the mercuric chloride method to increase the separation between the nuclear and mitochondrial bands. The ratio of mitochondrial to total DNA was estimated in each sample on the basis of the area under each of the peaks of densitometer tracings. As with chloroplast DNA of *Chlamydomonas,* mitochondrial DNA of yeast was shown to replicate semiconservatively and not at the same time in the cell cycle as nuclear DNA. An intact circular molecule of mitochondrial DNA from *S. lactis* is shown in Plate I.

Reich and Luck studied the pattern of replication of mitochondrial DNA in *Neurospora* by the ^{15}N–^{14}N transfer method (*63*). Their results are not clear, probably because of the presence of large pools of nitrogenous precursors. After one doubling of mass of the *Neurospora* mycelium, they found that the mitochondrial DNA was still about 80% ^{15}N, and after two doublings the DNA was about 55% ^{15}N with a considerable spread in densities of different molecules containing different proportions of ^{15}N–^{14}N. After three doublings, the DNA was still approximately 22% ^{15}N, again showing a broad band of heterogeneous densities. Thus the results do not resolve the question of whether all the mitochondrial DNA is replicated semiconservatively and whether all of the existing molecules are used as templates for replication in each successive doubling of the mycelial mass.

Parsons and Rustad (*55*) examined the replication of mitochondrial DNA in *Tetrahymena pyriformis* by an autoradiographic method. Cells were prelabeled with tritiated thymidine and the decrease in mitochondrial grain count was followed for four cell generations in unlabeled medium. They showed that the average mitochondrial grain count per cell was halved at each cell generation, and that the total mitochondrial label was conserved for at least four generations. These results are consistent with a pattern of semiconservative replication.

DNA replication in isolated rat liver mitochondria was studied by Karol and Simpson (*37*) using 5-bromodeoxyuridine as a density label to distinguish the newly synthesized DNA in a cesium chloride density gradient. Only a few percent of the DNA molecules incorporated the analog under these *in vitro* conditions. However, those DNA molecules in which synthesis did occur showed extensive replacement of thymine by bromouracil. The replacement was so extensive that the authors concluded that the mechanism was true replication rather than repair.

The replication of rat liver mitochondrial DNA has also been examined in electron micrographs (*40*). The illustration shown in Plate II represents a partially duplicated circular molecule identified by the presence of replicating forks. Although the frequency of occurrence of these molecules is hard to estimate from electron micrographic procedures, their occurrence is important independent evidence that replication of these circular molecules does occur.

Vinograd and his colleagues (*46b*) have now identified a 7 S fragment of mitochondrial DNA which is complementary to a region of the light strand of mitochondrial DNA from mouse L-cells and which seems to be part of a replicating intermediate form of the DNA. Electron micrographs of closed circular mitochondrial DNA (Plate II) show the 7 S fragment attached in place and indicated by arrows. A diagrammatic representation of this presumed replicating intermediate is shown in Fig. 2.10. The mitochondrial DNA is seen as opened to form a "displacement loop," within which a newly synthesized fragment, the displacing strand, is hydrogen-bonded to the preexisting light strand. The authors have computed that the 7 S length could be synthesized by a displacement reaction, unwinding the duplex without necessitating a single-strand break, i.e., without an initiating nick in the molecule. Evidence in line with this model comes from incorporation studies of mitochondrial DNA synthesis in isolated chick liver mitochondria (*67b*). Short fragments with a sedimentation coefficient "below 20 S with a peak around 10 S" were identified as newly synthesized DNA, hydrogen-bonded to covalently closed circular DNA molecules.

Taken together, these studies demonstrate the occurrence of semiconservative replication in chloroplast DNA of *Chlamydomonas* and mitochondrial DNA of yeast and suggest its occurrence in other systems as well. At a more quantitative level, we may ask whether every molecule

Fig. 2.10. Diagrammatic representation of closed circular D-mitochondrial DNA. The heavy line represents the "displacing strand" (i.e., the 7 S fragment) and the curved line with attached bars is the "displaced strand" in the region of the displacement loop, as described in the text. From Kasamatsu *et al.* (*46b*).

of organelle DNA is replicated regularly in synchrony with cell division and, more generally, how the replication of organelle DNA's is controlled.

CONTROL OF REPLICATION OF ORGANELLE DNA'S

The difference in time of replication of organelle and nuclear DNA's in *Chlamydomonas* and yeast, cited above, suggests that nuclear and organelle DNA's may be controlled by independent replication control systems. Supporting evidence for this view comes from recent experiments showing that the antibiotic cycloheximide blocks nuclear DNA replication preferentially while allowing some mitochondrial DNA synthesis in yeast (*30*), and similarly allowing some chloroplast DNA synthesis in *Chlamydomonas* (*24*).

Another approach to this question involves a comparison of the amounts of nuclear and organelle DNA's in cells under different physiological conditions. In *Chlamydomonas,* gametes contain half as much chloroplast DNA relative to the nuclear DNA content as do vegetative growing cells. Also, cells growing exponentially in the dark (heterotrophically) contain the gametic amount of chloroplast DNA in comparison to light-grown (phototrophic) cells (*43*).

In yeast, variations in the amount of mitochondrial DNA under different growing conditions have recently been examined very carefully in an effort to distinguish such variations from artifacts caused by differences in the difficulty of extracting mitochondrial DNA from cells in various physiological states (*25*). Using the double labeling concept, Fukahara standardized a method of examining the proportions of mitochondrial and nuclear DNA's in yeast (*25*). A standard yeast strain was tritium labeled in its DNA, and experimental cultures were unlabeled. The standard cells and experimental cells were mixed and the DNA was extracted together. The observed specific activity of the mitochondrial DNA was compared with that expected if both cultures contained equal ratios of nuclear to mitochondrial DNA. Anaerobically grown yeast was shown to have the same mitochondrial/nuclear DNA ratio as the aerobic control. Similarly, the mitochondrial DNA content of repressed and derepressed cultures was shown to be about the same. Finally, a strain known to contain a high concentration of mitochondrial DNA was found to have a threefold excess over that of a second strain known to be very low in mitochondrial DNA content.

Chloroplast DNA has been reported as replicating several times faster than the nuclear DNA in young tobacco seedlings, at a stage when chloroplasts are dividing faster than the leaf cells (*28, 69*). At such a time, the replication of chloroplast DNA faster than the cell cycle would seem not only likely, but essential, if every mature chloroplast is to contain a full genome of DNA.

LOSS OF ORGANELLE DNA

The irreversible disappearance of organelle DNA's has been reported in three systems: kinetoplast DNA of several species of trypanosomes (*70, 90*), chloroplast DNA of certain *Euglena* mutants (*21*), and, most recently, mitochondrial DNA of certain yeast mutants (*26a*). Although it is operationally difficult, if not impossible, to prove the total absence of DNA, DNA was undetected in these three systems under all conditions of cultivation.

The effect of acridine dyes in inducing akinetoplastic strains of trypanosomes was noted as early as 1910 (*90*). Following the demonstration that kinetoplasts contain Feulgen-staining material, the effect of acridine dyes on this material was reinvestigated (*33, 70, 74, 75*). The details of disappearance of kinetoplast DNA in organisms growing in the presence of acriflavine were studied by light and electron microscopy and by tritiated thymidine incorporation. It is now well established that acridine dyes block replication of the kinetoplast DNA under conditions not affecting nuclear DNA.

Euglena readily loses its chloroplast-forming ability following any one of numerous treatments including ultraviolet irradiation, heat shock, and growth in the presence of streptomycin (cf. Chapter 8). Some of the resulting colorless mutants have lost the ability to form chloroplasts, but have retained chloroplast DNA. Other strains have no detectable chloroplast DNA (*21*). This loss is not reflected in decreased growth rate, since *Euglena* grows as well heterotrophically as it does photosynthetically. Other algae, such as *Chlamydomonas,* seem to require the presence of chloroplast DNA and of a partly formed chloroplast at all times, even when growing heterotrophically in the dark; mutant strains without plastid structures or plastid DNA have never been found in *Chlamydomonas.*

Saccharomyces cerevisiae, bakers yeast, is a facultative anaerobe. Grown under anaerobic conditions, its mitochondrial functions are dispensable. Mutant strains (e.g., *petites*) blocked in mitochondrial function grow in the presence of oxygen as if they were anaerobic. Most *petite* strains retain mitochondrial DNA indefinitely, and there is evidence that this DNA is transcribed (cf. Chapter 7). However, Goldring *et al.* (*26a*) have found that yeast cells grown for some time in the presence of ethidium bromide lose all detectable mitochondrial DNA and this finding has been confirmed (*55a*).

The irreversible loss of an organelle DNA is evidence that the functions of the corresponding organelle are dispensable. In the three examples cited here, the organelle is known to be dispensable. Trypanosomes can grow parasitically without mitochondria, yeast can grow anaerobically, and *Euglena* can grow heterotrophically. In general however, the rarity and irreversibility of losses of organelle DNA provides evidence of the unique and indispensable role of these genetic molecules in the cell.

CONCLUDING REMARKS

This chapter has summarized the evidence for the presence of DNA in chloroplasts, mitochondria, and the kinetoplasts of trypanosomes. Are these organelle DNA's unique molecules in the genetic sense, carrying genetic information found nowhere else in the cell? This question cannot be answered from physical and chemical studies alone, but some suggestive evidence has been presented.

Differences in average base composition of nuclear and organelle DNA's provide only weak evidence of uniqueness, since the organelle DNA's might represent repetitive molecules copied from a small region of the nuclear genome. Stronger evidence of uniqueness, though still indirect, comes from nearest neighbor frequency analysis. This method has revealed large differences between the nuclear and chloroplast DNA's of *Chlamydomonas.*

Hybridization of nuclear and organelle DNA's should, in principle, provide a direct test of sequence homology. However, in practice, the method has not been definitive, primarily because the experimental technique of DNA–DNA hybridization is not sufficiently sensitive. Comparative hybridization of nuclear and organelle DNA's with organelle RNA's to be discussed in Chapters 7 and 8 provided extensive evidence of the sequence specificity of organelle DNA's.

Thus at present, the evidence that organelle DNA's carry unique genetic information comes not only from the physical studies reviewed in this chapter but largely from genetic and biochemical evidence to be considered in the succeeding chapters of this book. Correlations of genetic data with the behavior of organelle DNA's have provided the best experimental evidence for the localization of cytoplasmic genes in organelle DNA's. These correlations will be presented in the following four chapters in relation to the genetic data.

Suggested Review Articles

Borst, P., and Kroon, A. M. (1969). Mitochondrial DNA: Physicochemical properties, replication and genetic function. *Int. Rev. Cytol.* **26,** 107–190.

Kirk, J. T. O. (1971). Will the real chloroplast DNA please stand up? *In* "Autonomy and Biogenesis of Mitochondria and Chloroplasts," *Aust. Acad. Sci. Symp.* (N. K. Boardman, A. W. Linnane, and R. M. Smillie, eds.), pp. 267–276. North-Holland Publ., Amsterdam.

Linnane, A. W., and Haslam, J. M. (1971). Biogenesis of yeast mitochondria. *In* "Current Topics in Cellular Regulation" (B. L. Horecker and E. R. Stadtman, eds.), Vol. 2, pp. 101–172. Academic Press, New York.

Rabinowitz, M., and Swift, H. (1970). Mitochondrial nucleic acids and their relation to the biogenesis of mitochondria. *Physiol. Rev.* **50,** 376–427.

Vinograd, J., and Lebowitz, J. (1966). Physical and topological properties of circular DNA. *J. Gen. Physiol.* **49,** 103–125.

References

1. Arber, W., and Linn, S. (1969). DNA modification and restriction. *Annu. Rev. Biochem.* **38,** 647.
2. Baur, E. (1909). Das Wesen und die Erblichkeitsverhaltnisse der "Varietates albomarginatae hort" von Pelargonium zonale. *Z. Verebungslehre* **1,** 330–351.
3. Bernardi, G., Carnevali, F., Nicolaieff, A., Piperno, G., and Tecce, G. (1968). Separation and characterization of a satellite DNA from a yeast cytoplasmic "petite" mutant. *J. Mol. Biol.* **37,** 493–505.
4. Bernardi, G., Faures, M., Piperno, G., and Slonimski, P. P. (1970). Mitochondrial DNA's from respiratory-sufficient and cytoplasmic respiratory-deficient mutant yeast. *J. Mol. Biol.* **48,** 23–42.
5. Borst, P. (1970). Mitochondrial DNA: Structure, information content, replication and transcription. *In* "Control of Organelle Development," *Symp. Soc. Exp. Biol.* **24,** 201–226.
6. Borst, P., and Kroon, A. M. (1969). Mitochondrial DNA: Physicochemical properties, replication and genetic function. *Int. Rev. Cytol.* **26,** 107–190.
7. Brawerman, G., and Eisenstadt, J. M. (1964). Deoxyribonucleic acid from the chloroplasts of *Euglena gracilis. Biochim. Biophys. Acta* **91,** 477–485.
8. Britten, R. J., and Kohne, D. E. (1966). Nucleotide sequence repetition in DNA. *Carnegie Inst. Washington Yearb.* **65,** 78–106.
9. Chargaff, E., and Davidson, J. N., eds. (1955). "The Nucleic Acids," Vol. 3. Academic Press, New York.
10. Chiang, K.-S., and Sueoka, N. (1967). Replication of chloroplast DNA in *Chlamydomonas reinhardi* during vegetative cell cycle: its mode and regulation. *Proc. Nat. Acad. Sci. U. S.* **57,** 1506–1513.
11. Chun, E. H. L., Vaughan, M. H., and Rich, A. (1963). The isolation and characterization of DNA associated with chloroplast preparations. *J. Mol. Biol.* **7,** 130–141.

11a. Clark, T. B., and Wallace, F. G. (1960). A comparative study of kinetoplast ultrastructure in the Trypansomatidae. *J. Protozool.* **7,** 115.

12. Clayton, D. A., and Vinograd, J. (1967). Circular dimer and catenate forms of mitochondrial DNA in human leukaemic leucocytes. *Nature* (*London*) **216,** 652–657.
13. Clayton, D. A., and Vinograd, J. (1969). Complex mitochondrial DNA in leukemic and normal human myeloid cells. *Proc. Nat. Acad. Sci. U. S.* **62,** 1077–1084.
14. Corneo, G., Moore, C., Sanadi, D. R., Grossman, L. I., and Marmur, J. (1966). Mitochondrial DNA in yeast and some mammalian species. *Science* **151,** 687–689.
15. Correns, C. (1909). Vererbungsversuche mit blass (gelb) grunen und bluntblattrigen Sippen bei *Mirabilis, Urtica,* und *Lunaria. Z. Verebungslehre* **1,** 291.

16. Correns, C. (1937). "Nicht Mendelnde Vererbung." Gerbruder Borntraeger, Berlin.
17. Cummins, J. E., Rusch, H. P., and Evans, T. E. (1967). Nearest neighbor frequencies and the phylogenetic origin of mitochondrial DNA in *Physarum polycephalum. J. Mol. Biol.* **23,** 281–284.
18. Davidson, N., Widholm, J., Nandi, U. S., Jensen, R., Olivera, B. M., and Wang, J. C., (1965). Preparation and properties of native crab dAT. *Proc. Nat. Acad. Sci. U. S.* **53,** 111–118.
19. Dawid, I. B. (1965). Deoxyribonucleic acid in amphibian eggs. *J. Mol. Biol.* **12,** 581–599.
20. du Buy, H. G., Mattern, C. F. T., and Riley, F. L. (1965). Isolation and characterization of DNA from kinetoplasts of *Leishmania enriettii. Science* **147,** 754.
21. Edelman, M., Schiff, J. A., and Epstein, H. T. (1965). Studies of chloroplast development in *Euglena.* XII. Two types of satellite DNA. *J. Mol. Biol.* **11,** 769–774.
22. Edelman, M., Epstein, H. T., and Schiff, J. A. (1966). Isolation and characterization of DNA from the mitochondrial fraction of *Euglena. J. Mol. Biol.* **17,** 463–469.
23. Evans, T. E. (1966). Synthesis of a cytoplasmic DNA during the G_2 interphase of *Physarum polycephalum. Biochem. Biophys. Res. Commun.* **22,** 678.
24. Flechtner, V., and Sager, R. (1971). Preferential labeling of chloroplast DNA in cycloheximide-treated *Chlamydomonas. Abstr. 62nd Annu. Meeting Amer. Soc. Biol. Chemists,* June 1971.
25. Fukahara, H. (1969). Relative proportions of mitochondrial and nuclear DNA in yeast under various conditions of growth. *Eur. J. Biochem.* **11,** 135–139.
26. Gibor, A., and Izawa, M. (1963). The DNA content of the chloroplasts of *Acetabularia. Proc. Nat. Acad. Sci. U. S.* **50,** 1164–1169.
26a. Goldring, E. S., Grossman, L. I., Krupnick, D., Cryer, D. R., and Marmur, J. (1970). The petite mutation in yeast: Loss of mitochondrial deoxyribonucleic acid during induction of petites with ethidium bromide. *J. Mol. Biol.* **52,** 323.
27. Green, B. R., and Burton, H. (1970). *Acetabularia* chloroplast DNA: Electron microscopic visualization. *Science* **168,** 981.
28. Green, B. R., and Gordon, M. P. (1966). Replication of chloroplast DNA of tobacco. *Science* **152,** 1071–1074.
29. Green, B., Heilporn, V., Limbosch, S., Boloukhere, M., and Brachet, J. (1967). The cytoplasmic DNA's of *Acetabularia mediterranea. Proc. Nat. Acad. Sci. U. S.* **58,** 1351–1358.
30. Grossman, L., Goldring, E. S., and Marmur, K. (1969). Preferential synthesis of yeast mitochondrial DNA in the absence of protein synthesis. *J. Mol. Biol.* **46,** 367–376.
31. Guerineau, M., Grandchamp, C., Yotsuyanagi, Y., and Slonimski, P. P. (1968). Examen au microscope éléctronique du DNA mitochondrial de la levure. I. Molecules a deux éxtremités libres. II. Molecules circulaire. *C. R. Acad. Sci. Ser. D.* **266,** 1884–1887.
31a. Hickson, F. T., Roth, T. F., and Helinski, D. R. (1967). Circular DNA forms of a bacterial sex factor. *Proc. Nat. Acad. Sci. U. S.* **58,** 1731.
32. Hollenberg, C. P., Borst, P., Thuring, R. W. J., and Van Bruggen, E. F. J. (1969). Size, structure and genetic complexity of yeast mitochondrial DNA. *Biochim. Biophys. Acta* **186,** 417–419.
32a. Hudson, B., Upholt, W. B., Devinny, J., and Vinograd, J. (1969). The use of an ethidium analogue in the dye-buoyant density procedure for the isolation of closed circular DNA: The variation of the superhelix density of mitochondrial DNA. *Proc. Nat. Acad. Sci. U. S.* **62,** 813.
33. Inoki, S. (1957). Origin of the akinetoplastic strain of *Trypanosoma gambiense. Proc. Int. Genet. Symp. 1956.* (*Cytologia Suppl.* 1957), pp. 550–554.
34. Ishida, M. R., Tadatoshi, K., Matsubara, T., Hayashi, F., and Yokomura, E. (1969), Characterization of satellite DNA from the cells of *Porphyra tenera. Res. Reactor Inst. Kyoto Univ.* **2,** 73.

35. Iwamura, T. (1962). Characterization of the turnover of chloroplast deoxyribonucleic acid in *Chlorella. Biochim. Biophys. Acta,* **61,** 472–474.
36. Iwamura, T., and Kuwashima, S. (1969). Two DNA species in chloroplasts of *Chlorella. Biochim. Biophys. Acta* **174,** 330–339.
37. Karol, M. H., and Simpson, M. V. (1968). DNA biosynthesis by isolated mitochondria: A replicative rather than a repair process. *Science* **162,** 470–472.
38. Kirk, J. T. O. (1971). Will the real chloroplast DNA please stand up? *In* "Autonomy and Biogenesis of Mitochondria and Chloroplasts," *Aust. Acad. Sci. Symp.* (N. K. Boardman, A. W. Linnane, and R. M. Smillie, eds.), p. 267. North-Holland Publ., Amsterdam.
39. Kirk, J. T. O., and Tilney-Bassett, R. A. E. (1967). "The Plastids: Their Chemistry, Structure, Growth and Inheritance." W. H. Freeman, London.
40. Kirschner, R. H., Wolstenholme, D. R., and Gross, N. J. (1968). Replicating molecules of circular mitochondrial DNA. *Proc. Nat. Acad. Sci. U. S.* **60,** 1466–1472.
41. Kislev, N., Swift, H., and Bogorad, L. (1965). Nucleic acids of chloroplasts and mitochondria in Swiss chard. *J. Cell Biol.* **25,** 327–344.
42. Lane, D., and Sager, R. (Unpublished.)
43. Lane, D., Flechtner, V., Scheinbach, S., and Sager R. (Unpublished.)
44. Laurent, M., and Steinert, M. (1970). Electron microscopy of kinetoplastic DNA from *Trypanosoma mega. Proc. Nat. Acad. Sci. U. S.* **66,** 419–424.
45. Luck, D. J. L., and Reich, E. (1964). DNA in mitochondria of *Neurospora crassa. Proc. Nat. Acad. Sci. U. S.* **52,** 931–938.
46. Lyttleton, J. W., and Petersen, G. B. (1964). The isolation of deoxyribonucleic acid from platn tissues. *Biochim. Biophys. Acta* **80,** 391–398.
46a. Mandell, J. D., and Hershey, A. D. (1960). A fractionating column for analysis of nucleic acids. *Anal. Biochem.* **1,** 66–77.
46b. Manning, J. E., Wolstenholme, D. R., Ryan, R. S., Hunter, J. A., and Richards, O. C. (1971). Circular chloroplast DNA from *Euglena gracilis. Proc. Nat. Acad. Sci. U. S.* **68,** 1169.
47. Mehrota, B. D., and Mahler, H. R. (1968). Characterization of some unusual DNAs from the mitochondria from certain "petite" strains of *Saccharomyces cerevisiae. Arch. Biochem. Biophys.* **128,** 685–703.
48. Meselson, M., and Stahl, F. W. (1958). The replication of DNA in *Escherichia coli. Proc. Nat. Acad. Sci. U. S.* **44,** 671–682.
49. Meselson, M., Stahl, F. W., and Vinograd, J. (1957). Equilibrium sedimentation of macromolecules in density gradients. *Proc. Nat. Acad. Sci. U. S.* **43,** 581–588.
50. Meves, F. (1918). Die Plastosomen theorie der Vererbung. *Arch. Mikrosk. Anat.* **92,** 41.
50a. Miyazawa, Y., and Thomas, C. A., Jr. (1965). Nucleotide composition of short segments of DNA molecules. *J. Mol. Biol.* **11,** 223–237.
51. Mounolou, J. C., Jakob, H., and Slonimski, P. P. (1966). Mitochondrial DNA from yeast "petite" mutants: Specific changes of buoyant density corresponding to different cytoplasmic mutations. *Biochem. Biophys. Res. Commun.* **24,** 218–224.
52. Moustacchi, E., and Williamson, D. H. (1966). Physiological variations in satellite components of yeast DNA detected by density gradient centrifugation. *Biochem. Biophys. Res. Commun.* **23,** 56–61.
53. Nass, M. M. K., and Nass, S. (1963). Intramitochondrial fibers with DNA characteristics. I. Fixation and electron staining reactions. *J. Cell Biol.* **19,** 593–611.
54. Nass, S., and Nass, M. M. K. (1963). Intramitochondrial fibers with DNA characteristics. II. Enzymatic and other hydrolytic treatments. *J. Cell Biol.* **19,** 613–629.
55. Parsons, J. A., and Rustad, R. C. (1968). The distribution of DNA among dividing mitochondria of *Tetrahymena pyriformis. J. Cell Biol.* **37,** 683–693.

55a. Perlman, P. S., and Mahler, H. R. (1971). Molecular consequences of ethidium bromide mutagenesis. *Nature New Biol.* **231,** 12.
55b. Pettijohn, D. E., and Hanawalt, P. C. (1964). Evidence for repair-replication of ultraviolet damaged DNA in bacteria. *J. Mol. Biol.* **9,** 395.
56. Piko, L., Tyler, A., and Vinograd, J. (1967). Amount, location, priming capacity, circularity, and other properties of cytoplasmic DNA in sea urchin eggs. *Biol. Bull.* **132,** 68–90.
57. Rabinowitz, M., and Swift, H. (1970). Mitochondrial nucleic acids and their relation to the biogenesis of mitochondria. *Physiol. Rev.* **50,** 376–427.
58. Rabinowitz, M., Sinclari, J., DeSalle, L., Haselkorn, R., and Swift, H. H. (1965). Isolation of deoxyribonucleic acid from mitochondria of chick embryo heart and liver. *Proc. Nat. Acad. Sci. U. S.* **53,** 1126–1133.
59. Radloff, R., Bauer, W., and Vinograd, J. (1967). A dye-buoyant-density method for the detection and isolation of closed circular duplex DNA: The closed circular DNA in HeLa cells. *Proc. Nat. Acad. Sci. U. S.* **57,** 1514–1521.
60. Randall, J., and Disbrey, C. (1965). Evidence for the presence of DNA at basal body sites in *Tetrahymena pyriformis. Proc. Roy. Soc. Ser. B* **162,** 473–491.
61. Ray, D. S., and Hanawalt, P. C. (1964). Properties of the satellite DNA associated with the chloroplasts of *Euglena gracilis. J. Mol. Biol.* **9,** 812–824.
62. Ray, D. S., and Hanawalt, P. C. (1965). Satellite DNA components in *Euglena gracilis* cells lacking chloroplasts. *J. Mol. Biol.* **11,** 760–768.
63. Reich, E., and Luck, D. J. L. (1966). Replication and inheritance of mitochondrial DNA. *Proc. Nat. Acad. Sci. U. S.* **55,** 1600–1608.
64. Riou, G., and Delain, E. (1969). Electron microscopy of the circular kinetoplastic DNA from *Trypanosoma cruzi:* occurrence of catenated forms. *Proc. Nat. Acad. Sci. U. S.* **62,** 210–217.
65. Riou, G., and Paoletti, C. (1967). Preparation and properties of nuclear and satellite deoxyribonucleic acid of *Trypanosoma cruzi. J. Mol. Biol.* **28,** 377–382.
66. Ris, H., and Plaut, W. (1962). Ultrastructure of DNA containing areas in the chloroplast of *Chlamydomonas. J. Cell Biol.* **13,** 383–391.
67. Sager, R., and Ishida, M. R. (1963). Chloroplast DNA in *Chlamydomonas. Proc. Nat. Acad. Sci. U. S.* **50,** 725–730.
67a. Schatz, G., Haslbrunner, E., and Tuppy, H. (1964). Deoxyribonucleic acid associated with yeast mitochondria. *Biochem. Biophys. Res. Commun.* **15,** 127–132.
67b. Schegget, J., and Borst, P. (1971). DNA synthesis by isolated mitochondria. I. Effect of inhibitors and characterization of the product. II. Detection of product DNA hydrogen bonded to closed duplex circles. *Biochim. Biophys. Acta* **246,** 239.
68. Shapiro, L., Grossman, L. I., Marmur, J., and Kleinschmidt, A. K. (1968). Physical studies on the structure of yeast mitochondrial DNA. *J. Mol. Biol.* **33,** 907–922.
69. Shipp, W. S., Kieras, F. J., and Haselkorn, R. (1965). DNA associated with tobacco chloroplasts. *Proc. Nat. Acad. Sci. U. S.* **54,** 207–212.
70. Simpson, L. (1968). Effect of acriflavin on the kinetoplast of *Leishmania tarentolae. J. Cell Biol.* **37,** 660–682.
71. Smith, D., Tauro, P., Schweizer, E., and Halvorson, H. O. (1968). The replication of mitochondrial DNA during the cell cycle in *Saccharomyces lactis. Proc. Nat. Acad. Sci. U. S.* **60,** 936–942.
72. Smith-Sonneborn, J., and Plaut, W. (1967). Evidence for the presence of DNA in the pellicle of *Paramecium. J. Cell Sci.* **2,** 225–234.
73. Spencer, D., and Whitfeld, P. R. (1967). DNA synthesis in isolated chloroplasts. *Biochem. Biophys. Res. Commun.* **28,** 538–542.
74. Steinert, M., and Van Assel, S. (1967). The loss of kinetoplastic DNA in two species of Trypanosomatidae treated with acriflavine. *J. Cell Biol.* **34,** 489–503.

75. Stuart, K. D., and Hanson, E. D. (1967). Acriflavin induction of dykinetoplasy in *Leptomonas karyophilus*. *J. Protozool.* **14,** 39–43.
76. Stutz, E. (1970). The kinetic complexity of *Euglena gracilis* chloroplast DNA. *FEBS Lett.* **8,** 25–28.
77. Subak-Sharpe, J., and Sager, R. (Unpublished.)
78. Sueoka, N., Chiang, K-S., and Kates, J. R. (1967). Deoxyribonucleic acid replication in meiosis of *Chlamydomonas reinhardi*. I. Isotopic transfer experiments with a strain producing eight zoospores. *J. Mol. Biol.* **25,** 47–66.
79. Suyama, Y. (1966). Mitochondrial deoxyribonucleic acid of *Tetrahymena*. Its partial physical characterization. *Biochemistry* **5,** 2214–2221.
80. Suyama, Y., and Bonner, W. D., Jr. (1966). DNA from plant mitochondria. *Plant Physiol.* **41,** 383–388.
81. Suyama, Y., and Preer, J. R., Jr. (1965). Mitochondrial DNA from protozoa. *Genetics* **52,** 1051–1058.
82. Tewari, K. K., Votsch, W., Mahler, H. R., and Mackler, B. (1966). Biochemical correlates of respiratory deficiency. VI. Mitochondrial DNA. *J. Mol. Biol.* **20,** 453–481.
83. Tewari, K. K., and Wildman, S. G. (1970). Information content in the chloroplast DNA. *In* "Control of Organelle Development," *Symp. Soc. Exp. Biol.* **24,** 147–180.
83a. Trager, W. (1965). The kinetoplast and differentiation in certain parasitic protozoa. *Amer. Natur.* **99,** 255–266.
84. Van Bruggen, E. F. J., Runner, C. M., Borst, P., Ruttenberg, G. J. C. M., Kroon, A. M., and Schuurmans Stekhoven, F. M. A. H. (1968). Mitochondrial DNA. III. Electron microscopy of DNA released from mitochondria by osmotic shock. *Biochim. Biophys. Acta* **161,** 402–414.
85. Vinograd, J., and Lebowitz, J. (1966). Physical and topological properties of circular DNA. *J. Gen. Physiol.* **49,** 103–125.
86. Vogel, H. J., Thompson, J. S., and Shockman, G. D. (1970). Characteristic metabolic patterns of prokaryotes and eukaryotes. *Symp. Soc. Gen. Microbiol.*, No. XX, pp. 107–119.
86a. Weil, R., and Vinograd, J. (1963). The cyclic helix and cyclic coil forms of polyoma viral DNA. *Proc. Nat. Acad. Sci. U. S.* **50,** 730–738.
87. Wells, R., and Birnstiel, M. (1969). Kinetic complexity of chloroplastal deoxyribonucleic acid and mitochondrial deoxyribonucleic acid from higher plants. *Biochem. J.* **112,** 777–786.
88. Wells, R., and Ingle, J. (1970). The constancy of the buoyant density of chloroplast and mitochondrial DNA's in a range of higher plants. *Plant Physiol.* **46,** 178–179.
89. Wells, R., and Sager, R. (1971). Denaturation and the renaturation kinetics of chloroplast DNA from *Chlamydomonas reinhardi*. *J. Mol. Biol.* **58,** 611–622.
90. Werbitzki, F. W. (1910). Uber blepharophastlose *Trypanosomen*. *Zentralbl. Bakteriol. Parasitenk. Infektwnskr. Hyg. Abt. 1 Orig.* **53,** 303.
91. Wetmur, J. G., and Davidson, N. (1968). Kinetics of renaturation of DNA. *J. Mol. Biol.* **31,** 349–370.
92. Whitfeld, P. R., and Spencer, D. (1968). Buoyant density of tobacco and spinach chloroplast DNA. *Biochim. Biophys. Acta* **157,** 333–343.
93. Wolstenholme, D. R., and Dawid, I. B. (1967). Circular mitochondrial DNA from *Xenopus laevis* and *Rana pipiens*. *Chromosoma* **20,** 445.
94. Wood, D. D., and Luck, D. J. L. (1969). Hybridization of mitochondrial ribosomal RNA. *J. Mol. Biol.* **41,** 211–224.
95. Woodcock, C. L. F., and Bogorad, L. (1970). Evidence for variation in the quantity of DNA among plastids of *Acetabularia*. *J. Cell Biol.* **44,** 361–375.

GENETIC ANALYSIS OF CYTOPLASMIC SYSTEMS

Introduction

The next four chapters deal with the genetic evidence for the identification and properties of cytoplasmic genes. We will be concerned with methodological questions, such as the criteria for identifying cytoplasmic genes, and with experimental problems in the exploration of these new genetic systems.

Cytoplasmic genes were initially identified by their failure to follow Mendel's laws. In the past, when no independent supporting evidence (particularly the presence of cytoplasmic DNA's) was available, the criteria for recognizing cytoplasmic genes were often very indirect. As new cytoplasmic genetic systems become better known, more satisfactory methods of identifying cytoplasmic genes are becoming available. This introductory section examines criteria in general, and the succeeding chapters reconsider the methods being applied to each of the organisms under discussion.

Classically, the principal criteria for identification of cytoplasmic genes have been (*a*) differences in the results of reciprocal crosses; (*b*) non-Mendelian ratios; (*c*) extensive somatic segregation during vegetative or clonal growth; (*d*) infectivity in mycelial grafts in fungi; and (*e*) independence of nuclear and cytoplasmic gene assortment in suitable systems, e.g., heterokaryons. More recently, establishment of a cytoplasmic linkage group in *Chlamydomonas* has provided a powerful new means of identifying cytoplasmic genes by linkage relations; and in yeast, the physical alterations of mitochondrial DNA in *petite* mutants have provided a new means of identifying mitochondrial genes.

Having identified the cytoplasmic genes of interest, we will then turn to the study of their properties. What can we learn about cytoplasmic genetic systems from genetic analysis? For each genetic system (i.e., chloroplastal, mitochondrial) we may ask the same questions that motivated classical geneticists: How are the genes arranged, how many copies are present per organelle and per cell; how are they distributed at cell division; does recombination occur, and, if so, can the genes be mapped; are genes organized into operons or other regulatory groupings; and so forth.

In other words, everything that has been investigated in other genetic systems, nuclear, bacterial, or viral, can, in principle, be examined in cytoplasmic systems. As yet however, *Chlamydomonas* provides the only cytoplasmic genetic system sufficiently developed so that many of these questions can be approached experimentally. Let us start then with an examination of the methods and results of cytoplasmic genetic analysis in *Chlamydomonas.*

A note on terminology: non-Mendelian genes were originally identified by their failure to obey Mendel's laws. Subsequently the terms "cytoplasmic" and "nonchromosomal" (NC) were used. Now it seems likely that non-Mendelian genes are associated with specific cytoplasmic organelles. As this association becomes firmly established, it would seem appropriate to call them organelle genes. This term permits flexibility in denoting chloroplast genes, mitochondrial genes, etc., as new knowledge develops. In the balance of this chapter therefore the genes initially identified as non-Mendelian will be referred to as cytoplasmic or as organelle genes.

3

Cytoplasmic Genes in Chlamydomonas

INTRODUCTION

This chapter describes the research method and results which have led to the discovery of a cytoplasmic linkage group or "chromosome" in the sexual green alga *Chlamydomonas*. This linkage group is probably located in chloroplast DNA; the evidence for this association is summarized at the end of this chapter.

The story of this research begins with the choice of *Chlamydomonas* as a suitable organism for the genetic analysis of cytoplasmic heredity. Previous studies with higher plants had demonstrated the existence of cytoplasmic genes influencing many aspects of plant development and in particular of chloroplast development (*24*). However none of these extensive investigations had led to the establishment of a workable genetic system in which recombination and mapping could be carried out (cf. Chapter 6).

Following the precept that the best organism for solving a particular problem is the simplest one, I cast about for a suitable eukaryotic microorganism. Discussions with C. B. Van Niel and G. M. Smith led to the choice of *Chlamydomonas reinhardi* for a number of reasons. In the first place, it is a unicellular sexual microorganism with a simple sexual life cycle (*32, 58*), easy to control under laboratory conditions. Secondly, the organism grows well on a simple defined medium (*38*), can be grown in mass culture, and can be handled by standard bacteriological methods (*33*). Thirdly, of particular importance for our purposes, the organism contains one chloroplast and several mitochondria per cell, making available both of these organelle systems for investigations in the same organism.

The species we work with, *Chlamydomonas reinhardi,* is a facultative phototroph. It can be grown either photosynthetically or heterotrophically with acetate as its preferred carbon source (*38*). When we examined *Chlamydomonas* in the electron microscope (*43, 44*), we found yet another potential advantage of the organism. As shown in Plate III and discussed in the the next section, the cytoplasm is structurally complex with many of the components found in the cells of higher forms, such as endoplasmic reticulum, Golgi apparatus, microtubules, lysosomes, and ribosomes, as well as the complex basal body (blepharoplast) at the base of the flagella. The structural complexity is relevant to our interest in the genetic control of subcellular structure, as is the fundamental similarity of cytoplasmic organization in *Chlamydomonas* to that seen in higher organisms.

Chlamydomonas reinhardi is a soil organism, a member of a large genus including soil, fresh water, and marine forms. Closely related organisms include *Dunaliella,* which grows at high salt concentration (e.g., in the Great Salt Lake) and *Haematococcus,* which grows at high altitudes and is responsible for the "pink snow" seen on glaciers. Within the genus *Chlamydomonas* are species ranging from isogamous (both gametes of equal size) to heterogamous (one gamete small and the other large). The genus includes both homothallic forms in which one haploid cell can give rise to gametes which mate, form zygotes, and undergo meiosis; and heterothallic forms (like *C. reinhardi*) in which there are two genetically determined mating types or sexes.

This diversity of form and physiology suggests considerable genetic plasticity within the genus. In particular, the diversity in sexual differentiation is of interest in relation to the modes of transmission of cytoplasmic genes.

Of paramount importance for our investigation of cytoplasmic genetics, however, have been several unsuspected properties of the organism. With *Chlamydomonas* we have been able to overcome three of the principal obstacles to research on cytoplasmic heredity encountered with other organisms: inability to induce mutations, difficulty of distinguishing between nuclear and cytoplasmic mutations, and the absence of evidence of recombination or of linkage of cytoplasmic genes.

Chlamydomonas reinhardi was available as an experimental organism thanks to the prior work of G. M. Smith who isolated this species and worked out its sexual life cycle under laboratory conditions (*58, 59*). Fortunately for us, this species is heterothallic and isogamous: mating is controlled by a single nuclear gene, cells are either mt^+ or mt^-, and gametes contribute equal cell contents to the zygote. The sexual life cycle is shown in Fig. 3.1. Nuclear genes show 2:2 segregation among the four zoospores in this simple haploid life cycle.

We detected and identified the first cytoplasmic mutation in *Chlamydomonas* by its special pattern of inheritance (*31*). A mutant strain exhibiting a high level of resistance to streptomycin (to 500 μg/ml) was selected on streptomycin agar and test-crossed with a wild-type strain. The results diagrammed in Fig. 3.2 indicate that the allele conferring resistance (*sm-r*) was transmitted to all of the progency while that conferring sensitivity (*sm-s*) was not transmitted at all. We call this 4:0 inheritance, in contrast to 2:2 Mendelian ratios. In the F_1 backcross generation, the mt^+ parents transmitted *sm-r* to all progeny as in the original cross, but mt^- parents carrying the *sm-r* gene did not transmit it.

In an intensive study of the inheritance of *sm*, a series of four generations of backcrosses were carried out to look for evidence of multifactorial inheritance. No such evidence was found. On the contrary, the results showed that the uniparental pattern of transmission of *sr* was dependent upon mating type: whichever allele was carried by the mt^+ parent was transmitted to the progeny.

This pattern of 4:0 inheritance closely resembles classic maternal inheritance, the hallmark of cytoplasmic heredity in higher plants (*24*). Subsequently many other mutations were found which exhibited the same uniparental pattern of transmission. This pattern provides a convenient means of distinguishing between "Mendelian" (i.e., nuclear) and "non-Mendelian" (i.e., cytoplasmic) genes.

The recovery of this first cytoplasmic mutation following growth of a streptomycin-sensitive strain on streptomycin agar led us to examine streptomycin as a potential mutagen (*33, 34, 53, 54*). As will be discussed later in this chapter, we found that the drug was indeed mutagenic for many cytoplasmic genes, but not for nuclear ones. Thus streptomycin mutagenesis has been a powerful tool in the acquisition of our starting material, a collection of cytoplasmic gene mutations.

The fact that cytoplasmic genes followed the pattern of maternal inheritance in *Chlamydomonas* came as a distinct surprise, and aroused our interest in the mechanism by which it was accomplished. In higher plants maternal inheritance has been attributed to the unequal contributions to the fertilized egg of cytoplasm from the female and male (pollen) parents (*24*). In *Chlamydomonas reinhardi*, however, which is an isogamous species, both parents contribute equal amounts of cytoplasm to the zygote by complete fusion of the two isogametes.

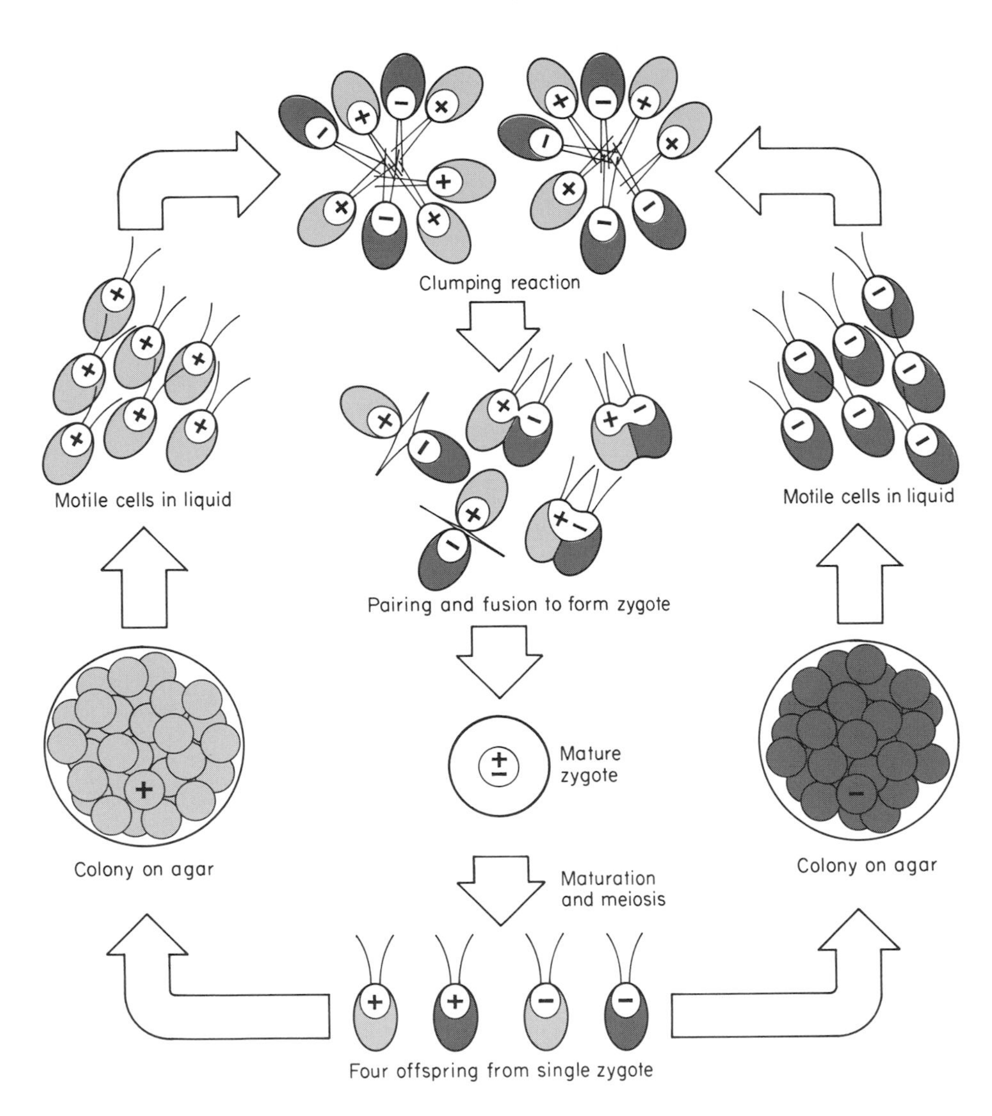

Fig. 3.1. The life cycle of *Chlamydomonas reinhardi,* showing the segregation of mating type, denoted by + and −, and of an unlinked nuclear gene pair denoted by light and dark shading. After a clumping reaction, pairs of cells of opposite mating type fuse to form zygotes, which are the only diploid stage of the usual life cycle (cf. however, exceptional diploid clones, p. 69). After a period of maturation (several days) zygotes germinate with the release of four zoospores, the four products of meiosis. From (32).

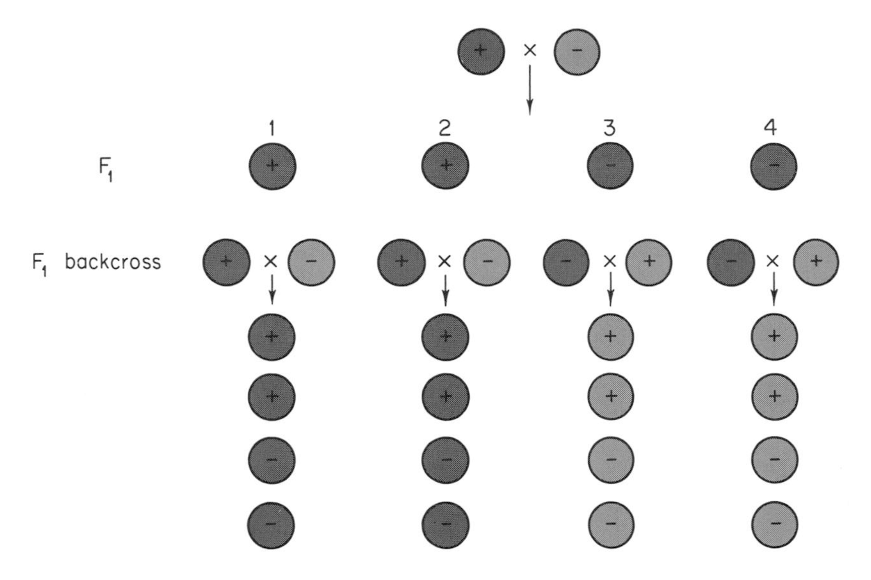

Fig. 3.2. Inheritance of cytoplasmic streptomycin resistance in *Chlamydomonas*. Plus and minus signs refer to mating type. In the initial cross, *sm-r* mt^+ $y1^+$ × *sm-s* mt^- $y1^-$, all progeny were *sm-r*, though zoospores of each zygote segregated 2:2 for the nuclear gene pairs mt^+/mt^- and yl^+/yl^-. F_1 clones of *sm-r* mt^+ backcrossed to *sm-s* mt^- produced all *sm-r* progeny (4:0 segregation) as in the initial cross, but F_1 clones of *sm-r* mt^- backcrossed to *sm-s* mt^+ produced only *sm-s* progeny (0:4 segregation). Dark shaded cells indicate streptomycin resistant cells and light shaded cells indicate streptomycin sensitive cells. From (*31*).

It was evident, therefore, that the widely accepted view of maternal inheritance as a simple consequence of cytoplasmic exclusion did not apply to *Chlamydomonas.* In examining the higher plant literature, we found considerable evidence there too that was not in keeping with the cytoplasmic exclusion hypothesis, in particular the occurrence of partially biparental as well as maternal inheritance of non-Mendelian genes (*24*). Evidently then, biparental inheritance could occur despite the miniscule contribution of cytoplasm from the male parent, and maternal inheritance could occur in the face of equal cytoplasmic contributions from the two parents. These considerations led us to look in *Chlamydomonas* for an enzymatic mechanism of maternal inheritance, based upon the destruction in the zygote of organelle DNA's from the male parent. These studies will be discussed later in this chapter.

The mechanism of maternal inheritance in *Chlamydomonas* was of special concern to us because uniparental transmission provides a drastic impediment to genetic analysis. Because of this phenomenon in other organisms no recombinants were found and consequently neither recombination analysis nor conventional mapping procedures could be applied.

Overcoming the obstacle of maternal inheritance represented a fundamental advance in cytoplasmic genetic analysis with *Chlamydomonas.* Mrs. Ramanis and I found that maternal inheritance could be converted to a form of biparental inheritance by a variety of means, most dramatically by UV irradiation of the mt^+ parent just before mating (*47*). Following this treatment, cytoplasmic genes are transmitted from the zygotes to all progeny, and recombination then proceeds during vegetative growth of progeny clones.

Thus the pattern of maternal inheritance provides the means by which we distinguish cytoplasmic genes from nuclear ones, but the pattern can be converted to biparental for purposes of genetic analysis.

Summarizing these introductory remarks, *Chlamydomonas* has been found to possess many characteristics which make it suitable for cytoplasmic genetic analysis. The principal aim of this chapter is to present our current understanding of the organization and behavior of cytoplasmic genes in *Chlamydomonas,* and the lines of evidence that have culminated in our identification of a linkage group in chloroplast DNA. Before turning to the genetic evidence, however, we will first consider some relevant properties of the organism.

SOME PROPERTIES OF CHLAMYDOMONAS

Morphology and Fine Structure

Studies of the fine structure of normal green and mutant yellow strains of *Chlamydomonas,* first reported by us (*43, 44*), have subsequently been extended, especially by Ohad *et al.* (*27*), Ringo (*29*), and Johnson and Porter (*21*). The principal features of the morphology of *Chlamydomonas,* based upon these studies, are shown diagrammatically in Fig. 3.3 and in Plates III, IV, and V. The organism is enclosed within a cell membrane, surrounded by a thin cell wall and an outer capsule of variable thickness. The single chloroplast is large and cup-shaped, encompassing about 50% of the volume of the cell. Within the chloroplast there are lamellar membranes where photosynthesis occurs, as well as a starch-synthesizing body called the pyrenoid, and the eyespot. The organism is motile, swimming by means of two anterior flagella; it is phototactic, detecting light by means of the eyespot located within the chloroplast. In the light microscope, one may also observe two anterior contractile vacuoles and the nucleus with a large nucleolus. The haploid chromosome number has been variously reported as 8–16. A recent cytological study (*25c*) as well as unpublished genetic data (*19, 25a*) sup-

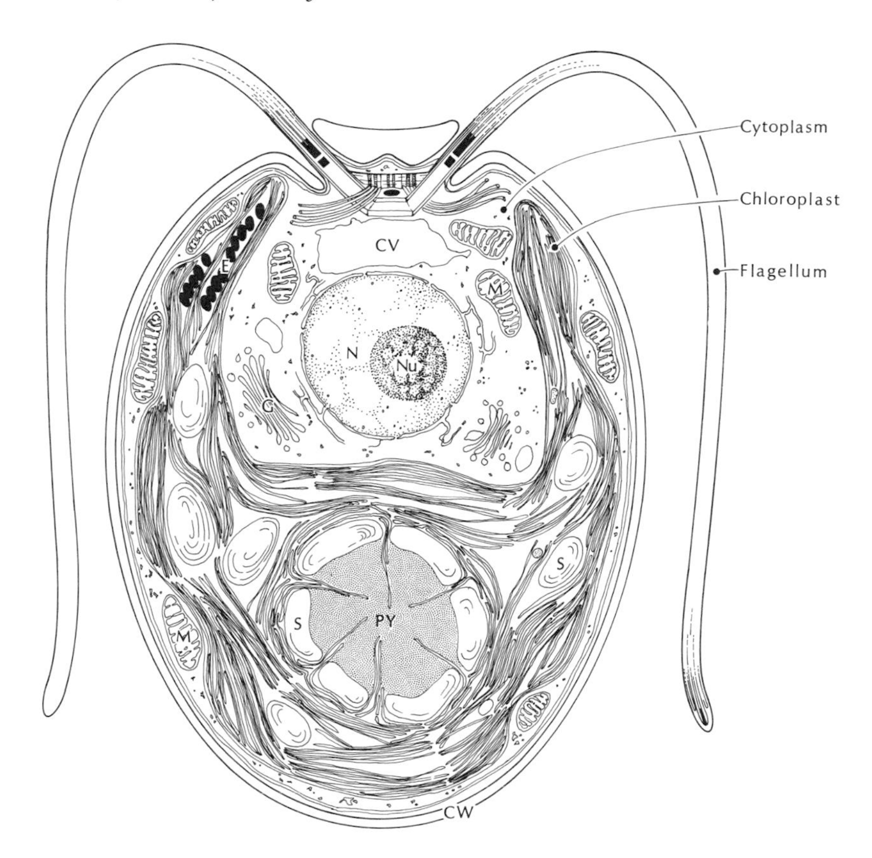

Fig. 3.3. Diagrammatic representation of *Chlamydomonas* based on electron micrographs. At the anterior end are two flagella which penetrate the cell wall (CW) and are associated with the basal body and microtubules. Two contractile vacuoles (CV) and mitochondria (M) are also found in that region. One large cup-shaped chloroplast contains eyespot (E), pyrenoid (PY), and starch grains (S), as well as lamellar membranes (ribosomes and DNA not seen at this magnification). Golgi membranes (G), mitochondria, cytoplasmic ribosomes, and endoplasmic reticulum surround the nucleus (N) with its chromosomes and nucleolus (Nu). (See Plates III, IV, V, XIII, and XIV for further details.)

port the higher value. The small size of the chromosomes shown in Fig. 3.4 indicates why establishing the correct chromosome number has been so difficult.

An electron microscope view of the normal green cell is shown in Plate III. Here one may see the complex structural organization of the cytoplasm, which apart from the chloroplast, resembles that of animal cells. Typical cytoplasmic components include numerous mitochondria, several sets of Golgi membranes, smooth and rough membranes of the endoplasmic reticulum, ribosomes, microtubules, microbodies, and

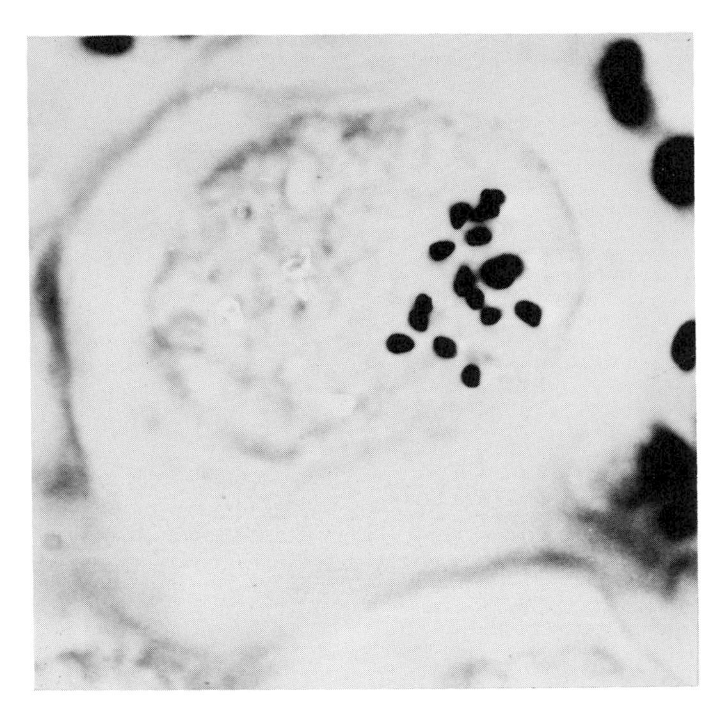

Fig. 3.4. Composite from photographs taken at two focal levels showing the 16 chromosomes of *Chlamydomonas reinhardi* at metaphase of mitosis. From McVittie and Davies (*25c*).

other components similar to those seen in other organisms. The fine structures of vegetative cells during cell division, with special emphasis on microtubules, was described by Johnson and Porter (*21*). Details of chloroplast structure will be discussed in Chapter 8 in relation to chloroplast biogenesis. Here we note that the chloroplast of the wild-type cell grown in the dark contains chlorophyll, lamellar membranes, and ribosomes, and looks very similar to that of the light-grown cell. We found a yellow mutant, *y-1* which behaves like a higher plant, in not being able to make chlorophyll in the dark (*43*). As shown in Plate IV, when grown in the dark, the yellow mutant has neither chlorophyll nor organized lamellar membrances, although chloroplast ribosomes are still present. When the dark-grown yellow mutant is exposed to light, chlorophyll synthesis begins and lamellar membranes form anew. Thus the yellow mutant provides an excellent system in which to investigate lamellar membrane formation (*27*). Abundant chloroplast ribosomes, estimated at about two-thirds the number present in the cytoplasm, are found both in light- and dark-grown cells of both wild-type and the yellow mutant.

The apparently smaller size of chloroplast ribosomes compared with those of the cytoplasm (*27, 44*) in electron micrographs led us to undertake a physical comparison of isolated ribosomes from the two cell compartments. We found (*40*) that cytoplasmic ribosomes had a sedimentation constant of 83 S at infinite dilution and contained RNA's of 24 S and 16 S as determined by sedimentation rate in the analytic ultracentrifuge. Thus, they resemble the 80 S ribosomes of animals and higher plants in S value. However, they resemble bacterial ribosomes in requiring high concentrations of Mg^{2+} (0.01 *M*) for monosome stability. When the Mg^{2+} concentration is lowered, the monosomes dissociate to 61 S and 41 S subunits. The chloroplast ribosomes were found to be approximately 70 S, highly Mg^{2+}-dependent, and easily dissociated to a stable 50 S and an unidentified smaller subunit. Subsequent studies by Hoober and Blobel (*20*) confirmed and extended these observations.

Mitochondria are present under all growth conditions (*27, 43, 44*). In dark-grown cells, the mitochondria are larger, have a denser matrix, and have more cristae than mitochondria of light-grown cells. Mitochondria are frequently seen lying between the chloroplast and the outer cell membrane, and a cluster of mitochondria are often seen close to the flagellar basal bodies. Mitochondrial ribosomes and DNA are presumed to be present here as they are in other organisms. However, as in other organisms, they are not evident in electron micrographs.

The intricate structure of the flagellar apparatus was beautifully described by Ringo (*29*) (Fig. 3.5 and Plate V). Extensive studies, both genetic and structural, were initiated by Randall and his students in an attempt to examine the genetic control of flagellar structure in *Chlamydomonas* (*28, 64*). More recently, further studies of flagellar development have been reported by Rosenbaum *et al.* (*30a*). Although only nuclear gene mutations were identified in these studies, some of the potentially most interesting mutants have not been studied genetically because they have abnormal flagella or none at all, and consequently cannot mate.

DNA can be detected in the chloroplast of *Chlamydomonas* by cytochemical means, as was first shown by Ris and Plaut (*30*). A few (frequently two) particles can be seen within the chloroplast in whole cell preparations, either by Feulgen staining or by acridine orange fluorescence. Chloroplast DNA has also been identified in electron micrographs of sectioned cells. The DNA fibers in *Chlamydomonas* chloroplasts, for example, have been identified not only by their appearance but particularly by their disappearance after treatment of the preparations with DNase (*30*).

If the chloroplast DNA of *Chlamydomonas* is localized in one or two

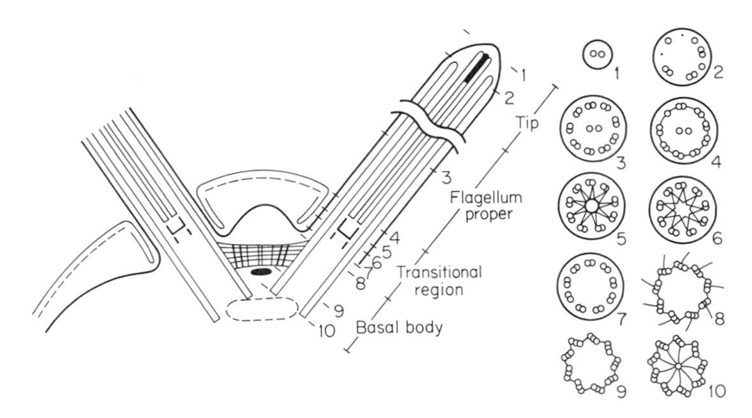

Fig. 3.5. A schematic drawing of an idealized longitudinal section through both basal bodies of *Chlamydomonas.* The tilt of the basal bodies with respect to one another is not shown, and the position of the two proximal connecting fibers, which would be out of the plane of the drawing, is indicated by a dashed line. The four regions of the flagellum are designated, and ten typical cross sections show the internal arrangement of fibers at the numbered points marked along the length of one flagellum. From *(29).*

regions within the chloroplast, it should be visible in the light microscope. The analytic amount of DNA per chloroplast is at least 4×10^9 daltons (cf. p. 22), a value which is above the minimal amount of 1×10^9 daltons visible by light microscopy after Feulgen staining. Even less DNA can be seen by acridine orange fluorescence. Thus, the presence of Feulgen staining particles in the chloroplast supports the hypothesis of DNA localization.

DNA has not been cytochemically detected elsewhere in the cytoplasm of *Chlamydomonas,* not even in mitochondria. The amount and arrangement of mitochondrial DNA in *Chlamydomonas* is not known, but may well be below the quantity needed for visualization by means of light microscopy.

Growth and Synchrony

Liquid cultures of *Chlamydomonas* grow exponentially at rates of one to four doublings in 24 hours depending on the culture medium and conditions, especially light intensity and temperature. As shown in Fig. 3.6, exponential growth can be maintained up to cell densities of 5×10^6 cells per milliliter.

With appropriate dilution, cultures may be maintained in synchronous growth indefinitely by means of diurnal light–dark cycles. The cell cycle in the unicellular algae, differs from that of bacterial and mammalian cells in the loose coupling between growth and cell division. In

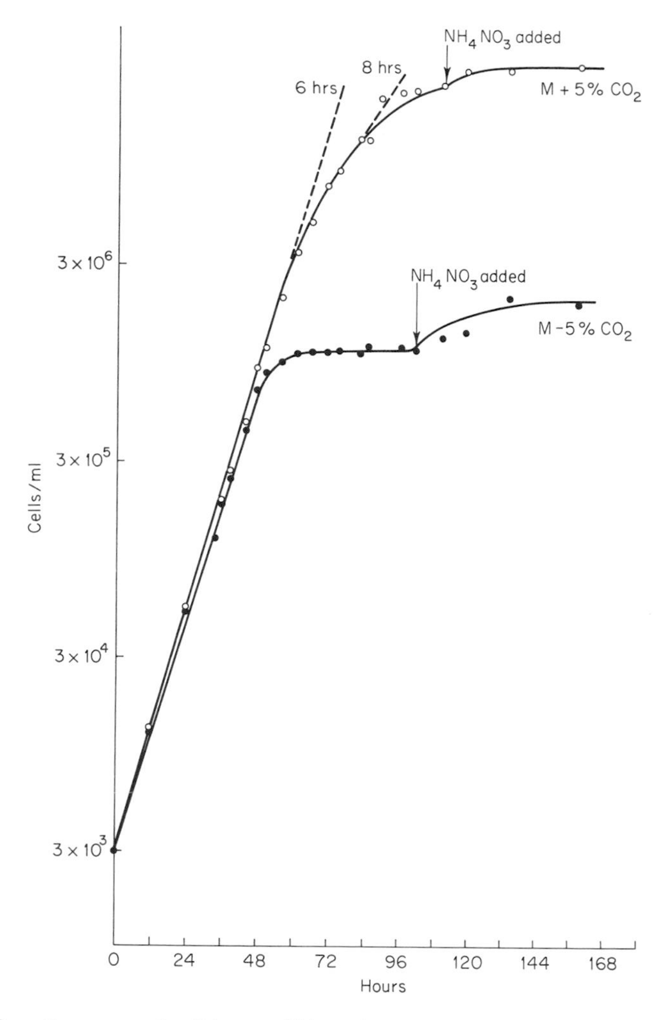

Fig. 3.6. Growth curves of wild-type *Chlamydomonas* (strain *21gr*) grown on minimal medium (M) with and without added 5% CO_2. Growth is exponential with a doubling time of 6 hours, stopping at about 1×10^6 cells/ ml in the absence of added CO_2 and continuing to higher densities with added CO_2. Growth stops for an unknown reason, not the lack of a nitrogen source (as shown by the results obtained with added NH_4NO_3). The dotted lines indicate the doubling time corresponding to the indicated slopes.

synchronous cultures of *Chlamydomonas,* cell division occurs just once in the 24-hour cycle, while within the enlarging cell mass, successive rounds of DNA replication and concomitant increases in other cell components occur, giving rise to four, eight, or even sixteen cells at the time of cell division.

Synchronous cultures are of particular value in studies of cytoplasmic genetics because the nuclear and chloroplast DNA's replicate on different time schedules (*5*). This temporal separation offers excellent experimental possibilities for studying replication of the two DNA systems independently.

DNA Replication in the Cell Cycle

As discussed in Chapter 2, only two classes of DNA have been satisfactorily identified in *Chlamydomonas* (Fig. 2.2)—nuclear DNA (α band) with a buoyant density of 1.724 gm/cm^3 in CsCl and chloroplast DNA (β band) banding at 1.694 gm/cm^3 (*41, 60*). In addition to these DNA's a special M band DNA seen only in zygotes has been described by Sueoka *et al.* (*60*). No such band has been found in the strains studied in our laboratory (*25*) and the identity of this component remains obscure. A band appearing at about 1.715 gm/cm^3 (called γ band) has been seen in most preparations of DNA from vegetative cells, gametes, and zygotes. Its identification as mitochondrial DNA has been proposed by Sueoka *et al.* (*60*) and Chiang (*4*) but the evidence as yet is circumstantial.

The behavior of nuclear and chloroplast DNA's in the cell cycle has been studied in several laboratories. Sueoka (*59a*) first demonstrated a pattern of semiconservative replication of nuclear DNA in exponentially growing cultures of *Chlamydomonas*. Under the growing conditions used, cells were dividing into four or eight daughters as well as into two. Consequently isotopic transfer experiments from ^{15}N to ^{14}N medium did not give the kind of clear-cut results obtained with *Escherichia coli* in which only hybrid density DNA was found after one doubling. Instead, heavy DNA (^{15}N–^{15}N) and light DNA (^{14}N–^{14}N) were seen as well as hybrid DNA (^{14}N–^{14}N) during the replication process. This finding probably resulted from initiation of a second round of DNA replication before completion of the first round. Subsequent studies with synchronous cultures have confirmed the semiconservative pattern of nuclear DNA replication in *Chlamydomonas* (*4*).

The replication of nuclear, chloroplast, and γ band DNA's was examined by Chiang and Sueoka (*5*) with an ^{15}N–^{14}N transfer experiment carried out with a synchronously growing culture (cf. Chapter 2). During the 12-hour light period of the light–dark cycle, the chloroplast DNA peak was seen to move from the fully heavy (^{15}N–^{15}N) to the hybrid position, representing one round of replication; then two peaks appeared, one hybrid and one fully light, resulting from further rounds of replication in the ^{14}N medium. The γ band also replicated during the light cycle. The nuclear DNA, on the other hand, remained in the fully heavy position throughout the light period and replicated during the dark period.

DNA synthesis has also been examined by means of incorporation of radioisotope-labeled adenine instead of by density shift (*4, 9a*). The synthesis of nuclear DNA in the dark period was confirmed, but incorporation of label into chloroplast DNA was seen both in the light and in the dark. Whether the dark incorporation represents replication or repair, and how extensive it may be, are questions which have not yet been clarified.

Gametogenesis and the Sexual Life Cycle

The sexual life cycle of *Chlamydomonas* is shown diagrammatically in Fig. 3.1. Early stages in mating are shown in Plate VI. In this isogamous species there are two mating types which, as vegetative cells, are morphologically indistinguishable. Smith and Regnery (*59*) first showed that mating type is determined by a pair of nuclear genes mt^+ and mt^-, which behave like alleles, segregating 2:2 in meiosis. In their studies, the mating reaction was not well controlled and the yield of zygotes was usually low.

Our earliest work with *Chlamydomonas* (*38, 39*) concerned the differentiation of gametes and control of the mating reaction. Exponentially growing cultures do not mate. Establishing conditions for effective mating with a high yield of zygotes was a necessary step in developing a successful methodology for genetic analysis with this organism, especially for cytoplasmic genetic analysis in which recovery of a random sample of zygotes from a random sample of gametes was important. Furthermore, all biophysical studies of DNA from zygotes have depended upon achieving close to 100% mating gametes.

Both the genetic and the biophysical studies were facilitated by our discovery that mating could be controlled by nitrogen starvation of the parental cultures (*39*). In a study comparing the effect upon mating efficiency of depleting each of the components of the medium, we found that nitrogen was the only component whose depletion enhanced the mating response. Indeed gametogenesis in *Chlamydomonas* appeared to be regulated by a balance between energy and nitrogen metabolism reminiscent of the carbon–nitrogen balance which regulates fruiting in higher plants. An example of our results is shown in Fig. 3.7. The detailed metabolic basis of this empirical finding is still unknown, but the phenomenon itself has been exploited by us and by all subsequent investigators working with *Chlamydomonas.*

Since those first studies we and other investigators (*4, 7, 9, 11, 13, 22, 23, 25, 37, 60*) have used nitrogen-depletion routinely to ensure high

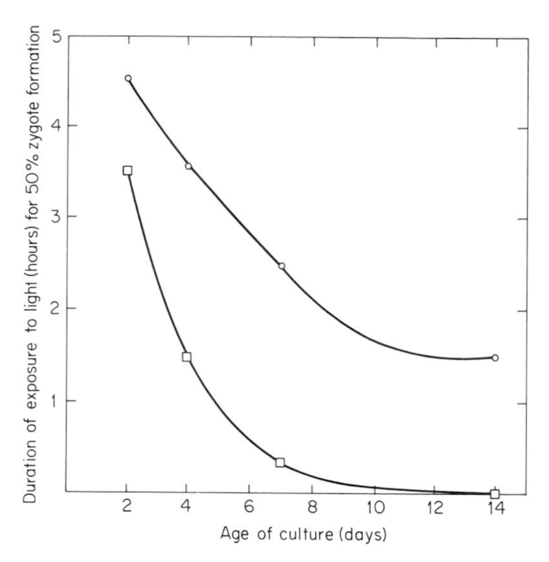

Fig. 3.7. Effect of age of culture and nitrogen depletion on zygote formation. Cells were grown either on a regular high-nitrogen medium (○, 0.03% NH_4NO_3) or on a special low-nitrogen medium (□, 0.003% NH_4NO_3) and tested at intervals for zygote forming ability. The test consisted of washing and resuspending equal numbers of cells of the two mating types in nitrogen-free medium in light and determining the time required for 50% zygote formation. Cells grown on low-nitrogen medium reached suitably gametic state much faster and more reproducibly than those grown on regular medium. No other component of the medium had this effect. Increasing age of the culture also increases zygote forming ability, probably by the same mechanism as nitrogen depletion. From (*39*).

yields of zygotes in genetic analysis and in biophysical studies of zygote DNA. Kates and Jones (*22*) incorporated nitrogen starvation into their regime for inducing synchronous mating of synchronous cultures grown in liquid medium on a 12-hour light/12-hour dark cycle. Kates *et al.* (*23*) found that during differentiation of gametes by nitrogen depletion, under synchronous conditions, the cells double twice, and the DNA content of gametes is the same as that of vegetative cells, indicating that the DNA (more precisely, the nuclear DNA) also replicates twice at this time. However, both Chiang (*4*) and we (*25*) found that chloroplast DNA apparently undergoes only one doubling during gametogenesis since gametes contain only 7% chloroplast DNA whereas the vegetative cells from which they are derived contain about 14%.

Thus, the physiology of the mating process appears to involve a number of regulatory events, determining the number of rounds of nuclear and chloroplast DNA replication and cell division, leading to the emergence of gametes capable of cell fusion. The gametes differ from vegetative cells in being smaller, unable to undergo further DNA replication or cell division without the addition of a nitrogen supply, and above all, in being capable of mating.

The first step in the mating process is the production, by mt^- cells, of a diffusable substance which attracts mt^+ gametes, leading to a clumping reaction (*66*). So far as is known, these "sexual" substances are involved only in the clumping process, and not in later stages of mating.

The fusion process itself has been described in an electron microscope study by Friedmann *et al.* (*10*) who noted a morphological difference between the mating types which develops during mating, i.e., a fertilization tubule produced by mt^+ cells (Plate VI). Further stages in zygote formation, described by Cavalier-Smith (*2*) involve cell fusion followed by fusion of the two chloroplasts and of the nuclei. Subsequent stages in internal reorganization leading to formation of four zoospores withing the zygote wall, have not been adequately described as yet. Of particular interest to us in terms of genetic analysis is the behavior of nuclear and chloroplast DNA's during zygote development and germination.

DNA Replication in Zygotes

The replication of nuclear DNA in zygotes was studied by Sueoka *et al.* (*60*) who found that only one round of replication occurred between mating and the emergence of eight zoospores, i.e., during two meiotic plus one mitotic division. This result implied that the octospores contained only half the nuclear DNA content of the gametes. Subsequently, it was shown (*4*) that the zoospores undergo an extra round of replication before the second mitotic division which brings their DNA content back to that of the gametes and vegetative cells.

The replication of chloroplast DNA in zygotes is a very complex phenomenon. As noted above, the cytoplasmic genes of *Chlamydomonas* show maternal inheritance, i.e., transmission from the female (mt^+) parent to all progeny, but little or no transmission from the male parent. What happens to the male cytoplasmic genome in the zygote which leads to its total disappearance as a genetic entity?

If maternally inherited genes are located in chloroplast DNA, as we think they are, then the behavior of chloroplast DNA molecules from the two parents should be different in the zygote, paralleling the behavior of the genetic markers. To investigate this possibility we have been following the fates during zygote development of parental DNA's distinguished by density labeling with ^{15}N and by radioisotope labeling with ^{3}H- and ^{14}C-adenine. Our experiments have revealed a complex process by which chloroplast DNA's from the two parents are distinguished and differently treated in the zygote (*25, 42*).

The first events occur soon after mating. As seen in CsCl gradients, chloroplast DNA from the female (mt^+) parent undergoes a density shift during the first 24 hours after zygote formation, becoming several density units lighter than it is in vegetative cells or in gametes (Figs. 3.8 and 3.9). In reciprocal crosses of ^{14}N- and ^{15}N-labeled parents no corresponding density shift is seen in the chloroplast DNA from the male (mt^-) parent. Thus the density labeling procedure clearly demonstrates the different paths followed by chloroplast DNA's from male and female parents during the first 24 hours in the zygote (*25*). The mechanism of this density shift is unknown but it could result from the covalent addition of new components, e.g., methyl groups, which are known to decrease the bouyant density of the DNA. Such a mechanism has been found in host modification and restriction in bacteria (*1*).

At later times in zygote maturation, the differences in the results of ^{14}N $\times$ ^{15}N reciprocal crosses are still evident in density gradients, as shown in Figs. 3.10–3.11. Thus it is evident that the two parental chloroplast DNA's are experiencing different fates in the zygote (*42*).

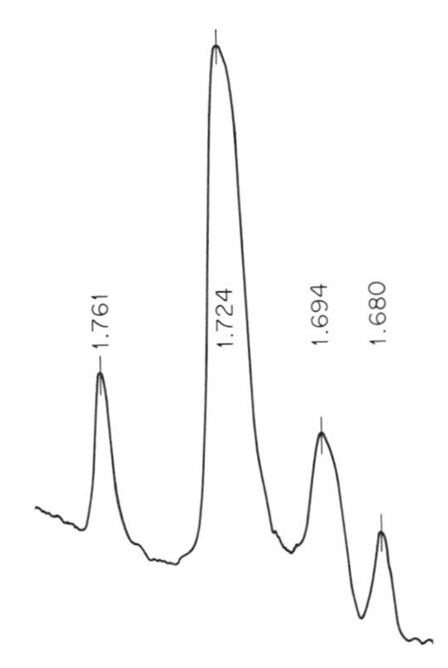

Fig. 3.8. Principal DNA's of *Chlamydomonas.* Microdensitometer tracing of DNA's from gametes (mating type plus) centrifuged to equilibrium in CsCl density gradient. Bands seen are: nuclear DNA at 1.724 gm/cm^3 chloroplast DNA at 1.694 gm/cm^3 as computed from markers at 1.761 gm/cm^3 (SP-15 phage DNA from Dr. Marmur) and 1.680 gm/cm^3 (crab poly dAT from Dr. Sueoka). In gametes, chloroplast DNA is 7% of total DNA, based on calibration with known amount of SP-15 DNA. From (*42*).

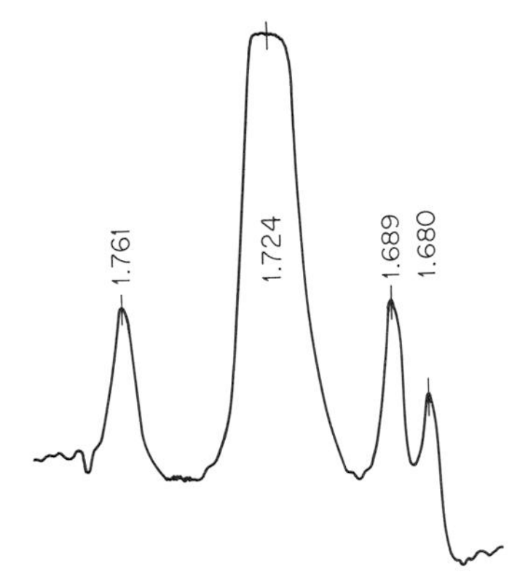

Fig. 3.9. DNA from zygotes of $^{14}N \times ^{14}N$ cross, extracted and centrifugated as in Fig. 3.8. Zygotes kept in nitrogen-free medium for 24 hours after mating and then harvested and DNA extracted. Nuclear DNA bands at 1.724 gm/cm^3 and chloroplast DNA at new density of 1.689 gm/cm^3. Makers as in Fig. 3.8 at 1.761 gm/cm^3 and 1.680 gm/cm^3.

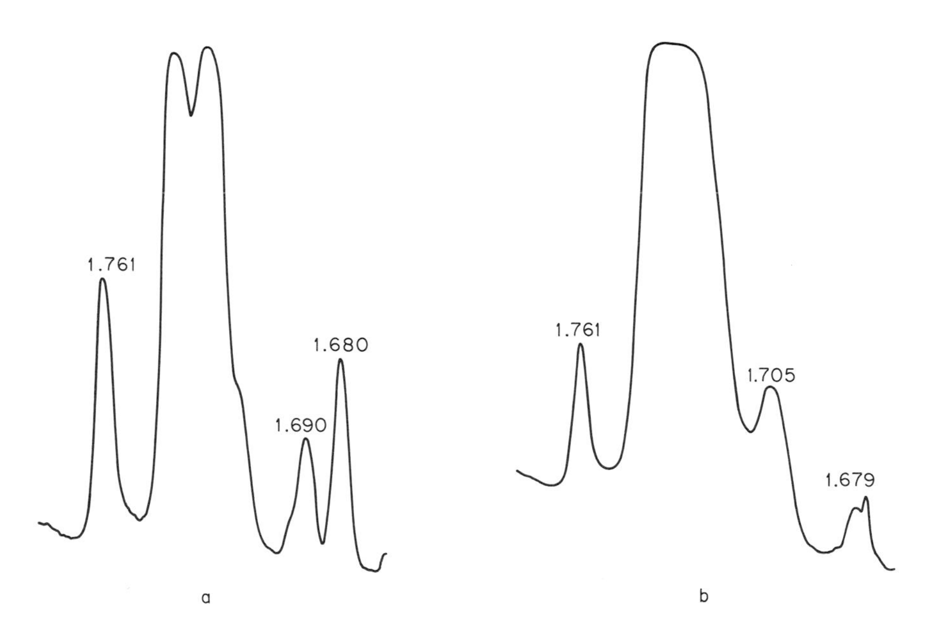

Fig. 3.10. (a) DNA from zygotes of $^{14}N \times ^{15}N$ cross, kept 1 day in nitrogen-free medium, 6 days in ^{14}N in dark, and 24 hours in light. DNA prepared and marker DNA's added as in Fig. 3.8. Nuclear DNA (overloaded) at 1.738 and 1.723 gm/cm^3. One chloroplast peak at 1.690 gm/cm^3. (b) DNA from zygotes of $^{15}N \times ^{14}N$ cross, kept 1 day in nitrogen-free medium, 6 days in ^{14}N in dark, and 24 hours in light. Nuclear DNA overloaded. One chloroplast peak at 1.705 gm/cm^3 (poly dAT shows two peaks).

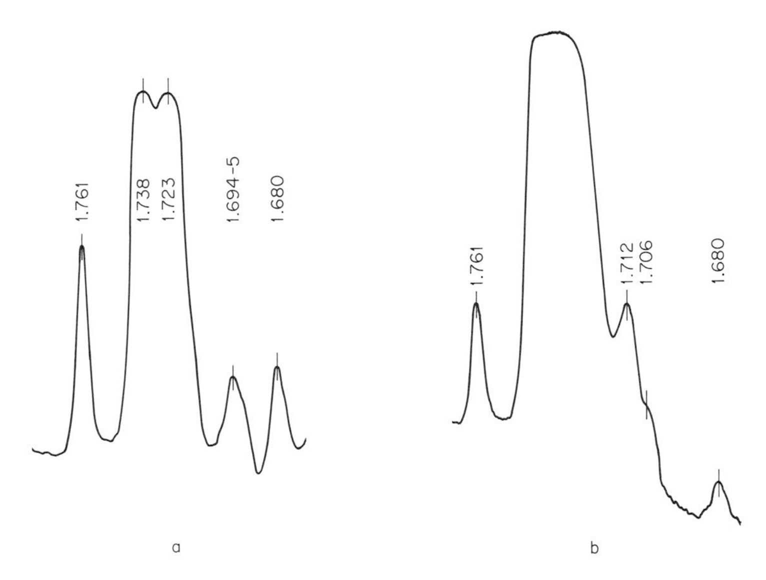

Fig. 3.11. (a) DNA from zygotes of $^{14}N \times {}^{15}N$ cross, kept 1 day in nitrogen-free medium, 6 days in ^{15}N in dark, and 24 hours in light. DNA prepared as in Figs. 3.8 and 3.9. Nuclear DNA (overloaded) peaks at 1.738 and 1.723 gm/cm^3, representing ^{15}N and ^{14}N components. One chloroplast peak at 1.694–1.695 gm/cm^3. (b) DNA from zygotes of $^{15}N \times {}^{14}N$ cross kept under the same regime as in (a). Nuclear DNA overloaded; chloroplast DNA seen as main band at 1.712 and shoulder at 1.706 gm/cm^3. Markers as in Fig. 3.8.

In similar experiments in which ^3H- and ^{14}C-adenine replaced the density labels, the results were less clear cut than with density labeling; some of the radio-activity from the male parent was found in the chloroplast DNA region. For example, after 24 hours in the zygote, when the chloroplast peak had shifted to its new light position, both ^3H (female) and ^{14}C (male) labels were found in the light peak. The ratio of $^{14}C/^3H$ was low compared with that of the nuclear region, indicating that most of the counts were from the female parent. Nonetheless, some ^{14}C counts were present. During further days of zygote maturation, this ratio gradually increased, approaching that of nuclear region.

We interpret these findings as indicating considerable turnover or repair synthesis. Radioisotope labeling is a much more sensitive indicator than is density labeling of random incorporation into a preexisting DNA. In these circumstances, the density label provides a clearer picture of the different paths followed by the two chloroplast DNA's than does the radioactive adenine label.

Chiang studied this problem initially with adenine labeling alone (*3*) and examined chloroplast DNA at a late time in zygote development and also in zoospores recovered after zygote germination. At these late times, he found that the ratio of $^{3}H/^{14}C$ in the chloroplast region was approximately the same as in the nuclear region. These findings are experimentally in line with our own observations. By using adenine labeling exclusively and by looking only at late times in zygote development, Chiang missed seeing the differential paths followed by chloroplast DNA's from the two parents, which we have found. More recently, Chiang (*4*) reported the results of a pair of reciprocal crosses involving both density and adenine labeling. Differences in the behavior of chloroplast DNA's from the two parents can be seen clearly in his Fig. 12.

A different line of evidence indicating that maternal inheritance is under enzymatic control comes from our discovery of the effectiveness of UV irradiation in converting maternal to biparental transmission of cytoplasmic genes (*47*). The details of this study will be discussed later in this chapter (p. 84).

Maternal inheritance is of the greatest interest and importance in the investigation of cytoplasmic genetics, since it is a widespread phenomenon, found not only in algae, but also in fungi and higher plants. Maternal inheritance appears to be one device by which preferential transmission to progeny of cytoplasmic genes from one parent is achieved. As we shall see in subsequent chapters, the phenomenon of preferential transmission is a fundamental feature of cytoplasmic genetic systems (cf. p. 219).

Nuclear Genetics of Chlamydomonas

When we began to work with *Chlamydomonas*, nothing was known of its nuclear genetics beyond the 2:2 segregation of mating type alleles in meiosis (*59*). A clear understanding of the behavior of nuclear chromosomes was essential as a control for the investigation of cytoplasmic genes, and consequently an extensive study of nuclear markers was undertaken. We established the regularities of a simple haploid life cycle, as shown in Fig. 3.1, utilizing a number of unlinked markers including mating type, nuclear streptomycin resistance, resistance to methionine sulfoximine, and four pigment mutants (*32*). Each allelic pair showed 2:2 segregation in meiosis, and gene–centromere distances were estimated for each of the markers. In subsequent studies by Ebersold *et al.* (*9*), six linkage groups were identified. In 1965, preliminary evidence for the existence of sixteen linkage groups was reported (*19*), but the supporting linkage data have not yet been published. These linkage groups are shown in Fig. 3.12 taken from a recent review (*25a*).

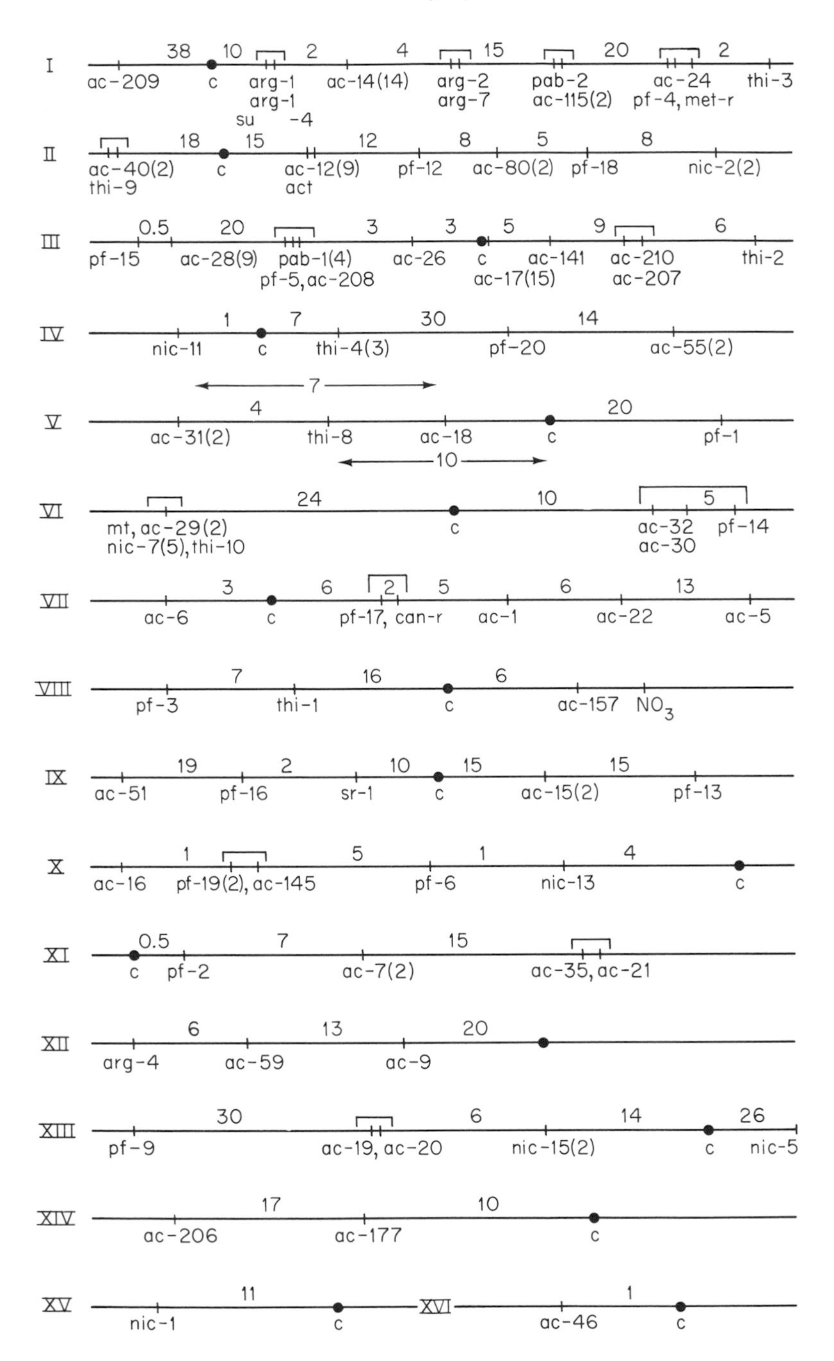
I
38 10 2 4 15 20 2
ac-209 c arg-1 ac-14(14) arg-2 pab-2 ac-24 thi-3
arg-1 arg-7 ac-115(2) pf-4, met-r
su -4
II
18 15 12 8 5 8
ac-40(2) c ac-12(9) pf-12 ac-80(2) pf-18 nic-2(2)
thi-9 act
III
0.5 20 3 3 5 9 6
pf-15 ac-28(9) pab-1(4) ac-26 c ac-141 ac-210 thi-2
pf-5, ac-208 ac-17(15) ac-207
IV
1 7 30 14
nic-11 c thi-4(3) pf-20 ac-55(2)
V
7
4 20
ac-31(2) thi-8 ac-18 c pf-1
10
VI
24 10 5
mt, ac-29(2) c ac-32 pf-14
nic-7(5), thi-10 ac-30
VII
3 6 2 5 6 13
ac-6 c pf-17, can-r ac-1 ac-22 ac-5
VIII
7 16 6
pf-3 thi-1 c ac-157 NO3
IX
19 2 10 15 15
ac-51 pf-16 sr-1 c ac-15(2) pf-13
X
1 5 1 4
ac-16 pf-19(2), ac-145 pf-6 nic-13 c
XI
0.5 7 15
c pf-2 ac-7(2) ac-35, ac-21
XII
6 13 20
arg-4 ac-59 ac-9
XIII
30 6 14 26
pf-9 ac-19, ac-20 nic-15(2) c nic-5
XIV
17 10
ac-206 ac-177 c
XV
11
nic-1 c
XVI
1
ac-46 c

Three strains of independent origin are now in common use. Our strain, received from G. M. Smith in 1949, and denoted *137(+)* and *137(−)* in his records, differs from the others in its ability to grow on nitrate as sole nitrogen source (*52b*), in its regular production of four zoospores from the zygote during germination, and in the absence of M band DNA (*25*). The strain used by Ebersold, Levine, and Gillham also came from Smith and was denoted *137c*. It differs from ours in not being able to grow on nitrate and in usually producing eight zoospores at germination. The strains used by Sueoka and Chiang containing an M band DNA in zygotes are 89 and 90 from the Indiana Collection. They have not been investigated genetically and carry no genetic markers other than the mating type alleles. In extensive crosses made in my laboratory between marked stocks of our strains and of Ebersold's, no evidence of chromosomal aberrations was detected on any of ten linkage groups, indicating close nuclear homology between the two strains.

Genetic evidence has not revealed any chromosomal abnormalities or complexities such as aneuploidy which is a serious technical problem in yeast. Nuclear gene conversion has been detected by us (*52b*) but no quantitative studies have been reported.

The exceptional occurrence of vegetative diploids was discovered by Ebersold (*8, 8a*). He found that a small fraction of zygotes do not undergo meiosis, but rather divide as diploids, giving rise to stable diploid clones which behave as mt^- and will mate with haploid mt^+ cells to form triploid zygotes. Diploids have many potential uses in genetic and physiological studies, but have hardly been exploited, except for a study by Gillham (*12*) to be discussed below (p. 79).

TRANSMISSION OF CYTOPLASMIC GENES IN CHLAMYDOMONAS: A SUMMARY

For purposes of clarity, the behavior of cytoplasmic genes in meiosis and mitosis in *Chlamydomonas*, as presently understood from our genetic studies, has been summarized in a set of didactic statements.

Fig. 3.12. The 16 nuclear linkage groups of *Chlamydomonas reinhardi*. Figures in parenthesis after the name of a locus indicate the number of alleles known at that locus. Numbers above the line are map distances. The bracket above the group of markers indicates that their relative positions are uncertain or unknown. Abbreviations: *c*, centromere; *mt*, mating type; *arg*, arginine requiring; $su^{arg\text{-}1}$, suppressor of *arg-1*; *ac*, acetate-requiring; *nic*, nicotinamide requiring; *pab*, *p*-aminobenzoate requiring; *thi*, thiamine requiring; *pf*, paralyzed flagellae; *sr*, streptomycin resistant; *can-r*, canavanine resistant; *met-r*, methionine sulfoxamine resistant; *act-r*, actidione resistant; NO_3, ability to grow on nitrate as sole nitrogen source. From (*25a* and *52b*).

The evidence in support of these statements will be presented in the subsequent sections of this chapter.

1. Streptomycin is an effective mutagen for cytoplasmic genes in *Chlamydomonas,* but is not detectably mutagenic toward nuclear genes.
2. Cytoplasmic genes of *Chlamydomonas* usually show maternal inheritance and can be identified by this transmission pattern.
3. Spontaneous exceptions to maternal inheritance occur with a frequency usually below 1%. These exceptional zygotes exhibit biparental inheritance giving rise to progeny containing a complete cytoplasmic genome from each parent. These progeny are therefore cytoplasmic heterozygotes or *cytohets.*
4. Yields of up to 50% *biparental* zygotes can be obtained by UV irradiation of the female (mt^+) parent immediately before mating. As with the spontaneous biparental zygotes, four cytohet progeny (zoospores) are produced.
5. In multiple marked crosses, all cytoplasmic genes show the same pattern of inheritance in individual zygotes i.e., all maternal or all biparental. In maternal zygotes, all markers from the female parent are transmitted to the progeny and none appear from the male. Recombination has not been observed in maternal inheritance. In biparental zygotes, all progeny are cytohets. Segregation and recombination are rare events during meiosis.
6. In cytohet progeny of biparental zygotes, whether of spontaneous origin or from UV treatment, segregation and recombination occur at each cell division beginning with the first mitotic doubling of zoospores, and are seen as long as any heterozygous markers remain to detect the process.
7. Segregation events are the result of exchanges that occur at a "four-strand" stage, when the DNA molecules are replicated and paired but before the cells have divided. Exchange events occur with the same probability at each cell division.
8. Two types of exchanges occur: reciprocal and nonreciprocal. The reciprocal event produces two daughter cells, each carrying one of the parental alleles. The nonreciprocal event, resembling gene conversion, produces one daughter cell which is a pure parental type (either type, on average 1:1) and the other which is still a cytohet (c.f. Fig. 3.19). Thus, parental alleles segregate 1:1 on the average in each zoospore clone.
9. Reciprocal exchange events provide evidence for the existence of a centromere-like attachment point which governs the distribution of the strands at cell division.
10. Genes can be mapped by three different parameters: (*a*) the

frequency of reciprocal exchange between gene and attachment point, leading to segregation; (*b*) the frequency of reciprocal recombination between genes; and (*c*) the frequency of coconversion (nonreciprocal exchange) of pairs or sets of genes.

11. The formal map which best describes the data generated by these three mapping procedures is circular. Genetic circularity is the consequence of even numbers of exchanges between strands and need not reflect the physical state of the DNA.

12. A special class of persistent cytohets exists in which segregation occurs very rarely during vegetative growth, but regularly at meiosis. These persistent cytohets exhibit maternal inheritance and the two copies of the cytogenome from the maternal parent segregate approximately 1:1 in the zoospore progeny.

GENETIC ANALYSIS OF CYTOPLASMIC GENES IN CHLAMYDOMONAS

The statements of the preceding section will now be documented and discussed, with the addition of information not explicitly noted above.

Mutagenesis of Cytoplasmic Genes

Streptomycin as a Mutagen. The deleterious effect of streptomycin on chloroplast development was discovered by von Euler (*62, 63*), who found that seedlings watered with a streptomycin solution developed colorless leaves. Subsequent studies with *Euglena* established that growth on streptomycin, as well as numerous other treatments, led to the irreversible loss of chloroplast-forming ability (cf. Chapter 2). Despite considerable investigation, the mode of action of streptomycin in these systems is still not known.

We chose streptomycin initially as a selective agent for the isolation of nuclear drug-resistant markers. However, we found that mutants appearing on streptomycin agar plates were of two types: one class were resistant to 100 μg of streptomycin per milliliter and the other were resistant to 500 μg/ml. The low-level resistant strains proved to be the result of a nuclear mutation, segregating 2:2 in crosses with wild type; and the high-level resistant strains all showed maternal inheritance and were classified as cytoplasmic (*31*).

With the discovery of these two classes of mutations to streptomycin resistance, nuclear and cytoplasmic, the means were at hand to compare their origin in the same experiment. Were both classes of spontaneous origin? Was streptomycin acting solely as a selective agent? The question was of great tactical importance to us. If we could establish the mode of origin of this first cytoplasmic mutation isolated in *Chlamydomonas,* perhaps we would know how to obtain additional cytoplasmic mutations. We therefore undertook to study the origin of these mutations (*33, 34*) using the method of fluctuation analysis (*25b*).

An example of the results, given in Table 3.1, showed clearly that the nuclear mutations were of spontaneous origin, preexisting in the population before they were identified by selection on streptomycin agar. The cytoplasmic mutations, on the other hand, were shown in the same experiments to be induced by streptomycin during residual growth of streptomycin-sensitive cells on drug agar plates. They were not preexisting in the population before exposure to the drug. These findings were subsequently confirmed by Gillham and Levine (*17*), who also suggested that a few of the mutations were of spontaneous origin.

Having established the mutagenicity of streptomycin for induction of streptomycin resistance, we then found that streptomycin was also an effective mutagen for other non-Mendelian genes. Indeed most of the mutations we have so far mapped and characterized (Table 3.2) were induced by growth of streptomycin-sensitive cells in the presence of a toxic but sublethal concentration of streptomycin.

Streptomycin is not mutagenic for nuclear genes in *Chlamydomonas* nor is it an effective mutagen in bacterial or viral systems. What is the mechanism of its mutagenic action and specificity? Since the drug binds strongly to DNA, its mutagenic action may result from a direct interaction with DNA. Another possibility is that the mutagenic action is a secondary effect resulting from streptomycin-induced miscoding leading to an altered DNA polymerase or repair enzyme.

Nitrosoguanidine. In contrast to streptomycin, *N*-methyl-*N*-nitrosoguanidine has been found to induce mutations in every system to which it has been applied. In *Chlamydomonas,* both nuclear and cytoplasmic gene mutations are induced by this drug (*14*). Its drawback as a mutagen of choice for cytoplasmic genes is the preponderance of nuclear mutations that are simultaneously induced.

Mutagenesis by Streptomycin Withdrawal. When cells of a cytoplasmic streptomycin-dependent (*sd*) mutant strain are grown in the absence of the drug, they undergo four to five doublings. In a study of reversion from *sd* to *ss* (streptomycin sensitivity), Ramanis and I found that streptomycin withdrawal was itself mutagenic in this strain (*52a*). When growth ceased in streptomycin-free medium, a burst of mutants appeared, including apparent revertants to sensitivity as well as other types of mutations, e.g., streptomycin resistance, loss of photosynthetic ability, and temperature sensitivity.

TABLE 3.1

Fluctuation Analysis: Origin of Nuclear (sr-100) and Cytoplasmic (sr-500) Mutations to Streptomycin Resistance [a]

	Liquid cultures grown on Ac and plated on Ac + 100 μg streptomycin per ml		Liquid cultures grown on Ac and plated on M + 300 μg/ml		Liquid cultures grown on M and plated on M + 100 μg/ml
No. of tubes	30	37	37	34	30
Cells/tube					
initial	200	82.5	37	270	800
final	5.5×10^{6}	9.9×10^{5}	5×10^{7}	3.8×10^{7}	1.5×10^{7}
Mean No. mutants per tube *sr-100*	1.4	3.0	2.1	5.3	5.3
Variance	6.0	39.0	27.0	250.0	48.0
P	0.1	0.005	0.005	0.005	0.005
Mutation rate/cell/division *sr-100* (P_0 method)	1.1×10^{-6}	8.2×10^{-7}	4.5×10^{-9}	9.2×10^{-9}	3.6×10^{-8} (partial plate suppression)
Mean No. mutants per tube *sr-500*	0.1	0	0.19	0.32	0.1
Variance	0.09	–	0.17	0.23	0.093
P	>0.9		0.75	0.2	>0.9
Mutation rate/cell/division *sr-500* (P_0 method)	1.4×10^{-7}		2.4×10^{-9}	3.6×10^{-9}	4.6×10^{-9}

Variation in the Number of Mutants in Aliquots Sampled from a Single Pooled Culture

	Mean	Variance	X^2	*P*
Experiment 1	1.4	1.6	0.0133	>0.9
2	0.71	0.83	0.009	>0.9
3	0.55	0.28	0.095	0.25
4	2.1	2.1	0	∞

[a] M = minimal medium; Ac = M plus sodium acetate. Fluctuation analysis distinguishes between spontaneous and induced mutations by examining their distribution in many small liquid cultures, each starting from a few wild-type cells. In these experiments, cells were grown and tested under three regimes previously shown to influence the yield of mutants. In each experiment, after growth in liquid, cells were plated on streptomycin–agar to count the mutants. Nuclear mutations to *sr-100* showed high variance from tube to tube, indicating spontaneous origin of mutations at random times. Cytoplasmic mutations to *sr-500* showed low variance typical of normal distribution, indicating induced mutation at a fixed probability characteristic for each regime. Control experiments in which sets of tubes were pooled before plating also showed low variance typical of a normally distributed population. From (*34*).

TABLE 3.2

Cytoplasmic Gene Mutations in Chlamydomonas

Gene	Origin[a]	Phenotype	Mapped[b]	Reference
ac1	SM induced	Requires acetate (*leaky*)	Yes	*45*
ac2	SM induced	Requires acetate (*stringent*)	Yes	*45*
ac3	SM induced	Requires acetate (*stringent*)	Yes	*52*
ac4	SM induced	Requires acetate (*leaky*)	Yes	*52*
tm1	SM induced	Cannot grow at 35°C	Yes	*51*
tm2	SM induced	Conditional: grows at 35°C only in the presence of streptomycin	No	*52*
Seven *tm* mutants	SM induced	Cannot grow at 35°C	No	*52*
ti1 thru *ti5*	NG	Tiny colonies on all media	No	*52*
ery1	SM induced	Resistant to 50 μg/ml erythromycin	Yes	*51*
kan1	SM induced	Resistant to 100 μg/ml kanamycin	No	*52*
spc1	SM induced	Resistant to 50 μg/ml spectinomycin	Yes	*51*
spi1 thru *5*	SM induced	Resistant to 100 μg/ml spiramycin	Yes	*51*
ole1 thru *3*	SM induced	Resistant to 50 μg/ml oleandomycin	Yes	*51*
car1	SM induced	Resistant to 50 μg/ml carbamycin	Yes	*51*
cle1	SM induced	Resistant to 50 μg/ml cleosine	Yes	*51*
ery3	SM induced	Resistant to erythromycin, carbamycin, oleandomycin, spiramycin (same concentrations as single mutations above)	No	*52*
ery11	SM induced	Same as for *ery3*	No	*52*
sm2	SM induced	Resistant to 500 μg/ml SM	Yes	*31*
sm3	SM induced	Resistant to 50 μg/ml SM	Yes	*46*
sm4	SM induced	SM dependent	Yes	*46, 49*
sm5	SM induced	Resistant to 500 μg/ml SM; recombines with *sm2*	Yes	*52*
D-371 and *D-310*	Induced by growth of strain *sm4* without SM	Resistant to 500 μg/ml SM	No	*52*
Four *D* mutants	Induced by growth of strain *sm4* without SM	Resistant to 500 μg/ml SM; segregate like persistant hets (sd/sr)	No	*51*
Eleven *D* mutants	Induced by growth of strain *sm4* without SM	Resistant to various low levels of SM: 20 μg/ml; 50 μg/ml; 100 μg/ml. Segregate like persistant hets (*sd*/low *sr*)	No	*52*

TABLE 3.2 (cont.)

Gene	Origin[a]	Phenotype	Mapped[b]	Reference
D-769	Induced by growth of strain *sm4* without SM	Conditional *sd*	Yes	*49*
Three *D* mutants	Induced by growth of strain *sm4* without SM	Conditional *sd*; segregate like persistant HETS (*sd*/cond. *sd*)	No No	*52* *52*
UV-16 *UV-17*	UV induced in strain *sm4*	Resistant to 500 μg/ml SM	No	*52*
Four *UV* mutants	UV induced in strain *sm4*	Resistant to 20 μg/ml SM	No	*52*
Three *UV* mutants	UV induced in strain *sm4*	Resistant to 20 μg/ml SM; segregate like persistant HETS (*sd*/low *sr*)	No	*52*
sr-2-1 *sr-2-60* *sr-2-280* *sr-2-218*	Spontaneous mutations selected on SM	Resistant to SM 500 μg/ml	No	*15* *61* *15* *15*
kan-1	Spontaneous; selected on kanamycin	Resistant to kanamycin 50 μg/ml	No	*61*
ery-2-y	NG	Resistant to erythromycin 100 μg/ml	No	*61*
ery-3-6	NG	Resistant to erythromycin 100 μg/ml	No	*61*
spr-1-27	NG	Resistant to spectinomycin 100 μg/ml	No	*15*
sd-3-18	NG	Dependent on at least 20 μg/ml SM	No	*61*
nea-2-1	NG	Resistant to 1 mg/ml neamine	Yes[b]	*15*

[a] SM = streptomycin; NG = nitrosoguanidine.
[b] Mapping references (*49, 51*).

All of these mutations show maternal inheritance, and are therefore cytoplasmic. However, two distinct patterns of transmission have been found. One class resembles those described in Fig. 3.2, and corresponds to the usual maternal pattern of 4:0 transmission. The other class is different from any we have previously seen (*52a*). For example, if the new mutant was phenotypically *sr* mt^+ and crossed with an *ss* mt^- test strain, the progeny segregated 2 *sr*: 2 *sd*—a non-Mendelian 2:2 ratio with both *sr* and *sd* coming from the mt^+ parent. Further study has established that these mutants represent a special class of cytohets, which we have called persistent cytohets (*52, 52a*). Their behavior in meiosis and mitosis will be discussed below (p. 96).

Other Mutagens. No systematic studies of the effects of other mutagens have been reported with *Chlamydomonas.*

Phenotype Classes of Cytoplasmic Mutants in Chlamydomonas

To date, four general phenotypic classes of cytoplasmic mutants have been described in *Chlamydomonas.* They are (*a*) nonphotosynthetic mutants, which have lost the ability to grow photosynthetically, and require a reduced carbon source, preferably acetate, for growth in light or dark; (*b*) temperature-sensitive mutants which show altered growth requirements at temperatures above or below the optimal 25°C; (*c*) mutants that are resistant to one or more antibiotics; and (*d*) slow-growing strains which form tiny colonies on supplemented agar media. Mutations thus far investigated are listed in Table 3.2.

The location of many of these mutations on a single genetically circular linkage group is discussed below. Here it should be noted that the *sd3* mutation which Schimmer and Arnold (*55–57*) have suggested as mitochondrial (c.f. p. 98) appears to be identical in behavior (*52*) with our *sd* gene (called *sm4*) which is linked to *sm2* and *sm3* on the chloroplast linkage group. Recently, Surzycki and Gillham (*61*) have tried to distinguish between chloroplast and mitochondrial genes on the basis of their responses to antibiotics. The results are preliminary and open to numerous interpretations. They do not provide an unambiguous localization of mutations. On the other hand, genetic data, as we shall see in this chapter, *do* provide an unambiguous basis for assigning individual mutations to a specific linkage group.

Spontaneous Exceptions to Maternal Inheritance: Early Studies

Spontaneous exceptions to the rule of maternal inheritance were noted in our first study of inheritance of streptomycin resistance (*31*) and subsequently noted by Gillham (*11*). Prior to 1966, all genetic studies of cytoplasmic genes in *Chlamydomonas* were undertaken with spontaneous exceptions to maternal inheritance. These studies, which provided the basic evidence of recombination and linkage, are reviewed in this section. Subsequently, with the discovery of the UV effect on maternal inheritance (cf. p. 84), it became possible to develop the quantitative mapping procedures now in use.

In our initial studies (*31*), about 1% of zygotes in crosses of *ss* mt^+ × *sr* mt^- transmitted the *sr* allele to the progeny. Subsequently with the isolation of a non-Mendelian mutation to streptomycin dependence (*sd*), it became possible to study exceptional zygotes in a systematic way (*45*).

As shown in Table 3.3, when the mt^+ parent is streptomycin dependent (sd mt^+) streptomycin must be present for growth of progeny from germinating zygotes. In the absence of streptomycin a small fraction of zygotes, approximately 0.1%, germinate. Similarly, in the reciprocal cross when the *sd* gene is present in the mt^- parent, a low of fraction of exceptional zygotes germinate and produce colonies in the presence of streptomycin.

The analysis of progeny from these exceptional zygotes showed that both streptomycin-dependent and streptomycin-sensitive cells were present in each zygote colony, indicating that both of the parental alleles had been transmitted. Furthermore the exceptional zygote colonies were found to contain eight genetically different progeny classes, as shown in Fig. 3.13. Each of the four distinguishable products of meiosis or zoospores gave rise to both streptomycin-dependent and streptomycin-sensitive subclones. Thus, *sd* and *ss* must have segregated from each other at postmeiotic divisions (45).

Further studies were carried out with a pair of reciprocal two-factor crosses.

Cross A: $sd\ ac^+\ mt^+ \times ss\ ac^-\ mt^-$
Cross B: $ss\ ac^-\ mt^+ \times sd\ ac^+\ mt^-$

In cross A exceptional zygotes were selected by plating on acetate agar without streptomycin thus selecting for the *ss* allele of the mt^- parent and permitting equal growth of ac^+- and ac^--containing cells.

TABLE 3.3

Selection of Exceptional Zygotes in Crosses between sd and ss[a]

	% Zygote colonies	
Cross[b]	Streptomycin agar	Minimal agar
1. $sd\ mt^+ \times ss\ mt^-$	100	0.07
2. $ss\ mt^+ \times sd\ mt^-$	0.08	100
3. $sd\ mt^+ \times sd\ mt^-$	100	<0.0001
4. $ss\ mt^+ \times ss\ mt^-$	<0.0001	100

[a] From Sager and Ramanis (45).

[b] In cross 1, the number of zygote colonies formed on streptomycin–agar was taken as 100% to compare with 0.07% of exceptional zygotes formed in the absence of the drug. In the reciprocal cross 2, 0.08% exceptional zygotes were identified. Crosses 3 and 4 are controls, showing no detectable exceptional zygotes in crosses of $sd \times sd$ or $ss \times ss$.

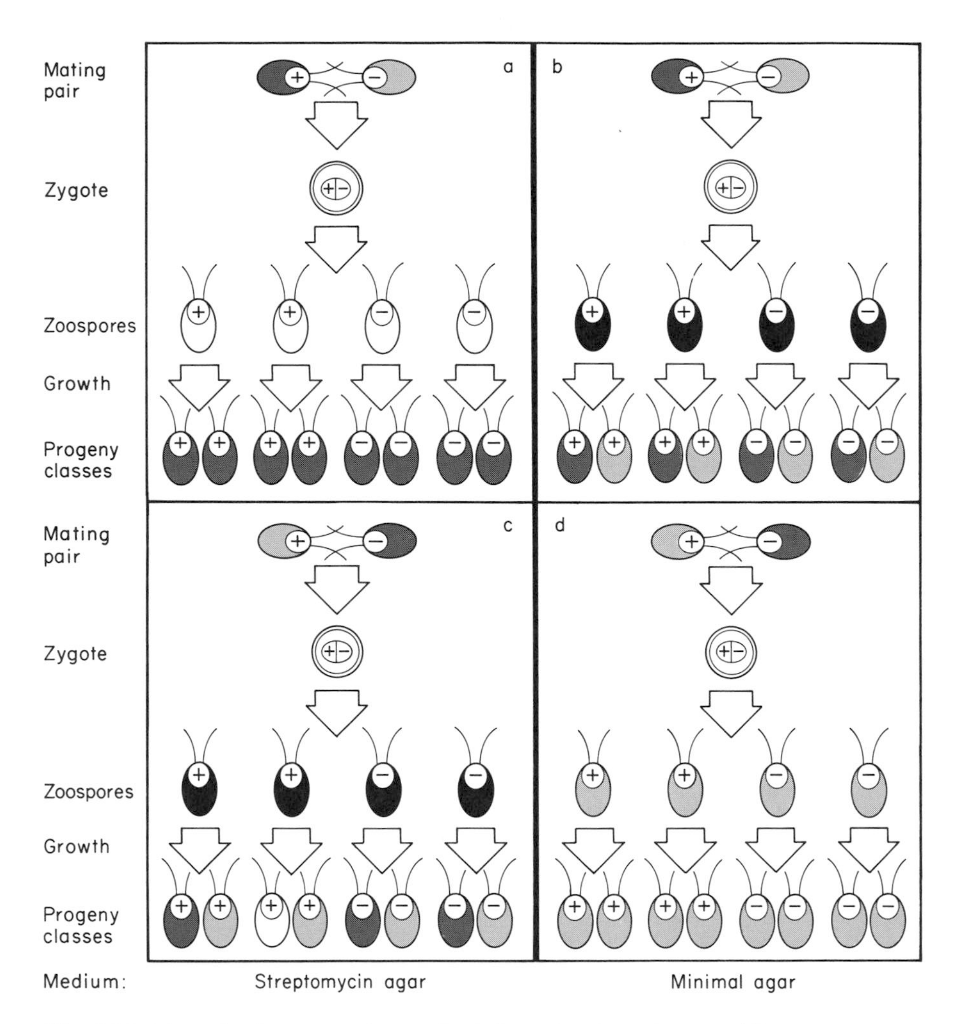

Fig. 3.13. (a) *sm-d* mt^+ × *sm-s* mt^- germinated on streptomycin agar. All zoospores are *sm-d*. (b) Same cross as (a) germinated on minimal agar; rare zygotes (approximately 0.1%) germinated; all zoospores are heterozygous, segregating *sm-d* and *sm-s* in clonal growth. (c) *sm-s* mt^+ × *sm-d* mt^- germinated on streptomycin agar; same result as in (b). (d) Same cross as (c) germinated on minimal agar; all zoospores are *sm-s*. Dark gray circles, streptomycin dependent (*sm-d*); black circles, streptomycin dependent/streptomycin sensitive (*sm-d*/*sm-s*); light gray circles, streptomycin sensitive (*sm-s*). mt^+ and mt^- are denoted as + and −, respectively. Based on data from reference 45.

The progeny were scored for segregation both of the *sd/ss* pair and of the ac^+/ac^- pair. In cross B three different selective conditions were compared: acetate-streptomycin agar which selects for *sd;* minimal agar which selects for ac^+; and the minimal streptomycin agar which selects for both ac^+ and *sd.*

About the same number of exceptional zygotes (0.02–0.1%) were found in both crosses. Selecting simultaneously for two cytoplasmic genes did not reduce the yield to the extent expected if the probabilities of transmission had been independent. In both crosses progeny were found to segregate independently for the *sd/ss* and the ac^+/ac^- pairs in postmeiotic divisions. When zygote colonies were analyzed after many doublings, sixteen genetically different meiotic products were recovered. The ratio of ac^+ to ac^- clones found among the progeny of crosses A and B was close to 1:1. The *sd/ss* ratio could not be measured with accuracy under the plating conditions employed.

Thus with the use of spontaneous exceptional zygotes it was possible to obtain qualitative evidence of postmeiotic segregation. These results provided the first evidence that the cytoplasmic genes under investigation shared with nuclear genes the properties of stability, mutability, maintenance of identity in heterozygotes, segregation of alleles, and the classical dichotomy between genotype and phenotype (*45*). This study also provided the first evidence of recombination between *ac* and *sd,* but did not distinguish between true recombination of linked genes and a sorting-out process.

Mitotic segregation of *sr* and *ss* has also been observed in vegetative diploids. Using Ebersold's selection method (*8, 8a*), Gillham crossed *arg2 sr* $mt^+ \times$ *arg1 ss* mt^- and examined the selected *arg1/arg2* diploid colonies for the presence of *sr* and *ss.* (*12*). He found that (*a*) most diploid clones contained both *sr* and *ss* cells, but that the ratios were far from 1:1, with the allele from the mt^+ parent always in excess; and (*b*) the *sr* and *ss* alleles segregated out rapidly during vegetative growth.

These results indicate that a mechanism for maternal inheritance operates in vegetative diploids as well as in zygotes, at least in the *sr* region. The findings also highlight the instability of heterozygotes. Presumably the basis of this instability is simply the continuing opportunities for segregation available to heterozygotes at each mitotic doubling. The diploid system clearly merits further investigation.

Our first evidence of linked recombination (*46*) came in the analysis of progeny from the following four-point cross: *ac1 sd* $mt^+ \times$ *ac2 sr* mt^-. The genotypes of the parents and progeny are shown in Fig. 3.14. Regular zygotes were recovered by plating on a nonselective medium containing streptomycin and acetate. Under these conditions all the

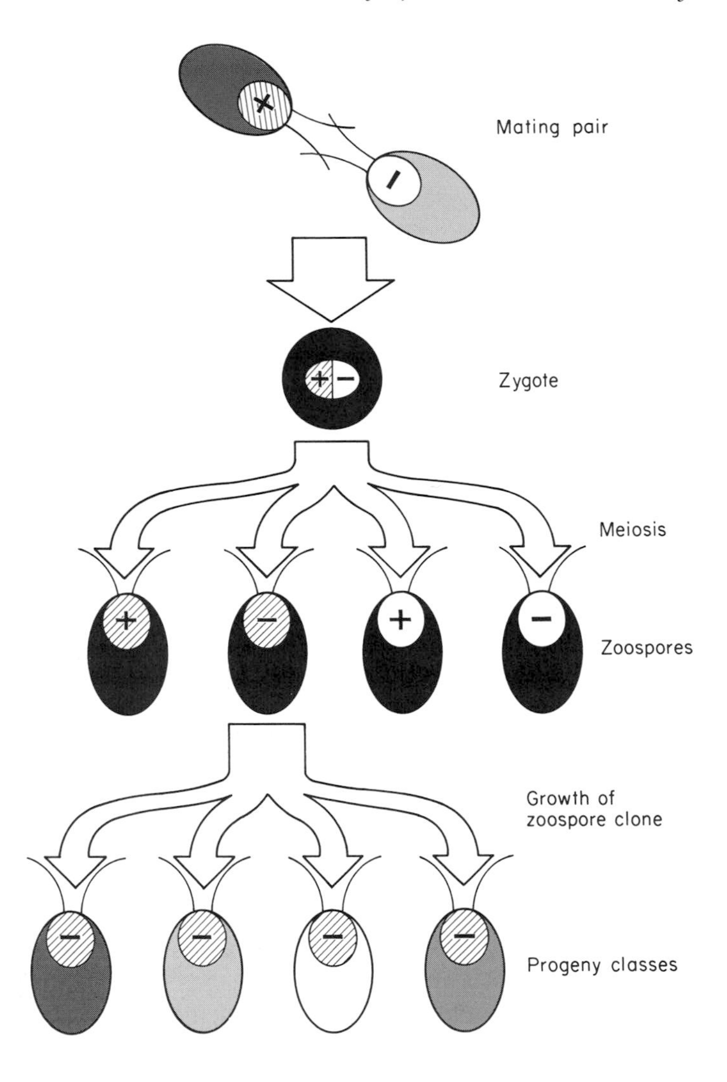

Fig. 3.14. Sixteen classes of progeny (only four are shown). Cross shows gene segregation in an exceptional zygote of *Chlamydomonas.* In this cross the female parent differs from the male by two pairs of unlinked nuclear genes shown by + and − for mating type and cross-hatching for an unlinked pair of alleles; and two pairs of cytoplasmic genes indicated by shading. The zygote is diploid, containing all genes from both parents. In meiosis the nuclear genes segregate as determined by chromosome behavior, giving rise to four genetically different products. The cytoplasmic genes do not segregate in meiosis, and in these exceptional zygotes the zoospores are heterozygous for the cytoplasmic genome. Cytoplasmic gene segregation occurs in the mitotic divisions of each zoospore clone after meiosis giving four progeny classes, two parental and two recombinant, in each zoospore clone. From (*46*).

progeny recovered were uniformly *ac1 sd* like the maternal (mt^+) parent, thus showing the usual maternal transmission pattern.

Exceptional zygotes were recovered by plating on acetate agar without streptomycin to select for the *sr* gene present in the mt^- parent. As shown in the figure, each of the four zoospores from exceptional zygotes gave rise to colonies containing the four classes of progeny types: *ac1 sd, ac2 sd, ac1 sr,* and *ac2 sr.* The ratio of *ac1*:*ac2* was approximately 1:1 whereas the ratio of *sr*:*sd* was about 1.5:1, probably because of the differential growth rates of the *sr* and *sd* types.

The results of this cross (Fig. 3.14) confirmed the results of the previous experiment (Fig. 3.13). (*a*) Segregation of cytoplasmic genes did not occur during meiosis as shown by the fact that each zoospore was initially heterozygous for both pairs of cytoplasmic genes though haploid for the nuclear complement. (*b*) Segregation of cytoplasmic genes did occur during vegetative growth. After four doublings there had been a substantial amount of segregation, approximately 60% for both the *ac1*/*ac2* pair and the *sr*/*sd* pair. (*c*) Recombination occurred between the *ac* markers and the *sd, sr* markers. (*d*) A new and unexpected feature of this cross was the appearance of two additional classes of progeny besides those shown in Fig. 3.14. Both new classes were phenotypically wild type, one being ac^+, and the other streptomycin sensitive, *ss*.

The *ss* and ac^+ progeny were shown to be recombinant. (*a*) They were not the progeny of revertant cells present in the parental population because plating of the parents revealed no such revertants; and secondly because they continued to arise anew during growth of the zoospore clones. (*b*) They did not arise by the sorting-out of wild-type genes present but unexpressed in the parental cells as shown by a series of homoallelic crosses designed to expose any wild-type genes present in the parental mutant cells; none were found. (*c*) The possibility that these wild-type progeny could have resulted from suppressor mutations, either nuclear or cytoplasmic, was excluded by a series of test crosses. Ten ac^+ and ten *ss* progeny, each from a different zygote, were backcrossed to standard wild-type strains and the progeny were examined for mutant phenotypes arising in either exceptional or nonexceptional zygotes. No mutant types were found. We concluded that the wild-type progeny appearing in this cross were true recombinants between linked markers.

A search was then instituted for reciprocal recombinant classes. Since recombination occurs during growth of zoospore clones, any cell which is a cytohet (e.g., *ac1*/*ac2*) can be used as a source of recombinants. In these experiments pairs of daughter cells were separated by

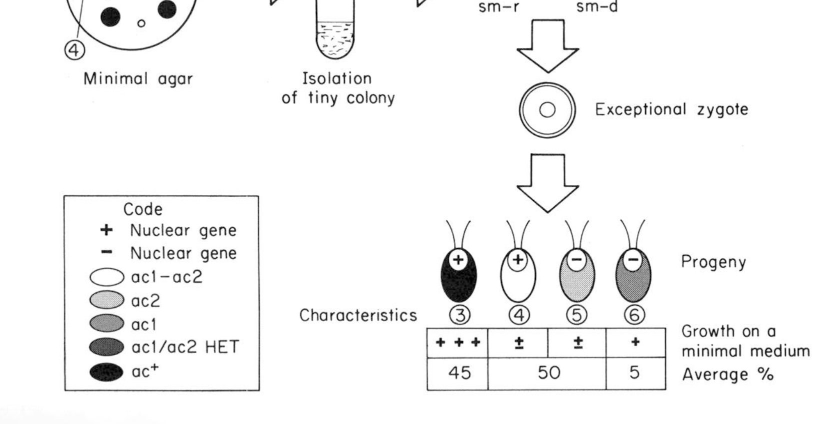

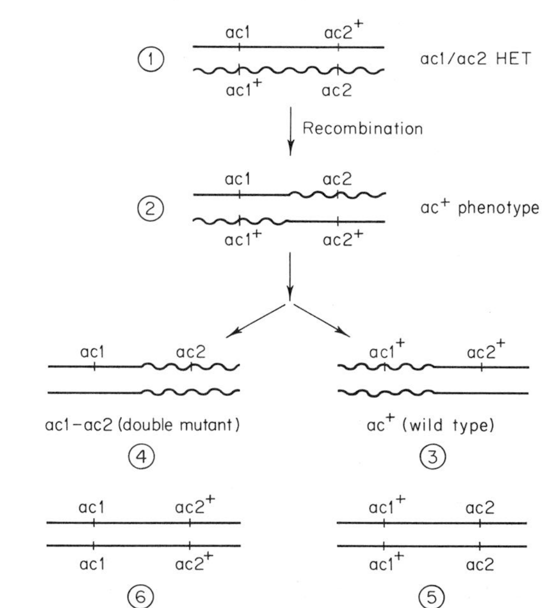

Fig. 3.15. (a) Identification of reciprocal recombinants from *ac1/ac2* HET. Phenotypically ac^+ clones were selected from pedigrees of *ac1/ac2* HETS and restreaked. Some ac^+ clones segregated *tiny* and *large* colonies on minimal agar. The large colonies were shown to be ac^+ by phenotype and test cross. The tiny colonies were shown to be *ac1-ac2* double mutants by test cross shown in (b) with ac^+ *sm-d* mt^+ selecting on acetate medium for exceptional zygotes. From (*36*).

dissection after one and two doublings, and the resulting sets of four pedigreed cells were allowed to grow and form colonies. The phenotype of the double mutant was unknown, but we expected a stringent acetate-requiring mutant similar to the stringent mutant *ac2*. We therefore looked for stringent mutants present in the same pedigree set with ac^+ recombinants. The experiment was performed as described in Fig. 3.15. The identification of the *ac1–ac2* recombinant was confirmed by test crosses with ac^+ wild-type cells.

Subsequently an analogous experiment was performed to look for the *sd-sr* reciprocal recombinant in crosses of *sd* × *sr*. Clones in which *ss* cells had been detected were examined for the presence of *sd-sr* double mutants. Since no class appeared which was phenotypically different from the parent, it was necessary to backcross a number of *sr* and *sd* clones chosen at random from colonies containing *ss* recombinants. Some clones of *sr* cells were found which segregated *sd* and *sr* in subsequent test crosses, whereas all the *sd* cells tested were uniformly *sd*. These test crosses identified and distinguished the *sd-sr* recombinant which were phenotypically indistinguishable from *sr* cells. These results established *ac1*, *ac1*, *sr*, and *sd* as separate genes, and the recombination frequencies observed in the cross provided the means for generating a simple map.

The need to distinguish among *ss* alleles at different loci led to a change in symbols. We now denote *sr* as *sm2*. A low-level streptomycin-resistant gene is denoted *sm3*, and *sd* has become *sm4*. Alleles at the *sm2* locus are denoted as *sm2-s* (sensitive) and *sm2-r* (resistant) and at the *sm4* locus as *sm4-s* (sensitive) and *sm4-d* (dependent).

Preliminary studies of recombination among the genes *sr2*, *nr2* (neamine-resistant), and *spr1*(spectinomycin-resistant) have been reported by Gillham (*13, 16*). In this work, zygotes were germinated and allowed to form colonies *in situ*. Exceptional zygotes were detected by replica plating to look for markers from the mt^- parent, and were classified by replica plating for the presence of some of the recombinant classes. In some instances, zygote colonies were subcultured and the relative proportions of the various genotypes were determined directly.

The results indicated rather loose linkage between *sr2* and *nr2*, and very tight linkage between *nr2* and *spr1*. These findings are difficult to evaluate quantitatively because (*a*) the parental types did not segregate 1:1 among the progeny; (*b*) colonies were examined after many rounds of recombination had occurred; (*c*) numbers of recombinants observed were low; and (*d*) not all classes of recombinants were recovered. Nonetheless the results are qualitatively in accord with our much more extensive data from UV-induced biparental zygotes. Before examining those data however, we must first consider the UV method itself.

The Effect of UV Irradiation upon Maternal Inheritance

The use of rare spontaneous exceptional zygotes for recombination analysis is at best a makeshift procedure. Quantitation is very difficult since the selection process itself selects against some classes of progeny. It seemed essential therefore to find a better method for routine recombination analysis. Experiments with temperature shifts and antibiotic treatments during zygote formation and maturation gave encouraging results. Ten-to twentyfold increases in yield of exceptional zygotes were obtained. However, these results did not prepare us for the dramatic effectiveness of UV irradiation *(47)*.

We found that a suitable exposure of mt^{+} cells to UV irradiation immediately before mating led to the recovery, depending on the UV dose, of exceptional zygotes equivalent to 100% of the zygotes present; in other words, there occurred a total conversion of maternal zygotes to exceptional ones. Further examination of exceptional zygotes showed that they were of two types: biparental zygotes transmitting cytoplasmic genes from both parents to progeny, and paternal zygotes transmitting cytoplasmic genes only from the unirradiated mt^{-} parent. Photoreactivation proved to be a very effective treatment in increasing the yield of biparental zygotes.

An example of results is shown in Fig. 3.16. In Fig. 3.16a, increasing doses of UV irradiation lead to a rapid conversion of maternal to exceptional zygotes in the absence of photoreactivation (curve D). In curves A and B one sees the results of two different regimes of photoreactivation. In both conditions the conversion to exceptional zygotes is considerably reduced by photoreactivation. However, photoreactivation primarily increases the proportionate yield of biparental zygotes at the expense of paternal zygotes. The effect of UV in the yield of biparental and paternal zygotes is shown in Fig. 3.16b and c. We have interpreted these data as evidence for two distinct processes affected by UV irradiation:

$$1.\ \text{Maternal} \underset{\text{PR}}{\overset{\text{UV}}{\rightleftharpoons}} \text{biparental}$$

$$2.\ \text{Biparental} \underset{\text{PR}}{\overset{\text{UV}}{\rightleftharpoons}} \text{paternal}$$

Process 1 can be reversed by visible light much more effectively before mating than afterward; we view this step as the one that regulates maternal inheritance, perhaps by interfering with the transcription of a messenger RNA required for synthesis of a restriction enzyme. Process 2, on the other hand, may be involved directly with the replication of

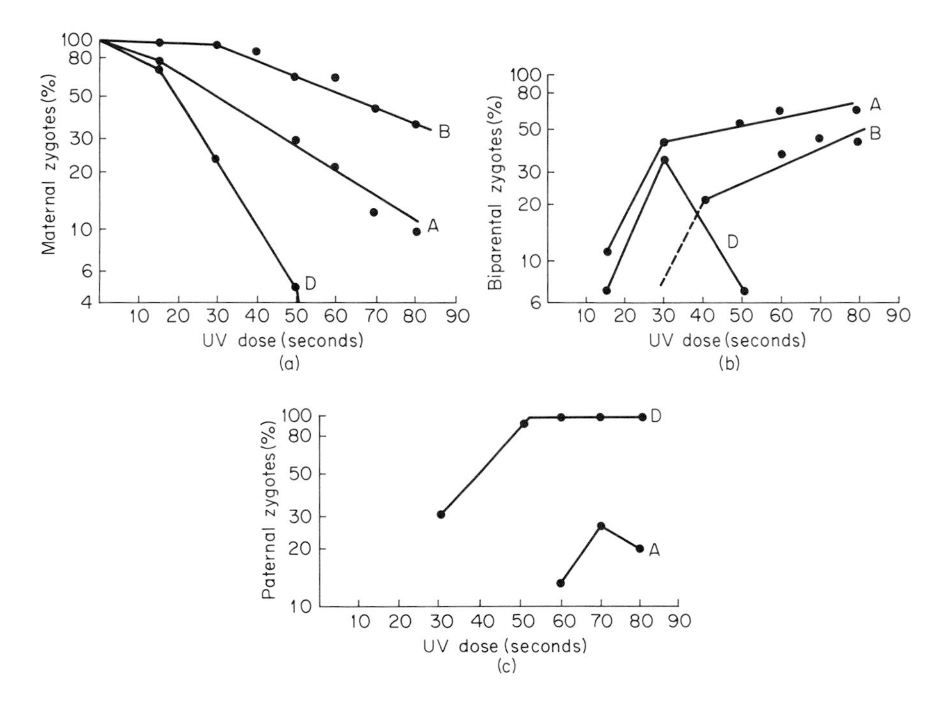

Fig. 3.16. Effectiveness of UV irradiation in converting maternal zygotes to exceptional ones. (a) Conversion of maternal to exceptional zygotes as a function of UV dose. (b) Yield of biparental zygotes as a function of UV dose. (c) Yield of paternal zygotes as a function of UV dose. Female (mt^+) gametes were irradiated, mated with unirradiated males (mt^-), and kept in dark until zygote formation was completed. After 2 hours, zygotes were diluted and plated; plates were incubated in dark (D) or in light (A) for photoreactivation. In (B) mating gametes were exposed to light during and after zygote formation. After 24 hours in light, A and B were incubated in dark with D series for one week, then all plates were exposed to light to induce germination of zygotes. From (*47*).

chloroplast DNA as influenced by the presence of pyrimidine dimers. In paternal zygotes, presumably the mt^+ chloroplast DNA has been too heavily damaged to be replicated; this damage is partially reparable by photoreactivation. The fact that the UV target is subject to photoreactivation indicates that it is DNA. Irradiation of the male parent before mating has no effect on maternal inheritance and irradiation of zygotes after mating is highly lethal.

A detailed study was carried out of segregation and recombination in spontaneous biparental zygotes in comparison with UV-induced biparental zygotes in the same cross. No special effect of UV irradiation upon recombination frequencies was noted under our experimental conditions: namely, UV irradiation plus photoreactivivation go give a final yield of approximately 50% biparental zygotes. At higher UV doses and in the absence of photoreactivation, distorted segregation and recombination ratios have been observed.

Segregation Patterns of Individual Gene Pairs

When biparental zygotes are produced under our experimental conditions, the four zoospores formed upon germination are each hybrid for all of the cytoplasmic genes by which the parents differ, although the zoospores are normal haploids for their nuclear genome. The hybrid state of the zoospores does not persist. At each successive doubling in the formation of zoospore clones, some pure types arise which resemble each of the parents.

This segregation process was initially studied with the pair of closely linked genes, *ac1* and *ac2*. We found that the pure types appear at a constant rate per doubling for many doublings and that eventually no hybrids remain. The combined results of many experiments involving

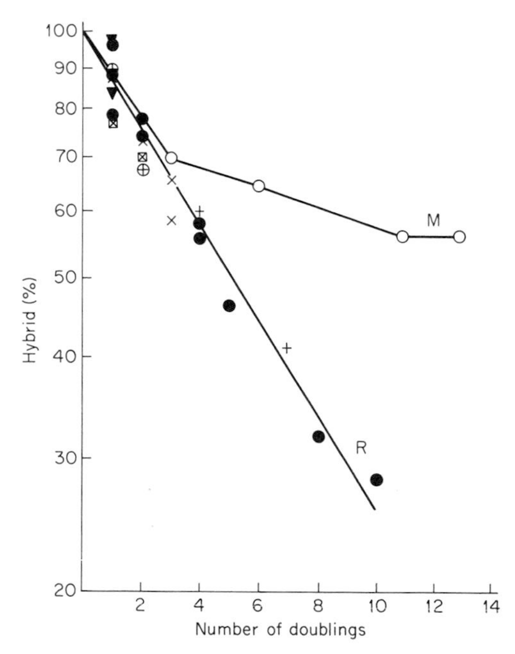

Fig. 3.17. Segregation of *ac1*/*ac2* during growth of heterozygous zoospores obtained with and without UV treatment. Curve R: cultures grown in acetate medium that is nonselective for acetate mutants. Curve M: cultures grown in minimal medium, selecting for hybrids and ac^+ recombinants. Symbols denote different experiments. Curve R shows exponential decay of heterozygotes as a result of segregation, indicating a constant rate of segregation per doubling. From (*48*).

these two genes are shown in Fig. 3.17. New ac^+ recombinants also arise during these doublings (*48*).

In subsequent studies we found that the rate of segregation of pure types is characteristically different for each gene observed and that these differences provide the basis for a mapping procedure. An example of this procedure is shown in Fig. 3.18 in which the segregation rates for four genes were compared in the progeny of a single cross. The mapping procedure utilizes the slope of each curve as the basis for computing the linear distance from gene to gene.

The rationale of this mapping procedure is based upon our analysis of the segregation process. In Mendelian systems the allelic segregation ratios of 1:1 seen in meiosis are the consequence of the mechanics of chromosome replication and distribution on the meiotic spindle. In classical cytogenetics allelic pairs served as markers whose segregational patterns were correlated with the behavior of physical markers on the chromosome such as knobs and small discontinuities. In non-Mendelian systems previously studied there was little evidence of allelic segregation. Typically the pattern has been maternal; and in the exceptions to maternal inheritance, segregation has occurred in an irregular fashion during clonal or vegetative growth. Attempts to study somatic segregation in higher plants have been thwarted primarily by the complexity of cell lineage patterns.

In *Chlamydomonas* the behavior of cytoplasmic gene pairs in meiosis and in the subsequent clonal multiplication of zoospores has been examined by pedigree analysis. In preliminary studies we examined colonies grown from single zoospores and found that each pair of alleles segregated on the average 1:1. However when individual zygotes were examined at the 16–32 cell stage, we found that single gene pairs did not segregate 1:1 in the first few doublings. The pedigree analyses were undertaken to investigate the deviation from 1:1.

In the pedigree studies involving heterozygotes taken either at the first, second, or later doubling of zoospores, three patterns of segregation were seen.

	HET ↓	$\frac{ac1}{ac2}$ ↓
Type I:	HET + HET	$\frac{ac1}{ac2} + \frac{ac1}{ac2}$
Type II:	HET + P_1 *or* HET + P_2	$\frac{ac1}{ac2} + \frac{ac1}{ac1}$ *or* $\frac{ac1}{ac2} + \frac{ac2}{ac2}$
Type III:	$P_1 + P_2$	$\frac{ac1}{ac1} + \frac{ac2}{ac2}$

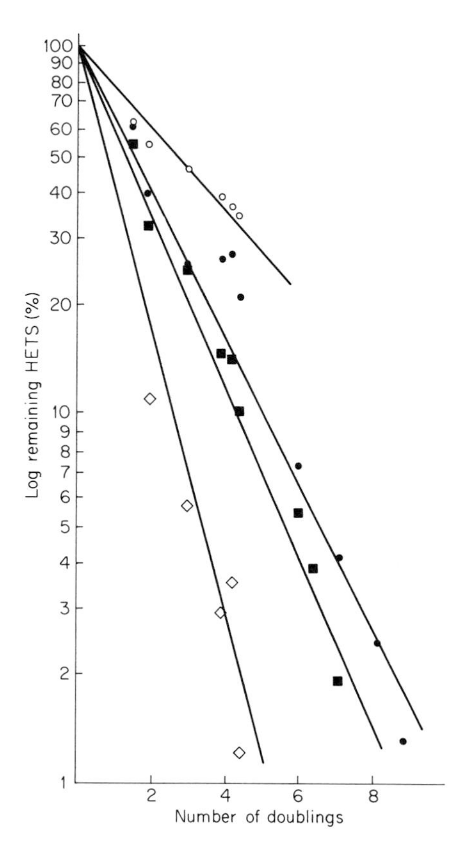

Fig. 3.18. Comparison of segregation rates in liquid culture of four genes in the progeny of a single cross. Symbols: ○, temperature sensitivity (*tm1*); ■, streptomycin resistance (*sm2*); ●, spectinomycin resistance (*spc*); ◇, erythromycin resistance (*ery*).

Type I segregation, which preserves the heterozygous condition, is the prevalent pattern. Type II segregation is the result of a nonreciprocal event and may be thought of as analogous to gene conversion. The type III pattern is the result of a reciprocal exchange event. Schematic diagrams shown in Fig. 3.19 indicate how these segregation events may be visualized at the DNA level. It is the type II events that distort the 1:1 ratio temporarily in individual zygotes. However on the average the type II events produce equal numbers of the two parental alleles.

In multifactor crosses involving several genes, the frequency of type II events is approximately the same for each gene. Detailed evidence of this will be presented later. However the frequency of type III segregation shows a definite polarity. That is, different genes show different

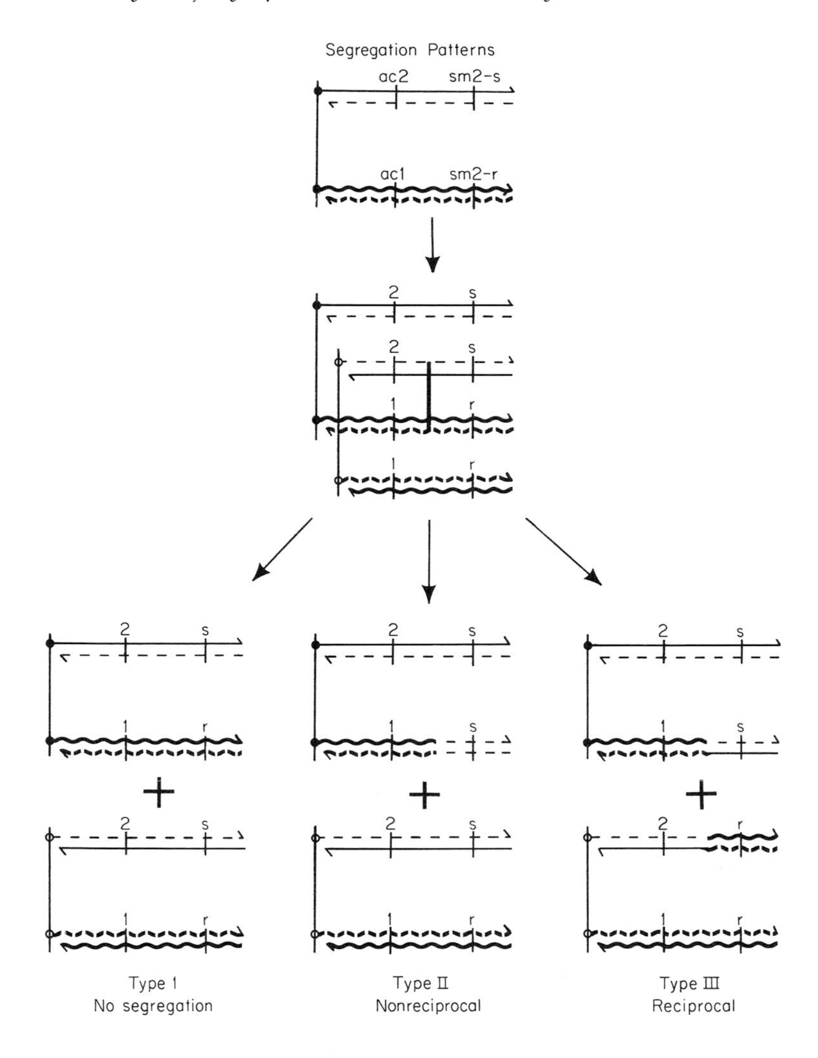

Fig. 3.19. Segregation patterns. The homologous DNA's from the two parents are shown as straight (*ac2* parent) and wavy (*ac1* parent) solid lines. The complementary replicated strands are shown as dashed lines. The homologous DNA's are shown attached to a hypothetical membrane by one of the complementary strands. The model predicts that the 'old' attachment points go to one daughter at cell division and the 'new' points to the other daughter, thus determining the regularity of distribution following semiconservative replication. Type II segregation is pictured as a double-stranded loss and replacement of a segment, and type III is shown as a double-stranded exchange. This model is consistent with the genetic data but otherwise speculative. From (*37*).

and characteristic frequencies of type III segregation. These differences provide an order on the basis of which genes can be mapped. The map order and relative distances based upon the polarity of type III segregation scored after two doublings is very similar to the map produced from the rate of overall segregation scored after eight to ten doublings as shown in Fig. 3.17 and 3.18. Since the liquid culture method shown in Fig. 3.18 does not discriminate between segregants from type II and from type III events, it is evident that the type II events which occur with equal frequency for each gene do not contribute to the differential slope. Thus the kinetic data can be compared directly with the pedigree data in which type II and type III events are enumerated separately.

We have assumed, as shown in Fig. 3.19 that type III segregation results from a reciprocal exchange event occurring between two nonhomologous strands at a four-strand stage, i.e., after replication of the DNA's but before cell division. The polarity of type III segregation is taken as evidence of a centromere-like "attachment point" (*ap*) governing the distribution of these molecules to daughter cells at cell division. In Fig. 3.19 a reciprocal exchange occurring between the genes *ac* and *sm*2 results in type III segregation for *sm*2 which is beyond the exchange point with respect to *ap*. Mapping by the frequency of type III segregation is based on the assumption that exchanges occur at random along the molecule and that the farther a gene lies from *ap* the more frequently it will undergo type III segregation.

The success of this mapping procedure provides evidence that type III segregation is not the result of a sorting-out process but rather that it is the result of a true recombinational event. Analysis of progeny recovered, discussed below, provides evidence that these cells are permanently diploid for this linkage group and that recombination occurs at a four-strand stage, after replication and before cell division.

Recombination between Linked Genes and Mapping Analysis

Estimation of recombination frequencies is complicated by two special features of this genetic system. One is the continuing occurrence of recombinational events at successive mitotic doublings. Recombination rates are so high that exchange patterns can rarely be interpreted after more than three or four doublings. Loosely linked genes soon appear unlinked, as for example the genes *ac*1 and *sm*3 as shown in Table 3.4. The second complication is the high frequency of nonreciprocal exchanges (resembling gene conversion) that accompany the reciprocal exchanges. We have developed methods to utilize both classes of events for mapping purposes, but they must be separately enumerated. For these reasons we found it necessary to develop a method of pedigree

TABLE 3.4

Linkage of Acetate and Streptomycin Regions[a]
(cross: *ac2*$^+$ *ac1 sm3-s sm2-s* × *ac2 ac1*$^+$ *sm3-r sm2-r*)

Progeny class 1	Progeny class 2	Progeny class 3	Recombination region	Numbers of progeny[b] Expt. 1	Expt. 2	Expt. 3
ac2$^+$ *ac1*	*sm3-s*	*sm2-s*	Parental	75	32	66
ac2 ac1$^+$	*sm3-r*	*sm2-r*	Parental	64	36	52
ac2$^+$ *ac1*	*sm3-r*	*sm2-r*	2	13	7	71
ac2 ac1$^+$	*sm3-s*	*sm2-s*	2	20	15	57
ac2$^+$ *ac1*	*sm3-r*	*sm2-s*	2 + 3[c]	13	0	ns[d]
ac2 ac1$^+$	*sm3-r*	*sm2-s*	3	15	6	ns
ac2$^+$ *ac1*$^+$	*sm3-r*	*sm2-r*	1	7	2	7
ac2$^+$ *ac1*$^+$	*sm3-s*	*sm2-s*	1 + 2	5	3	8
ac2$^+$ *ac1*$^+$	*sm3-r*	*sm2-s*	1 + 3	2	0	ns
Total progeny segregated:				214	101	261
Recombination in region						
1				6.5	5.0	5.8
2				23.9	25.0	52.2
3				14.0	6.0	ns

[a] From Sager and Ramanis (*49*).
[b] In experiments 1 and 2, the progeny were scored after two mitotic doublings and linkage was seen in regions 1, 2, and 3. In experiment 3, the progeny were scored after four to six doublings and the linkage in region 2 was lost.
[c] Position of *sm3*- between *ac1* and *sm2* assigned on basis of other data.
[d] ns = not scored.

analysis, using only the first two mitotic doublings after meiosis for routine mapping purposes. Our method of pedigree analysis is shown in Fig. 3.20 (*49*).

We classify the progeny of individual zygotes at the sixteen-cell stage, when each of the four zoospores has divided twice. The initial classification of the data establishes for each heterozygous gene pair whether segregation occurred at either the first or second doubling, and whether it was a reciprocal or nonreciprocal event. If segregation occurred at neither doubling, the gene is classified as "HET" and no further doublings are analyzed. (The "HET" classification is based upon subsequent segregations occurring during growth of the zoospore colony which is eventually scored for all markers, but we do not know at which doublings the later segregations occurred.)

Our first linkage map, shown in Fig. 3.21 was based upon recombination fractions obtained by pedigree analysis of five different crosses involving a set of eight genes: *ac2, ac1, sm4, sm3, sm2, csd, nea, csd,* and

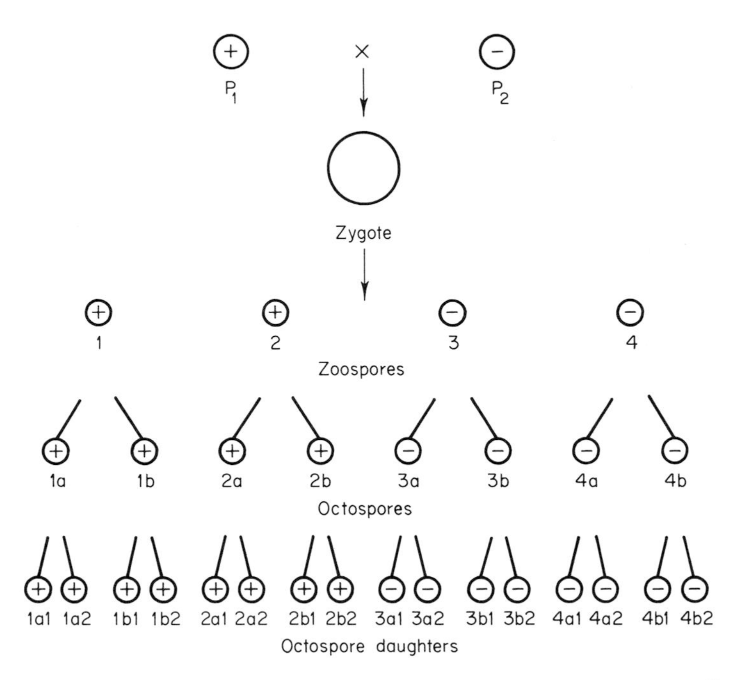

Fig. 3.20. Procedure for pedigree analysis. After germination, zoospores are allowed to undergo one mitotic doubling and then the eight cells (octospores) are transferred to a fresh Petri plate and respread. After one further doubling, each pair of octospore daughters is separated and allowed to form colonies. The sixteen colonies, derived from the first two doublings of each zoospore are then classified for all segregating markers. From (*49*).

ery, as well as the attachment point, *ap*. The arrangement and relative distances between genes, estimated from recombination fractions, agreed well with the results obtained from frequencies of reciprocal recombination of each gene, i.e., mapping by distances of each gene from the attachment point. These data have been discussed in several reviews (*27a*, *37*, *50*).

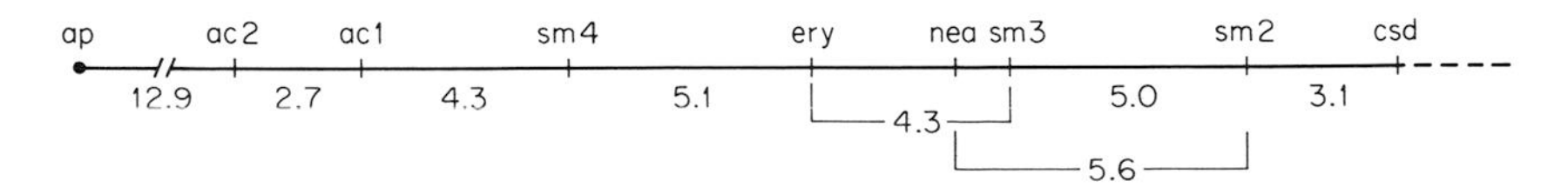

Fig. 3.21. Composite map of chloroplast linkage group I, based on published data. Symbols are: *ap*, attachment point; *ac2* and *ac1*, acetate requirement; *sm4*, streptomycin dependence; *ery*, erythromycin resistance; *nea*, neamine resistance; *sm3*, low level streptomycin resistance; *sm2*, high level streptomycin resistance; *csd*, conditional streptomycin dependence. From (*50*).

More recently, this set of genes has been found to lie on a segment of a larger linkage group, which appears from genetic data to be circular (*51*). By examining the recombination of markers located all around this map, we have considerably increased our understanding of the exchange processes, and improved our mapping procedures.

The gene order and relative distances shown in the circular map in Fig. 3.22 are based upon recombination and co-segregation frequencies found in several five- and six-factor crosses each involving one new marker as well as the genes *tm*, *spc*, *ery*, *sm2*, and *ac* present in all of the crosses (*51*, *52b*). The evidence of circularity is supported not only by the data on which Fig. 3.22 is based, but also by segregation rates in liquid culture of the sort shown in Fig. 3.18. For example, in all crosses examined in liquid cultures, the gene *ery* shows the highest rate of segregation, and therefore we have assigned it a position 180° from the attachment point, *ap*. The genes *spc* and *sm2* with similar rates of segregation, have been found by recombination analysis to be more closely linked to *ery* than to each other. In other experiments, *ac* and *tm* exhibit similar segregation rates in liquid culture experiments, indicating that they are similar distances from *ap*, but they are not closely linked in recombination analysis.

The map shown in Fig. 3.23 incorporates data from these and earlier crosses (*46*, *49*, *51*, *52b*). We have not yet been able to establish a linear

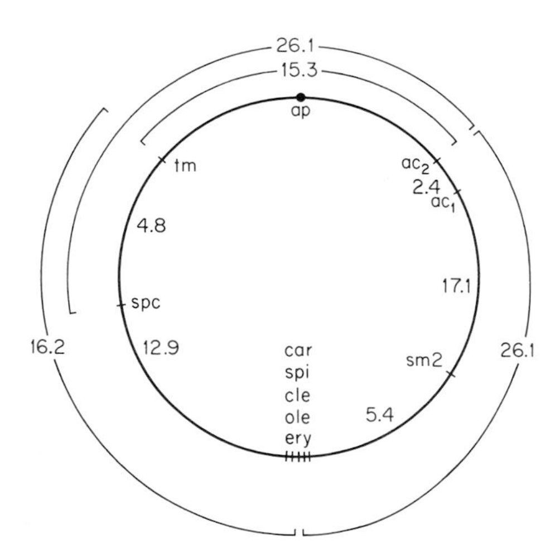

Fig. 3.22. Gene order and intergenic distances given in the figure are average values based on recombination and cosegregation frequencies in progeny of four crosses, each segregating for one of the markers *car*, *cle*, *ole*, and *spi*, in addition to all six markers *ac2*, *ac1*, *sm2*, *ergy*, *spc*, and *tm*. From (*51*).

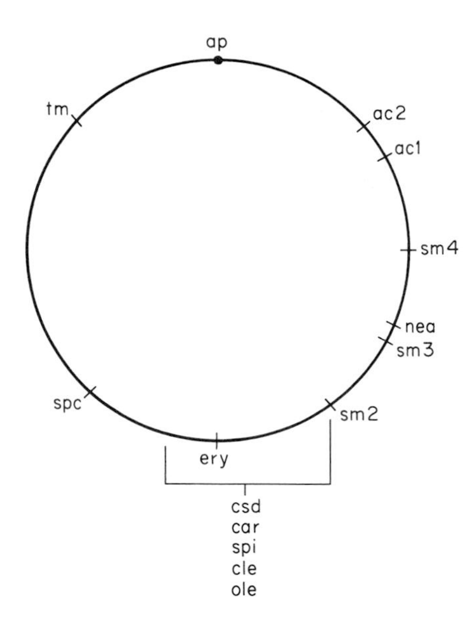

Fig. 3.23. Circular map incorporating data from several crosses. The qualitative evidence of gene order, relative distances, and circularity are well supported by the data but quantitative map distance need further clarification. The bracketed genes all map close to *ery* and no linear order has been established. From (*49*, *51*, and *52*). Symbols; *ap*, attachment point; *ac2* and *ac1*, acetate-requirement; *sm4*, streptomycin dependence; *nea*, neamine resistance, *sm3*, low level streptomycin resistance; *sm2*, high level streptomycin resistance, *ery*, erythromycin resistance; *csd*, conditional streptomycin dependence; *car*, carbomycin resistance; *spi*, spiramycin resistance; *cle*, cleosine resistance; *ole*, oleandomycin resistance; *spc*, spectinomycin resistance; and *tm*, temperature sensitivity.

order for the markers closest to *ery*. Some discrepancies in the quantitative data and other questions, like the extent of negative interference, await further analysis with the aid of additional markers.

The Four-Strand Stage

The most extensive evidence that recombination in this linkage group occurs at a four-strand stage, similar to that pictured in meiotic and mitotic recombination of nuclear chromosomes, comes from strand analysis. The procedure involves assigning the location of each exchange event on the basis of pedigree data. The assumed patterns of distribution are shown in Fig. 3.24.

The fact that recombination in this linage group is occurring at a four-strand stage was already evident from the pedigree data (*49*, *51*, *51a*). Before circularity had been established, we were able to identify

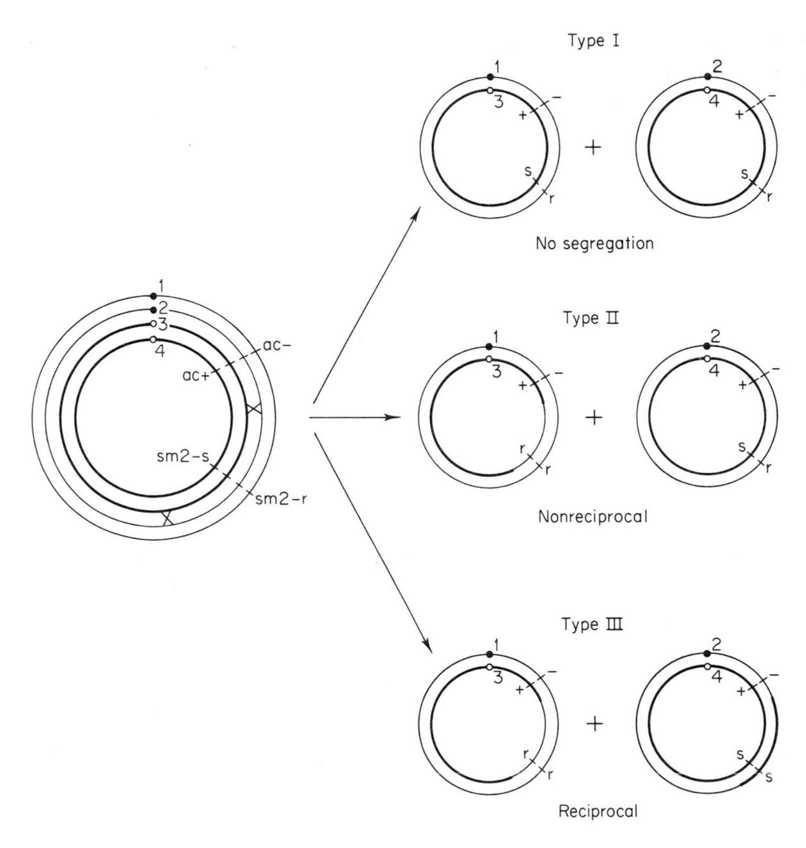

Fig. 3.24. Segregation patterns in circular molecules. The homologus DNA's from the two parents are shown as thick and thin lines, each representing double-stranded molecules with corresponding attachment points. After replication but before cell division, the four molecules are arranged so that exchange events can occur. The attachment points are oriented so that 1 and 2 go to separate cells at cell division, as do 3 and 4. If no exchanges occur, both daughter cells are fully heterozygous (type I). Nonreciprocal (type II) exchange is viewed as the miscopying of a region. Reciprocal (type III) segregation requires an exchange event between the genes and attachment point, as well as an exchange beyond the gene to insure proper separation of the circles. "X" indicates point of exchange between strands 2 and 3. From *(51)*.

four distinct strands in pedigrees. However, with the recognition of circularity and the development of regular strand analysis, vastly more evidence has become available about recombination patterns. We are now in a position to examine the recombination process in this system, and these studies are now in progress. We wish to investigate the relationship, if any, between reciprocal (type III) and nonreciprocal (type II) exchanges. Are these events correlated in any way? Can their relative frequencies be altered experimentally or by mutation? Above all, can the genetic map be related to the physical size and structure of chloroplast DNA?

PERSISTENT HETEROZYGOTES

Beginning about 1965, evidence has been accumulating in our laboratory of a second pattern of transmission of cytoplasmic genes in *Chlamydomonas:* maternal transmission, but with 2:2 segregation ratios in meiosis. This novel behavior was initially studied with a special set of progeny from crosses with an *sd* mutant that was phenotypically *sr,* but which in crosses of *sr* mt^+ × *ss* mt^- produced zygotes segregating 2 *sr*:2 *sd* progeny in meiosis. Subsequently, the same pattern was seen with *sr* mutants recovered after streptomycin starvation of an *sd* mutant strain (cf. page 72). Here again, the newly arisen *sr* mutants, in crosses with *ss* mt^- testers, gave rise to progeny segregating approximately 2 *sr*:2 *sd* in meiosis.

The evidence of maternal inheritance in these crosses was provided by the fact that cytoplasmic genes from the mt^- parent were not transmitted. Thus the *sr* mt^+ parent was carrying both the *sr* and the *sd* genes, and transmitted both genes at meiosis, with a segregation process sufficiently regular to produce, in the main (but not exclusively) 2:2 ratios.

These observations raised many questions. Were the *sr* and *sd* genes seen in these experiments identical with those we had been studying before? Could the same genes exhibit both 4:0 and 2:2 segregation ratios in meiosis? If the *sr* mutants were actually *sr/sd,* why didn't they segregate in mitosis?

To answer these questions, we have taken advantage of our newly developed set of linked markers. We set out to test the following hypothesis: that the same genes segregate either 2:2 or 4:0 in meiosis, depending upon whether the female (mt^+) parent is homozygous or heterozygous for the cytoplasmic genome. How does one obtain a persistent heterozygote to start with? Our previous studies had demonstrated that cytohets do not persist, but rather that they segregate during the first few mitotic doublings after meiosis. We decided to search for rare persistent cytohets (*52, 52a*).

Using selective techniques, we reexamined the progeny from the cross shown in Fig. 3.25, and found a small fraction which we suspected might be persistent cytohets. They were resistant to four antibiotics, having received the genes *sm2-r, ery-r* and *car-r* from one parent, and

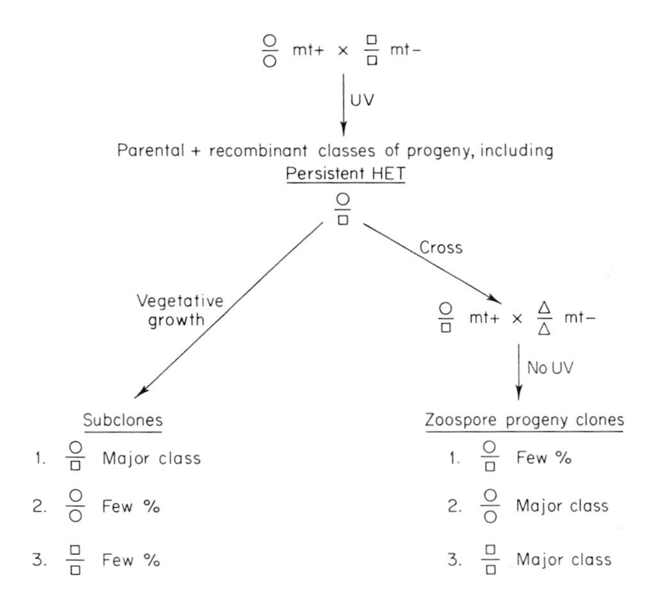

Fig. 3.25. Origin and transmission of a persistent cytohet. Key, ○, *sm2-s ery-s car-s spc-r tr;* □, *sm2-r ery-r car-r spc-s ts;* △, *sm2-s ery-s car-s spc-s ts.* Note that (1) the persistent HET rarely segregates during vegetative growth, but usually segregates in meiosis; (2) persistent HET shows maternal inheritance; and (3) recombination is very rare.

spc-r from the other parent. During vegetative growth, these clones were found to segregate out the two parental genotypes at a low frequency of 1% or less.

When these cytohets were crossed, as shown in Fig. 3.25, the progeny were found to segregate for the parental types that had given rise to the cytohet strain (i.e., the mt^+ parent in this cross). Thus, most progeny were either *sm-r, ery-r, car-r, spc-s,* or *sm-s, ery-s, car-s spc-r.* In addition, some progeny were still cytohets, but segregated out in the first few doublings.

Thus, persistent cytohets have been shown to behave in a unique way. They segregate principally in meiosis and rarely in mitosis, whereas conventional cytohets arising as progeny from biparental zygotes segregate primarily in vegetative growth and rarely in meiosis. The property of persistence as a cytohet is not itself inherited.

The use of the set of markers shown in Fig. 3.25, coming from a well-analyzed cross in which most of the progeny behaved in the conventional manner previously described, has established beyond doubt that the same genes, and by inference, the same DNA, may exhibit two distinct modes of behavior in meiosis and in mitosis.

The opposite conclusion was reached by Schimmer and Arnold (*55–57*) in a series of papers in which they examined one aspect of the behavior of persistent cytohets. They recovered *ss* revertants from an *sd* strain, and studied in great detail the frequencies of appearance of *sd* subclones during growth of the new *ss* mutant strains. They found that different mutant clones exhibited different frequencies of segregation of *sd* cells, and concluded that they were observing the results of a random segregation process. Schimmer and Arnold proposed that the *sd* gene that they had been studying was probably mitochondrial, since the chloroplast genome had been shown (*49*) to be diploid.

It now seems likely that Schimmer and Arnold were in fact examining one of the properties of persistent cytohets, namely their ability to segregate with a low frequency in mitosis. Had they made crosses with these strains, they would have seen the maternal 2:2 ratios described above, and would have been led to discard the mitochondrial hypothesis.

What is the molecular basis for the special behavior of the persistent cytohets? Most cytohets arising in crosses segregate rapidly in mitosis during the growth of zoospore clones. By the time they are ready to be crossed again, they are fully homozygous. Persistent cytohets, on the other hand, do not segregate rapidly during clonal growth, and clones of persistent cytohets are still primarily heterozygous when they are crossed. Thus, the origin of persistent cytohets seems to depend upon some event which blocks segregation and occurs in zoospores or their immediate clonal descendants.

What is this event? As will be recalled (p. 87) segregation involves pairing and strand exchange, i.e., recombination. We have no evidence of segregation by a sorting out process apart from recombination. Consequently, we assume that persistent cytohets result from an event that blocks recombination. Since rare segregation events do occur, the block evidently is not absolute.

The nature of this block is not known. We have postulated a change in orientation of the chloroplast DNA molecules to inhibit recombination but the mechanism of this process remains to be investigated.

CONCLUDING REMARKS

This chapter has described the development of methods of cytoplasmic genetic analysis in the alga *Chlamydomonas* (*31, 32, 34, 38, 39, 45–51*). This organism was initially chosen for the problem because it was the best sexual photosynthetic eukaryotic microorganism available. Among its many advantages are ease of growth and control of the sexual cycle under laboratory conditions; absence of nuclear chromosomal aberrations; ease of inducing and identifying cytoplasmic gene mutations; presence of only one chloroplast per cell, greatly simplifying the analysis of the chloroplast genome; a large difference in the average base composition of nuclear and chloroplast DNA, leading to ease of identification and separation of these components in cesium chloride gradients.

With this organism we have established the presence of a cytoplasmic genetic linkage group with numerous genes affecting growth (temperature-sensitive), photosynthetic capacity (acetate-requiring), and antibiotic resistance (*49–51*). All cytoplasmic genes so far mapped have been located within this single linkage group.

Recombination analysis of crosses involving multiply-marked parental stocks has revealed some behavioral features of this linkage group. These features were summarized earlier in the chapter (page 70) and include: (*a*) production of biparental zygotes by UV irradiation, which blocks the events leading to loss of cytogenes from the mt^- parent; (*b*) reciprocal and nonreciprocal exchange events occurring at a four-strand stage prior to cell division; (*c*) presence of a centromere-like attachment point that governs the distribution of the linkage groups at cell division; and (*d*) evidence from mapping data that the linkage group under consideration is genetically circular.

We identify this cytoplasmic linkage group with chloroplast DNA for the following reasons. First, let us consider the evidence that the linkage group consists of DNA. The principal evidence comes from recombination analysis: the regularities of exchanges at the four-strand stage generating a unique linear or circular map are properties unique to DNA. Supporting evidence comes from studies of photoreactivation following UV irradiation of mt^+ gametes to produce biparental zygotes; at high UV doses, the mt^+ cytogenes are not transmitted unless they are rescued by photoreactivation.

Accepting then that this linkage group consists of DNA, we may ask: which DNA? Genetic evidence has clearly distinguished the behavior of the cytoplasmic linkage group from that of nuclear chromosomes on the basis of (*a*) maternal inheritance; (*b*) biparental inheritance after UV irradiation without recombination in meiosis; (*c*) regular patterns of recombination in vegetative growth; and (*d*) genetic circularity. The genetic evidence thus excludes nuclear DNA and leads us to choose among cytoplasmic DNA's for the location of this linkage group.

The only cytoplasmic DNA so far identified in *Chlamydomonas* is chloroplast DNA. We assume that mitochondrial DNA is also present, but its identity has not yet been established. For the sake of discussion, we will compare chloroplast and mitochondrial DNA's as possible carriers of the cytoplasmic linkage group.

1. The genetic evidence has demonstrated that the cytoplasmic linkage group is functionally diploid (*49, 51*). Diploidy is consistent with the presence of one chloroplast per cell and would be difficult to reconcile with the presence of numerous mitochondria per cell.
2. Diploidy is consistent with the demonstration by Ris and Plaut of the presence of two Feulgen-positive bodies in the chloroplast (*30*).
3. Most of the cytogenes so far mapped in *Chlamydomonas* have been mutagenized by streptomycin, a drug known to affect chloroplast development, but having little or no mutagenicity in any other system (*34*).
4. The loss of cytogenes from the mt^- parent in maternal inheritance is reflected in the different fates of chloroplast DNA's from the mt^+ parent and the mt^- parent seen during zygote maturation (*4, 25, 42*).
5. The phenotypic properties of cytogenes of this linkage group, including loss of photosynthetic capacity and resistance to antibiotics that affect the functioning of chloroplast ribosomes, are consistent with the location of these genes in chloroplast DNA (*18, 35–37, 61*).

If indeed, this linkage group is located in chloroplast DNA, we need to reconcile the genetic evidence with data from reannealing kinetics concerning the physical organization and genomic complexity of chloroplast DNA. Reannealing kinetics has revealed the presence of one component with a size of about 2×10^8 daltons, and a minor component in the size range of 10^6–10^7 daltons (*65*). If these are the only components, then the larger one would be present in about twenty copies in gametes. For technical reasons it has not yet been possible to detect a postulated third component consisting of DNA present in only two copies. Thus it is not yet known whether the linkage group which is genetically diploid is physically diploid or multiploid. Also the arrangement of the components, whether in one or many separate molecules, has not been ascertained.

It seems evident that genetic and biophysical studies of cytoplasmic DNA's in *Chlamydomonas* have opened new avenues for investigation. Recently, a cytoplasmic genetic system very similar to that of *Chlamydomonas* has been described in *Eudorina,* a colonial green alga related to *Chlamydomonas.* Both streptomycin resistance and a morphological property, the compactness of the colony, showed maternal inheritance (*26*).

Further research problems of general interest for which *Chlamydomonas* and related algae are uniquely suited include control of replication and transcription of organelle DNA's in the cell cycle; mechanisms of recombination and repair of organelle DNA's; molecular basis of maternal inheritance; physical organization of organelle DNA's; and above all, the functions of cytoplasmic genes and their interactions with the nuclear and mitochondrial genomes. Our present understanding of the role of organelle genes in biogenesis will be discussed in the latter half of this book.

Suggested Review Articles

Chiang, K. S. (1971). Replication, transmission and recombination of cytoplasmic DNAs in *Chlamydomonas reinhardi*. *In* "Autonomy and Biogenesis of Mitochondria and Chloroplasts," *Austr. Acad. Sci. Symp.* (N. K. Boardman, A. W. Linnane, and R. M. Smillie, eds.), p. 235. North-Holland, Publ., Amsterdam.

Gillham, N. W. (1969). Uniparental inheritance in *Chlamydomonas reinhardi*. *Amer. Natur.* **103,** 355.

Sager, R. (1970). Genetic studies of chloroplast DNA in *Chlamydomonas*. *In* "Control of Organelle Development," *Symp. Soc. Exp. Biol.* 24 (P. L. Miller, ed.), p. 401. Cambridge Univ. Press, London and New York.

Sager, R. and G. E. Palade (1957). Structure and development of the chloroplast in *Chlamydomonas*. I. The normal green cell *J. Biochem. Cytol.* **3,** 63.

Sager, R. and Z. Ramanis (1971). Methods of genetic analysis of chloroplast DNA in *Chlamydomonas*. *In* "Autonomy and Biogenesis of Mitochondria and Chloroplasts," *Austr. Acad. Sci. Symp.* (N. K. Boardman, A. W. Linnane, and R. M. Smillie, eds.), p. 250. North-Holland, Publ., Amsterdam.

References

1. Arber, W. and Linn, S. (1969). DNA modification and restriction. *Amer. Rev. Biochem.* **38,** 467.
2. Cavalier-Smith, T. (1970). Electron microscopic evidence for chloroplast fusion in zygotes of *Chlamydomonas reinhardi*. *Nature (London)* **228,** 333.
3. Chiang, K. S. (1968). Physical conservation of parental cytoplasmic DNA through meiosis in *Chlamydomonas reinhardi*. *Proc. Nat. Acad. Sci. U. S.* **60,** 194.
4. Chiang, K. S. (1971). Replication, transmission and recombination of cytoplasmic DNAs in *Chlamydomonas reinhardi*. *In* "Autonomy and Biogenesis of Mitochondria and Chloroplasts," *Aust. Acad. Sci. Symp.* (N. K. Boardman, A. W. Linnane, and R. M. Smillie, eds.), p. 235. North-Holland, Publ. Amsterdam.
5. Chiang, K. S. and Sueoka, N. (1967). Replication of chloroplast DNA in *Chlamydomonas reinhardi* during vegetative cell cycle: It's mode and regulation. *Proc. Nat. Acad. Sci. U. S.* **57,** 1506.
6. Chiang, K. K., Kates, J. R., Jones, R. F., and Sueoka, N. (1970). On the formation of a homogenous zygote population in *Chlamydomonas reinhardi*. *Develop Biol.* **22,** 655.
7. Ebersold, W. T. (1962). Biochemical genetics. *In* "Biochemistry and Physiology of Algae" (R. A. Lewin, ed.), p. 731. Academic Press, New York.
8. Ebersold, W. T. (1963). Heterozygous diploid strains of *Chlamydomas reinhardi*. *Genetics* **48,** 888.

8a. Ebersold, W. T. (1967). *Chlamydomonas reinhardi:* Heterozygous diploid strains. *Science* **157,** 447.

9. Ebersold, W. T., Levine, R. P., Levine, E. E., and Olmsted, M. A., (1962). Linkage maps in *Chlamydomonas reinhardi. Genetics* **47,** 531.
9a. Flechtner, V. and Sager, R. (1972). In preparation.
10. Friedmann, I., Colwin, A. L., and Colwin, L. H. (1968). Fine-structural aspects of fertilization in *Chlamydomonas reinhardi. J. Cell Science* **3,** 115.
11. Gillham, N. W. (1963) The nature of exceptions to the pattern of uniparental inheritance for high level strepomycin-resistance in *Chlamydomonas reinhardi. Genetics* **48,** 431.
12. Gillham, N. W. (1963) Transmission and segregation of a nonchromosomal factor controlling streptomycin resistance in diploid *Chlamydomonas. Nature* (*London*) **200,** 294.
13. Gillham, N. W. (1965). Linkage and recombination between nonchromosomal mutations in *Chlamydomonas reinhardi. Proc. Nat. Acad. Sci. U. S.* **54,** 1560.
14. Gillham, N. W. (1965). Induction of chromosomal and nonchromosomal mutations in *Chlamydomonas reinhardi* with *N*-methyl-*N*-nitrosoguanidine. *Genetics* **52,** 529.
15. Gillham, N. W. (1969). Uniparental inheritance in *Chlamydomonas reinhardi. Amer. Natur.* **103,** 355.
16. Gillham, N. W., and Fifer, W. (1968). Recombination of nonchromosomal mutations: A three-point cross in the green alga *Chlamydomonas reinhardi. Science* **162,** 683.
17. Gillham, N. W., and Levine, R. P. (1962). Studies on the origin of streptomycin resistant mutants in *Chlamydomonas reinhardi. Genetics* **47,** 14631.
18. Gillham, N. W., Boynton, J. E., and Burkholder, B. (1970). Mutations altering chloroplast ribosomes phenotype in *Chlamydomonas.* I. Non-Mendellian mutations. *Proc. Nat. Acad. Sci. U. S.* **67,** 1026.
19. Hastings, P. J., Levine, E. E., Cosbey, E., Hudock, M. O., Gillham, N. W., Surzycki, S. J., and Levine, R. P. (1965). The linkage groups of *Chlamydomonas reinhardi. Microbial Genet. Bull.* **23,** 17.
20. Hoober, J. K., and Blobel, G. (1969). Characterization of the chloroplastic and cytoplasmic ribosomes of *Chlamydomonas reinhardi. J. Mol. Biol.* **41,** 121.
21. Johnson, U. G., and Porter, K. R. (1968). Fine structure of cell division in *Chlamydomonas reinhardi.* Basal bodies and microtubules. *J. Cell Biol.* **38,** 403.
22. Kates, J. R., and Jones, R. F. (1964). The control of gametic differentiation in liquid cultures of *Chlamydomonas. J. Cell. Comp. Physiol.* **63,** 157.
23. Kates, J. R., Chiang, K. S., and Jones, R. F. (1968). Studies on DNA replication during synchronized vegetative growth and gametic differentiation in *Chlamydomonas reinhardi. Exp. Cell Res.* **49,** 121.
24. Kirk, J. T. O., and Tilney-Bassett, R. A. E. (1967). "The Plastids." Freeman, London and San Francisco, California.
25. Lane, D., and Sager, R. (1972). In preparation.
25a. Levine, R. P., and Goodenough, U.W. (1970). *Annu. Rev. Genet.* **4,** 397.
25b. Luria, S., and Delbruck, M. (1943). Mutations of bacteria from virus sensitivity to virus resistance. *Genetics* **28,** 491.
25c. McVittie, A., and Davies, D. R. (1971). The location of the linkage groups in *Chlamydomonas reinhardi.* Submitted for publication.
26. Mishra, N. C., and Threlkeld, S. F. H. (1968). Genetic studies in *Eudorina. Genet. Res.* **11,**21.
27. Ohad, I., Siekevitz, P., and Palade, G. E. (1967). Biogenesis of chloroplast membranes. I. Plastid dedifferentiation in a dark-grown algal mutant (*Chlamydomonas reinhardi*) *J. Cell Biol.* **35,** 521. II. Plastid differentiation during greening of a dark-grown algal mutant (*Chlamydomonas reinhardi*) *J. Cell Biol.* **35,** 553.
27a. Preer, J. R., Jr. (1971). Extrachromosomal inheritance: Hereditary symbionts, mitochondria, chloroplasts. *Annu. Rev. Genet.* **5,** 626.

28. Randall, J. T., Cavalier-Smith, T., McVittie, S., Warr, J. R., and Hopkins, J. M. (1967). Development and control processes in the basal bodies and flagella of *Chlamydomonas reinhardi. Develop. Biol. (Suppl.)* **1,** 43.
29. Ringo, D. L. (1967). Flagellar motion and fine structure of the flagellar apparatus in *Chlamydomonas. J. Cell Biol.* **33,** 543.
30. Ris, H. and Plaut, W. (1962). Ultrastructure of DNA-containing areas in the chloroplast of *Chlamydomonas. J. Cell Biol.* **13,** 383.
30a. Rosenbaum, J. L., Moulder, J. E., and Ringo, D. L. (1969). Flagellar elongation and shortening in *Chlamydomonas.* The use of cycloheximide and colchicine to study the synthesis and assembly of flagellar proteins. *J. Cell Biol.* **41,** 600.
31. Sager, R. (1954). Mendelian and non-Mendelian inheritance of streptomycin resistance in *Chylamydomonas reinhardi. Proc. Nat. Acad. Sci. U. S.* **40,**356.
32. Sager, R. (1955). Inheritance in the green alga *Chlamydomonas reinhardi. Genetics* **40,** 476.
33. Sager, R. (1960). Genetic systems in *Chlamydomonas. Science* **132,** 1459.
34. Sager, R. (1962). Streptomycin as a mutagen for nonchromosomal genes. *Proc. Nat. Acad. Sci. U. S.* **48,** 2018.
35. Sager, R. (1965). On non-chromosomal heredity in microorganisms. *In* "Function and Structure in Micro-organism," *Symp. Soc. Microbiol. XV,* p. 324.
35a. Sager, R. (1966). Mendelian and non-Mendelian heredity: A reappraisal. *Proc. Roy. Soc. Ser. B* **164,** 290.
36. Sager, R. (1968). Cytoplasmic genes and organelle formation. In "Formation and Fate of Cell Organelles," *Ann. Symp. Int. Soc. Cell Biol.* (K. G. Warren, ed.), p. 317. Academic Press, New York.
37. Sager, R. (1970). Genetic studies of chloroplast DNA in *Chlamydomonas. In* "Control of Organelle Development," *Symp. Soc. Exp. Biol.* (P. L. Miller, ed.), Vol. 24, p. 401. Cambridge Univ. Press, London and New York.
38. Sager, R., and Granick, S. (1953). Nutritional studies with *Chlamydomonas reinhardi. Ann. N. Y. Acad. Sci.* **56,** 831.
39. Sager, R., and Granick, S. (1954). Nutritional control of sexuality in *Chlamydomonas reinhardi. J. Gen. Physiol.* **37,** 729.
40. Sager, R., and Hamilton, M. G. (1967). Cytoplasmic and chloroplast ribosomes of *Chlamydomonas:* Ultracentrifugal characterization. *Science* **157,** 709.
41. Sager, R., and Ishida, M. R. (1963). Chloroplast DNA in *Chlamydomonas. Proc. Nat. Acad. Sci. U. S.* **50,** 725.
42. Sager, R. and Lane, D. (1969). Replication of chloroplast DNA in zygotes of *Chlamydomonas. Fed. Proc. Fed. Amer. Soc. Exp. Biol.* **38,** 347.
43. Sager, R., and Palade, G. E. (1954). Chloroplast structure in green and yellow strains of *Chlamydomonas. Exp. Cell Res.* **7,** 584.
44. Sager, R., and Palade, G. E. (1957). Structure and development of the chloroplast in *Chlamydomonas.* I. The normal green cell. *J. Biophys. Biochem.* **3,** 463.
45. Sager, R., and Ramanis, Z. (1963). The particulate nature of nonchromosomal genes in *Chlamydomonas. Proc. Nat. Acad. Sci. U. S.* **50,** 260.
46. Sager, R., and Ramanis, Z. (1965). Recombination of non-chromosomal genes in *Chlamydomonas. Proc. Nat. Acad. Sci. U. S.* **53,** 1053.
47. Sager, R., and Ramanis, Z. (1967). Biparental inheritance of non-chromosomal genes induced by ultraviolet irradiation. *Proc. Nat. Acad. Sci. U. S.* **58,** 931.
48. Sager, R., and Ramanis, Z. (1968). The pattern of segregation of cytoplasmic genes in *Chlamydomonas. Proc. Nat. Acad. Sci. U. S.* **61,** 324.
49. Sager, R., and Ramanis, Z. (1970). A genetic map of non-Mendelian genes in *Chlamydomonas. Proc. Nat. Acad. Sci. U. S.* **65,** 593.

50. Sager, R., and Ramanis, Z. (1971). Methods of genetic analysis of chloroplast DNA in *Chlamydomonas*. *In* "Autonomy and Biogenesis of Mitochondria and Chloroplasts," *Aust. Acad. Sci. Symp.* (N. K. Boardman, A. W. Linnane, and R. M. Smillie, eds.) p. 250. North-Holland Publ., Amsterdam.
51. Sager, R., and Ramanis, Z. (1972). Genetic circularity of chloroplast DNA in *Chlamydomonas*. (Manuscript in preparation.)
51a. Sager, R. and Ramanis, Z. (1971). Formal genetic analysis of organelle genetic systems. *Stadler Symp.* **1** and **2,** 65.
52. Sager, R. and Ramanis, Z. (1971). Persistent cytoplasmic heterozygotes in *Chlamydomonas*. *Genetics* **68,** 356.
52a. Sager, R. and Ramanis, Z. (1972). Mutagenesis of cytoplasmic genes in a streptomycin-dependent strain of *Chlamydomonas*. I. Induction of mutations by streptomycin-withdrawal. II. A novel pattern of inheritance. (Manuscripts in preparation.)
52b. Sager, R. and Ramanis, Z. (1972). (Manuscript in preparation.)
53. Sager, R., and Tsubo, Y. (1961). Genetic analysis of streptomycin-resistance and -dependence in *Chlamydomonas*. *Z. Vererbungslehre* **92,** 430.
54. Sager, R., and Tsubo, Y. (1962). Mutaganic effects of streptomycin in *Chlamydomonas*. *Arch. Mikrobiol.* **42,** 159.
55. Schimmer, O. and Arnold, C. G. (1970). Untersuchungen uber Reversions- und Segregationsverhalten eines ausserkaryotischen Gens bei *Chlamydomonas reinhardi*. *Mol. Gen. Genet.* **107,** 281.
56. Schimmer, O., and Arnold, C. G. (1970). Uber die Zahl der Kopien eines ausserkaryotischen Gens bei *Chlamydomonas reinhardi*. *Mol. Gen. Genet.* **107,** 366.
57. Schimmer, O., and Arnold, C. G. (1970). Hin- und Rucksegregation eines ausserkaryotischen Gens bei *Chlamydomonas reinhardi*. *Mol. Gen. Genet.* **108,** 33.
58. Smith, G. M. (1948). Sexuality in *Chlamydomonas*. *Science* **108,** 680.
59. Smith, G. M., and Regnery, D. C. (1950). Inheritance of sexuality in *Chlamydomonas reinhardi*. *Proc. Nat. Acad. Sci. U. S.* **36,** 246.
59a. Sueoka, N. (1968). Mitotic replication of deoxyribonucleic acid in *Chlamydomonas reinhardi*. *Proc. Nat. Acad. Sci. U. S.* **46,** 83.
60. Sueoka, N., Chiang, K. S., and Kates, J. R. (1967). Deoxyribonucleic acid replication in meiosis of *Chlamydomonas reinhardi*. *J. Mol. Biol.* **25,** 47.
61. Surzycki, S. J., and Gillham, N. W. (1971). Organelle mutations and their expression in *Chlamydomonas reinhardi*. *Proc. Nat. Acad. Sci. U. S.* **68,** 1301.
62. Von Euler, H. (1949). Einfluss des Streptomycins auf die chlorophyllbildung. *Kem. Arb. II* **9,** 1.
63. Von Euler, H., Bracco, M., and Heller, L. (1948). Les actions de la streptomycine sur les graines en germination des plantes vertes et sur les polynucleotides. *C. R. Acad. Sci.* **227,** 16.
64. Warr, J. R., McVittie, A., Randall, J. T., and Hopkins, J. M. (1966). Genetic control of flagella structure in *Chlamydomonas reinhardi*. *Genet. Res.* **7,** 335.
65. Wells, R., and Sager, R. (1971). Denaturation and renaturation kinetics of chloroplast DNA from *Chlamydomonas reinhardi*. *J. Mol. Biol.* **58,** 611.
66. Wiese, L. (1965). On sexual agglutination and mating type substances (gamones) in isogamous heterothallic *Chlamydomonas*. I. Evidence of the identity of the gamones with the surface components responsible for sexual flagellar contact. *J. Phycol.* **1,** 46.

4

Mitochondrial Genetics of Yeast

Yeast has played a central role in the study of mitochondrial genetics, beginning with the discovery and subsequent investigation of a class of *petite* mutations by Ephrussi (*7*) and his collaborators. At the genetic level, some of the *petite* mutations were shown to be non-Mendelian. At the biochemical level these mutants were shown to lack respiratory capacity, being deficient in cytochrome activity and growing only on fermentable substrates. The non-Mendelian *petite* mutation was postulated to be mitochondrial because of its phenotype—a block in mitochondrial activity. But until recently no experimental evidence established the non-Mendelian *petites* unequivocally as mitochondrial mutations.

Subsequently, with the discovery that some *petite* mutants contain an altered mitochondrial DNA, the mitochondrial location of the *petite* mutation was substantially established. The critical evidence associating the *petite* mutation with mitochondrial DNA is based on two discoveries: (*a*) that some *petites* contain a physically altered mitochondrial DNA, and (*b*) that mutations to *petite* can also alter other non-Mendelian genes thought to be mitochondrial. In the past, a serious limitation to genetic analysis of the mitochondrial system was the absence of any mutations other than *petite.* The recent demonstration of non-Mendelian mutations to drug resistance in yeast has made it possible to examine mitochondrial DNA by genetic methods.

This chapter begins with the initial evidence of the non-Mendelian nature of the *petite* mutation and the discovery of suppressiveness. The phenomenon of suppressiveness, a kind of dominance of the *petite* phenotype over the wild type, will be considered in some detail because of its importance in the analysis of interactions between different mitochondrial DNA's in the same cell. Finally, new experiments demonstrating recombination of mitochondrial genes and illuminating the relationship between the *petite* phenotype and mitochondrial DNA will be considered. Genetic analysis of this system is just at its inception. One may anticipate substantial progress in the understanding of mitochondrial genetics in the next few years.

SEXUAL LIFE CYCLE OF SACCHAROMYCES CEREVISIAE

Applied yeast genetics began in prehistoric times with the selection of strains which were particularly suitable for leavening bread and for fermenting sugar to produce alcoholic beverages. (Ceres was the Roman goddess of the harvest.) There was probably little change in methodology from then until 1935 when Winge showed that haploid strains could be crossed and that classical genetic analysis was feasible (*46*). Actually, the first account of the sexual cycle of yeast under laboratory control was published in 1918 by Kruis and Satava (*22*). However, this paper was lost in the literature for many years, during which time the genetic analysis of yeast was developed by Winge (*48*) and subsequently by Lindegren (*24a*).

The life cycle of the principal yeast strain (*Saccharomyces cerevisiae*) used in genetic investigations is a simple one, as shown in Fig. 4.1. Haploid strains are differentiated by a pair of chromosomal alleles *a* and α which control the mating system. When mixed under appropriate conditions, individual cells of *a* and α fuse to form the diploid zygote which can then divide mitotically to give rise to clones of diploid cells. Single diploid cells can be induced to sporulate, a process involving meiosis and the formation of an unordered tetrad consisting of the four meiotic products called ascospores. The alleles *a* and α segregate 2:2 in each tetrad, and each ascospore can grow, giving rise to a haploid clone.

When haploid cells are crossed, genetic analysis can be carried out both at the level of the resulting diploid clone and by tetrad analysis of diploid cells induced to sporulate. As with *Chlamydomonas,* nuclear genes, by which the parents differ, segregate 2:2 among the ascospores, and deviations from 2:2 segregation provide the first line of evidence in the recognition of cytoplasmic genes.

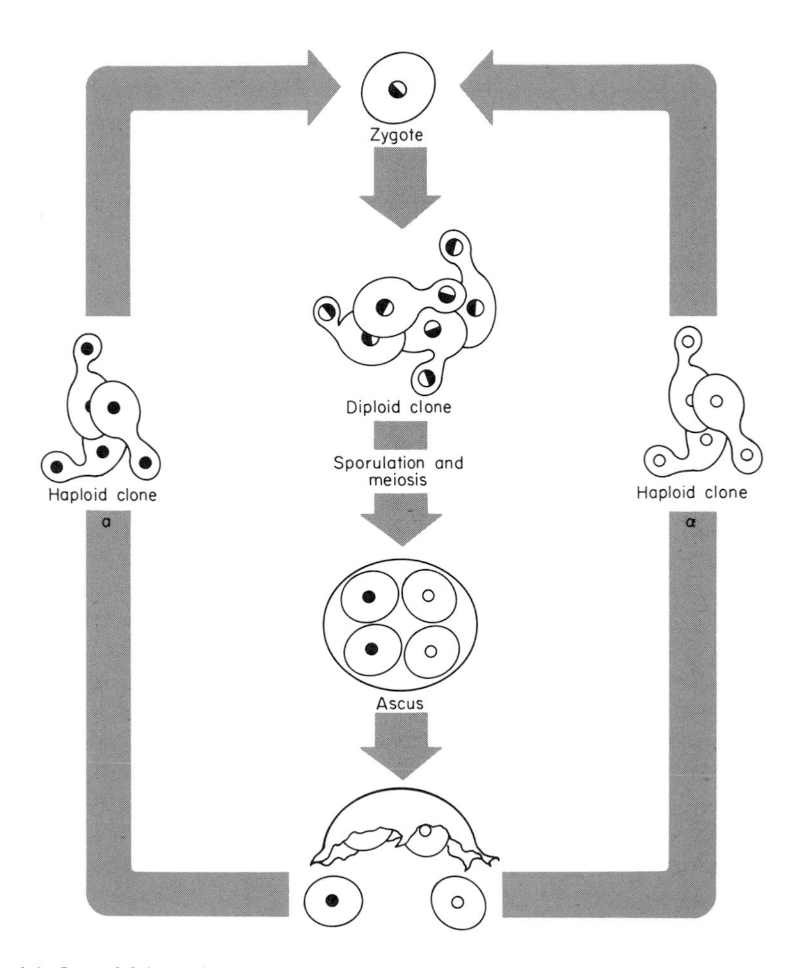

Fig. 4.1. Sexual life cycle of yeast (*Saccharomyces cerevisiae*). Haploid cells of opposite mating types, *a* and α, fuse to form the diploid zygote, which grows by budding to form a diploid colony. Diploid cells can be induced to undergo meiosis and sporulation, producing four ascospores (two are mating type *a* and two are α), which are the four products of meiosis, in an unordered ascus. Each ascospore grows to form a haploid clone of cells which can fuse with cells of opposite mating type and repeat the cycle.

EARLY HINTS OF MITOCHONDRIAL HEREDITY

Probably the first evidence of an extranuclear system in yeast was reported by Winge and Laustsen (*47*). They were studying strains of homozygous diploid yeast which had formed by diploidization of haploids. In previous studies they had shown that haploid yeast could form diploids by two alternate routes: (*a*) cell diploids, formed by cell fusion of two haploid cells, as shown in Fig. 4.1, and (*b*) spore diploids, formed by nuclear fusion within the ascus immediately after meiosis and before ascospores have formed. If the parents were isogenic except for mating type no genetic variability would be expected in either of these homozygous diploid lines except for random changes owing to new mutations. However, they found evidence for what they called an inbreeding degeneration, based upon a decline in germination and a precipitous decline in the amount of dry matter produced by spore diploids but not by cell diploids. They emphasized that these effects were permanent over many years.

Winge and Laustsen interpreted their results as a decrease in the chondriosome (as mitochondria were then called) component of the cytoplasm. They suggested that the degeneration resulted from a decrease in the number of mitochondria per cell, owing to the manner of formation of diploids; that in cell diploids the correct mitochondrial complement was maintained, but that in spore diploids, the nuclear genome diploidized properly but the mitochondria did not. This hypothesis is illustrated in Fig. 4.2 taken from their paper. Translated into molecular terms, this hypothesis predicts that spore diploids contain less mitochondrial DNA than cell diploids. Only now have methods become available to test this hypothesis. Probably Winge and Laustsen were thinking in terms of cytological studies which had demonstrated regular and oriented distribution of mitochondria at cell division, such as those of E. B. Wilson.

In "The Cell," Wilson pointed out that Benda and his successors had emphasized the fact that the chondriosomes were distributed to the daughter cells with approximate equality. In ciliates, Faure-Fremiet reported that the numerous scattered mitochondria divide synchronously with the nucleus. Orientation of the chondriosome with respect to the spindle had been described in spermatocytes of *Ascaris* and in mitotic divisions of higher plants. Indeed, a regular orientation of the chondriosomes in sperm formation has been described in a great number of organisms (*45*).

In 1916 Wilson (*45*) had described the distribution of mitochondria during spermatogenesis in two scorpions, *Centrurus* and *Opisthacanthus*. He stated that with respect to *Centrurus*, "It is possible to conclude with certainty that the chondriosome-material is divided with exact equality among all the spermatozoa. . . . Before the maturation divisions take place the whole of the chondriosome-material in the primary spermatocytes is concentrated in a single, ring-shaped body which is then equally divided in such a way that each spermatid receives one-quarter of the product, the process taking place with a

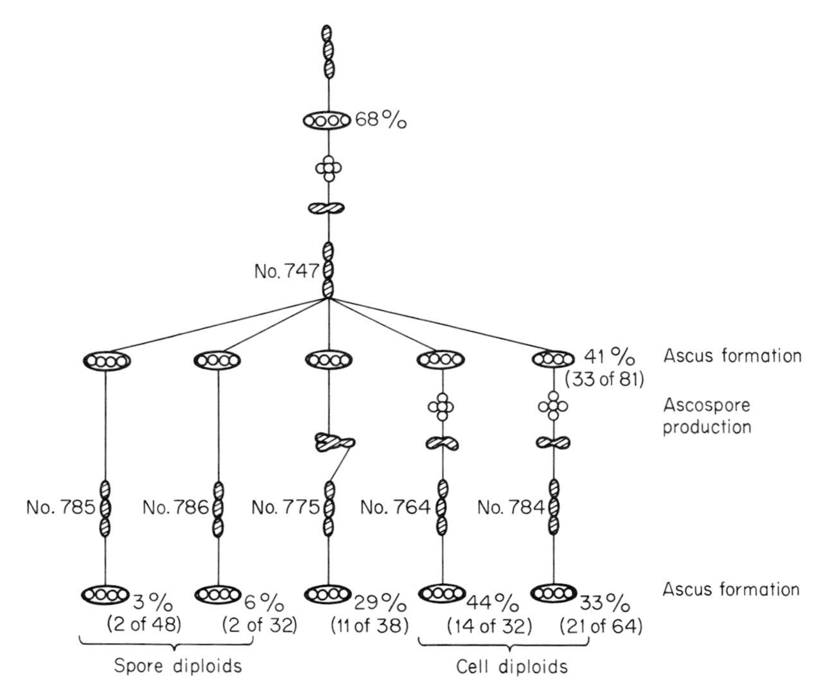

Fig. 4.2. Spore diploids and cell diploids of yeast. Comparative viability assessed as sporulating ability of spore diploids (Nos. 785 and 786) formed by nuclear fusion in the ascus just following meiosis; and of cell diploids (Nos. 747, 764, and 784) formed by the fusion of haploid cells. From Winge and Laustsen (*47*).

precision that is comparable to that seen in the distribution of the chromosome material." From E. B. Wilson "The Cell in Development and Heredity." Copyright 1925 by The Macmillan Company, New York.

CYTOPLASMIC PETITES

The first unequivocal demonstration of a cytoplasmic hereditary determinant in yeast came in a series of papers by Ephrussi and coworkers describing the "*petite*" mutations (*9, 11, 12, 13, 40*). *Petite* cells are respiratory deficient; they require a fermentable sugar such as glucose for growth and contain no detectable amounts of cytochromes a, a_3, or b. They do contain cytochrome c.

The first paper (*11*) on the origin of the *petite* mutation showed that an entire population of cells could be transformed from normal to *petite* by growth in the presence of the acridine dye, acriflavine; and that the susceptibility of the cells to acriflavine was the same for haploid and for diploid strains. This observation provided the first hint that the effect, though permanent and hereditary, might not be occurring at the level of nuclear gene mutation, because of its independence of gene dosage. [Much later, Pittman (*31*) extended this observation to tetraploid yeast as well.] Some yeast strains contained spontaneous mutants, present at a frequency of about 1%, which appeared to resemble the acriflavine-induced ones.

Subsequent papers in the series reported a comprehensive genetic analysis of *petite* mutants, as well as physiological and biochemical studies comparing the *petite* mutant with the wild type (*3, 6, 10, 15*). Genetic studies of *petites* will be considered in this chapter; discussion of physiological and morphogenetic studies will be deferred to Chapter 7.

The genetic analysis was straightforward. When *petite* cells of either mating type were crossed with the wild type, diploid cells were produced which were normal in respiration like the wild-type parent. When these diploid cells were sporulated, they produced the haploid products of meiosis, the ascospores which were all wild type with rare exceptions. The *petite* phenotype had vanished as shown in Fig. 4.3.

Wild-type cells of the F_1 generation were chosen at random and backcrossed to the original *petite* parent; all the progeny were normal. This procedure was carried out for four backcross generations; the results are shown in Table 4.1. In these crosses the parents differed by three pairs of segregating unlinked nuclear genes: mating type, a requirement for adenine, and a requirement of thiamine. Each of the Mendelian gene pairs segregated 2:2 with rare exceptions. But the petite phenotype appeared in only 5 out of 596 ascospores.

Because of the high spontaneous mutation rate from the wild type to *petite*, it seemed likely that these five exceptional *petite* progeny were new mutants, but the possibility that they were segregants of a multigenic system had to be considered. The problem was dealt with in an appendix by L'Heritier (*12*). He developed a statistic treatment to test the null hypothesis that the five *petite* progeny arising in backcross generations were the result of segregation of many recessive nuclear mutant genes simultaneously induced by acriflavine in the original mutagenic exposure.

He showed that in this instance, with five *petites* out of 596 progeny, it would require more than twenty independently segregating nuclear genes to give as rare a segregation as was observed. Since the probability is remote that acriflavine had simultaneously induced mutations of more than twenty unlinked nuclear genes, the backcross data provide strong evidence against the null hypothesis.

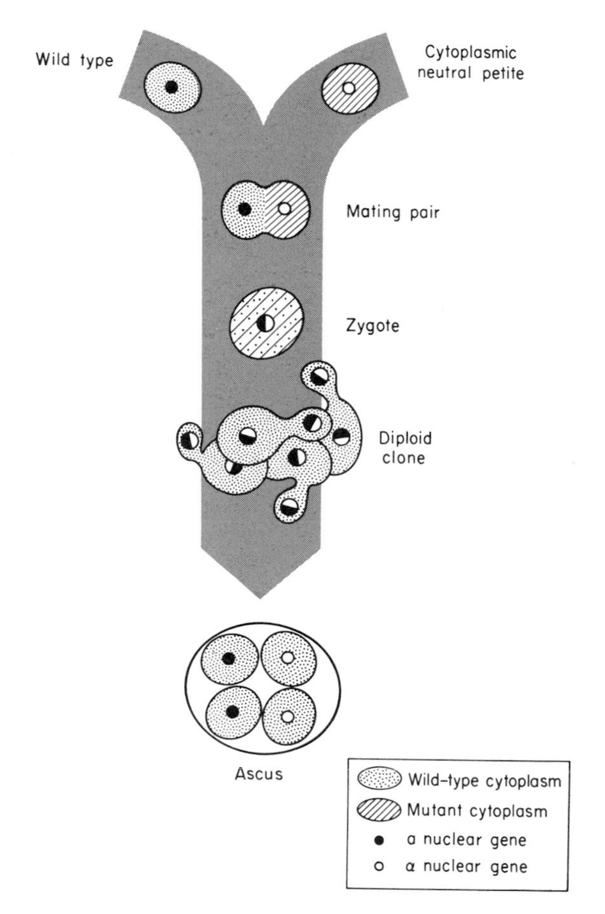

Fig. 4.3. Cross of wild type × neutral *petite*. Zygote is initially a cytohet, then gives rise to diploid colonies that have totally lost the *petite* phenotype. After sporulation, ascospores are all wild type.

The paper concludes that the genetic analysis is not compatible with a Mendelian pattern of inheritance and that the *petite* mutation is an example of cytoplasmic inheritance. Subsequent studies have fully substantiated this conclusion.

These studies were reported in 1949 and some of the original *petite* strains are still being cultivated. No reversions have ever been found despite the great selective advantage that wild-type revertants would have by virtue of a much faster growth rate. The original strains are now referred to as neutral or recessive *petites* to distinguish them from a second class of cytoplasmic *petites*, suppressive or dominant, which were discovered later (*14*) and which will be discussed below (p. 120).

TABLE 4.1

Progeny of Yeast Crosses and Backcrosses: Neutral Petite × Wild Type[a]

	First cross	Backcross 1	Backcross 2	Backcross 3	Backcross 4	Total
Segregation of nuclear genes						
Mating type						
2a:2α	28[b]	23	18	20	21[c]	110
Adenine requirement						
2A:2*a*	28	23	18	20	55	144
1A:3a	0	0	0	0	1	1
3A:1a	2	0	0	0	1	3
4A:0a	1	0	0	0	0	1
Thiamine requirement						
2T:2t	22	16	12	19	56	125
3T:1t	5	0	0	0	0	5
1T:3t	3	7	5	0	1	16
0T:4t	1	0	1	1	0	3
Number of asci analyzed	31	23	18	20	57	149
Segregation of cytoplasmic *petite*						
grande (wild type)	121	91	72	80	227	591
petite (mutant)	3	1	0	0	1	5
Number of progeny analyzed	124	92	72	80	228	596

[a] Adapted from (*12*), Table VII.
[b] Three diploid progeny not included.
[c] Mating type was determined in only twenty-one asci of backcross 4.

The original experiments on the induction of *petites* were carried out with acriflavine. Marcovich (*27*) compared the effects of proflavine and euflavine, the two acridine dyes present in a ratio of 1:2 in acriflavine. He found that proflavine had no effect upon the production of *petites* when tested at the highest nontoxic concentration. On the other hand, the euflavine component was extremely active in *petite* formation; one could transform the population completely to *petite* at nonlethal concentrations. This study showed that the toxicity and mutagenecity of the two acridine dyes were different, that euflavine was the effective component and that its effectiveness as a mutagen occurred at sublethal concentrations. Under conditions of maximal mutagenicity, it was further noted that only the buds were *petites*, the mother cells remaining phenotypically wild type. The significance of this differential effect on mother cell and bud will be discussed below (page 119).

The mutagenic effect of purified euflavine was directly demonstrated

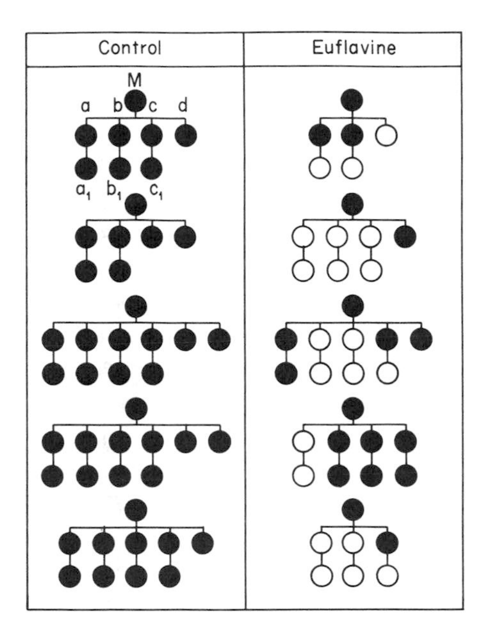

Fig. 4.4. Induction of *petite* mutations by euflavine. Pedigree analysis of single yeast cells and their descendants grown in the presence and absence of euflavine. The first line of descendants represents the buds produced by each treated cell, and the second line represents one bud from each of the first generation of buds. The phenotype of each cell was determined, by tests of the clone developing from it, as the presence or absence of cytochrome oxidase activity. Black circles represent wild-type clones; open circles represent *petite* mutants. From (*9*).

by Ephrussi and Hottinguer (*9*) in a study of hand-dissected pedigrees. These pedigrees, shown in Fig. 4.4, demonstrated the origin of *petites* in buds removed from the mother cell when cells were grown in media containing 10^{-6} *M* purified euflavine. A mutagen with close to 100% effectiveness in one doubling was a totally new phenomenon in genetics and lent further weight to the idea that the target genes were not nuclear.

An independent demonstration of the cytoplasmic nature of the *petite* determinants was given by Wright and Lederberg (*49*). They took advantage of a phenomenon first described by Fowell: the existence of transient heterokaryons during zygote formation in yeast. When haploid strains and mated there is a period during cell fusion, before the nuclei have fused, during which buds are produced. Sometimes buds arise containing only one of the unfused parental nuclei in a common cytoplasm. They form haploid clones of cells with nuclei from just one parent and cytoplasm from both. These yeast heterokaryons are for-

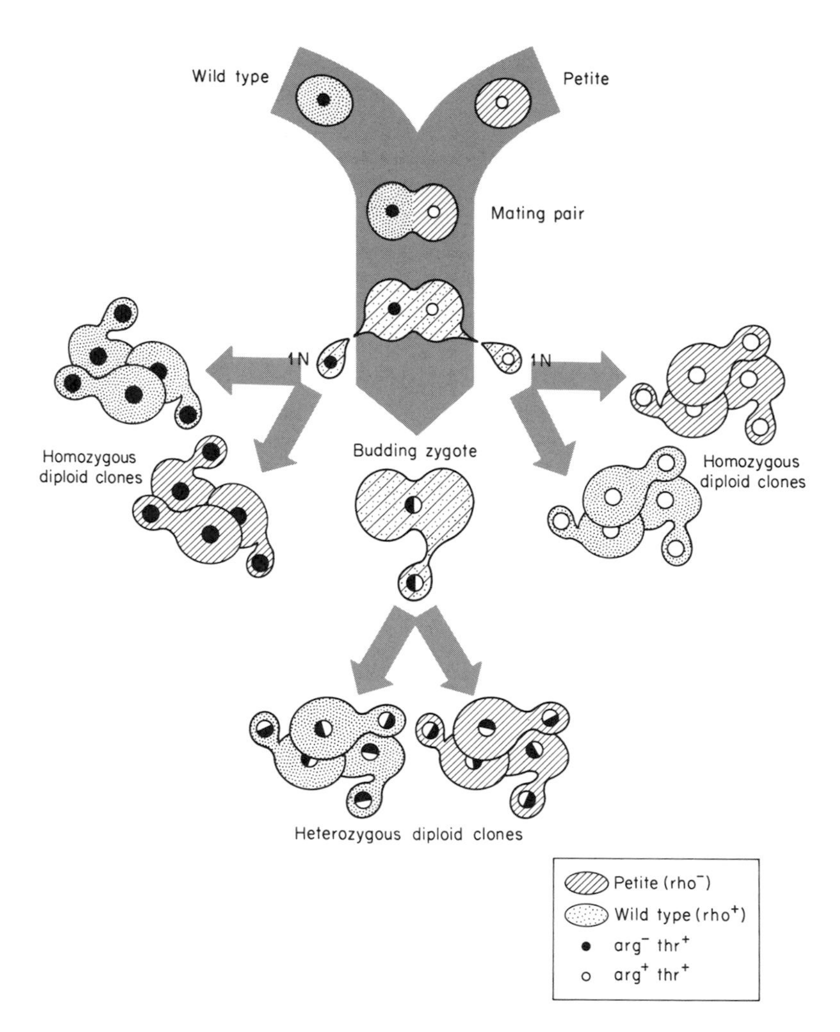

Fig. 4.5. Wright and Lederberg experiment: transmission of *petite* genotype through yeast heterokaryons. Mating pairs were hand dissected to obtain rare homozygous buds which received cytoplasm from both parents but nuclei from only one. Initially haploid, they spontaneously diploidize during growth. Since the parental *rho*$^-$ strain was suppressive (p. 120), both *rho*$^+$ and *rho*$^-$ clones were obtained from the homozygous, as well as heterozygous, diploids. Adapted from (*49*).

mally analogous to the fungal heterokaryons used to detect somatic segregation of cytoplasmic genes in *Aspergillus* (Chapter 5).

In the Wright and Lederberg study diagrammed in Fig. 4.5, the parental nuclei were marked by auxotrophic mutations, one strain being arginine-minus and the other threonine-minus. The strains were mated and mating pairs were isolated, allowed to form twenty to one hundred cells, each of which was replated and allowed to form a colony which was then classified for markers. The purpose was to find cells in which the parental nuclei had sorted out in buds before nuclear fusion. In crosses of *petite* × wild type, a total of ninety-one appropriate clones (derived from unfused parental nuclei) were isolated from 530 mating pairs. Six of the clones contained nuclei from the wild-type (non-*petite*) parent but were *petite* in phenotype, and five of the clones contained nuclei from the *petite* parent, but were wild type with respect to respiratory ability. Thus the *petite* property was shown to segregate independently of the nucleus. The authors concluded that the respiratory phenotype is subject to extranuclear genetic transmission in accordance with the conclusions of Ephrussi *et al.* (*12*).

In 1950 Chen *et al.* (*3*) reported the isolation of a series of mutants with a *petite* phenotype, but with a Mendelian pattern of genetic segregation. When these strains were crossed with wild type, the diploid zygotes were all normal and the ascospores produced after sporulation each segregated two *petites* to two wild types. In the early literature, these *petites* were called segregational, but we shall refer to them as nuclear. Since then, more than twenty different nuclear genes have been located, for the most part unlinked, each having an effect upon mitochondrial development resulting in a mutant phenotype similar to that of the cytoplasmic *petites* (*1, 39*). These strains and their biochemical interactions with the mitochondrial *petites* will be discussed in Chapter 7.

As shown in Fig. 4.6, crosses between nuclear and cytoplasmic *petites* give the following characteristic result: the zygotes develop normal respiration and after sporulation of the zygotes the ascospores are found to segregate two *petites* to two normal. In contrast to this result, in crosses between different cytoplasmic (neutral) *petites* the diploids are always *petites*.

Thus the nuclear and cytoplasmic *petites* complement one another. In formal genetic terms, complementation indicates nonidentity of genes. On this basis we may conclude that the cytoplasmic *petite* differs in its biochemical effect from all the nuclear *petite* loci, which also complement among themselves.

In summary, these early genetic studies of *petite* mutants established the existence of both nuclear (i.e., Mendelian) and cytoplasmic (i.e., non-Mendelian) genes influencing mitochondrial biogenesis. One method of distinguishing the two classes was clearly established; 4:0 segregation of alleles after sporulation of diploids arising from crosses of mutant × wild type. The heterokaryon experiment of Wright and

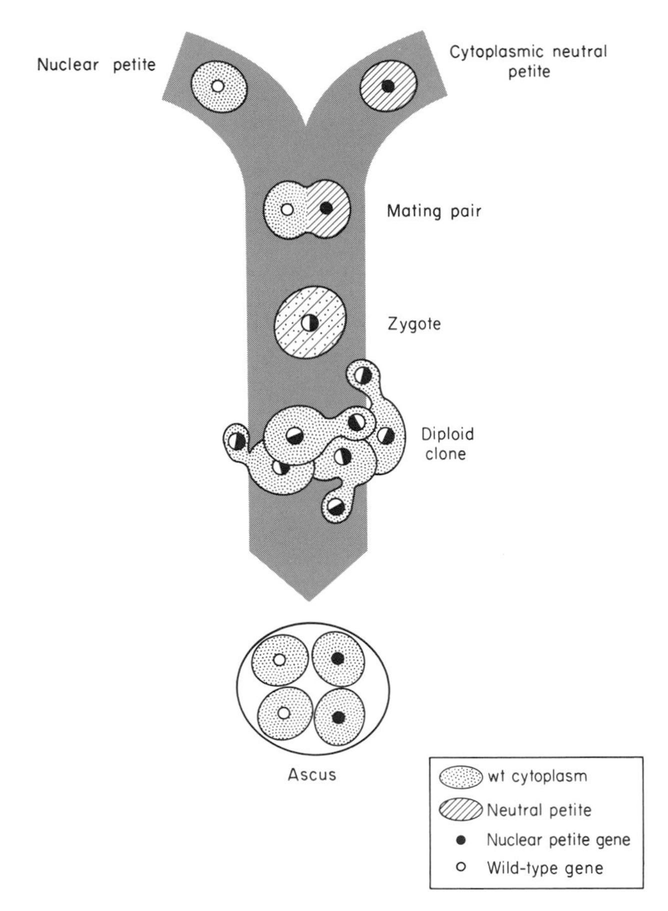

Fig. 4.6. Cross of nuclear *petite* × cytoplasmic neutral *petite*. Diploid clones are phenotypically wild type. After sporulation, ascopores segregate: two wild type, two nuclear *petite;* all contain wild-type cytoplasm.

Lederberg anticipated a second method: somatic segregation of cytoplasmic genes during vegetative growth.

INDUCTION OF CYTOPLASMIC PETITES BY OTHER METHODS

Once the occurrence of both cytoplasmic and nuclear gene *petites* had been firmly established, it became possible to compare the effects of various mutagens on the two genetic systems.

Pittman pointed out two simple methods of distinguishing cytoplasmic from nuclear *petites* (*31*). Nuclear *petites* have a measurable reversion rate to wild type, and if one grows up a large enough population of cells on glucose and then replates on a nonfermentable medium it is possible to detect the occurrence of revertants. As is known from the studies of Ephrussi (*7*), cytoplasmic *petites* never revert. Pittman confirmed this for a number of his induced *petite* strains by plating procedures which surveyed about 10^9 cells per mutant strain. He also showed that intercrosses of different cytoplasmic haploid *petites* gave diploid clones which were 100% *petite*, whereas crosses of different nuclear *petite* mutants gave wild-type clones, unless of course the parental haploid strains were mutant in the same cistron. As he pointed out, a known cytoplasmic *petite* can be used to distinguish a cytoplasmic from a nuclear mutation since cytoplasmic × nuclear mutants always give diploids which will grow on nonfermentable carbon sources.

A complication in this simple test arises from the fact that some nuclear mutations either induce or allow the selection of cytoplasmic mutations (called secondary *petites*). In mutation studies, it is necessary to distinguish between primary and secondary *petites* because the latter may not result directly from the mutagenic treatment. Thus, the *petites* which are recovered after a mutagenic treatment should first be crossed with a cytoplasmic *petite* tester stock to find out whether the cytoplasmic *petite* mutation is present. If so, the strains should then be crossed with wild type and sporulated to find out if a nuclear *petite* mutation is also present.

Analysis of the response to mutagens has been one of the most powerful methods of studying the nature of genetic material at the cellular level. Patterns of mutagenic response can be highly instructive. Table 4.2 provides a brief summary of the available information from the *petite* system of yeast.

The effects of ultraviolet light and of x-ray were first investigated by Caroline Raut (*34*). She showed that cytoplasmic *petites* could be induced by UV light. Later Raut and Simpson (*35*) showed that the action spectrum of cytoplasmic *petite* induction by ultraviolet light followed the nucleic acid absorption spectrum with a pronounced peak in the region of 2600 Å. The UV effect was very strong, since at most survival levels there were about 25% *petites* among the survivors. There are also another 25–30% of mixed colonies which could be classified as *petites* or not. In either case, their frequency as a function of dose also followed the nucleic acid absorption curve, as did survival curves for both the wild type and stable *petite* cells. Thus UV light induces *petites* with an effectiveness paralleling nucleic acid absorption.

TABLE 4.2

Effects of Mutagenic Agents on Nuclear and Cytoplasmic Genes in Yeast

	Effect of mutagenic agent on:[a]				
	Nuclear gene	Cytoplasmic *petite*	Buds	Mother	Reference
Euflavine	−	+	+	−	*27*
Proflavine	−	−	−	−	*27*
5-Fluorouracil	NT	+	+	+	*22a, 23, 27*
Ethidium bromide	NT	+	+	+	*41*
2,3,5-Triphenyl-tetrazolium	NT	+	+	+[b]	*24*
Nitrosomethyl urethane	+	+	+	NT	*37*
HNO_2 (nitrous acid)	+	−	NT	NT	*23, 37*
Dithranol (1,8,9-trihydroxy-anthracene)	NT	+	NT	NT	*18*
N-Methyl-*N*′-nitro-*N*-nitrosoguanidine	+	+	NT	NT	*30*
Mn, Co, Cu, Ni	NT	+	NT	NT	*25*
UV	+	+	+	NT	*34*
x-Ray	+	−	−	−	*35*
Anaerobiosis	−	−	−	−	*21*
Growth at 40°C	−	+	+	−	*38*
Heat shock at 54°C	−	+	+	−	*38*

[a] Symbols: − = no effect; + = mutagenic effect; NT = not tested.
[b] Slight effect.

In contrast to these results, x-irradiation showed very little increment, at most 2%, in the frequency of *petites* among the survivors at a series of x-ray doses. Similar results were reported by Pittman (*32*).

Harris (*21*) compared the rate of *petite* formation in anaerobic versus aerobic yeast. He grew normal yeast for long periods under conditions of rigorous anaerobiosis, and then tested at intervals for the frequency of *petites*. He showed unequivocally that the mutation rate from neutral to *petite* is independent of the presence of oxygen and not influenced by anaerobiosis.

The effectiveness of 5-fluorouracil as a mutagen for cytoplasmic *petites* was shown by Moustacchi and Marcovich (*29*) and by Lacroute (*23*). The yield of cytoplasmic mutants can be raised to 100% at concentrations of 5-fluorouracil that produce about 1% survival. An interesting aspect of the 5-fluorouracil effect is that the treated mother cells, as well as the buds, are mutated to *petites*, whereas in the case of acriflavine the mother cell remains unaffected and only the buds are *petite* (cf. p. 113).

Laskowski (*24*) studied the mutagenic effect of 2,3,5-triphenyltetrazolium (TZ) on the induction of *petites*. At concentrations of 2.5×10^{-5} gm/liter or higher, TZ induced 100% *petites* in growing yeast, but at lower concentrations there was no effect.

Growth at an elevated temperature was first shown to induce *petite* formation in yeast by Ycas (*50*) and subsequently studied in more detail by Sherman (*38*). Ycas found that yeast would grow slowly at 40°C, and that at this temperature one could get 100% conversion to *petites*. In following up this work, Sherman found that either growth at 40°C or heat shock, in which cells were pregrown at 30°C and then shifted for a short time to 54°C, was effective in inducing *petites*. He also noted that cells must be growing for the high-temperature induction to be effective, and that usually only the buds, not the mother cell, would mutate to *petite*.

The results summarized in Table 4.2 provide substantial evidence for the differences in response to mutagens of nuclear genes and cytoplasmic *petites*. The molecular bases of these differences are not understood at this time. Ethidium bromide and 5-fluorouracil are particularly noteworthy for inducing mutations not only in yeast buds, which contain primarily newly replicated DNA, but also in the mother cells, which presumably retain the parental strands. Further study of these systems may lead to a more detailed understanding of the distribution of mitochondrial DNA's in the budding process.

The differential effects of particular acridine dyes are not peculiar to yeast; similar differences have been noted in many systems. For example, proflavine is an effective frame-shift mutagen for phage but not for bacteria; acridine orange blocks replication of the sex factor, F, but not of bacterial DNA in the same cell; several acridine dyes block replication of kinetoplast DNA in trypanosomes without affecting nuclear DNA.

Perhaps the most perplexing feature of the cytoplasmic *petite* mutation is the very high spontaneous mutation rate. In many strains *petites* are present at frequencies around one per hundred, corresponding to even higher mutation rates.

In general, high mutation rates are not characteristic of cytoplasmic genes. Mitochondrial drug resistant mutants in yeast (p. 131) are found at conventionally low frequencies (below 10^{-6}), as are chloroplast mutations in *Chlamydomonas* (p. 72). However, as we have already seen in Chapter 2, cytoplasmic *petite* mutations sometimes (always?) involve gross changes in mitochondrial DNA, and are probably not to be considered in the same class as small nucleotide substitutions, additions, or deletions.

These considerations help to set the *petites* apart from other kinds of mutations, but throw no light on the mechanism of their origin. Recent studies of the effects of ethidium bromide (*20, 30a*) have shown that this dye causes major changes in the replication of mitochondrial DNA. Even this important finding provides little insight into the mechanism of the high spontaneous rate of *petite* production, beyond suggesting that the effect is somehow upon the replication process itself.

A further puzzle concerns the number of replicating copies of mitochondrial DNA. In experiments involving pedigree analysis after mutagenesis (*23, 27, 41*), the high frequency of mutant clones originating from individual buds was seen directly. If the treated cells contain many copies of replicating mitochondrial DNA, as is likely from the observed size, amount per cell, and semiconservative pattern of replication (Chapter 2) of yeast mitochondrial DNA, it is very hard to imagine how simultaneous mutagenesis of all these copies may occur. Similar difficulties arise from estimates of the number of genetic copies of mitochondrial DNA derived from mutation experiments, estimated in the range of two to six genetic copies (*38, 41*). These numbers are much lower than the estimate of twenty copies based on 5×10^7 daltons per copy and approximately 1×10^9 daltons of mitochondrial DNA per cell.

SUPPRESSIVE PETITES

In 1955 Ephrussi *et al.* reported the existence of a second class of cytoplasmic *petites* called suppressive (*14*). They chose to describe the phenomenon first with an hypothetical experiment (generalized from actual experiments) comparing the results of two crosses: (*a*) a neutral *petite* × wild type, and (*b*) a suppressive *petite* × wild type. In the first cross with a neutral *petite*, as was seen in Fig. 4.3, the diploid clones arising from zygotes all respired normally, and after sporulation the asci all segregated 4:0; the F_1 progeny were wild type as previously described.

In the cross of suppressive *petite* × normal, the results depended very much upon when sporulation occurred, as shown in Fig. 4.7. If the diploids were sporulated immediately after mating, most of the tetrads segregated 0:4, that is, the F_1's were *petite*. If the mating pairs were immediately plated out to form zygote colonies they gave rise mostly to *petite* colonies, a result consistent with the sporulation data. However, if the mating mixture was subcultured in liquid and then either plated out

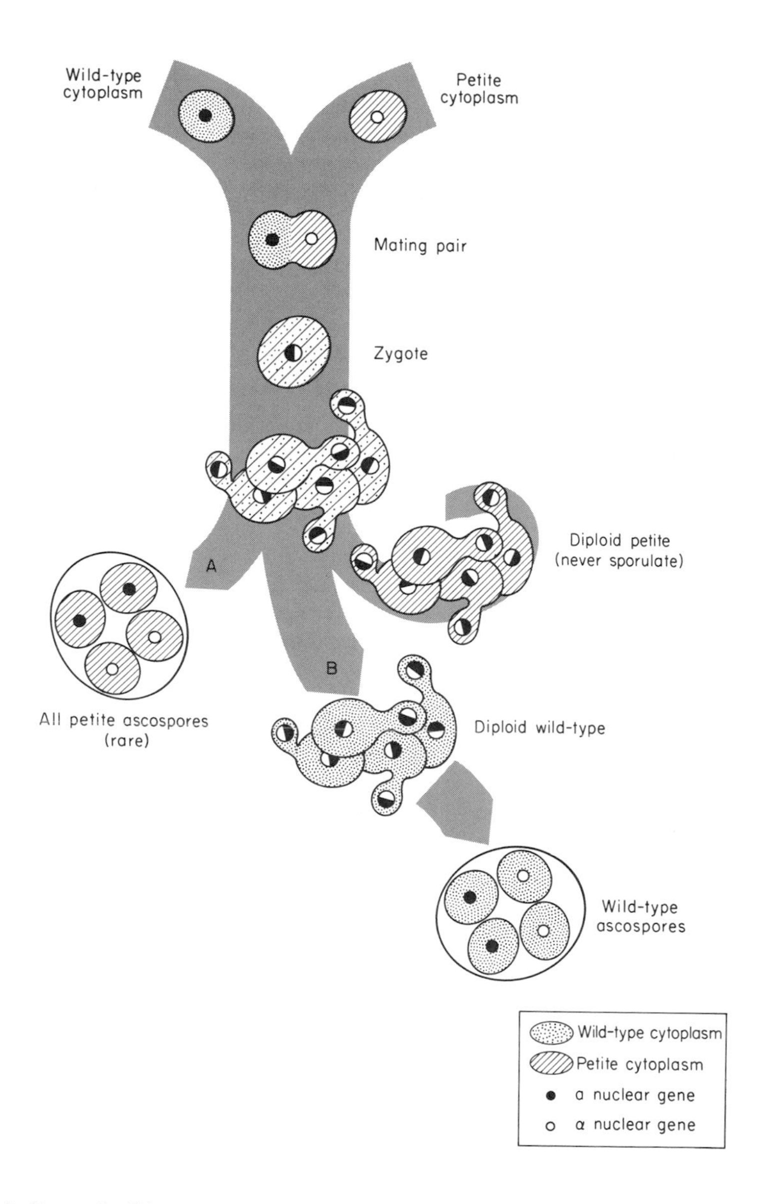

Fig. 4.7. Cross of wild type × suppressive cytoplasmic *petite*. If the zygotes were immediately sporulated (A), they would produce *petite* ascospores; if allowed to form diploid colonies, only those with wild-type cytoplasm (B) could sporulate, producing all wild-type progeny.

for diploid colonies or sporulated, the results were strikingly different—the diploids were mainly wild type and after sporulation the ascopores were all wild type, the segregation being 4:0.

These results were interpreted by Ephrussi *et al.* (*14*) as follows. The difference in outcome between immediate sporulation on the one hand, and intervening liquid culture on the other hand, was explained in a simple way. In liquid cultures, the wild-type cells which multiply faster than the *petites,* rapidly take over. Thus the suppressive *petites* are gradually lost if even a few percent of wild-type cells are initially present. Furthermore, sporulation requires respiratory metabolism and cannot be induced in established *petite* strains, whether neutral or suppressive. Thus, the 4:0 segregation ratios seen in sporulated diploids from suppressive × wild-type crosses result from the sporulation of the wild-type cells present in the mixture.

However, if the mating mixture is induced to sporulate immediately after zygote formation, suppressive *petite* zygotes will sporulate, presumably because the respiratory enzymes of the wild-type parent needed for sporulation are still present in the cytoplasm of the zygote. These zygotes give rise to tetrads showing 0:4 segregation ratios, the progeny all being haploid suppressive *petites.* This 0:4 ratio provides a direct demonstration of the non-Mendelian nature of the suppressive *petite* determinant.

Now let us consider the difference between neutral and suppressive *petites.* Suppressiveness is seen as the 0:4 ratio of wild-type to *petite* progeny in tetrads from freshly formed diploids; and as the presence of diploid *petite* colonies formed from zygotes in crosses of suppressive *petite* × wild type.

Suppressiveness has been viewed as a kind of dominance. However, the total disappearance of the wild-type genes as indicated by 0:4 segregation ratios (as well as the disappearance of neutral *petite* genes in the 4:0 segregations of neutral *petite* × wild-type crosses) shows that dominance in the sense of *phenotypic dominance* is not an appropriate descriptive term. The allele from one parent is not simply inactive and not expressed; it has vanished and never reappears among the progeny. For this reason it seems appropriate to view suppressiveness as *replicative dominance.*

In some *petite* strains, the mitochondrial DNA is considerably higher in AT content than the wild type. Some investigators (*2, 33*) suggested that these high AT molecules may replicate faster than the wild-type molecules, and on this basis may become the predominant form. This explanation does not account for those suppressive *petites* which have a base composition very similar to that of the wild type. A more general hypothesis might regard suppressiveness as the result of preferential replication of the mutant mitochondrial DNA without specifying the mechanism. Various mechanisms might be involved: more rounds of replication of shorter molecules, preferential distribution to the bud of mitochondrial DNA of the mutant type, or a recombination process.

As yet, the molecular mechanism of suppressiveness is unknown. Nonetheless, considerable information about the phenomenon at the cellular level is available. A detailed descriptive study of suppressiveness was undertaken first by Milkman (*28*) and then by Ephrussi and collaborators (*8, 16*) to clarify some features of the phenomenon at the cellular level. Because of its potential importance and unsolved status, we will consider the evidence from these studies in some detail, and then return to the molecular interpretation.

STUDIES OF SUPPRESSIVENESS AT THE CELLULAR LEVEL

Most suppressive *petites* are not fully suppressive. The degree of suppressiveness is measured by a standard test: a haploid suppressive *petite* strain is crossed with a haploid wild type tester stock, and the resulting zygotes are plated on a selective medium that supports growth only of the zygotes, and not of either parental strain. The resulting colonies are tested for presence of *petites* by staining with tetrazolium. The percent of *petite* colonies is taken as the measure of the percent suppressiveness (*8*).

Ephrussi and Grandchamp (*8*) in an elaborate series of cloning experiments showed that in a population of *petites,* the degree of suppressiveness is itself inherited. They found, for example, that a population which is 50% suppressive transmits that intermediate condition to its clonal descendants. Their procedure was to plate out a haploid suppressive strain, pick individual colonies, allow them to grow for a limited number of doublings, and then divide the culture in half. One half was subcultured to keep the haploid clone available and the other half was crossed with wild type to measure the percent of suppressiveness. An example of the results is shown in Fig. 4.8.

Since clones of intermediate suppressives give rise after mating to two kinds of diploid colonies, one class being wild type and the other being *petite,* one might have predicted the presence of two classes of cells in the haploid suppressive strain, one giving rise to normal colonies and one giving rise to *petites.* (The homogeneity of the wild-type parent was shown in control crosses.) But this clonal study demonstrates that the suppressives themselves have a hereditary property which is expressed in terms of the percent of suppressives which they

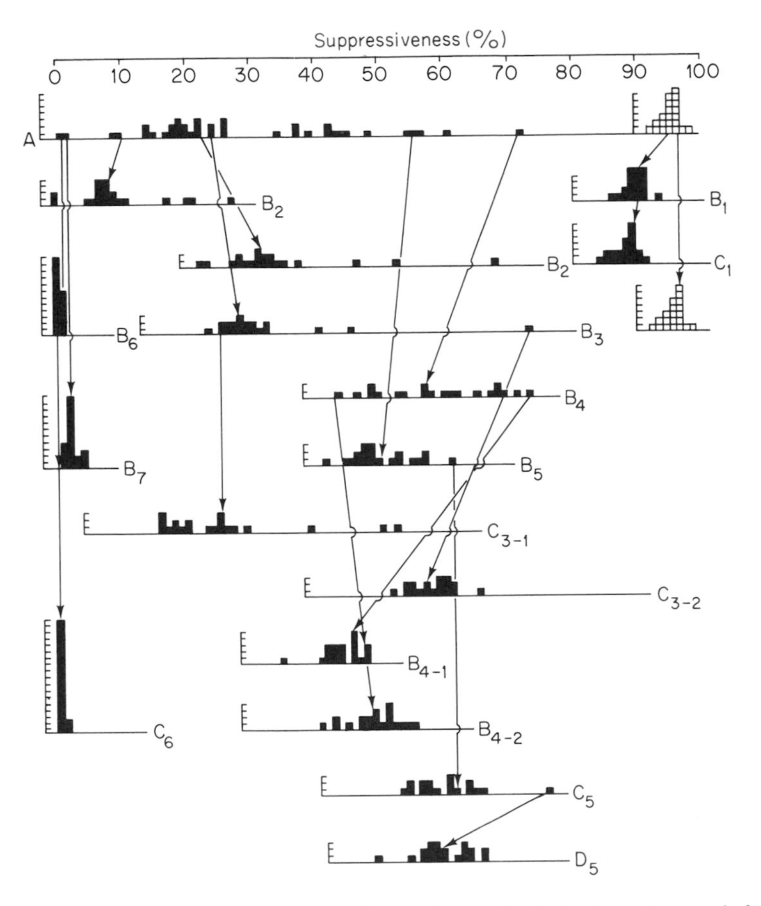

Fig. 4.8. Cloning of two suppressive *petite* strains and subcloning of several derived strains, showing the extent of clonal inheritance of percent suppressiveness. Each box represents one subclone. Solid blocks represent strain 1, open blocks represent strain 2. From (*8*).

will generate predictably in the future. Thus the intermediate suppressives are not mixtures of two cell types, extreme suppressives and neutrals, but rather they represent a distinctive genetic class of cells. Similarly, clones were characterized which regularly show a high percent of suppressiveness and others which show a low percent, e.g., 20–30% suppressiveness.

In discussing these results, Ephrussi proposed the term "suppressive factor" and defined it as something which suppresses the perpetuation of the wild-type cytoplasmic factor, perhaps by destroying it or by interfering with its multiplication or with its distribution into the diploid buds. According to this hypothesis, different intermediate suppressive strains would have different suppressive factors. In molecular terms, the suppressive factor could be the altered mitochondrial DNA itself, and the degree of suppressiveness would be the probability that the mutant DNA would displace the wild type.

The nature of suppressiveness was further analyzed by Ephrussi *et al.* (*16*) in a subsequent paper with two stated objectives: (*a*) to establish more conclusively whether or not partially suppressive clones are homogeneous with respect to the suppressive factor and give rise initially to a homogeneous population of zygotes; and (*b*) to examine the effects of environmental variables occurring in early stages of growth on the ultimate extent of suppression.

In any cross of normal × suppressive *petite*, eventually two types of colonies are recovered: those which are typically normal and those which are typically *petite*. A number of strains of suppressive *petites* were described, each having maintained a characteristic range of suppressiveness over a number of years of subculture. The relative stability of these lines is shown in Table 4.3.

In the initial plating of the mating cells, mixed colonies were also seen with enough of both types of cells present so that they could not be classified as either wild type or *petite*. Two principal types of intermediates were described: (*a*) a scalloped colony type and (*b*) so-called "abcessed *petites*." The scalloped colonies result from sectors in which the wild types tend to overgrow the *petites*. When replated, these colonies give a

TABLE 4.3

Relative Frequencies of Different Types of Zygote Colonies from Crosses of Various Petite Strains with a Standard Wild-Type Tester Strain[a]

Petite strain	% *Grandes*[b]		% Abscessed *petites*[b]		% *Petites*[b]	
2/16	2.2	(1–3.6)	2.8	(0.4–7.6)	95.0	(91–98)
20/9[c]	9.0	(1.1–17)	23.0	(9.6–36.4)	68.0	(47-81)
12[c]	39.5	(20–59)	15.5	(10.5–28)	45.0	(27.8–67)
2	74.7	(62–85)	9.0	(3.5–14)	16.2	(9.8–26.8)
A_{19}	78.0	(67–87)	7.0	(4–14)	15.0	(10–22.8)
9	87.7	(78–95)	5.2	(0.3–11.4)	7.2	(3.7–11.4)
$19dA_1$ (p_n)[d]	98.2		0.3		1.5	
19d(G)[d]	99.0		0		1.0	

[a] Adapted from (*16*), Table 1.
[b] Limits of variation shown in parentheses.
[c] Results shown in Fig. 4.9.
[d] (p_n) is neutral *petite* control; (G) is *grande* control.

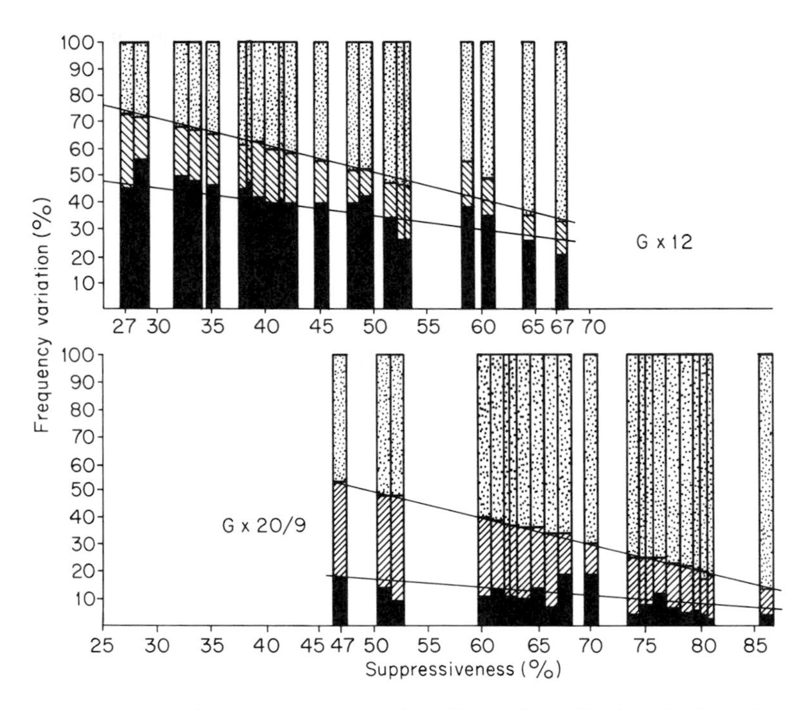

Fig. 4.9. Variation in the frequency of *grande* (solid portion of bar), *petite* (dotted portion of bar), and abscessed *petite* (i.e., mixed colonies) (striped portion of bar) zygote colonies arising from crosses between a wild type (G) and two suppressive *petite* haploids: 12 and 20/9 (see Table 4.3 for properties of *petites*). From (16).

high proportion of wild type. At the other extreme are the so-called "abcessed *petites*" which are tiny colonies. Late in colony formation they develop a central papilla or little mound of cells which stain with tetrazolium dyes and by this test are wild type. These colonies sometimes produce a mixture of *petites* and normals on subculture, but under some conditions only *petites* are found. The relative ease of classification suggests that by the time the original zygote colonies have matured, they contain predominantly wild-type cells as in the case of the scalloped colonies or predominantly *petites* as in the case of the abcessed *petites*.

Zygote colonies progress during their growth from being predominantly *petite* toward wild type. Part of the shift results from selection at the cellular level, since the wild-type cells grow much faster than do *petites*, but the authors suggest that other factors are involved as well. In general, the strains that are more highly suppressive have a lower percent of the mixed colony types. At the other extreme, zygotes from crosses of wild type × neutral *petite* rarely contain colonies of the mixed type.

These observations suggest that the diploid cells, which are initially heterozygous for the *petite* and wild-type genotypes, produce segregants that are pure wild type or pure *petite* during clonal growth to form colonies. However segregation is not the full explanation since pure *petite* colonies contain no wild-type cells. Thus there must be at least two processes taking place in diploid clones: segregation of mitochondrial genomes and outright destruction of the wild-type mitochondrial DNA.

To examine suppressiveness further at the cellular level, experiments were designed to examine the influence of environmental factors. The effect of temperature was studied by plating freshly mated zygotes at 18°C and at 30°C. The results in Fig. 4.9 show clearly that for a number of different suppressive strains, raising the temperature shifts the clones toward the wild type. In other words, at higher temperatures there is less suppressiveness.

The effects of five media were compared: (*a*) low glucose (0.13%), (*b*) glycerol alone, (*c*) glycerol plus low glucose, (*d*) sodium acetate alone, and (*e*) sodium acetate plus low glucose. Glucose and glycerol plus

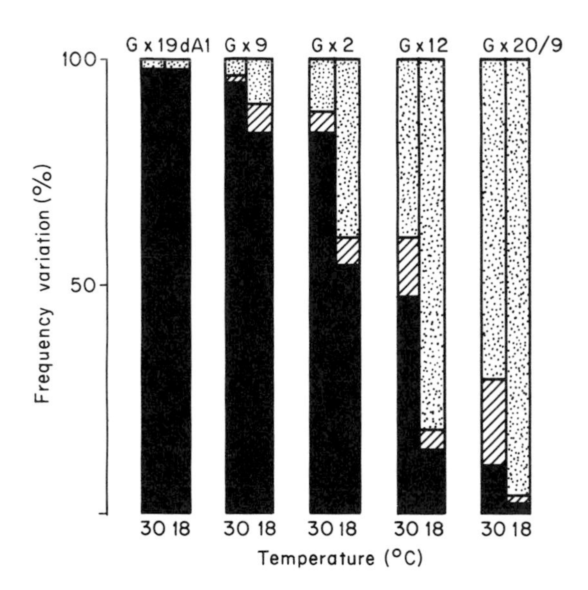

Fig. 4.10. Frequencies of *grande* (solid portion of bar), abscessed *petite* (striped portion of bar), and *petite* (dotted portion of bar) zygote colonies formed in crosses of wild type (G) × (five) suppressive *petite* strains at two temperatures, 30°C and 18°C. From (*16*).

glucose gave essentially the same proportion of the various cell types, but the medium containing sodium acetate gave sharply increased amounts of mixed and of wild-type colonies. By increasing the concentration of sodium acetate, the percent of *petite* colonies in a mating mixture from a highly suppressive strain could be reduced from about 75% down to about 25%. (In subsequent studies it was found that the acetate effect seen with sodium acetate was not reproduced with potassium acetate. This observation was not pursued and the authors refer only to the "acetate effect.")

The added acetate increases the fraction of abcessed *petites* which produce wild-type cells on subculture more than it initially increases the fraction of pure wild type. Presumably this means that the acetate effect takes time for expression. How many generations are involved cannot be computed from the data given. However, the effect is not entirely due to selection at the population level because the wild-type cells grow much more slowly on acetate than they do on glycerol, but more wild-type colonies are recovered when a mating mixture is plated on acetate than when it is plated on glycerol.

The variation in suppressiveness for a particular strain under different environmental conditions cannot be explained simply as a function of the growth rate. In the temperature experiment, raising the temperature increased the growth rate and also increased the percent of wild types. However, increasing the concentration of sodium acetate decreased the growth rate, while at the same time increasing the percent of wild types.

Thus the percent of *petites* finally produced in a given cross can be varied widely depending upon the environmental conditions employed. A strain which under standard conditions is 70% S (suppressive) can be varied between 30 and 100% *petite* colony formation. Similarly, weakly suppressive strains can be varied between 35 and 0% *petites* depending upon the growing conditions.

To clarify the events occurring in the first few generations of growth of the zygotes, four of the suppressive *petites* and a neutral *petite* were crossed with a standard wild-type strain. Buds were removed by micromanipulation from fresh zygotes and compared in a number of ways.

1. The zygotes were grown on glycerol medium without glucose and buds were removed into the same medium with and without added glucose. The results (Table 4.4) showed, first, that the ability to form buds on pure glycerol medium varied with the suppressive *petite* strain, the more suppressive *petites* forming fewer buds than the less suppressive strains. But even strain A-19, characterized by only about

TABLE 4.4

Average Number of Buds Formed by Zygotes from Crosses of *Petite* Strains × a Standard Grande[a,b]

Crosses	%S	Average no. buds	Frequency of *petite* zygote colonies (%)	Frequency of *petite* bud colonies (%)
Controls[c]				
G × G	0	17.6	0	0
G × p_n	0	16.6	1	1
p_n × p_n	0	0.6	100	100
G × suppressive *petites*				
2/16	97	5.7	100	100
20/9	70	9.75	100	100
12	45	10.3	73	91
A-19	13	12.2	34.5	74

[a] From *(16)*, Table 5.
[b] Buds were classified after plating on glycerol-glucose agar.
[c] G is grande; P_n is neutral *petite.*

13% S, made fewer buds than did the wild-type control. The total number of cells formed by all the buds indicates an average of seven doublings per bud for the controls, about four to five doublings per bud for the weakly suppressive, and two to three doublings per bud for the zygotes from the highly suppressive strains. The zygotes and the isolated buds were subsequently grown with glucose and classified for suppressiveness. Under these conditions, the more highly suppressive strains gave progeny which were 100% *petite* and even the low suppressive strains gave rise to a much higher percent of *petites* than under the usual testing conditions, that is, in the presence of glucose from the outset. These results showed that suppressiveness was not immediately established, even in strains which would eventually be close to 100% *petite* since a number of buds were formed in glycerol.

2. The wild-type parent was grown under strictly anaerobic conditions, mated anaerobically with *petites,* and then compared with aerobically grown cells. The results were indistinguishable; zygotes from anaerobic and from aerobically grown parents produced the same number of buds. Thus, neither synthesis nor function of the respiratory enzymes required for growth of the zygotes on glycerol were blocked by the presence of the suppressive factor immediately after zygote formation.

3. Wild-type cells can be identified by their ability to grow on nonfermentable carbon sources like glycerol and acetate, and by their ability to reduce tetrazolium dyes. In a number of experiments it was noted that the tetrazolium reaction and ability to grow on glycerol did not always coincide. For example, buds from a cross of wild type with a low suppressive stopped growing in glycerol as if they were *petite* and when glucose was added they continued to grow, but eventually they gave rise to wild-type colonies. In the opposite type of experiment, cells taken from the center of abcessed colonies which stain with tetrazolium, and therefore presumably are wild type, do not grow on glycerol. In a third case, buds from a cross of wild type × highly suppressive can give clones of 100–1000 cells on glycerol, but if buds are picked while they are still growing they will still be 100% *petite*. Thus the *petite* state was being established even while cells were behaving like *grandes*. These and other examples indicate a discrepancy between genotype and phenotype in cells taken from mixed colonies.

The authors conclude that the zygotes and their immediate descendants may remain in a so-called premutational state for some time before the final decision is made to stabilize as wild type or as *petite*. Each suppressive strain is characterized by some fraction of the zygote population which will inevitably become *petites*, and another fraction which will become wild type, almost irrespective of manipulation. However, in all strains some zygotes can be pushed in either direction.

The metabolic pathway by which glycerol or other environmental conditions influence the selection of mutant or wild-type mitochondrial DNA in these zygotes remains unknown. The results do suggest that preferential replication and transmission may not fully account for the findings. The total absence of wild-type cells in *petite* zygote colonies indicates at the cellular level that the wild-type mitochondrial DNA is probably destroyed or converted to *petite*. Indeed, it may be that the factor initially responsible for the *petite* mutation in the *petite* haploid strain is at work in the zygote as well. It would seem appropriate to follow up these cellular studies of suppressiveness with molecular studies of the mitochondrial DNA in these newly formed zygotes.

RECENT DEVELOPMENTS IN MITOCHONDRIAL GENETICS OF YEAST

In the years since Ephrussi's studies of suppressiveness, two discoveries have transformed the analysis of *petites* into a more accessible problem: the discovery of physical alterations in mitochondrial DNA of *petite* strains (cf. Chapter 2); and the acquisition of a new class of cytoplasmic mutations in yeast, conferring resistance to various antibiotics. This section describes genetic properties of the new drug-resistant mutants, summarizes evidence of recombination among them, and considers the relation of these mutations to mitochondrial DNA, to the *petite* mutation, and to suppressiveness.

Isolation and Properties of Cytoplasmic Mutations to Drug Resistance

The discovery of cytoplasmically inherited streptomycin resistance in *Chlamydomonas* raised the possibility that similar mutations might be found in yeast. With the recognition that mitochondria contain a protein-synthesizing system different from that of the cytoplasm and sensitive to the antibiotics that affect 70 S ribosomes as in the chloroplast (cf. Chapters 7 and 8), it seemed plausible that mutant strains resistant to these drugs might carry mutations in mitochondrial DNA. Accordingly, spontaneous mutations to drug resistance were selected on antibiotic-containing agar. The first cytoplasmically inherited mutations that were reported conferred resistance to erythromycin (*26, 43*).

In crosses of $ER^r \times ER^s$ (erythromycin-resistant and erythromycin-sensitive, respectively), diploid clones growing from zygotes were shown to contain mixtures of both cell types; on subculture, pure ER^s and ER^r clones were obtained. When these pure diploid clones were sporulated, they gave only 4:0 segregation (Fig. 4.11). [Thomas and Wilkie (*42*) reported 3:1 and 1:3 ratios as well, but it is not stated how many doublings the diploids underwent before sporulation. Some may still have contained both ER^r and ER^s alleles when sporulated.]

Evidence of reassortment of erythromycin resistance and respiratory competence was reported by Gingold *et al.* (*19*) in studies that followed the progeny of crosses through sporulation. These studies showed unambiguously that *ER* could be sorted out and transmitted independently of rho^- but the relationship between *ER* and *rho* was not fully clarified (see p. 138 for definition of *rho*).

Subsequently an extensive collection of erythromycin-resistant and chloramphenicol-resistant strains was described (*4*). Seven independently arising ER^r* strains of spontaneous origin, and fourteen chloramphenicol-resistant (C^r) strains, two of them UV-induced and the rest spontaneous, were shown to behave similarly in crosses. Each of them gave rise to mixed zygote clones in crosses with wild-type sensitive strains; and the stable resistant diploids recovered after subcul-

* Coen *et al.* (*4*) denote their erythromycin-resistant strains as E^r but we denote them as ER^r for consistency.

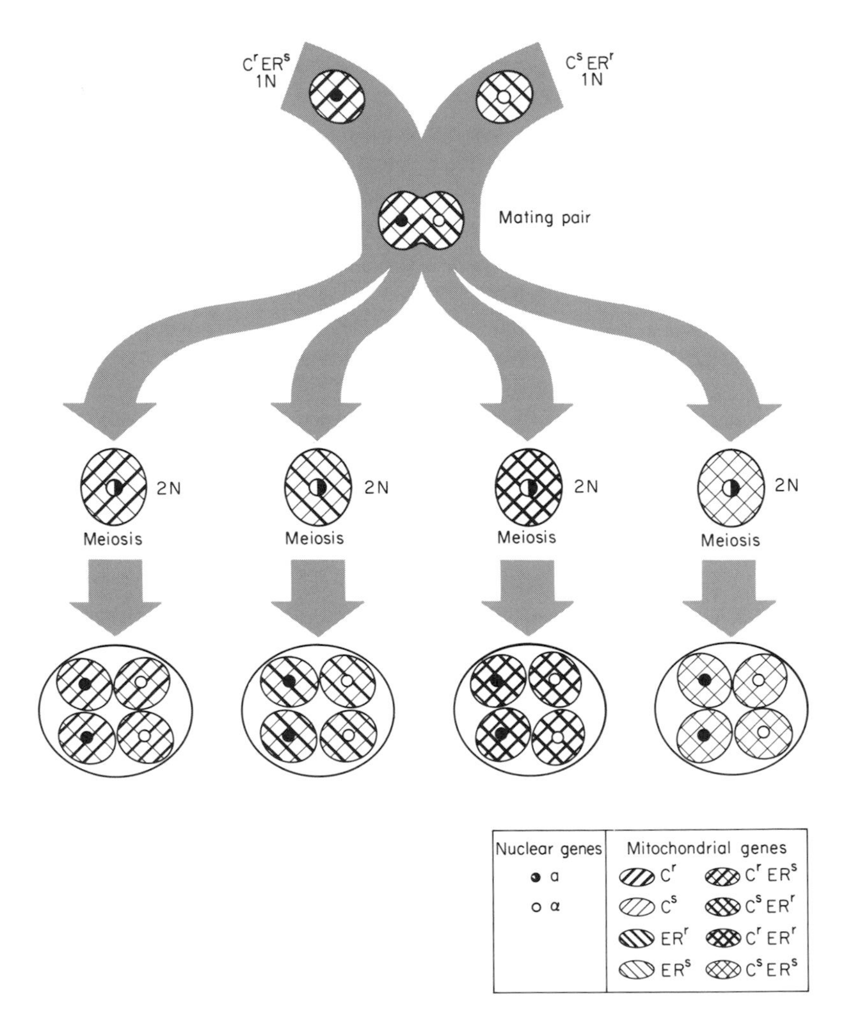

Fig. 4.11. Segregation of mitochondrial drug-resistance markers in the life cycle of yeast. The haploid parental strains differ in the nuclear mating type alleles *a* and α, and in two pair of mitochondrial genes: C^r/C^s and ER^r/ER^s. Zygote formation is followed by segregation of the parental and recombinant types during vegetative growth of diploid colonies. After one subculture each diploid cell gives rise to a uniform clone, either parental or recombinant. When these pure clones are induced to sporulate, they produce asci in which the four products segregate 4:0 for the mitochondrial genes and 2:2 for the mating type alleles *a* and α.

ture of zygote colonies all sporulated with typical 4:0 non-Mendelian ratios. This pattern in shown in Fig. 4.11. Recombination studies with these mutant strains will be discussed below.

Recombination of Cytoplasmic Genes in Yeast

In a short note reporting preliminary evidence of recombination, Thomas and Wilkie utilized strains resistant to spiramycin (SP^r) and paramomycin (P^r) as well as erythromycin (ER^r) (*42*). The results of their crosses indicate the recovery of recombinant clones. For example, in the cross $ER^s\ SP^s\ P^r \times ER^r\ SP^r\ P^s$, the following zygote clones were found:

$ER^s\ SP^s\ P^r - 49$
$ER^r\ SP^r\ P^s - 25$
$ER^s\ SP^s\ P^s - 14$
$ER^r\ SP^r\ P^r - 23$
$ER^s\ SP^r\ P^s - 3$
$ER^s\ SP^r\ P^r - 4$
$ER^r\ SP^s\ P^s - 1$

These preliminary data are suggestive of linkage between *ER* and *SP*, and looser linkage of these markers to *P*, but as yet no experimental details or further evidence have been reported.

A serious criticism of this report is the absence of a sporulation test, which is necessary to find out whether or not the recombination is at the level of DNA molecules per se. Perhaps the recombinant progeny contain copies of both parental DNA's rather than recombinant molecules. If so, one would expect segregation to occur at meiosis and this possibility could easily be tested by sporulation.

Extensive genetic studies of cytoplasmic drug-resistance markers in yeast have been reported by Coen *et al.* (*4*) and Bolotin *et al.* (*1a*). First they established the non-Mendelian nature of the chloramphenicol- and erythromycin-resistant mutations by crosses of the type shown in Fig. 4.11. They further demonstrated that recombination had occurred at the DNA level by sporulation tests of recombinant progeny. Major emphasis of their studies has been toward establishing the rules of segregation and recombination for mitochondrial genes. This ambitious aim has not yet been achieved, but a number of important facts have been established and these will now be discussed.

We must start with a consideration of experimental procedures, because of their decisive influence on the data. The overriding difficulty in this yeast system is the growth of clones by budding rather than by fission. Thus, the zygote produces buds sequentially, and the buds themselves produce clones by the same budding procedure. A zygote may produce as many as twenty sequential buds, and each bud then grows at about the same rate as the mother cell. Thus the buds produce progeny while the zygote is still producing buds. The consequences of this include unequal partitioning of cytoplasm and overlapping vegetative generations.

Both of these effects cause severe difficulties in the analysis of segregation and recombination especially in the first few cell divisions of the zygote. Two theoretical distributions are shown in Fig. 4.12,

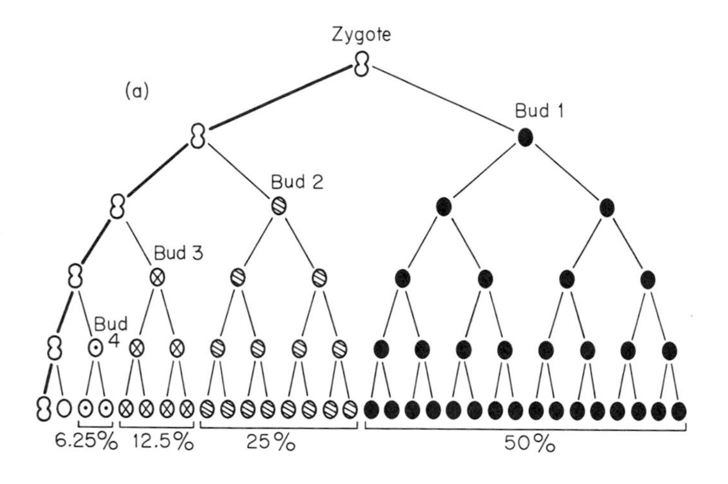

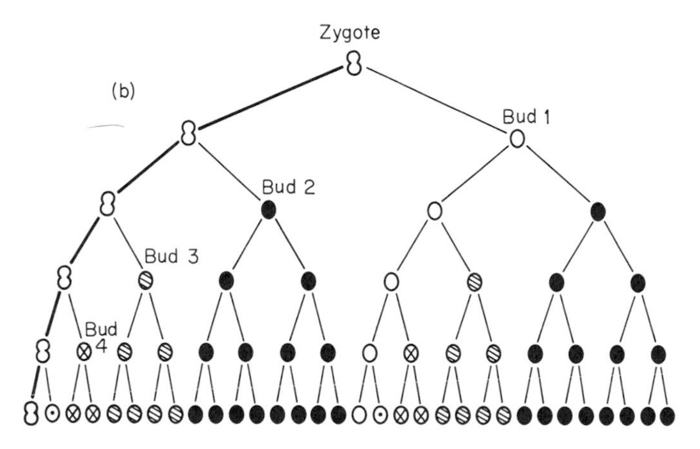

Fig. 4.12. Distribution of segregating parental and recombinant cell types in budding zygote clones of yeast. Diagrams show two theoretical distribution patterns. In (*a*), a different cell type was transmitted to each of the first four buds. In (*b*), the same order of transmission occurred in two branches of the pedigree. In both (*a*) and (*b*), the four genotypes accounted for 50%, 25%, 12.5%, and 6.25% of the total population, their frequency depending upon the time of origin from the budding zygote. From (4).

which highlight the skewed ratios of progeny classes that result from the budding pattern. Bud number 1 contributes 50% of the cells to the zygote colony, number 2 contributes the next 25% of the cells, etc.

Coen *et al.* (4) have tried to compensate for the skewing effect by sampling from large populations. For example, the data on which the evidence of polarity is based comes from classifying a mixture of cells taken from many diploid colonies each grown for about twenty doublings on a nonselective medium. In other experiments, the skewing effect has been examined by subculturing individual zygote colonies to determine the proportions of the various genotypic classes in each zygote colony. With these methods, certain regularities have been shown.

1. The parental genotype influences the recovery of alleles from the two parents; the allele coming from one parent predominates among the progeny. Coen *et al.* (4), examining several independent mutations to chloramphenicol resistance and to erythromycin resistance found that each had a characteristic ratio of allelic transmission. The maps shown in Fig. 4.13 are based upon the deviation from 1:1 of alleles recovered. These deviations can be viewed as polarity and used as the basis for sequencing of genes as shown in the figure. In monofactorial crosses,

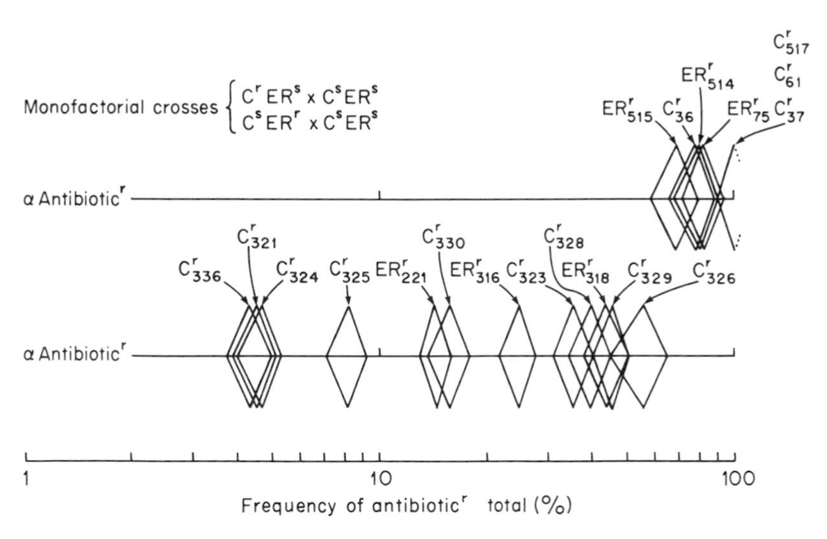

Fig. 4.13. Degree of transmission of the antibiotic-resistant mitochondrial alleles ER^r and C^r in one-factor crosses. The diamonds indicate the range of variation in % transmission of each allele tested. The effect of "mitochondrial sex" is shown by the higher transmission of *resistant* alleles carried in the α strain later shown to be ω^+ than in the α strain later shown to be ω^-. The spread of % transmission especially when resistance was carried in the ω^- parent provides a polarity used for mapping. From (4).

frequencies of $C^r/C^r + C^s$ and $ER^r/ER^r + ER^s$ ranged from 5% to almost 100%.

2. In two factor crosses, $C^s\ ER^r \times C^r\ ER^s$, a similar polarity was seen in the relative frequencies of $C^s\ ER^s$ and of $C^r\ ER^r$ recombinants recovered. Polarity is defined as the deviation from 1:1 of the two classes of recombinants. An example of the data obtained is given in Table 4.5. Here the ratio of $C^s\ ER^s$ to $C^r\ ER^r$ recombinants is given for ten different C^r mutations in crosses with two (and in two cases with four) different ER^r mutations. The ratios ranged from 1.5 to 10,000 and could be grouped. Five of the C^r mutants gave ratios between 1 and 10, four of

TABLE 4.5

Frequencies of Reciprocal Recombinants in Diploid Cells from Crosses of aC^{r}ER^s × αC^{s}ER [a,b]

			αC^{s}ER^r			
			650-2C		DP1-1B	
			ER^r_3	ER^r_{75}	ER^r_{514}	ER^r_{515}
aC^{r}ER^s	55R5-3C	C^r_{321}	5.1/0.12 = [43]	6.5/0.054 = [120] *	26.5/0.080 = [330] *	16/0.2 = [80] *
		C^r_{323}		2.3/0.85 = [2.7]	3.7/0.57 = [6.5] *	
		C^r_{324}		7.3/0.056 = [130]	15.6/0.093 = [170]	
		C^r_{325}		6.8/0.053 = [130]	16.3/0.105 = [160]	
		C^r_{326}		2.8/0.75 = [3.7]	3.7/0.53 = [6.9]	
		C^r_{327}	1.5/0.62 = [2.4]	1.5/0.87 = [1.7] *	2.8/0.21 = [13]	2.6/0.7 = [3.7] *
		C^r_{328}			2.8/0.46 = [6.1]	
		C^r_{329}		1.2/0.78 = [1.5]	1.6/0.52 = [3.1]	
		C^r_{330}		19.5/0.082 = [240]	34.6/0.053 = [650]	
		C^r_{336}		7.4/0.000 > [10000]	10.8/0.000 > [10000]	

[a] From (4), Fig. 13.

[b] In each square, the numerator is % C^sER^s, denominator is % C^rER^r and boxed value is the ratio (% C^sER^s)/(% C^rER^r).

* Indicates 2–6 independent crosses.

them had ratios between 100 and 500, and one had the extreme value of 10,000. As shown in Fig. 4.14, this classification fits quite well with the order seen in Fig. 4.13, based upon the polarity of segregation of genes in monogenic crosses.

The basis of this polarity has been attributed by Bolotin *et al.* (*1a*) to a pair of genetic factors called ω^+ and ω^- responsible for "mitochondrial sexuality," i.e., the preferential transmission of mitochondrial genes from ω^+ parents to progeny. Mitochondrial sexuality segregates independently of the mating type genes *a* and α. The effects of ω^+ and ω^- are shown in Table 4.6. They appear to determine the polarity of transmission in heterosexual crosses ($\omega^+ \times \omega^-$). In homosexual crosses, ($\omega^+ \times \omega^+$ and $\omega^- \times \omega^-$) no polarity is seen; the parental and recombinant clones segregate 1:1.

It is difficult to interpret these findings in the absence of pedigree data for the first few doublings. Without such data it has not proved possible to assess the mode of action of ω^+ in the distribution of mitochondrial DNA from the zygote to the buds, or to establish when molecular recombination occurs, whether exclusively in the zygote or also in the buds, or to distinguish between recombination and sorting out, or to determine how many copies are involved in the recombination process. Nonetheless, these two studies (*1a, 4*) have provided a solid basis for genetic analysis and mapping of mitochondrial genes in yeast.

The availability of these drug-resistance markers has also provided material for a further study of the molecular basis of the *petite* mutation and of suppressiveness. These new developments will now be discussed.

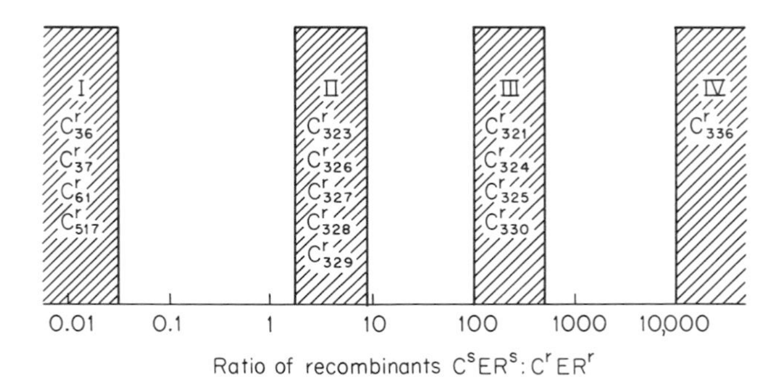

Fig. 4.14. Classification of C^r mitochondrial mutants into groups on the basis of polarity measured by ratio of C^sER^s/C^rER^r recombinants in a series of two-factor crosses. From (*4*).

TABLE 4.6

Consequences of Heterosexual and Homosexual Crosses of Mitchondrial Genes[a]

ER^r \ C^r	C^r_{321} ω^-	C^r_{36} ω^+
ER^r_{514} ω^+	8 54 [290] 38 0.13	44 44.5 [1.6] 7 4.5
ER^r_{221} ω^-	47.5 41.5 [1.2] 6 5	79 3.5 [0.02] 0.4 17.5

$C^r ER^s$	$C^s ER^r$
[$C^s ER^s$ / $C^r ER^r$]	
$C^s ER^s$	$C^r ER^r$

[a] Numbers represent % of progeny in each phenotypic class.

Molecular Genetics of Petites

The term *rho*$^-$ was initially introduced to denote the cytoplasmic gene associated with the *petite* phenotype. However, it has now become evident that many different mitochondrial mutations may give rise to *petites* (as will be discussed below and in Chapter 7), and consequently there is some confusion in the literature in the use of the term *rho*$^-$. In this book *rho*$^-$ will be used to denote any mitochondrial mutation giving rise to the *petite* phenotype; and *rho*$^+$ will then denote the corresponding wild-type state.

Several studies of the relation between *petite* mutations and cytoplasmic drug resistance have recently been reported (*1a, 4, 36, 43*). The presence of mitochondrial mutations to drug resistance cannot be demonstrated directly in *petite* strains, because these drugs act only on the mitochondrial system. Thus for example erythromycin resistance in yeast is defined as the ability of a cell to grow on glycerol in the presence of the drug; *petite* mutants do not grow on glycerol at all (cf. Chapter 7), and therefore cannot be classified for resistance to erythromycin or any other drugs that block mitochondrial activities. The effect of mutation from *rho*$^+$ to *rho*$^-$ on mitochondrial genes for drug resisatance can be seen by crossing such strains with *rho*$^+$ drug-sensitive testers, and looking for drug resistance in the zygote clones, as shown in Fig. 4.15.

Initially Thomas and Wilkie (*43*) reported that when a *rho*$^+$ ER^r strain is converted to *rho*$^-$ by treatment with euflavine, and then crossed with a strain that is *rho*$^+$ ER^s, all the zygote colonies are uniformly ER^s, the ER^r gene having totally disappeared. Subsequently, however, Saunders *et al.* (*36*), Coen *et al.* (*4*), and Rank (*33*) all found that some *petite* strains do transmit mitochondrial genes in test crosses. This finding has led to the development of the genetic method, shown in Fig. 4.15, for comparing different *petite* strains based on their retention or loss of markers.

For example, consider a double mutant strain, ER^r C^r *rho*$^+$. *Rho*$^-$ mutants, either spontaneous or induced, can be selected from this strain and tested for retention of markers by crossing with a double sensitive *rho*$^+$ tester strain, as follows: *rho*$^-$ ER^r C^r × *rho*$^+$ ER^S C^s. We must also

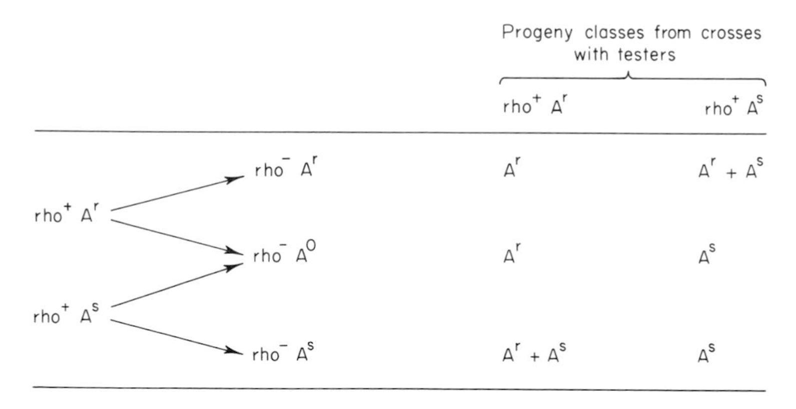

Fig. 4.15. Effect of mutation from wild type to *petite* upon the survival of various mitochondrial genes. Test crosses designed to determine whether mitochondrial genes present in a *rho*$^+$ strain are still present after mutation to *rho*$^-$. Since mitochondria are inactive in *rho*$^-$ cells, the markers on mitochondrial DNA of *rho*$^-$ cells must be transferred into *rho*$^+$ strain to assess their activity. A^r = antibiotic resistant; A^s = antibiotic sensitive; A^0 = absence of A gene.

test the possibility that mutation from *rho*$^+$ to *rho*$^-$ was accompanied by mutation of the marker genes from resistance to sensitivity. The appropriate test cross would be: *rho*$^-$ ER^s C^s $\times$ *rho*$^+$ ER^r C^r. The loss of one marker but not the other would provide a way of mapping the location of the alterations in mitochondrial DNA leading to the *petite* phenotype. Studies of this kind are just beginning to be done.

Drug-resistance markers have been used to study suppressiveness. Saunders *et al.* (*36*) correlated the percent retention of ER^r by individual *rho*$^-$ mutants with their percent suppressiveness. They found that neutral *petites* all become ER^0, retaining no resistance, and that highly suppressive *petites* are also entirely ER^0. The strains retaining some ER^r were limited to intermediate suppressives in the range of 5–50%. Also, in one of these strains, the ER^r gene continued to be lost, slowly, in subsequent subcultures.

Apparently then, using *ER* as an indicator, the intermediate suppressives were the least altered in their mitochondrial DNA, compared with neutrals and with highly suppressives, but they were unstable. These findings are consistent with the evidence of Ephrussi and Grandchamp (*8*) that intermediate suppressives show the widest variation in cloning experiments. What is the molecular basis of this variability?

The hypothesis of preferential replication of mitochondrial DNA from suppressive *petites* discussed earlier in the chapter (p. 122) was examined experimentally by Rank (*33*). First he correlated the percent suppressivity of *rho*$^-$ mutants with the percent transmission of ER^r in experiments similar to those of Saunders *et al.* (*36*). He found comparable results—that the higher the suppressivity, the lower the transmission of the *ER* gene. He then tested the hypothesis of selection at the level of mitochondrial DNA replication by crosses between high and low suppressive strains. His results showed unambiguous dominance of high suppressiveness over low, at the genetic level; but as yet, the mitochondrial DNA's of his strains have not themselves been examined.

CONCLUDING REMARKS

This chapter has summarized the evidence which associates mitochondrial DNA with various non-Mendelian mutations in yeast affecting mitochondrial function: loss of respiratory activity and resistance to several antibiotics. In Chapter 2, the physical alterations in mitochondrial DNA's of several *petite* mutants were described. In this chapter the genetic evidence has been presented that establishes these mutations as cytoplasmic rather than nuclear. The evidence rests primarily upon 4:0 segregation ratios of wild type:neutral *petites* in tetrads after meiosis, and upon the somatic segregation of suppressive *petites* in zygote clones.

Mutations conferring resistance to a number of antibiotics have been localized in mitochondrial DNA by genetic tests of two types: (*a*) somatic segregation in diploids and subsequent 4:0 segregation in tetrad analysis; and (*b*) frequent loss of marker genes associated with mutations from rho^+ to rho^-. These losses are interpreted as genetic evidence of extensive changes in nucleotide sequence in mitochondrial DNA, paralleling the changes seen in nucleotide composition (p. 31). Independently arising rho^- mutants may differ considerably in the extent and nature of these changes.

The occurrence of segregation and recombination of mitochondrial genes in yeast zygotes (and perhaps in diploid buds as well) has provided a basis for genetic mapping. Preliminary mapping efforts have been reported and rapid progress is to be anticipated.

The mechanism of suppressiveness, discussed at length in this chapter, remains a puzzle. Suppressiveness is seen as replicative dominance of the rho^- DNA over the wild type in zygotes and their diploid clonal descendants. One mechanism which has been proposed to account for suppressiveness is competition between wild-type and mutant DNA molecules at the level of replication resulting from differences in the primary structure of the DNA, e.g., higher percent of AT pairs or shorter molecules. However the pronounced effects of growing conditions upon the outcome of this competition, demonstrated by Ephrussi *et al.* (*16*), suggests that other factors are involved in suppressiveness as well as the structure of the DNA itself.

The phenomenon of suppressiveness may be viewed as a form of preferential transmission in which one of the parental cytoplasmic genomes takes over. The mechanism of the take-over process probably differs in different organisms, but analogous results are seen. In *Neurospora* for instance (cf. Chapter 5), cytoplasmic hybrids are unstable during mycelial growth; one or the other of the parental types eventually takes over. In higher plants, somatic segregation of cytoplasmic genes, seen, for example, as sectors of green and white tissue, provides evidence of a preferential segregation process. Maternal inheritance, as seen, for example, in *Chlamydomonas,* represents one extreme in which paternal chloroplast genes are not transmitted to progeny. The phenomenon of preferential transmission and its possible evolutionary significance will be discussed at the end of this section on genetic analysis of various cytoplasmic systems.

Suggested Review Articles

Coen, D., Deutsch, J., Netter, P., Petrochilo, E., and Slonimski, P. P. (1970). Mitochondrial genetics: I. Methodology and phenomenology. *In* "Control of Organelle Development," *Soc. Exp. Biol. Symp.* (P. L. Miller, ed.), Vol. 24, pp. 449–496. Cambridge Univ. Press, London.

Linnane, A. W., and Haslam, J. M. (1970). Biogenesis of yeast mitochondria. *In* "Current Topics in Cellular Regulation" (B. L. Horecker and E. R. Stadtman, eds.), Vol. 2, pp. 102–172. Academic Press, New York.

References

1. Beck, J. C., Parker, J. H., Balcavage, W. X., and Mattoon, J. R. (1971). Mendelian genes affecting development and function of yeast mitochondria. *In* "Autonomy and Biogenesis of Mitochondria and Chloroplasts," *Aust. Acad. Sci. Symp.* (N. K. Boardman, A. W. Linnane, and R. M. Smillie, eds.). North-Holland Publ., Amsterdam.

1a. Bolotin, M., Coen, D., Deutsch, J., Dujon, B., Netter, P., Petrochilo, E., and Slonimski, P. P. (1971). La Recombinaison des mitochondries chez *Saccharomyces cerevisiae. Bull. Inst. Pasteur* **69,** 215.

1b. Bunn, C. L., Mitchell, C. H., Lukins, H. B., and Linnane, A. W. (1970). Biogenesis of mitochondria xviii. A new class of cytoplasmically determined antibiotic resistant mutants in *Saccharomyces cerevisiae. Proc. Nat. Acad. Sci. U. S.* **67,** 1233.

2. Carnevali, F., Morpurgo, G., and Tecce, G. (1969). Cytoplasmic DNA from petite colonies of *Saccharomyces cerevisiae:* A hypothesis on the nature of mutation. *Science* **163,** 1331–1333.

3. Chen. S-Y., Ephrussi, B., and Hottinguer, H. (1950). Nature genetique des mutants à deficience respiratoire de la souche B-II de la levure de boulangerie. *Heredity* **4,** 337–351.

4. Coen, D., Deutsch, J., Netter, P., Petrochilo, E., and Slonimski, P. P. (1970). Mitochondrial genetics. I. Methodology and phenomenology. *In* "Control of Organelle Development," *Soc. Exp. Biol. Symp.* (P. L. Miller, ed.), Vol. 24, pp. 449–496. Cambridge Univ. Press, London and New York.

5. Criddle, R. S., and Schatz, G. (1969). Promitochondria of anaerobically grown yeast. I. Isolation and biochemical properties. *Biochemistry* **8,** 322–334.

6. Ephrussi, B. (1952). The interplay of heredity and environment in the synthesis of respiratory enzymes in yeast. *Harvey Lect. Series XLVI, 1950–1951,* pp. 45–67.

7. Ephrussi, B. (1953). "Nucleo-Cytoplasmic Relations in Micro-Organisms." Clarendon Press, Oxford.

8. Ephrussi, B., and Grandchamp, S. (1965). Études sur la suppressivité des mutants à déficience respiratoire de la levure. I. Existence au niveau cellulaire de divers "digres de suppressivité." *Heredity* **20,** 1–7.

9. Ephrussi, B., and Hottinguer, H. (1950). Direct demonstration of the mutagenic action of euflavin on Baker's yeast. *Nature (London)* **166,** 956.

10. Ephrussi, B., and Hottinguer, H. (1951). Cytoplasmic constituents of heredity. On an unstable state in yeast. *Cold Spring Harbor Symp. Quant. Biol.* **16,** 75–85.

11. Ephrussi, B., Hottinguer, H., and Chimenes, A.-M. (1949). Action de l'acriflavine sur les levures. I. La mutation "petite colonie." *Ann. Inst. Pasteur Paris* **76,** 351–367.

12. Ephrussi, B., Hottinguer, H., and Tavlitzki, J. (1949). Action de l'acriflavine sur les levures. II. Etude genetique du mutant "petite colonie." (Appendice par P. L'Heritier.) *Ann. Inst. Pasteur Paris* **76,** 419–450.

13. Ephrussi, B., L'Heritier, P., and Hottinguer, H. (1949). Action de l'acriflavine sur les levures. VI. Analyse quantitative de la transformation des populations. *Ann. Inst. Pasteur Paris* **77,** 64–83.
14. Ephrussi, B., Margerie-Hottinguer, H. de, and Roman, H. (1955). Suppressiveness: A new factor in the genetic determinatism of the synthesis of respiratory enzymes in yeast. *Proc. Nat. Acad. Sci. U. S.* **41,** 1065–1071.
15. Ephrussi, B., Slonimski, P. P., Yotsuyanagi, Y., and Tavlitzki, J. (1956). Variations physiologiques et cytologiques de la levure au cours du cycle de la croissance aerobie. *C. R. Trav. Lab. Carlsberg, Ser. Physiol.* **26,** 87–102.
16. Ephrussi, B., Jakob, H., and Grandchamp, S. (1966). Études sur la suppressivité des mutants à déficience respiratoire de la levure. II. Étapes de la mutation grande en petite provoquée par le facteur suppressif. *Genetics* **54,** 1–29.
17. Fukahara, H. (1969). Relative proportions of mitochondrial and nuclear DNA in yeast under various conditions of growth. *Eur. J. Biochem.* **11,** 135–139.
18. Gillberg, B. O., Zetterberg, G., and Swanbeck, G. (1967). Petite mutants induced in yeast by Dithranol (1,8,9-trihydroxy-anthracene), an important therapeutic agent against Psoriasis. *Nature (London)* **214,** 415.
19. Gingold, E. B., Saunders, G. W., Lukins, H. B., and Linnane, A. W. (1969). Biogenesis of mitochondria. X. Reassortment of the cytoplasmic genetic determinants for respiratory competence and erythromycin resistance in *Saccharomyces cerevisiae. Genetics* **62,** 735–744.
20. Goldring, E. S., Grossman, L. I., Krupnick, D. Cryer, D. R., and Marmur, J. (1970). The petite mutation in yeast. Loss of mitochondrial deoxyribonucleic acid during induction of petites with ethidium bromide. *J. Mol. Biol.* **52,** 323–336.
21. Harris, M. (1956). Occurrence of respiration deficient mutants in baker's yeast cultivated anaerobically. *J. Cell. Comp. Physiol.* **48,** 95.
22. Kruis, K., and Satava, J. (1918). Ovyvoji a kliceni spor i sexualite kvasinek. *Nakl. C. Akad. Praha.*
22a. Lacroute, F. (1963). Genetique de la resistance au 5-fluorouracil chez la levure. *C. R. Acad. Sci. Ser. D.* **257,** 4213.
23. Lacroute, F. (1966). Regulation de la chaine de biosynthese de l'uracile chez *Saccharomyces cerevisiae.* Thesis, Paris University.
24. Laskowski, W. (1954). Induction par le chlorure de tetrazolium de la mutation "petite colonie" chez la levure. *Heredity* **8,** 79.
25. Lindegren, C. C., Nagai, S., and Nagai, H. (1958). Induction of respiratory deficiency in yeast by manganese, copper, cobalt and nickel. *Nature (London)* **182,** 446–448.
26. Linnane, A. W., Saunders, G. W., Gingold, E. B., and Lukins, H. B. (1968). The biogenesis of mitochondria. V. Cytoplasmic inheritance of erythromycin resistance in *Saccharomyces cerevisiae. Proc. Nat. Acad. Sci. U. S.* **59,** 903–910.
27. Marcovich, H. (1951). Action de l'acriflavine sur les levures. VIII. Determination du composant actif et étude de l'euflavine. *Ann. Inst. Pasteur Paris* **81,** 452–468.
28. Milkman, R. (1963). Suppressiveness in petite yeast. *Microbial Genet. Bull.* **19,** 16.
29. Moustacchi, E., and Marcovich, M. H. (1963). Induction de la mutation "petite colonie" chez la levure par le 5-fluorouracile. *C. R. Acad. Sci. Ser. D.* **256,** 5646–5648.
30. Nordstrom, K. (1967). Induction of the petite mutation in *Saccharomyces cerevisiae* by N-methyl-N'-nitro-N-nitrosoguanidine. *J. Gen. Microbiol.* **48,** 277–281.
30a. Perlman, P. S., and Mahler, H. R. (1971). Molecular consequences of ethidium bromide mutagenesis. *Nature New Biol.* **231,** 12.
31. Pittman, D. D. (1957). Induction of respiratory deficiency in tetraploid *Saccharomyces* by ultraviolet radiation. *Exp. Cell Res.* **11,** 654.

32. Pittman, D. D. (1959). Ultraviolet induction of respiration deficient variants of *Saccharomyces* and their stability during vegetative growth. *Cytologia* **24,** 315.
33. Rank, G. H. (1970). Genetic evidence for 'Darwinian' selection at the molecular level. I. The effect of the suppressive factor on cytoplasmically-inherited erythromycin-resistance in *Saccharomyces cerevisiae.* II. Genetic analysis of cytoplasmically-inherited high and low suppressitivity in *Saccharomyces cerevisiae. Can. J. Genet. Cytol.* **12,** 129–136, 340–346.
34. Raut, C. (1954). Heritable non-genic changes induced in yeast by ultraviolet light. *J. Cell. Comp. Physiol.* **44,** 463–475.
35. Raut, C., and Simpson, W. L. (1955). The effect of x-rays and of ultraviolet light of different wavelengths on the production of cytochrome-deficient yeasts. *Arch. Biochem. Biophys.* **57,** 218–228.
36. Saunders, G. W., Elliot, B., Trembath, M. K., Lukins, H. B., and Linnane, A. W. (1971). Mitochondrial genetics in yeast. *In* "Autonomy and Biogenesis of Mitochondria and Chloroplasts," *Aust. Acad. Sci. Symp.* (N. K. Boardman, A. W. Linnane, and R. M. Smillie, eds.) North-Holland Publ., Amsterdam.
37. Schwaier, R., Nashed, N., and Zimmerman, F. K. (1968). Mutagen specificity in the induction of karyotic versus cytoplasmic respiratory-deficient mutants in yeast by nitrous acid and alkylating nitrosamides. *Mol. Gen. Genet.* **102,** 290–300.
38. Sherman, F. (1959). The effects of elevated temperatures on yeast. II. Inducation of respiratory-deficient mutants. *J. Cell. Comp. Physiol.* **54,**37–52.
39. Sherman, F. (1963). Respiration-deficient mutants of yeast. I. Genetics. *Genetics* **48,** 375.
40. Slonimski, P. P., and Ephrussi, B. (1949). Action de l'acriflavine sur les levures. V. Le systeme des cytochromes des mutants "petite colonie." *Ann. Inst. Pasteur Paris* **77,** 47–63.
41. Slonimski, P. P., Perrodin, G., and Croft, J. H. (1968). Ethidium bromide-induced mutation of yeast mitochondria: Complete transformation of cells into respiratory-deficient nonchromosomal "petites." *Biochem. Biophys. Res. Commun.* **30,** 232–239.
42. Thomas, D. Y., and Wilkie, D. (1968). Recombination of mitochondrial drug-resistance factors in *Saccharomyces cerevisiae. Biochem. Biophys. Res. Commun.* **30,** 368–372.
43. Thomas, D. Y., and Wilkie, D. (1968). Inhibition of mitochondrial synthesis in yeast by erythromycin: Cytoplasmic and nuclear factors controlling resistance. *Genet. Res.* **11,** 33–41.
44. Wilkie. D., Saunders, G. W., and Linnane, A. W. (1967). Inhibition of respiratory enzyme synthesis in yeast by chloramphenicol: Relationship between chloramphenicol tolerance and resistance to other antibacterial antibiotics. *Genet. Res.* **10,** 199–203.
45. Wilson, E. B. (1925). "The Cell in Development and Heredity." Macmillan, New York.
46. Winge, O. (1935). On haplophase and diplophase in some Saccharomycetes. *C. R. Trav. Lab. Carlsberg Ser. Physiol.* **21,** 77–109.
47. Winge, O. & O. Laustsen (1940). On a cytoplasmatic effect of inbreeding in homozygous yeast. *C. R. Trav. Lab. Carlsberg Ser. Physiol.* **23,** 17–39.
48. Winge, O., and Roberts, C. (1958). Life history and cytology of yeasts. Yeast genetics. *In* "The Chemistry and Biology of Yeasts" (A. H. Cook, ed.), pp. 93–156. Academic Press, New York.
49. Wright, R. E., and Lederberg, J. (1957). Extranuclear transmission in yeast heterokaryons. *Proc. Nat. Acad. Sci. U. S.* **43,** 919–923.
50. Ycas, M. (1956). A hereditary cytochrome deficiency appearing in yeast grown at elevated temperature. *Exp. Cell Res.* **10,** 746.

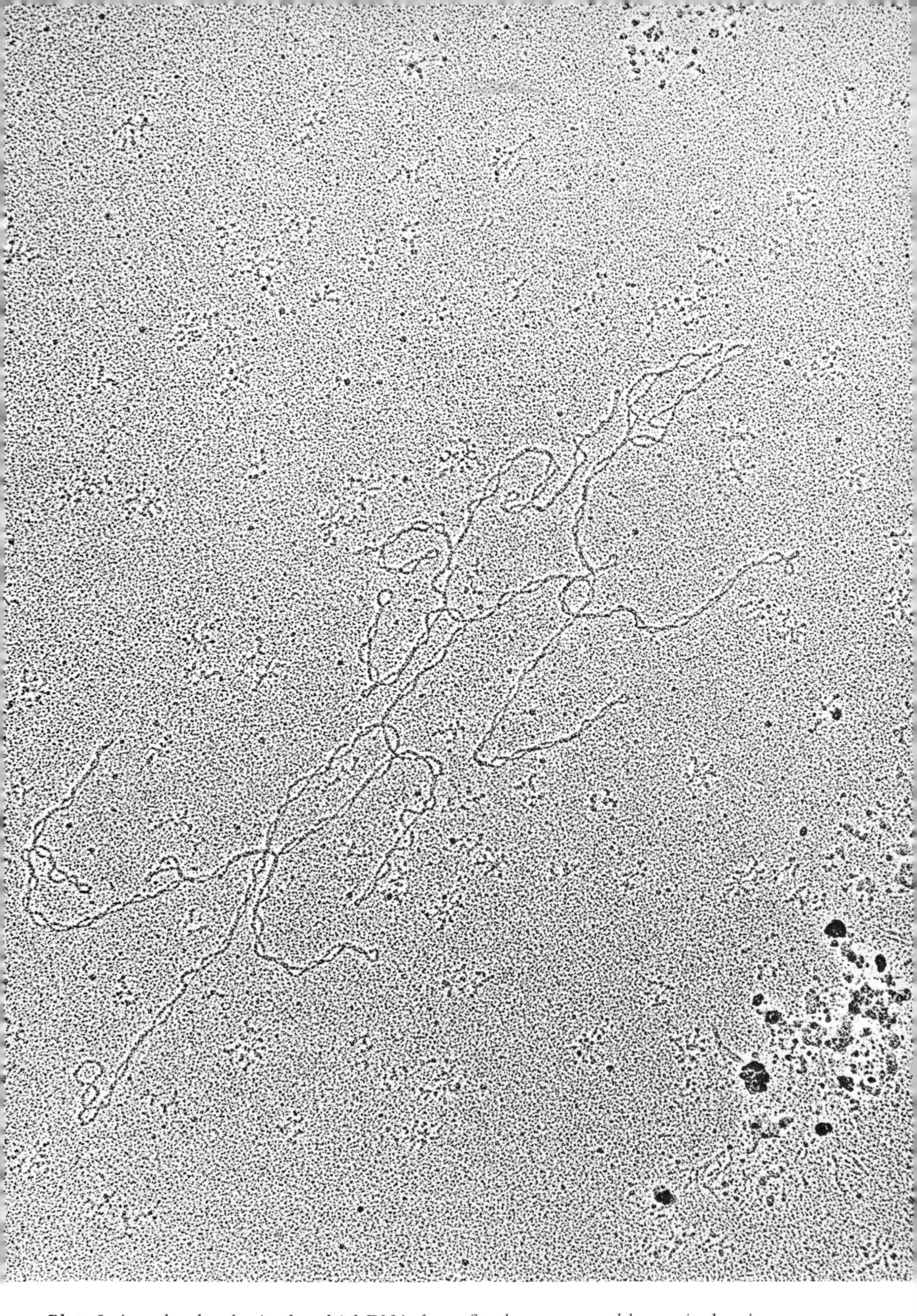

Plate I. A molecule of mitochondrial DNA from *Saccharomyces carlsbergensis* showing supercoiled configuration of the covalently closed circular molecule. 55,000×. Courtesy Dr. E. F. J. Van Bruggen, University, Groningen, Netherlands.

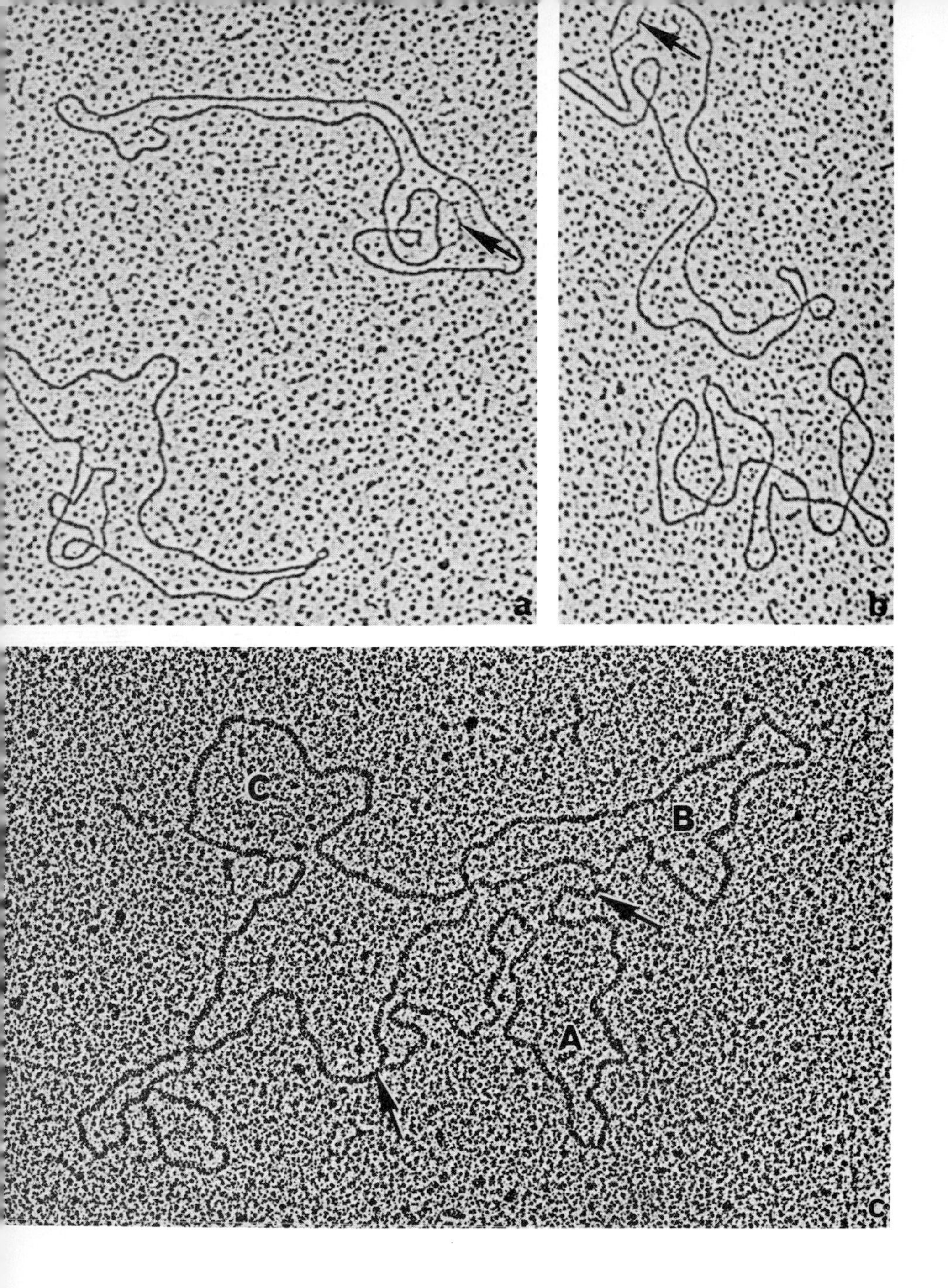

Plate II. (a) and (b). Electron micrographs of closed-circular mitochondrial DNA from mouse L-cells prepared for electron microscopy by the formamide technique, stained, and shadowed. Displacement loops are indicated by arrows. A single-stranded $\phi\chi$ DNA is present in a. Courtesy Dr. Vinograd (*46b* in Chapter 2). (c). A molecule of replicating mitochondrial DNA from rat liver. On one interpretation, the two replicating forks are indicated by arrows, dividing the molecule into two replicated loops of equal length A and B, and an unreplicated region C. 77,480×. Courtesy Dr. Kirschner (*40* in Chapter 2).

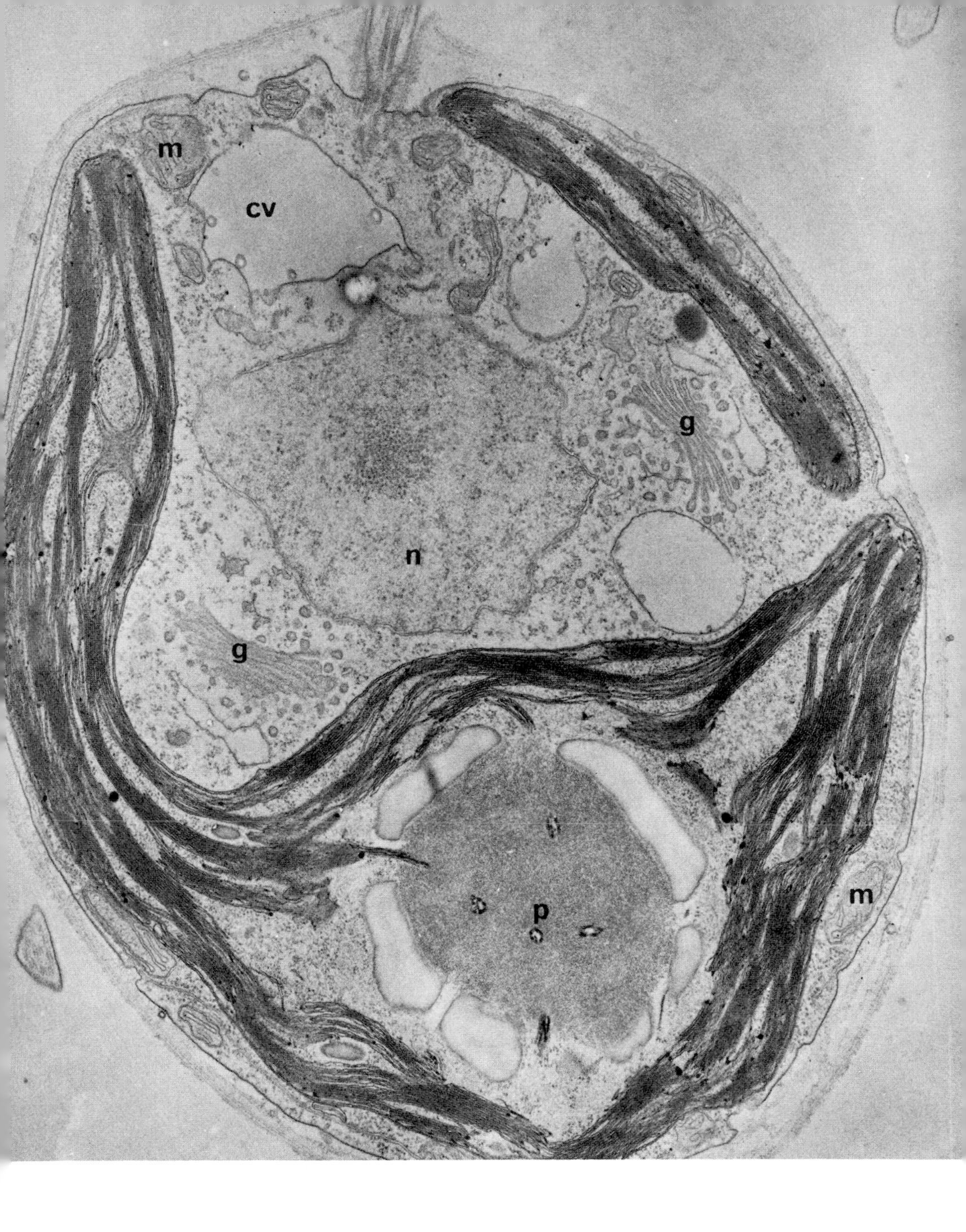

Plate III. Light-grown, fully greened *Chlamydomonas* strain *y-1* in longitudinal section. At anterior end, one of the two flagella is seen, as is one of the two contractile vacuoles (cv). Two sets of Golgi (g) vesicles are seen on either side of the nucleus. In the posterior region, the pyrenoid (p) is seen surrounded by starch plates and embedded in the cup-shaped chloroplast. Mitochondria (m) are located between the chloroplast and plasma membrane, as well as in the anterior region. Numerous ribosomes are seen in the cell sap and in the matrix of the chloroplast. 17,670×. Courtesy Dr. G. E. Palade, unpublished. For detailed description and preparative methods, see next and refs. *83* and *98* of Chapter 3.

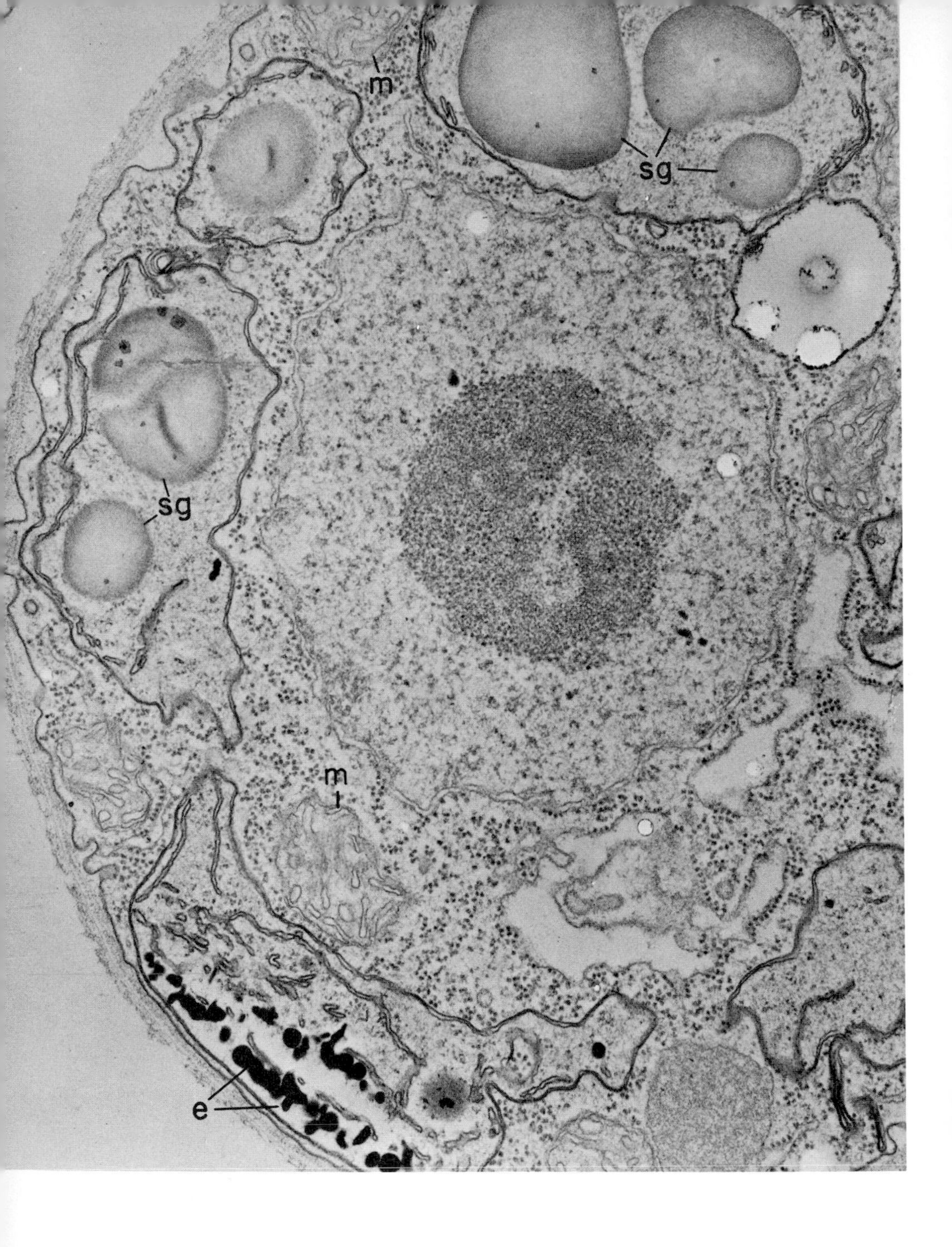

Plate IV. Dark-grown *Chlamydomonas* strain *y-1*. Section through nucleus and eyespot, not including pyrenoid or anterior region. Eyespot (e) is grossly disorganized, lamellar membranes are gone, but few scattered vesicles can be seen within the plastid. Several sections through the single chloroplast are seen, as well as starch (Sg). Mitochondria (m) are larger and denser than in light-grown cells; numerous ribosomes are present in the plastid and in cell sap, as in light-grown cells. 25,000×. Courtesy Dr. G. E. Palade, unpublished. For further details see text and ref. *83*, Chapter 3.

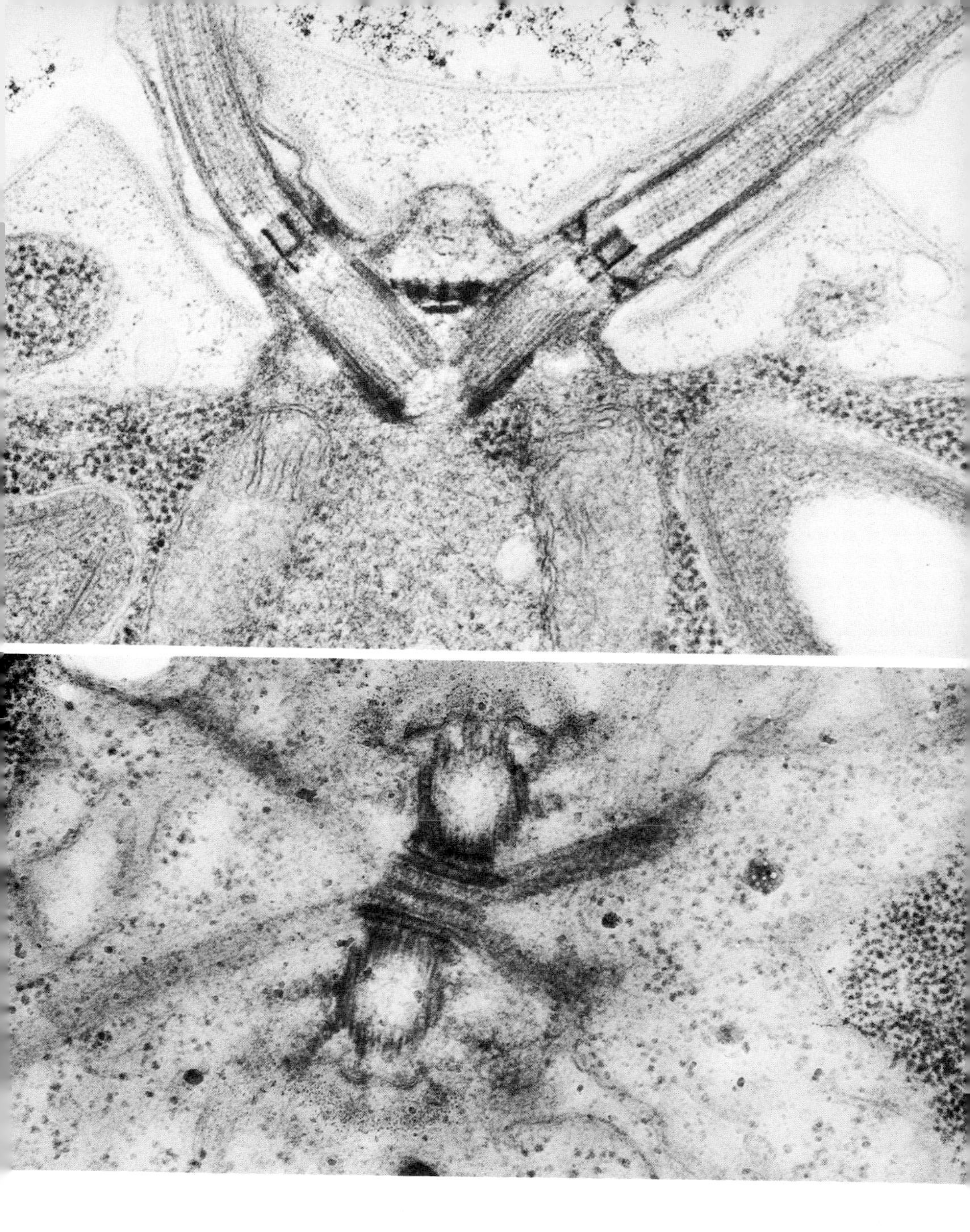

Plate V. Flagellar apparatus of *Chlamydomonas.* (a). A view of both flagella in longitudinal section showing the orientation of the basal bodies. The distal striated fiber connects the basal bodies; an electron-dense plate lies below the fiber. No microtubules are present in the region between the basal bodies. The central fibers of both flagella are visible in this section. 63,000×. (b). A section perpendicular to that of (a), showing the distal striated fiber in longitudinal section and the basal bodies in oblique section. Four bands of tubules approach the region between the basal bodies in an X-shaped configuration. 63,000×. Courtesy Dr. Ringo (*29* in Chapter 3).

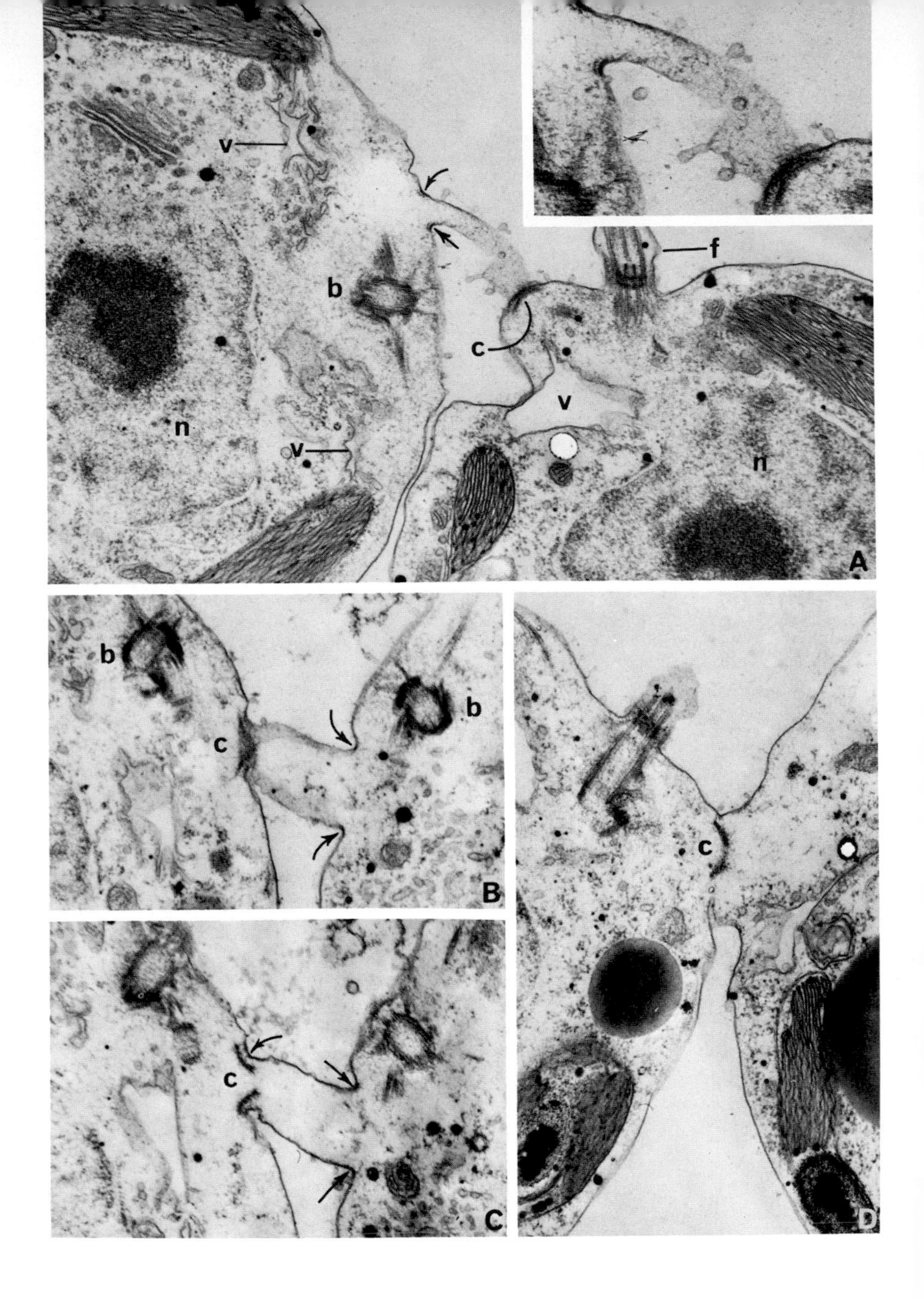

Plate VI. The mating process in *Chlamydomonas.* (a). The apical ends of two gametes are connected by a narrow fertilization tubule. In the gamete on the right side lies the edge of a well-developed choanoid body (c). In the gamete on the left, at the end of the tubule, lies a structure which resembles a choanoid body but is less dense and much smaller (b, basal body; f, flagellum; n, nucleus; v, contractile vacuole). 27,400×. The insert shows at higher magnification the fertilization tubule in continuity with both gametes. 54,000×. (b), (c). Consecutive sections showing a later stage in gamete association. 27,400×. (d). A later stage when the fertilization tubule is no longer present as such. The former gametes lie closer to each other and are in communication over a greater area than during earlier stages. One continuous plasma membrane bounds the two original gametes 27,400×. Courtesy Dr. Friedmann *et al.* (*10* in Chapter 3).

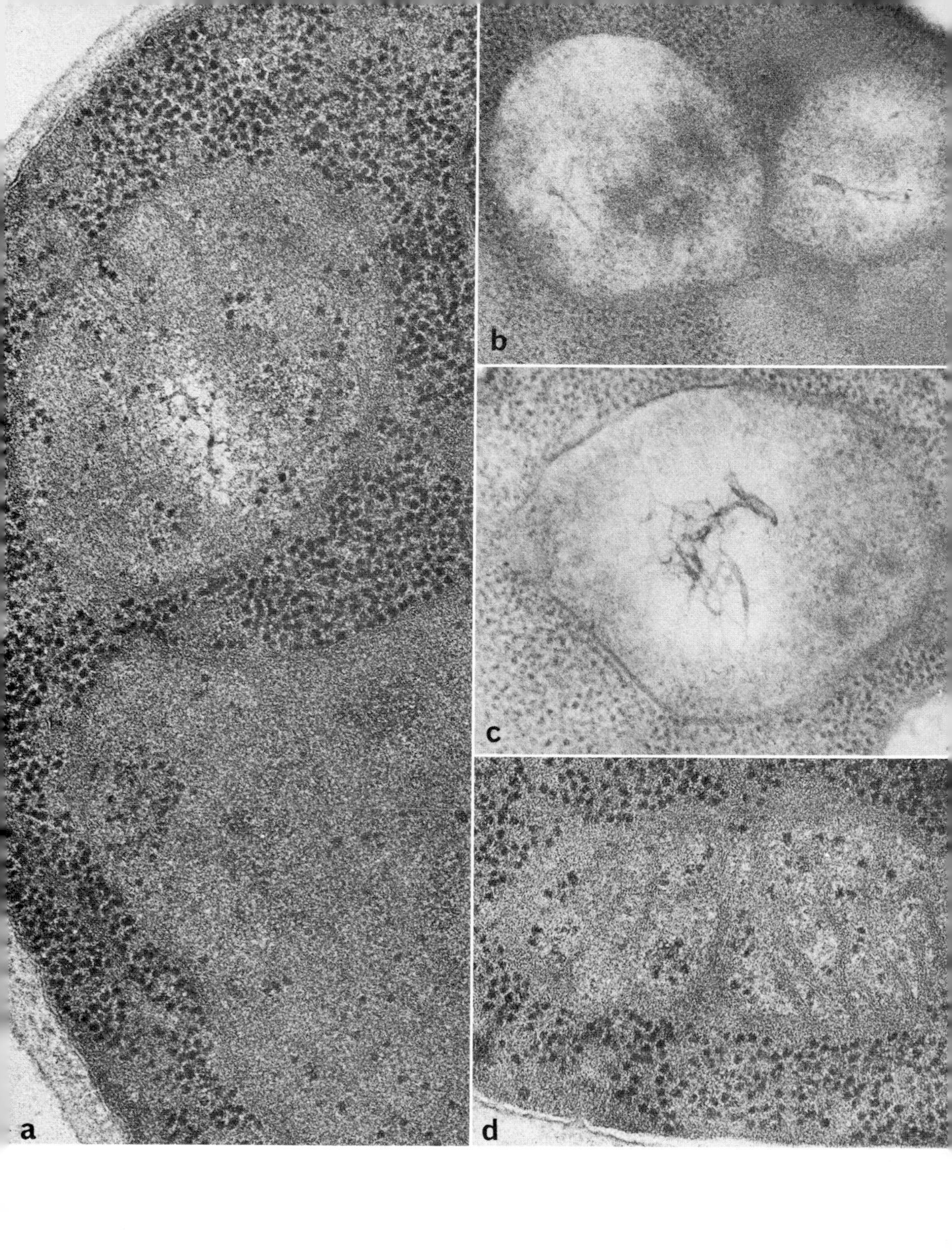

Plate VII. Mitochondria of aerobic and anaerobic yeast, *Saccharomyces cerevisiae,* strain $R_1\ \rho^+$ (a). Aerobically grown cell. Nucleus below, mitochondrion with small area of DNA filaments and a few scattered ribosomes above. 120,000×. (b). The same strain grown anaerobically *without* ergosterol or Tween 80. Mitochondria contain DNA but no cristae. 70,000×. (c). The same strain grown anaerobically *with* ergosterol and Tween 80 showing a prominent central area of DNA filaments but no cristae. 96,000×. (d). Another mitochondrion from an aerobic culture. 120,000×. Courtesy Dr. H. Swift, unpublished.

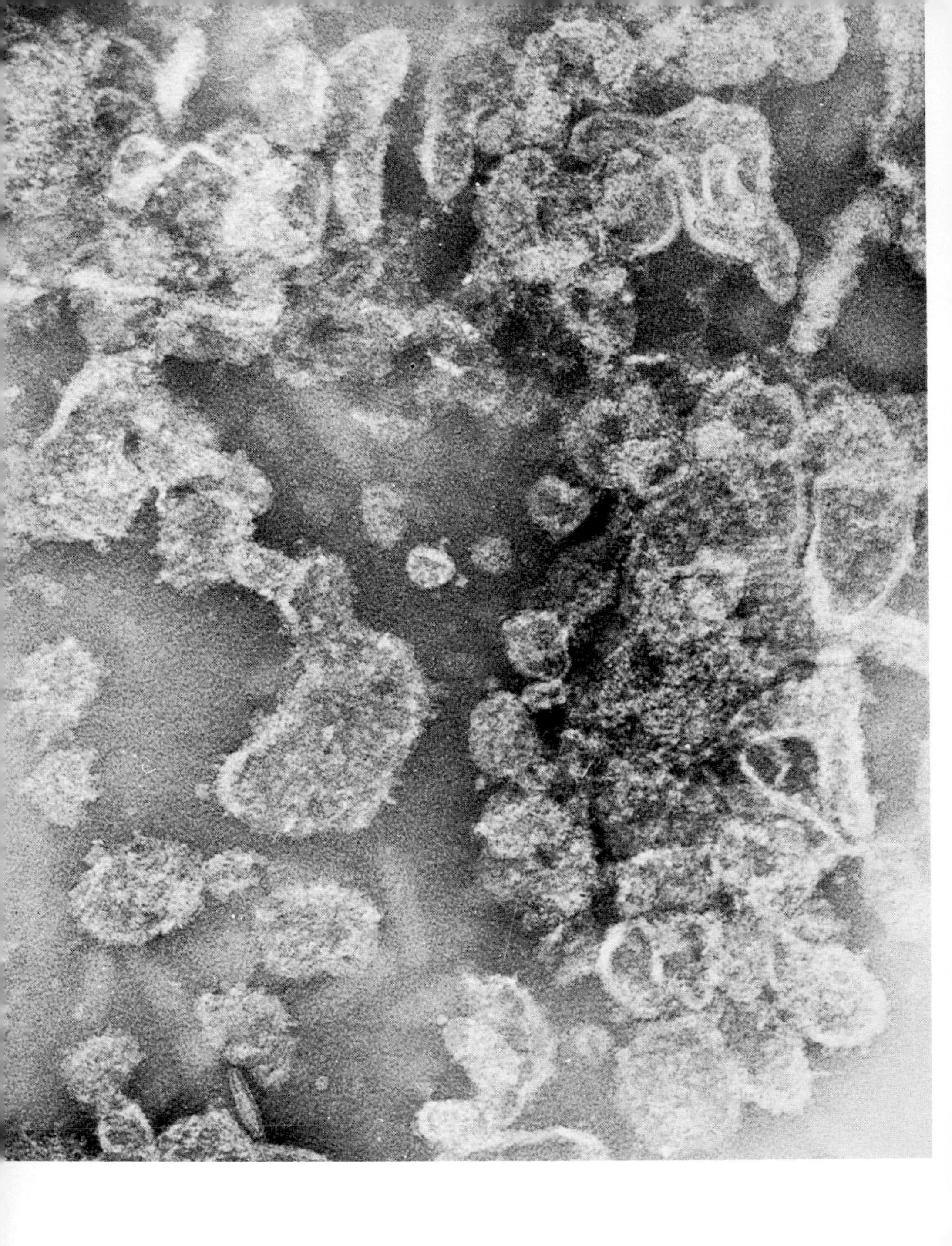

Plate VIII. Electron micrograph of CF_0. 250,000×. Courtesy Dr. E. Racker (*75* in Chapter 7).

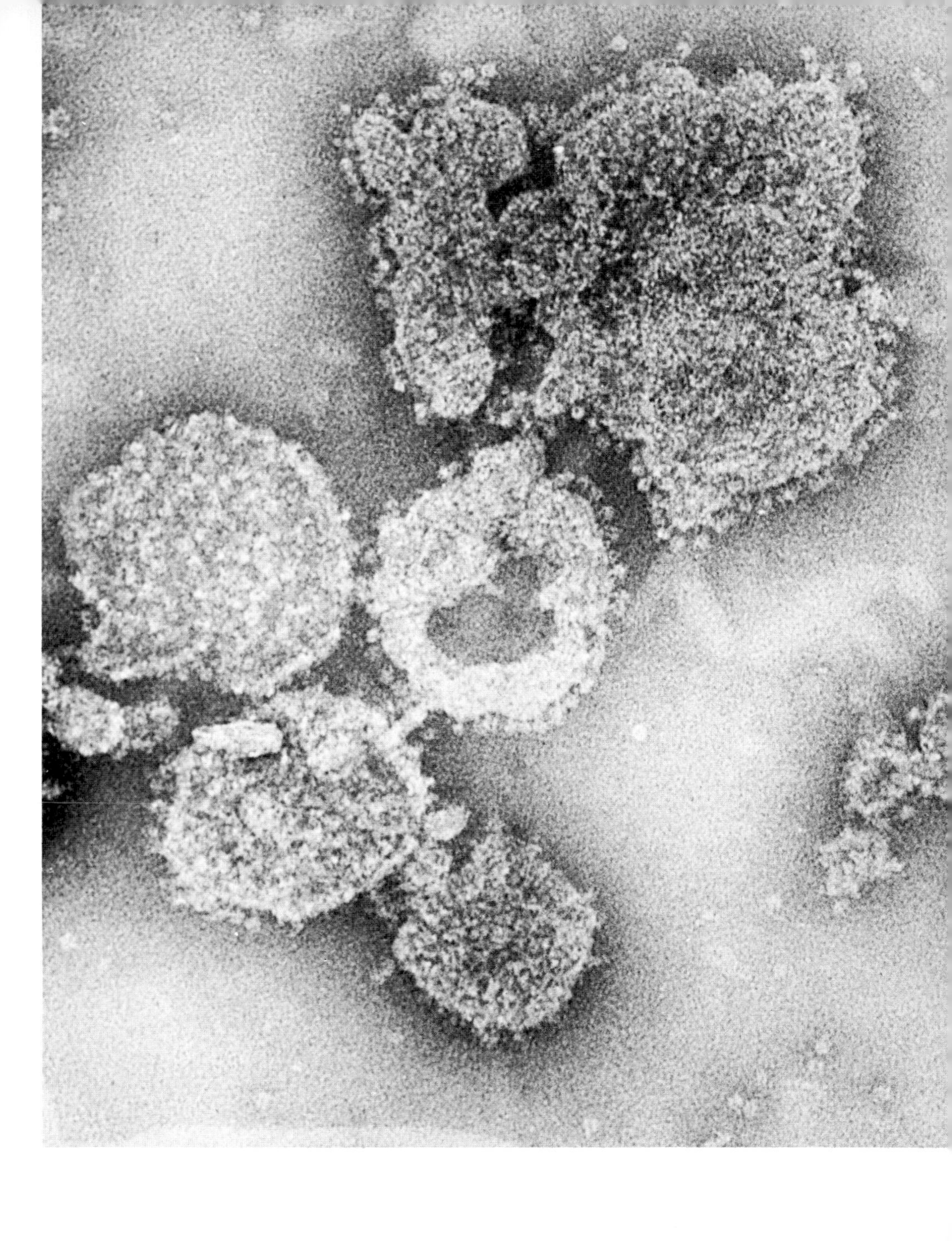

Plate IX. Electron micrograph of CF_0 + ATPase (F_1). 250,000×. Courtesy Dr. Racker (*75* in Chapter 7).

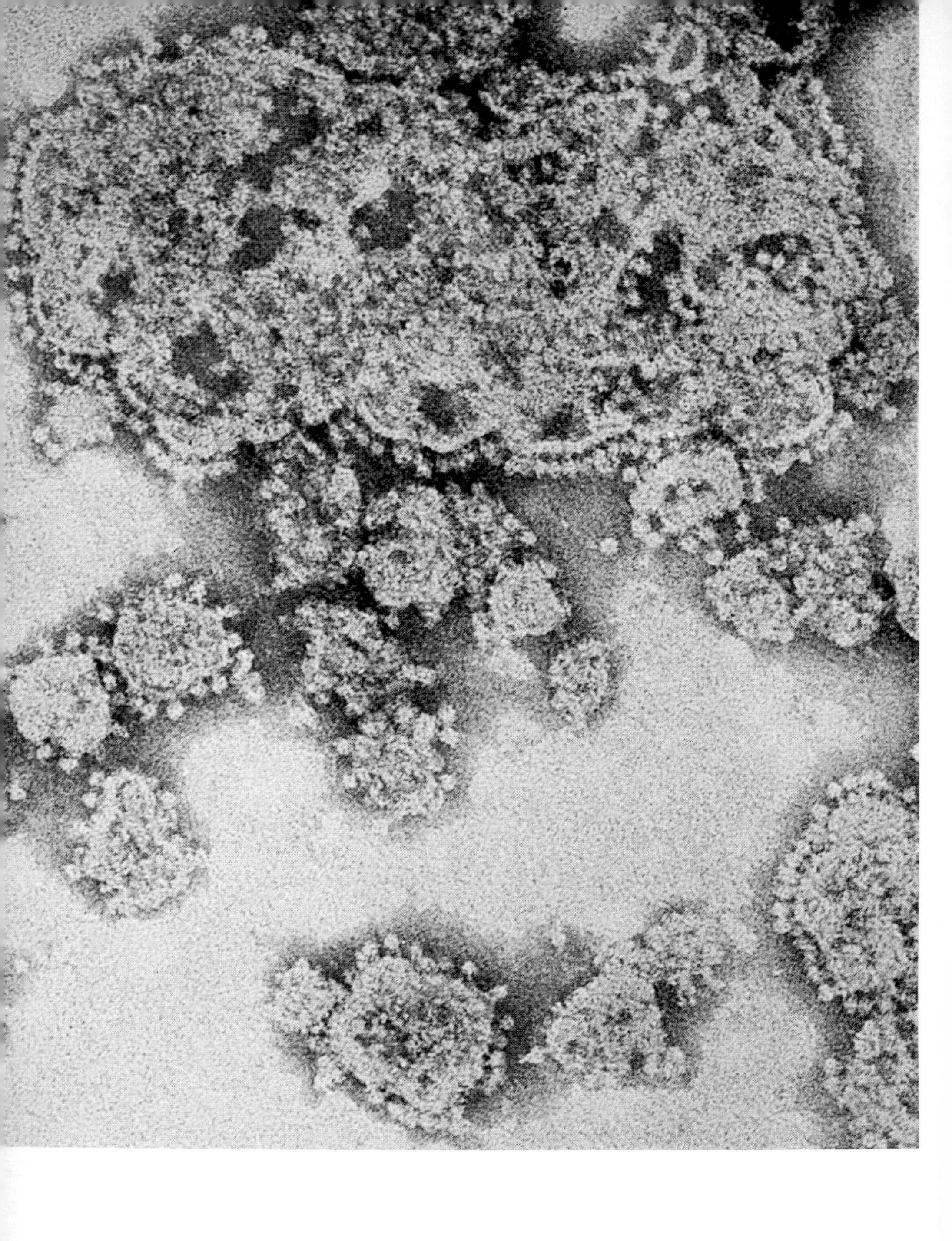

Plate X. Electron micrograph of CF_0 + Phospholipids × ATPase (F_1). 250,000×. Courtesy Dr. E. Racker (*75* in Chapter 7).

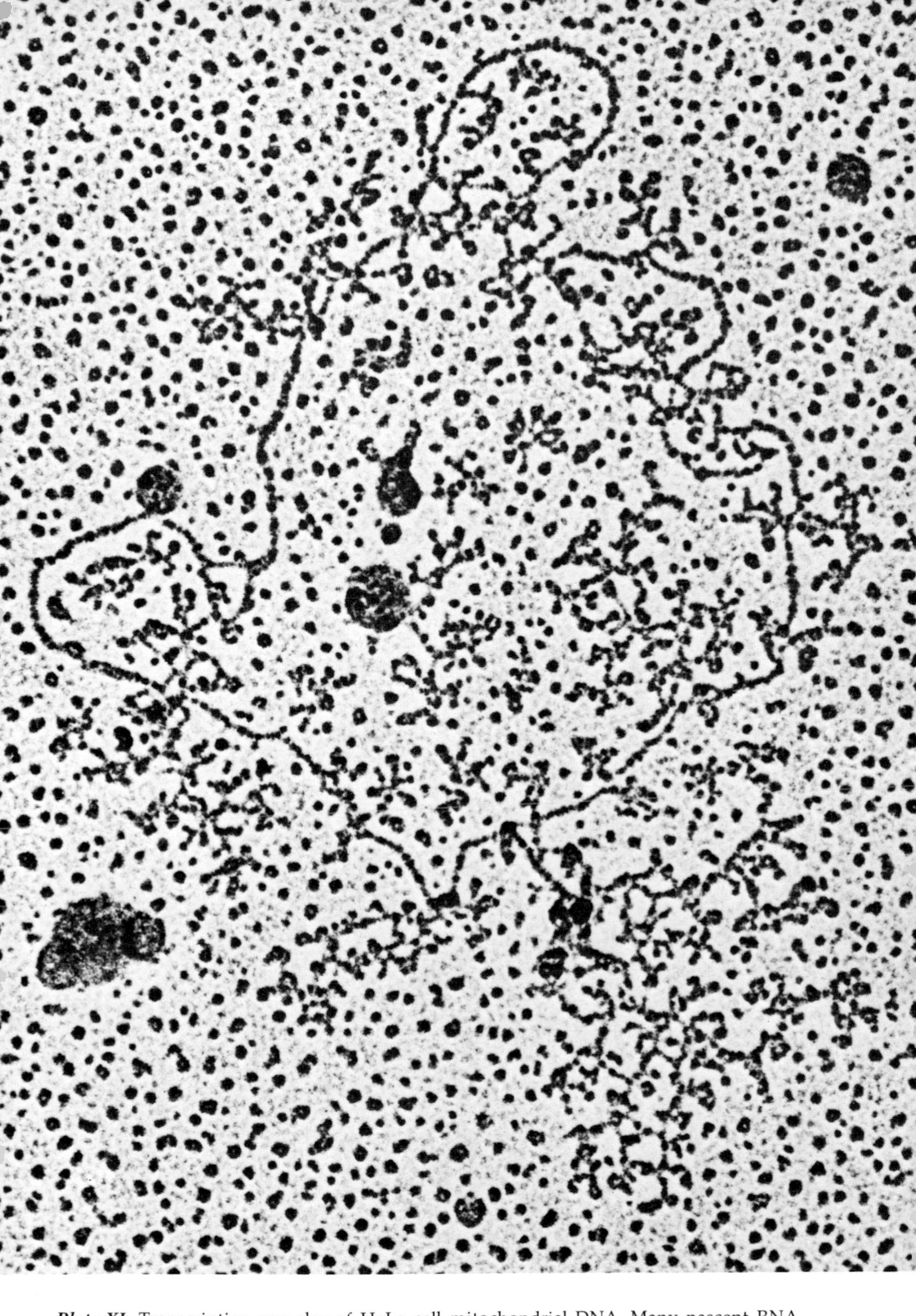

Plate XI. Transcription complex of HeLa cell mitochondrial DNA. Many nascent RNA chains can be seen attached to the mitochondrial DNA molecule. Under the aqueous mounting conditions used here, the RNA molecules are collapsed and appear as bushes. 107,500×. Courtesy Drs. Aloni and Attardi (*1* in Chapter 7).

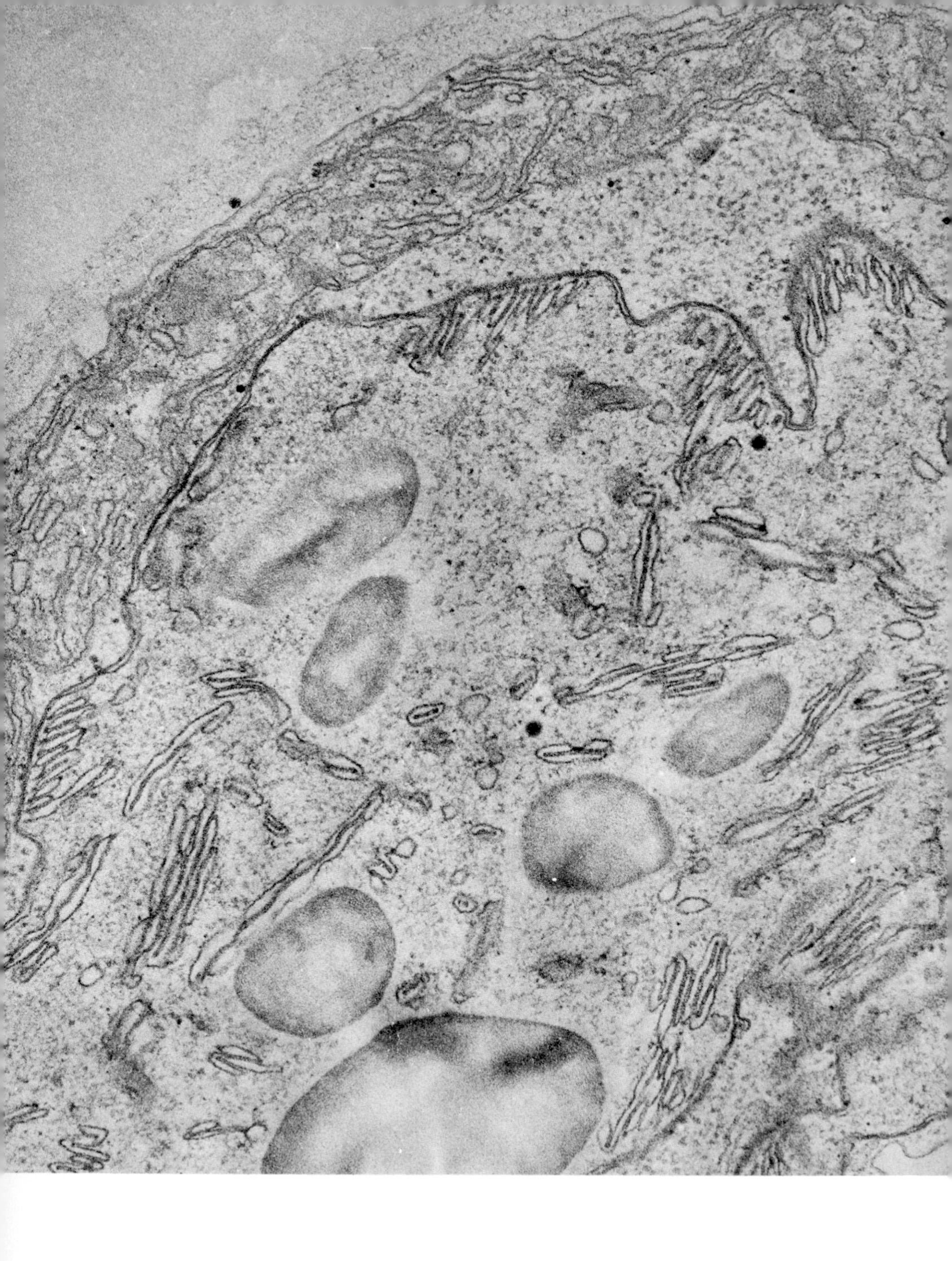

Plate XII. Degreening of *Chlamydomonas* strain *y-1*. The plastid envelope is intact and surrounds the remnants of lamellar membranes. The stacks have disappeared, the discs are found mainly in pairs or single and they have become very short. Numerous starch grains and ribosomes are present. A huge mitochondrion lies between the plastid and the plasma membrane. 90,000×. Courtesy Dr. G. E. Palade, unpublished.

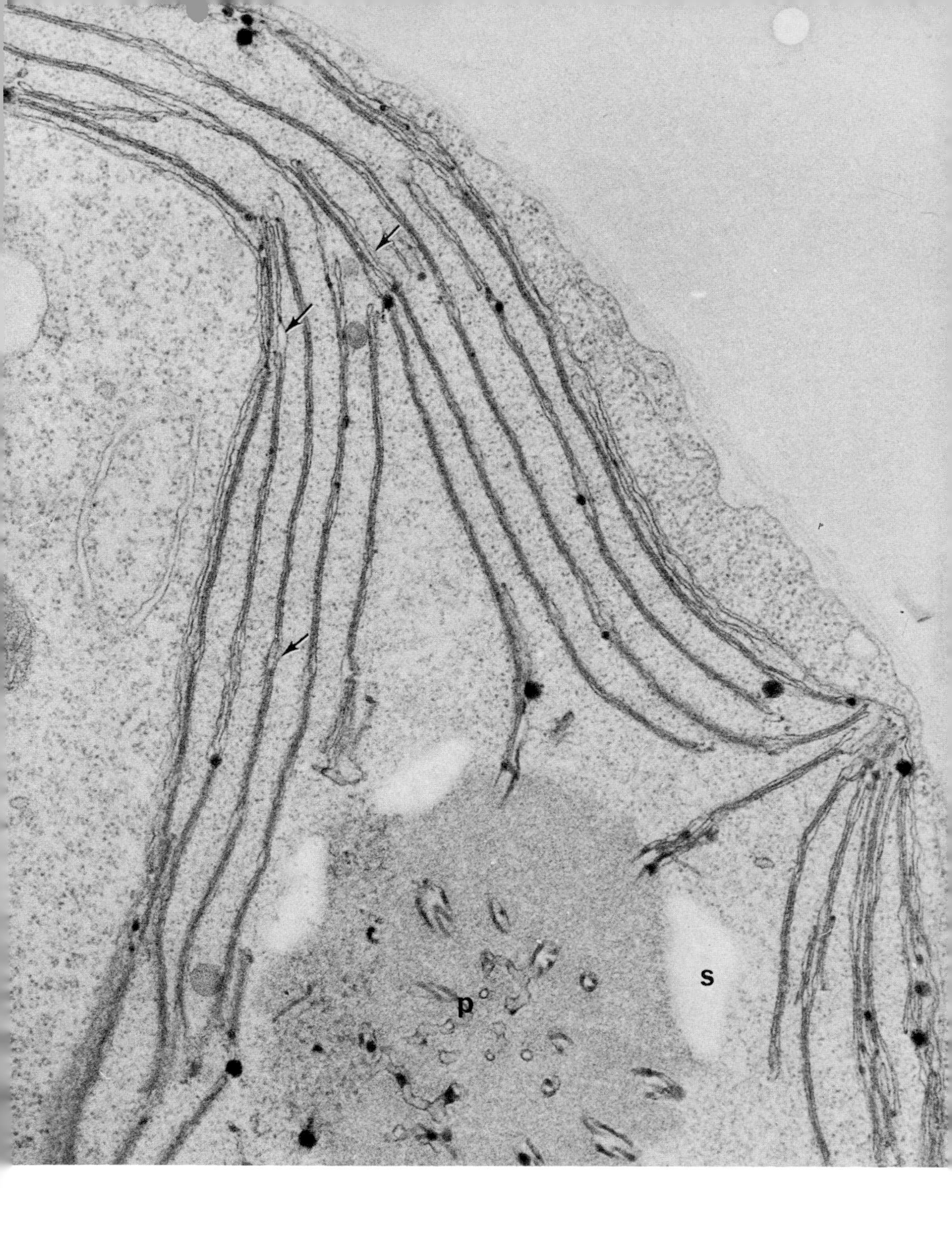

Plate XIII. Normal green strain of *Chlamydomonas* from synchronous culture growing on 12 hour light:12 hour dark regime. Harvested after 3 hours light. Lamellar membranes are unstacked, mostly in pairs with frequent unpaired regions (see arrows). Pyrenoid (p), starch grains (s), and ribosomes are present. 45,000×. Courtesy Dr. G. E. Palade, unpublished.

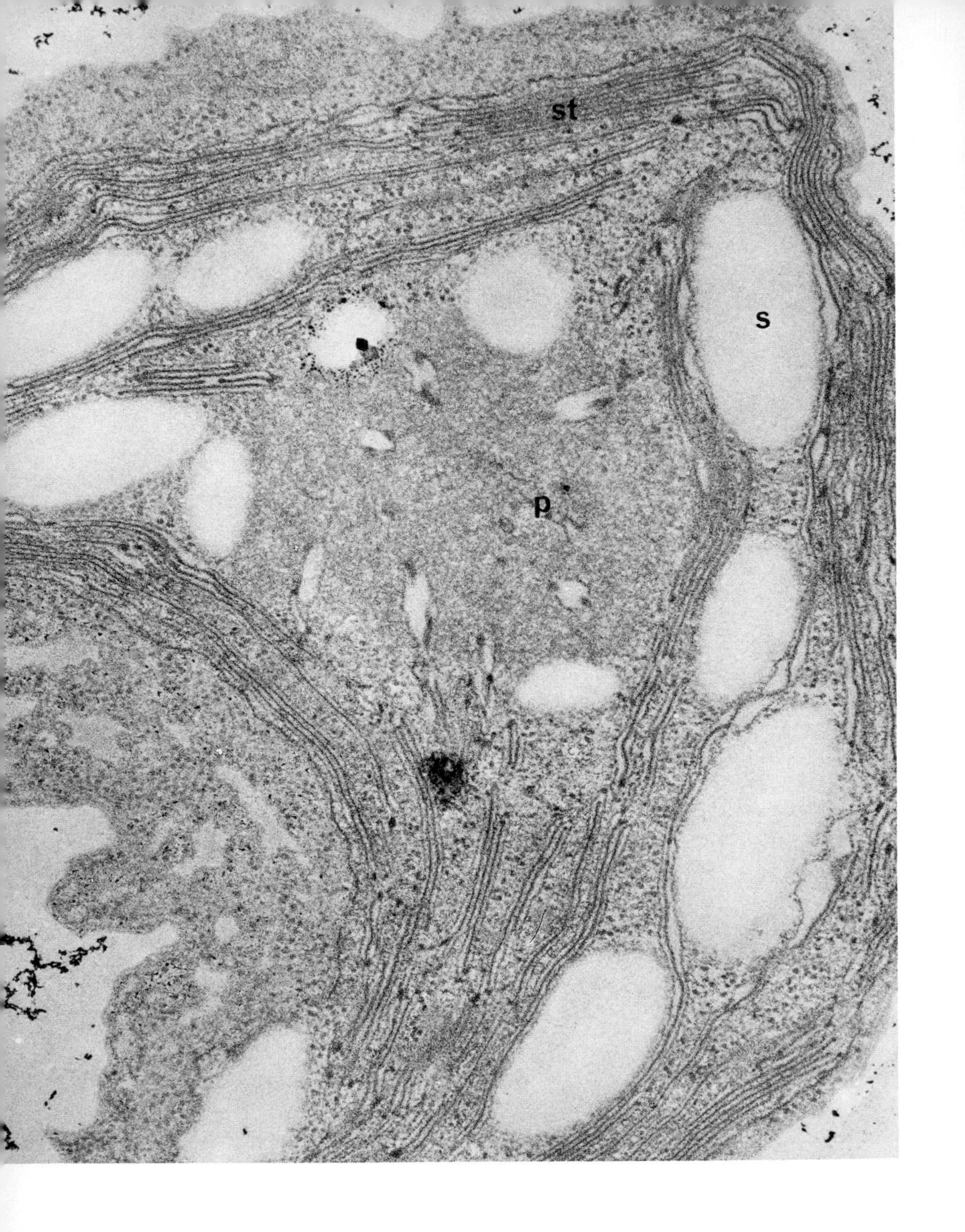

Plate XIV. Synchronously grown *Chlamydomonas* from same culture as in Plate III. Harvested after 12 hours dark. Lamellar membranes are stacked (st) as in light grown cultures (Plate IV). Pyrenoid (p), starch (s), and ribosomes are present as at other stages of growth. See text for further details. 50,000×. Courtesy Dr. G. E. Palade, unpublished.

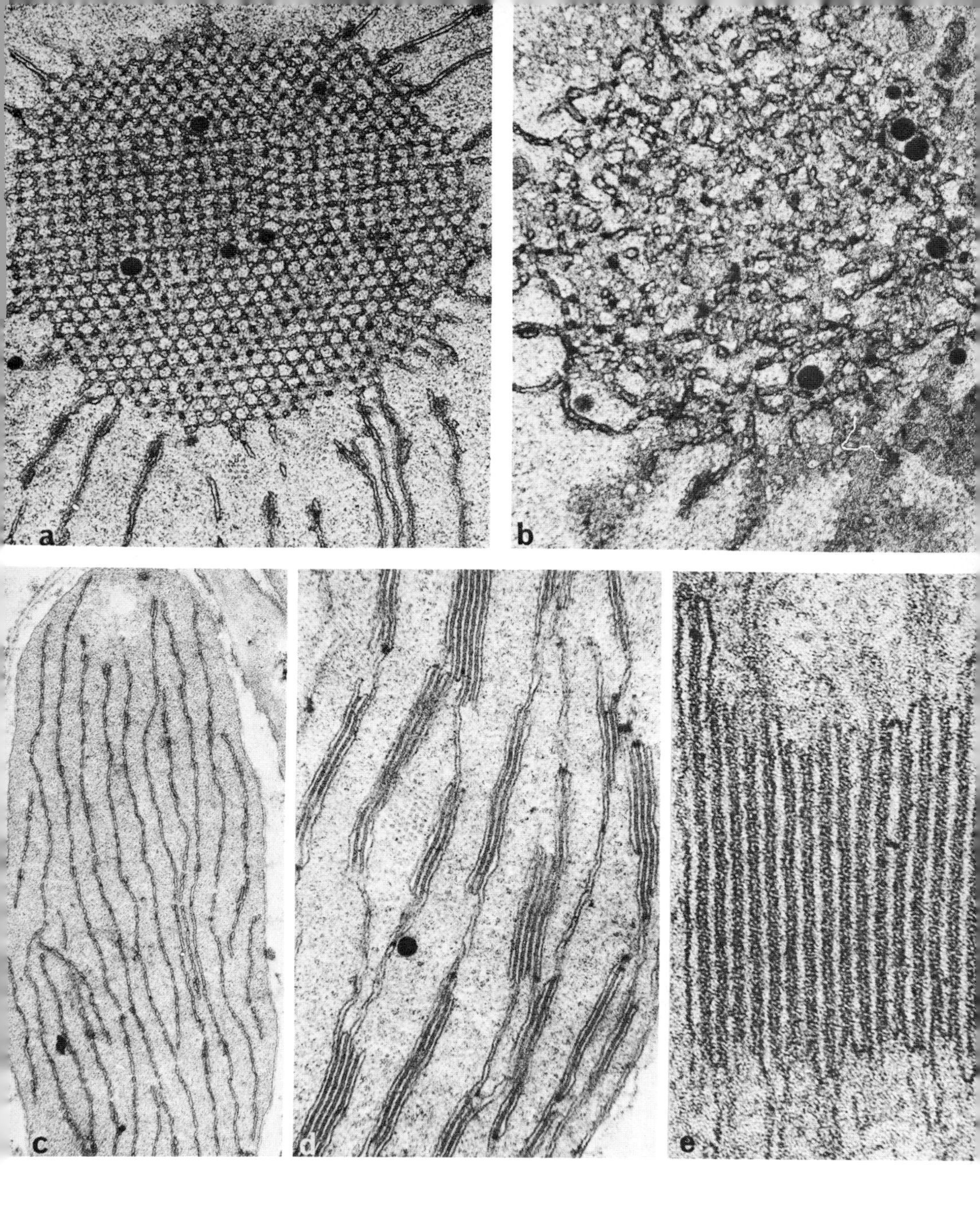

Plate XV. Development of etioplasts in 7-day-old dark-grown barley seedlings illuminated with white light. (a) Prolamellar body in etioplast. 45,000×. (b) Tube transformation: prolamellar body after 1 minute illumination. 59,000×. (c) Dispersal of prolamellar body into primary lamellar layers: after 1 minute illumination followed by 15 minutes in darkness. 22,000×. (d) Formation of grana: after 24 hours continuous illumination. 49,000×. (e) Granum of mature chloroplast. 160,000×. Courtesy K. W. Henningsen, J. E. Boynton, O. F. Nielsen, and D. von Wettstein, Institute of Genetics, University of Copenhagen.

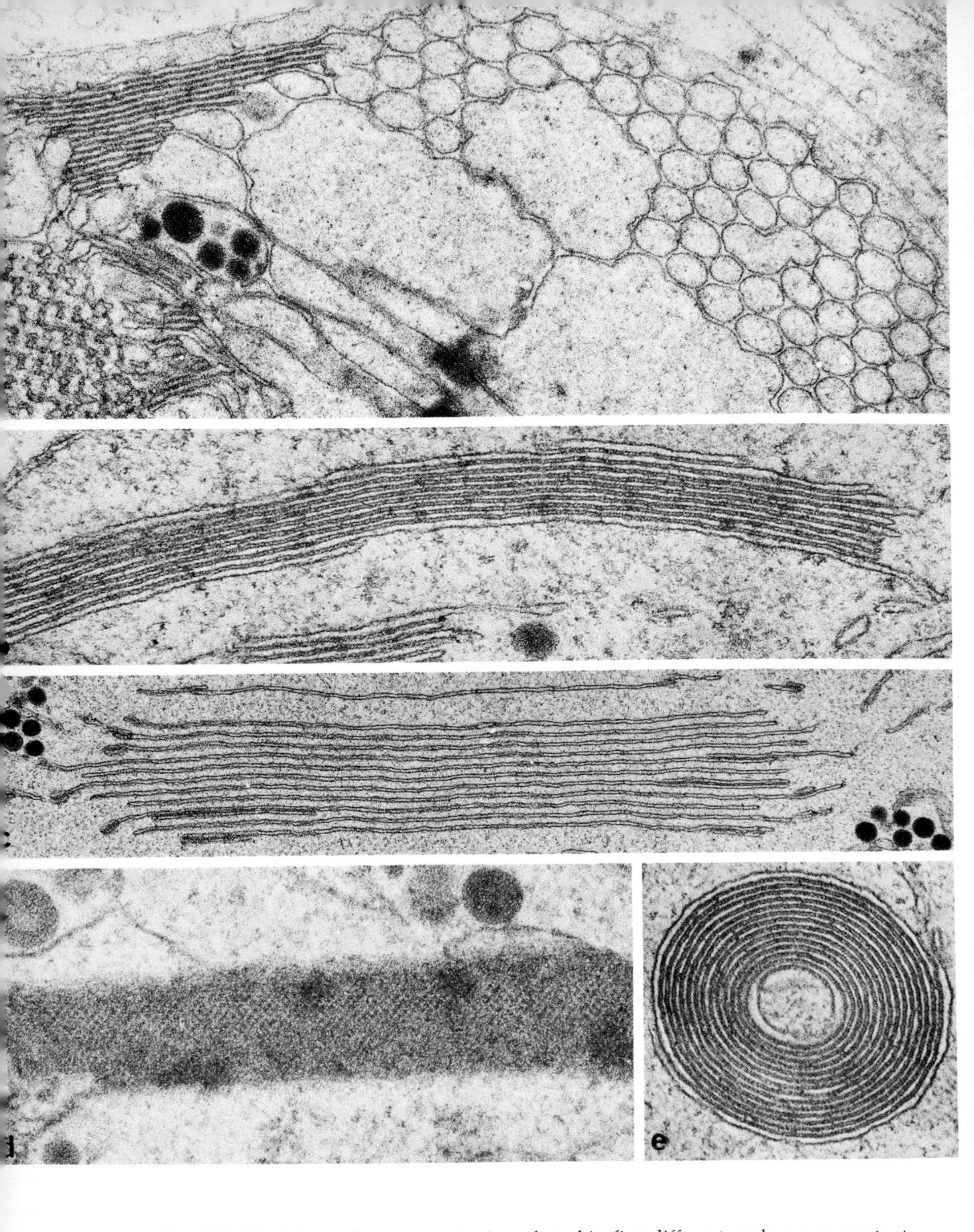

Plate XVI. Plastid membrane organizations found in five different nuclear gene mutants of barley after exposure of dark-grown seedlings to light. (a) *xantha-a*^{63}: Uncontrolled synthesis of chloroplast specific lipids leads to membranes in honey comb configuration together with grana and prolamellar bodies. 46,000×; (b) *xantha-d*^{31}: Membrane discs (thylakoids) aggregate into giant grana. 90,000×; (c) *xantha-f*^{60}: Mutant blocked in chlorophyll synthesis does not aggregate its discs into grana. 38,000×; (d) *infrared-5:* Mutant defective in the regulation of chlorophyll synthesis produces abnormally structured grana. 108,000×; (e) *xantha-b*^{18}: Mutant with spheroidal granum. 64,000×. Courtesy K. W. Henningsen, J. E. Boynton, O. F. Nielsen, and D. von Wettstein, Institute of Genetics, University of Copenhagen.

5

Cytoplasmic Genes in Neurospora and Other Fungi

Studies of cytoplasmic inheritance in the fungi began with the discovery in 1952 of *poky*, a slow-growing mutant strain of *Neurospora* which appeared spontaneously in a stock culture in the laboratory of Mitchell and Mitchell (*23*). They found that the *poky* phenotype was inherited maternally, and that the property of slow growth was associated with the absence of cytochromes a and b of mitochondrial electron transport.

Biochemical and genetic studies of the *poky* mutation and of other slow-growing mutant strains of *Neurospora* have been pursued in a number of laboratories. Some ten to fifteen different cytoplasmic genes have been identified (*2a, 19a, 19c, 33, 36*) and distinguished from one another by their phenotypes and in some cases by complementation in heterokaryons (discussed on p. 150). No single definitive piece of evidence establishes their location in mitochondrial DNA, but all of the available evidence supports this widely held assumption.

Thus, as we shall see in this chapter, the findings with *Neurospora* are consistent with the much more extensive yeast data (Chapter 4) in delineating a mitochondrial genetic system.

In marked contrast to the *Neurospora* work, studies of cytoplasmic inheritance in other fungi have not clearly implicated the mitochondria, nor have the genetic findings been unambiguous. In this chapter we will examine some of these studies, in particular the intensive investigations carried out with *Podospora* (*2, 22*) and with some species of *Aspergillus* (*18*).

These investigations are important because they represent unsolved puzzles in cellular heredity. Whether or not they are examples of cytoplasmic inheritance attributable to organelle DNA is an open question at this time. Two of the principal phenotypes involved are sexual differentiation (i.e., development of protoperithecia and conidia) and aging of vegetative mycelia (senescence). Both are of great intrinsic interest, and it would be important to establish the modes of control of their inheritance.

GENETIC ANALYSIS OF POKY AND RELATED STRAINS OF NEUROSPORA

The life cycle of *Neurospora* is shown in Fig. 5.1. *Neurospora crassa* is heterothallic with two mating types determined by the nuclear alleles A and a. The mycelium originating from a single asexual haploid conidium (of either mating type) gives rise to protoperithecia (fruiting bodies in which fertilization occurs) and to conidia which serve as male gametes. When a nucleus of opposite mating type enters the protoperithecium, a dikaryon is formed, which reproduces mitotically to form many dikaryotic cells. In each dikaryon, the nuclei fuse to produce a zygotic nucleus which will later undergo meiosis. The protoperithecium develops into a mature perithecium, containing many asci, each with one zygotic nucleus. In each ascus, meiosis occurs with the formation of four haploid nuclei, the four meiotic products. In *Neurospora crassa*, each nucleus divides before the ascus matures, liberating eight ascospores. Each nuclear gene pair segregates 4:4 among the eight haploid ascospores.

The first cytoplasmic gene described in *Neurospora* was called *poky* and later *mi-1* (*23*). (The gene symbol *mi* refers to maternal inheritance.) The crosses which established the cytoplasmic and maternal pattern of inheritance are shown in Fig. 5.2. *Poky* was crossed with a wild-type strain which also differed from it by a number of nuclear gene markers. In reciprocal crosses, when the protoperithecial or female parent was *poky*, all the progeny were *poky;* and in the reciprocal cross, when the female parent was wild type, all the progeny were wild type. In a series of backcrosses, the *poky* progeny (as protoperithecial parent) always gave rise only to *poky* progeny.

Subsequently, three other slow-growing strains were isolated and found to be phenotypically similar to *poky: mi-3, C-115,* and *C-117* (*25*). Of these, *mi-3* proved to behave genetically just like *poky* and was

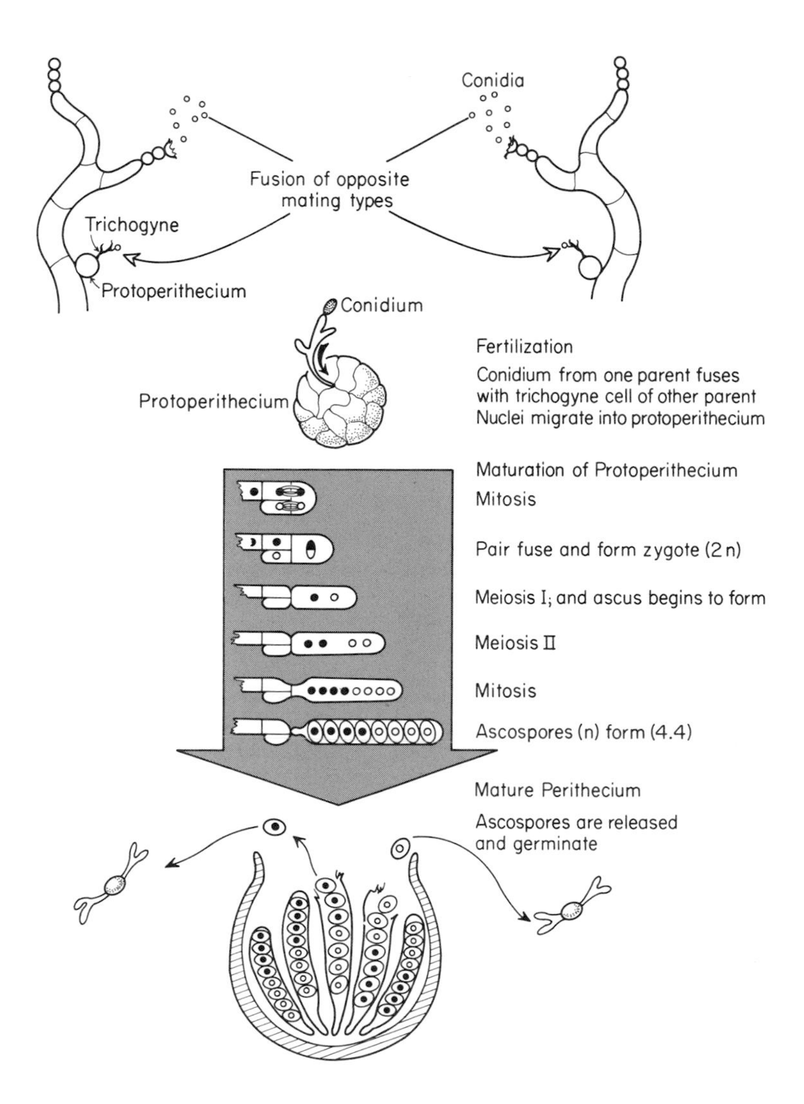

Fig. 5.1. Life cycle of *Neurospora crassa.* Single ascospores of both mating types, *a* and *A*, produce haploid mycelia which differentiate to form protoperithecia (female) and conidia (male). Fertilization takes place by entrance of a nucleus from conidia or mycelium of one mating type into a protoperithecium of the opposite mating type. The resulting dikaryotic cell divides mitotically to form many dikaryotic cells, each of which becomes a zygote when the two haploid nuclei fuse. Each fused nucleus undergoes meiosis within the ascus, producing four haploid nuclei, each of which divides once mitotically. Cell walls are then formed and eight haploid ascospores are released from the ascus. Nuclear alleles, like *A* and *a*, segregate 4:4 in the ascus.

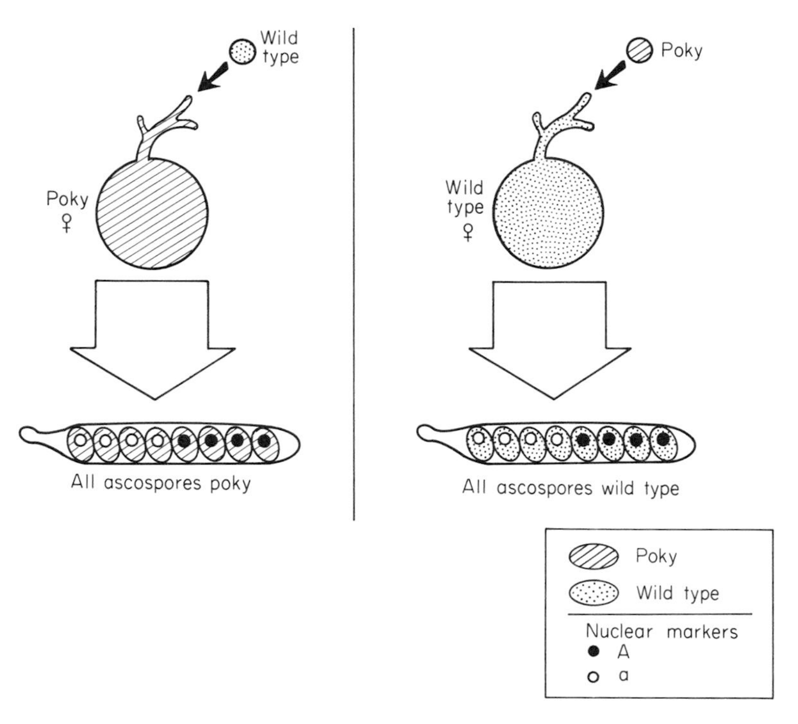

Fig. 5.2. Maternal inheritance of *poky* (*mi-1*). In reciprocal crosses between wild-type and *poky* strains, all ascospores produce *poky* progeny when the protoperithecial parent is *poky;* and wild-type progeny when the protoperithecial parent is wild type. Nuclear alleles used as markers segregate 4:4.

therefore classified as cytoplasmic. The other two strains were found to carry different single nuclear gene mutations each producing a *poky*-like phenotype. These results, of nuclear and cytoplasmic genes affecting the same phenotype are comparable to those found in other organisms, e.g., nuclear and cytoplasmic *petites* in yeast, and nuclear and cytoplasmic genes affecting chloroplast formation in *Chlamydomonas* and in higher plants.

Phenotypically, *poky* was originally identified by its slow growth. Wild-type ascospores complete their growth on slants in 3–4 days while *poky* ascospores under similar conditions require 10–12 days. A chromosomal gene *f* (for *fast-poky*) was later found which had the property of increasing the growth rate of strains carrying *mi-1* without affecting the growth rate of strains carrying *mi-3*. Gene *f* does not interact phenotypically either with *mi-3* or with *C-115* or *C-117*. A suppressor gene restoring wild-type growth to strain *C-115* was also found, and it in turn had no effect on *mi-1*, *mi-3*, or *C-117* (24).

Crosses were carried out to study interactions of the cytoplasmic genes, *mi-1* and *mi-3*, with the nuclear genes, *C-115* and *C-117* as shown in Fig. 5.3. For example, in a cross of *mi-1* × *C-115*, ascospores were recovered at random and backcrossed to wild type (as female parent) to determine whether they were carrying the *C-115* mutation or not. The backcrosses were necessary since phenotypically the progeny were indistinguishable. The backcross progeny no longer carried *mi-1* and were segregating 1:1 for the nuclear mutant gene *C-115*. Thus, although the progeny of the initial cross were all *poky*, half of them were double mutants carrying both a nuclear and a cytoplasmic determinant of slow growth. By backcrossing to a wild-type female, the cytoplasmic and chromosomal factors were separated. These results are comparable with crosses of nuclear × cytoplasmic *petites* in yeast (p. 115).

Neurospora hyphae will undergo fusion to form heterokaryons, mycelia containing two different kinds of nuclei. With cytoplasmic mu-

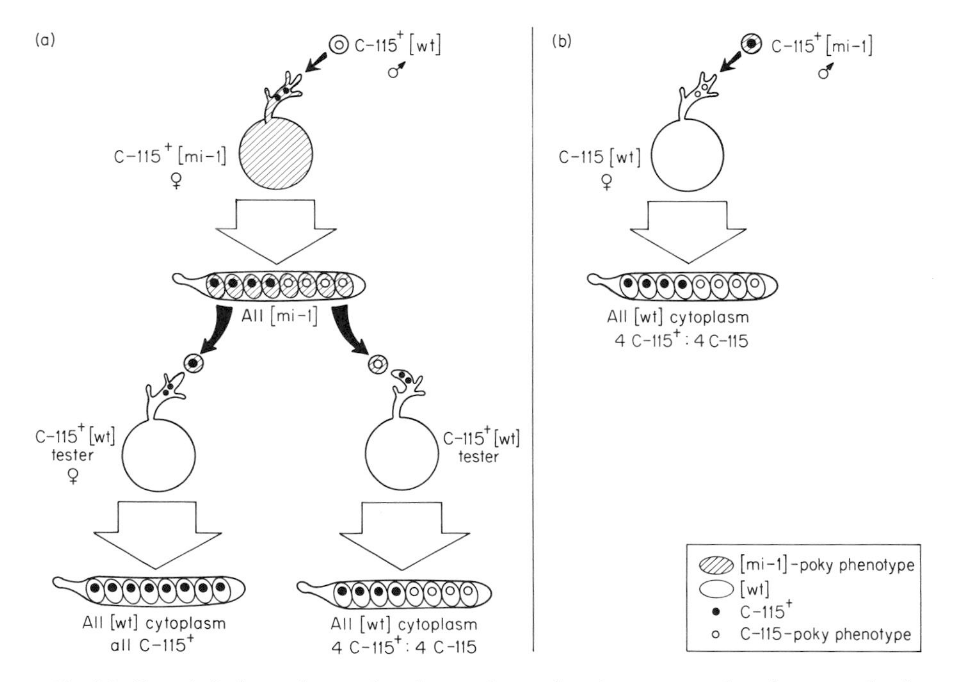

Fig. 5.3. Genetic independence of nuclear and cytoplasmic genes causing slow growth of *Neurospora* mycelia. Reciprocal crosses were made between a nuclear *poky* (*C-115*) and a cytoplasmic *poky* [*mi-1*]. (a) When *C-115*$^{+}$ [*mi-1*] was the female parent, the progeny were all *poky* [*mi-1*] regardless of the nuclear genotype. (b) When *C-115* was female parent, the *mi-1* gene of the male parent was not transmitted and ascospores segregated 4:4 for the nuclear gene *C-115*.

tants, another kind of hybrid mycelium can be constructed carrying different cytoplasmic genomes. Cytoplasmic hybrids of *Neurospora*, called heterocytons or more succinctly cytohets, were first studied by Gowdridge and by Pittenger.

Gowdridge constructed cytohets between *mi-1* and wild type, *mi-1* and *mi-3*, and *mi-3* and wild type (*8*). She found that the *mi-1* wild-type combination always gave wild-type growth rates, but that the *mi-3* wild-type cytohets sometimes gave *mi-3* growth and sometimes gave wild-type growth. Cytohets containing both *mi-1* and *mi-3* cytogenes usually grew at the *mi-3* rate, but sometimes behaved as pure *mi-1*.

Pittenger studied heterocytons between *mi-1* and another cytoplasmic gene which he had isolated, *mi-4* (*26*). The *mi-4* gene differs from those previously studied in that it never produces protoperithecia and therefore can not be used as the female parent in crosses. Conidia produced from *mi-4* mycelia never transmit the *mi-4* property through crosses. This pattern of nontransmission through the male parent is consistent with maternal inheritance, but, in the absence of positive transmission, does not represent a full analysis. It was therefore of particular interest to see whether *mi-4* could be transmitted by mycelial fusion to other strains.

A number of strains were established with *mi-4* cytoplasms. Then, cytohets between *mi-1* and *mi-4* were formed which initially grew at wild-type rate, but subsequently slowed down to growth rates typical either of *mi-4* or of *mi-1*. Thus the two cytoplasmic types showed complementation, i.e., growth at wild-type rate, followed by segregation of the individual types. These results, summarized in Table 5.1 and Fig. 5.4, show clearly that *mi-1* and *mi-4* represent different mutations.

A mutant strain of *Neurospora* called *stopper* (*stp*) arose after several successive exposures of macroconidia to UV irradiation (*19c*). This mutant grows erratically and does not produce protoperithecia. The principal evidence that *stp* is cytoplasmic comes from heterokaryon tests in which the *stp* phenotype was transferred from one nuclear genetic background to another by mycelial fusion. In this system, the relative abnormality of growth varied from one segregant to another, suggesting that the phenotype might be controlled by the ratio of wild-type to mutant alleles of *stp* present.

In further studies of stopper strains, Bertrand and Pittenger (*2a*) isolated several mutant strains (eg., *stp-A* and *stp-B*) of spontaneous origin from continuously growing cultures. In Fig. 5.5, the mutant growth property of two of the *stp* strains has been transferred to a wild-type strain by mycelial fusion. Since neither *stp-A* nor *stp-B* form protoperithecia, the evidence of their cytoplasmic nature was shown by: (*a*) nontransmission of the *stp* phenotype through conidia in sexual

TABLE 5.1

Summary of the Behavior of a Number of Combinations of Three Cytosomes (Normal, mi-4, and Poky) and Two Nuclear Types, Pan and Lys[a]

Genotype			
Nucleus	Cytosome	Phenotype[b]	Stability of phenotype
pan	*poky*	Mutant	Stable
lys	*poky*	Mutant	Stable
pan + *lys*	*poky*	Mutant	Stable
pan	*mi*-4	Mutant	Stable
lys	*mi*-4	Mutant	Stable
pan + *lys*	*mi*-4	Mutant	Stable
pan	*mi*-4 + *poky*	Wild	Becomes mutant
lys	*mi*-4 + *poky*	Wild	Becomes mutant
pan + *lys*	*mi*-4 + *poky*	Wild	Becomes mutant
pan	Normal	Wild	Stable
lys	Normal	Wild	Stable
pan + *lys*	Normal	Wild	Stable
pan + *lys*	Normal + *mi*-4	Wild	Becomes mutant
pan + *lys*	Normal + *poky*	Wild	Becomes mutant

[a] From (26), Table 1.

[b] Phenotype refers to initial growth rate and ability of conidia to form normal colonies on sorbose medium but not to spectroscopic analysis.

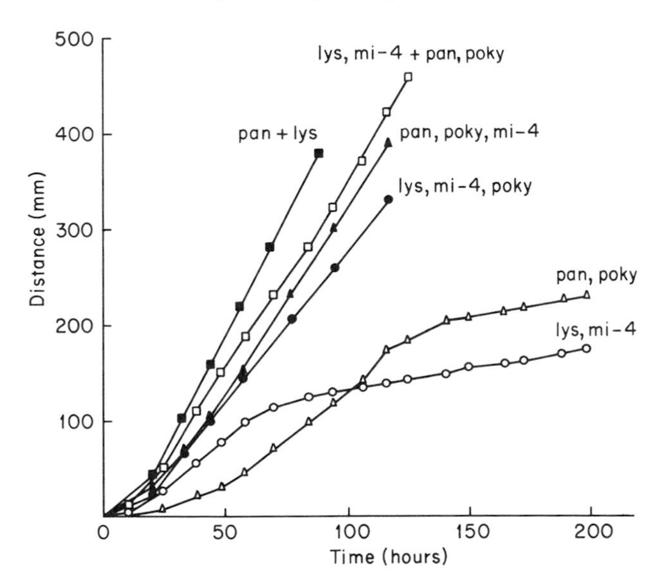

Fig. 5.4. Characteristic growth curves on supplemented medium of identical nuclear types, *pan* and *lys*, with various combinations of normal, *mi-4*, and *poky* cytosomes. From (26).

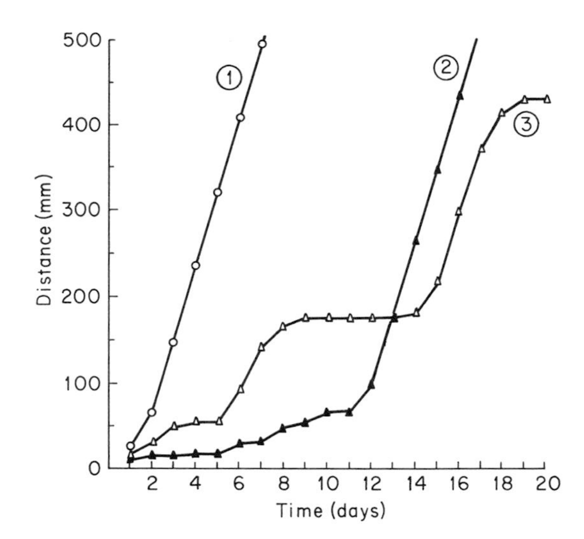

Fig. 5.5. Growth curves showing the advance of the mycelial front as a function of the time during stop-and-go growth in cultures maintained by continuous hyphal elongation. Curve 1 (○) is a wild-type culture and curves 2 (▲) and 3 (△) are two mutant isolates in which (*stp*) cytoplasm was transferred into a wild-type strain by mycelial fusion. From (*2a*).

crosses, and (*b*) transfer of the *stp* phenotype (and gene) from one nuclear homokaryon to another through heterokaryons, as shown in Fig. 5.5. The recovery of mycelia growing at the *stp* rate, out of cytohets containing both *stp* and wild-type alleles, indicates that *stp* may be suppressive, taking over the mycelial growing front from the wild type.

In general, cytohets in *Neurospora* are not stable mixtures; the pure parental types do segregate out during mycelial growth. However, *Neurospora* does not provide a very good system for the quantitative study of cytohets because cell partitions are virtually absent, and nuclei and other organelles intermix freely. Sectors of the kind commonly seen in *Aspergillus,* for example, are rare in *Neurospora.* Consequently, the take-over of a mycelial front by one or another parental genotype is difficult to assess in quantitative terms. The emergence of pure *mi-1, mi-4,* or *stp* hyphae from *Neurospora* cytohets may be comparable to suppressiveness but the hypothesis is difficult to test. It may be that the segregation seen in *Neurospora* cytohets represents an expression of replicative dominance, but no method has yet been devised to study suppressiveness quantitatively in *Neurospora.*

If the *mi* and *stp* genes are located in mitochondrial DNA, then segregation implies a sorting out process at the level of mitochondrial

DNA. If segregation occurs, what about recombination? Cytohets could be used to look for recombination.

Recently, Munkres and his students (*11*) have isolated in *Neurospora* a number of respiratory deficient mutants (R^-) from a selected strain carrying the An^+ nuclear gene for facultative anaerobic growth. These R^- strains also carry other mutations which arose concurrently and affect growth rate (S^-) and mycelial morphology (N^-). In crosses with wild type, the $R^-N^-S^-$ factors segregate with a biparental but non-Mendelian pattern. Preliminary results indicate a polarity of transmission, perhaps like that seen in yeast. The most important aspect of these preliminary findings is the presence of recombinant types. If this system can be developed, it will represent a breakthrough in *Neurospora* mitochondrial genetics, paralleling the recent findings in yeast.

A different approach to examining the stability of the *poky* phenotype was undertaken by Silagi (*31*). She attempted to modify the strain genetically by growing it at the fastest possible growth rate, by introducing the modifier gene *f* and by continual subculture over a long period of time. Strains carrying *mi-1* and gene *f* were maintained in a state of continuous growth by repeated transfer for 180 days. At intervals, samples were taken and allowed to conidiate in order to assess the presence or absence of *mi-1*.

During the first 40–50 days of continuous vegetative growth cultures were found to revert to the *poky* condition on germination. After this time conidia germinated with approximately wild-type lag and wild-type growth rate, but if such cultures were allowed to form protoperithecia and crossed to wild type, the progeny were again *poky* as judged from rate of germination and rate of vegetative growth. This experiment provides an additional kind of evidence that *mi-1* is a true genetic determinant and that its stability is not subject to influence by changes in conditions of growth at the organism level.

Studies of *Neurospora* mutants were carried out by Srb and his students using acriflavine-induced, slow-growing strains (*12, 19a, 27a, 32, 33*). In a summary of their genetic studies (*33*), Srb noted that six *Neurospora* mutants had been distinguished on the basis of clearly different growth patterns, out of a total of fifty-one cytoplasmic variants of independent origin. He suggested that since the means of discrimination were rather crude, the number of mutations isolated may have been much larger than six. This comment points up the difficulty of distinguishing different mutations in the absence of recombination.

The *SG* strains showed the same pattern of maternal inheritance previously reported for *poky* (Fig. 5.2). The *SG* property was transmitted through the maternal or protoperithecial parent and not through the conidial or male parent. The original *SG* parent was backcrossed for twenty generations to a normal *Neurospora* male, thus effectively introducing the male nuclear genetic material completely into the female cytoplasm with no change in the *SG* phenotype. Thus the *SG* phenotype was expressed independently of the nucleus with which the trait was originally associated, and the inheritance of *SG* was shown to be relatively permanent. *SG* was also outcrossed to linkage testers and no evidence was found for association of *SG* with any of the seven nuclear chromosomes of *Neurospora crassa*.

Reciprocal crosses of *SG* with *poky* showed maternal inheritance. When *SG* was the female parent, all progeny behaved like *SG*, whereas in the reciprocal cross in which *poky* was the maternal parent, all the progeny gave growth patterns typical of *poky*.

One of the difficulties in distinguishing different cytoplasmic genes on the basis of their phenotypes is the influence of the nuclear genome on the observations. To minimize this problem, Littlewood (*19a*) transferred the four cytogenes—*SG-1*, *SG-3*, *mi-1*, and *mi-3*—onto an essentially isogenic nuclear background. He then compared the four strains with respect to a number of parameters including linear growth rate, effect of the nuclear gene *f*, optimal temperature for maximal linear growth, and the effect of chloramphenicol on growth and differentiation of protoperithecia.

Littlewood showed that the four cytoplasmic mutations were distinguishable from one another by a number of criteria. The cytogenes *SG-1*, *SG-3*, and *mi-1* were partially suppressed by *f*, whereas, *mi-3* was unaffected by *f*. Young cultures of *SG-3* were much more thermolabile than the other three strains. The optimal temperature for growth of *mi-1* was several degrees lower than that for the other strains. Of particular interest in relation to mitochondrial function was the differential response of these strains to growth in the presence of chloramphenicol. Wild-type and *mi-3* strains were unaffected in their rate of linear growth by 4.0 mg/ml chloramphenicol, whereas the *mi-1* strain was almost totally inhibited by this concentration. *SG-1* and *SG-3* were intermediate in their response to the drug. Some of these results are shown in Figs. 5.6 and 5.7.

In a further study of the heritability and stability of *SG*, Infanger and Srb (*12*) reported studies of heterokaryons produced by fusion of hyphal tips between wild-type and *SG* strains. They found that the cytoplasmic traits, *SG* and wild type, were mutually exclusive, with either *SG* or wild type taking over, irrespective of the nuclear constitution of the mycelium. These results are similar to those of Gowdridge (*8*) and of Pittenger (*26*) described above.

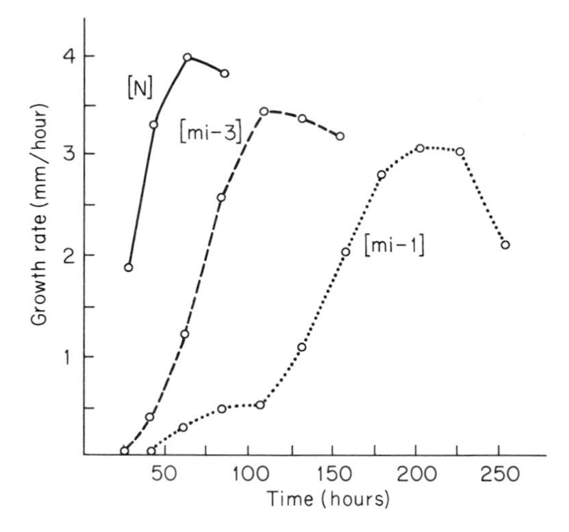

Fig. 5.6. Growth curves for three isogenic strains of *Neurospora:* wild type (*N*); mutants (*mi-1*) and (*mi-3*) grown on minimal medium. From (*19a*).

Puhalla and Srb (*27a*) carried the analysis of *SG* cytohets one step further by improving the technique for observing the movement of nuclear and cytoplasmic genes after mycelial fusion. They found that the nuclei in these heterokaryons moved faster than did the *SG* genes. They also found that *SG* could be identified by the acriflavine-sensitivity it conferred in contrast to the wild type.

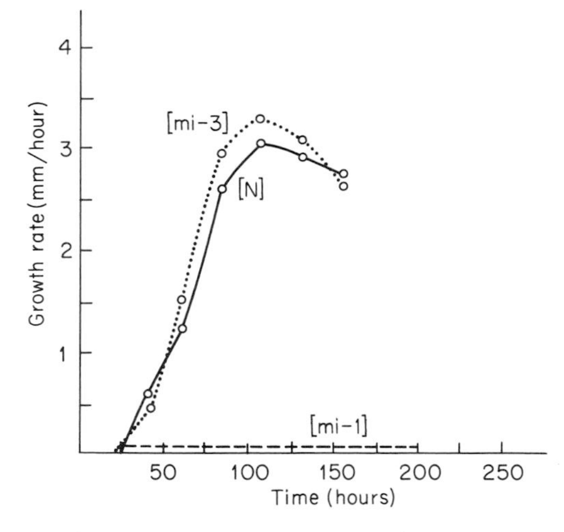

Fig. 5.7. Growth curves for three strains grown on minimal medium plus chloramphenicol (same strains and same experiment as Fig. 5.6). From (*19a*).

A particularly interesting cytoplasmic mutant, *ac-7*, has been described by Srb in which the pattern of genetic transmission is not maternal (*32*). Reciprocal crosses of *ac-7* × wild type gave only wild-type progeny, regardless of the polarity of the cross. Considered alone, this result has numerous interpretations. However, the situation is considerably clarified by further crosses made between *ac-7* and *SG*, as shown in Fig. 5.8. When *SG* is the female parent, all the progeny are *SG*, as in

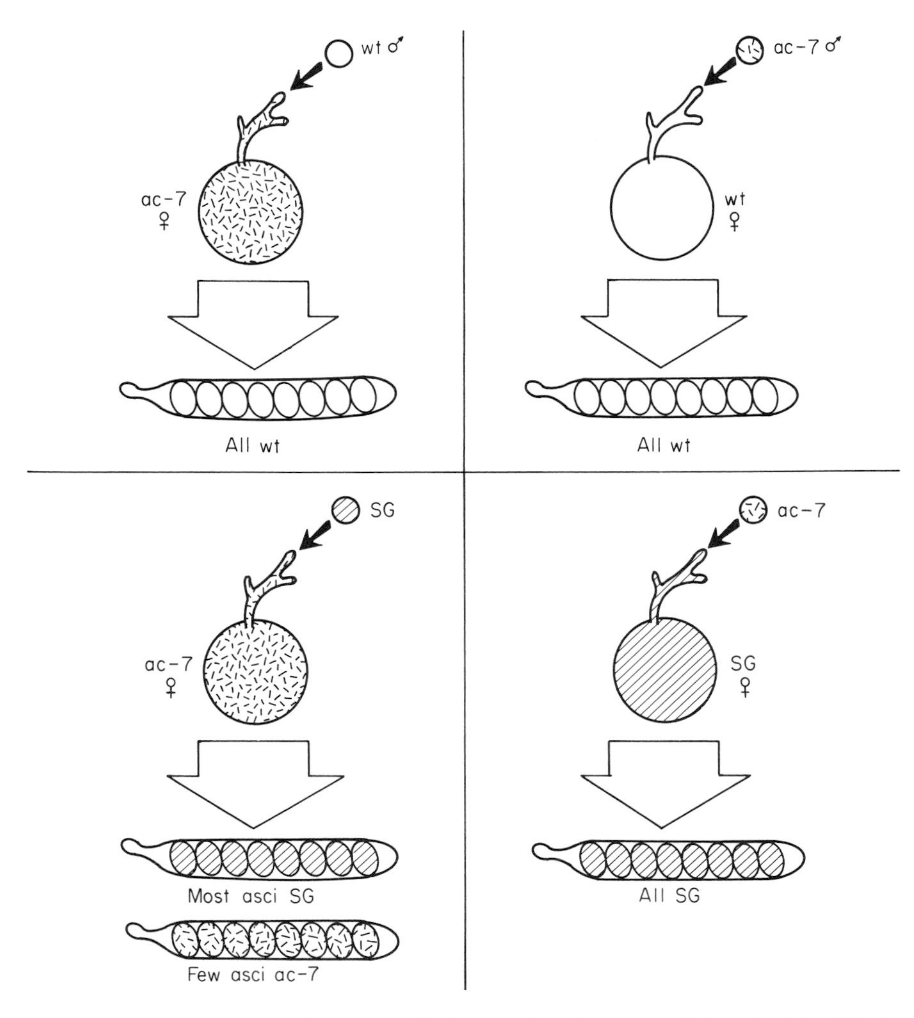

Fig. 5.8. Results of reciprocal crosses with the cytoplasmic mutant allele *ac-7*. This allele is not transmitted by the conidial and rarely by the protoperithecial parent. Based on (*33*).

typical maternal inheritance of *SG*. When *SG* is the conidial parent, however, most of the progeny are *SG*; a minority are *ac-7*.

These results suggest that *ac-7* is a permissive female parent, accepting the cytoplasmic genes contributed by the conidial (male) parent. However, the common interpretation of maternal inheritance in *Neurospora* has been that no cytoplasmic genes are contributed by the conidia, which like sperm, contain very little cytoplasm. The occurrence of paternal inheritance in these crosses is good evidence that the conidia do transmit cytoplasmic genes in crosses.

If this interpretation is correct, then a further mechanism is required in normal crosses to eliminate the male genome. Evidence for such a process has been discovered in *Chlamydomonas*, and was discussed in Chapter 3. The possibility that all organisms contain special systems to distinguish cytoplasmic genetic contributions of the two parents and to eliminate one of them finds further support in the finding of polarity of transmission in yeast mitochondrial DNA and in recent studies with the higher plant *Pelargonium*, to be discussed in Chapter 6.

In addition to these genetic studies of cytoplasmic mutations, the study of mitochondrial heredity was approached directly by Reich and Luck (*28*), using the buoyant density of mitochondrial DNA itself as the phenotype to be observed. *Neurospora crassa*, the species used in most genetic studies, contains two mitochondrial DNA's, one banding in CsCl density gradients at 1.698 and the other at 1.702 gm/cm^3. The related species, *N. sitophila*, has three mitochondrial DNA's: 1.692, 1.698, and 1.702 gm/cm^3. In a cross of *N. crassa* female × *N. sitophila* male, the mitochondrial DNA's from four sets of ascospores were examined in CsCl gradients as shown in Fig. 5.9. None contained any DNA of density 1.692 gm/cm^3, a result suggesting maternal inheritance of this DNA. In the reciprocal cross of *N. sitophila* female × *N. crassa* male, the results were less definitive: one set of ascospores contained the DNA of density 1.692 gm/cm^3 in all four progeny, while the other three sets contained no DNA of this density at all. Thus, the unique DNA of *N. sitophila* was not regularly transmitted in this cross. Although the results are difficult to assess, the method is interesting, because of the direct examination of the inheritance of mitochondrial DNA itself.

In summary, the genetic experiments reviewed in this section provide unambiguous evidence of a cytoplasmic genetic system in *Neurospora*. Are the various *mi* and *SG* mutations located in mitochondrial DNA? There is as yet no definitive evidence on this point, but biochemical studies to be described now show that some of these mutations alter mitochondrial function.

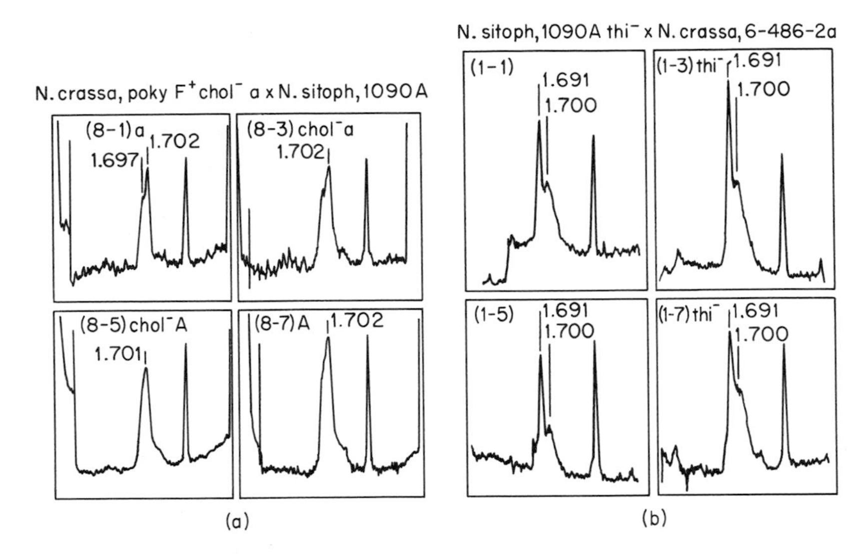

Fig. 5.9. Equilibrium density gradient centrifugation in CsCl of mitochondrial DNA isolated from *Neurospora* cultures derived from one member of each spore pair from single asci. The reference marker is bacteriophage SP-8 DNA. From (*28*). (a) *N. crassa* (♀) × *N. sitophila* (♂) (b) *N. sitophila* (♀) × *N. crassa* (♂).

BIOCHEMICAL STUDIES OF POKY

The phenotypic differences between wild-type and *mi* mutants were studied by Mitchell and collaborators over a period of several years (*10, 10a, 25, 37*). Initially they found that the slow-growing strains, both cytoplasmic and chromosomal mutants, had altered cytochrome activity, as shown in Fig. 5.10 (*37*). All four mutants lack cytochrome a. Strains carrying *mi-1* have an excess of cytochrome c and no detectable cytochrome b, whereas strains carrying *mi-3* have an excess of cytochrome c, some cytochrome b, and no a. (In subsequent studies, a small amount of a_1 was detected.) The cytochromes of *C-115* are similar to *mi-3*, whereas *C-117* has no cytochrome c, but does contain b.

Strains containing both nuclear and cytoplasmic mutations showed complex interactions which have not been fully analyzed. For example, the double mutant containing *C-117* and *mi-3* has a cytochrome spectrum very similar to that of the nuclear *C-117* and shows no effect of *mi-3*. On the otherhand, a double mutant containing *C-115* and *mi-1* has a spectrum like that of *mi-1* and one sees no effect of *C-115*.

The cytochrome patterns were correlated with studies of the respiratory enzymes, in particular succinic acid oxidase, cytochrome oxidase,

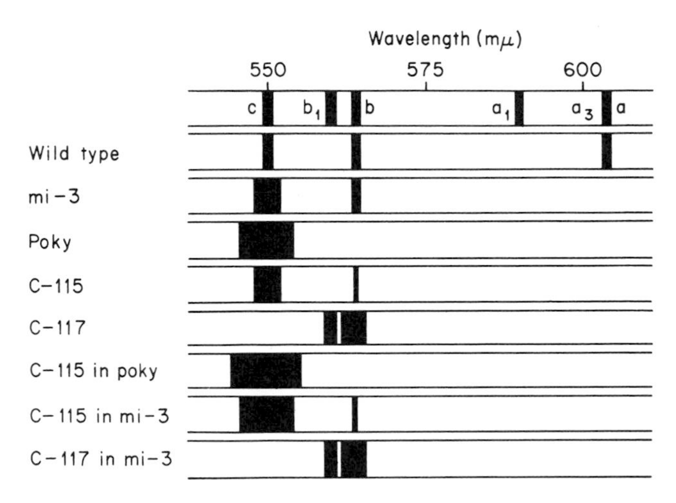

Fig. 5.10. Diagrammatic representation of cytochrome absorption bands of normal and mutant strains of *Neurospora*. Relative concentrations of cytochromes are indicated by the width of the bands, e.g., the ratio of cytochrome c concentration in wild type, *mi-3* and *"poky"* is approximately 1:5:15. From (*25*).

and succinic dehydrogenase. As shown in Fig. 5.11, correlated with the slow growth of *mi-1* is slow appearance of a low amount of cytochrome oxidase activity and the subsequent appearance of succinic oxidase activity. Succinic dehydrogenase is present in *poky* strains, and is initially more active than in the wild type. It is partially destroyed during growth of the *poky* mycelia. The deficiencies of cytochromes a and b seen in the absorption spectra correlate well with the low activities of cytochrome oxidase and of succinic acid oxidase.

Respiration in young *poky* mycelia does not depend upon the cytochromes; there is another terminal oxidase system which is cyanide- and azide-insensitive. Strains carrying *mi-3* have wild-type oxidase activity in cell-free preparations, and this activity correlates with the presence of cytochrome a_1 instead of $a + a_3$.

A number of studies were carried out in an attempt to analyze and understand the cytochrome pattern in the *poky* mutant. Hardesty and Mitchell (*10*) found that 3-day-old *poky* cultures in which the *poky* phenotype is most extreme contain about twenty times the wild-type concentration of free fatty acids and of cytochrome c, as well as a great excess of total lipid and phospholipid. Most of the lipid is in the form of free fatty acids, principally linolenic, linoleic, and palmitic acids.

Edwards and Woodward (*5*) have reported a difference in the absorption spectra of cytochrome oxidase preparations from wild-type and

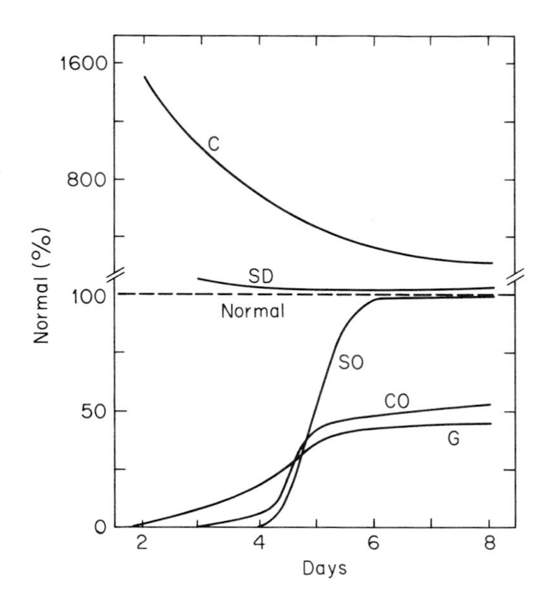

Fig. 5.11. Enzyme activities of *Neurospora* "*poky*" (*mi-1*) mutant during growth. Diagram shows relative growth (G), cytochrome c content (C), succinic dehydrogenase activity (SD), succinic acid activity (SO), and cytochrome oxidase activity (CO) of "*poky*" as compared to wild type. The latter is represented by the horizontal dotted line at the level of 100%. From (*10a*).

poky mitochondria, as shown in Fig. 5.12. The interpretation is complicated by the lack of a definitive means of identifying the apoprotein. *Poky* membranes from young cultures are deficient in cytochrome oxidase activity, but whether they lack the apoenzyme is still unresolved.

In a recent study of mitochondrial ribosomes from *poky* (*mi-1*) and wild-type strains of *Neurospora,* Rifkin and Luck (*29*) found a striking deficiency of monomers in mitochondrial lysates. As shown in Fig. 5.13 and Table 5.2, young *poky* mycelia contain principally the larger subunit of the mitochondrial ribosome, little or none of the smaller subunit, and a small amount of the monomer. Older *poky* mycelia, with increased cytochrome activity were found to contain more of the monomeric form than did the young (20-hour) cultures.

The authors suggest that the cytochrome defect in *poky* is a direct consequence of the ribosome deficiency. The deficiency is a partial one, from which the *poky* strains may recover almost completely at the end of growth. Thus it would seem likely that the underlying mutation involves a regulatory rather than a structural gene. An analogous mutation leading to a decrease in the number of chloroplast ribosomes has

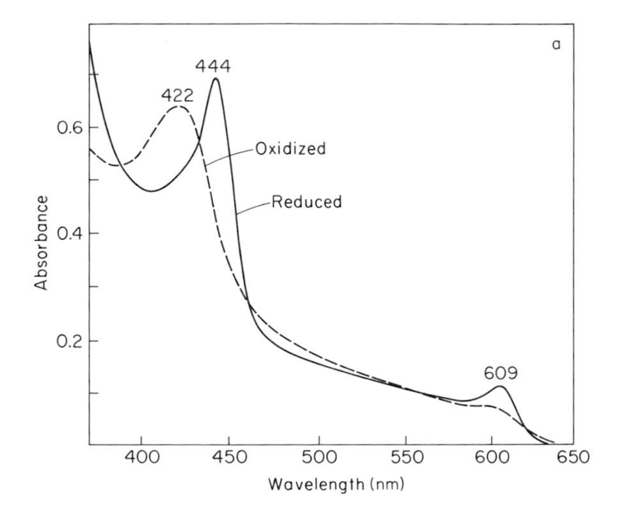

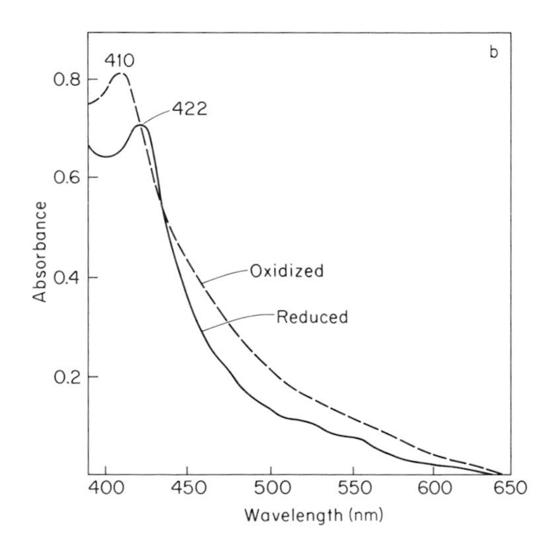

Fig. 5.12. Comparison of cytochrome a spectra of purified enzyme preparation from (a) wild type, and (b) "*poky*" (*mi-1*) mutant strains of *Neurospora crassa*. From (*5*).

been described in *Chlamydomonas* in which the regulatory gene *ac-20* is nuclear (cf. Chapter 8).

Thus, despite a great deal of effort expended over many years to establish the primary lesion responsible for the *poky* phenotype, that goal has not yet been reached; the primary action of the *poky* cytogene remains unknown.

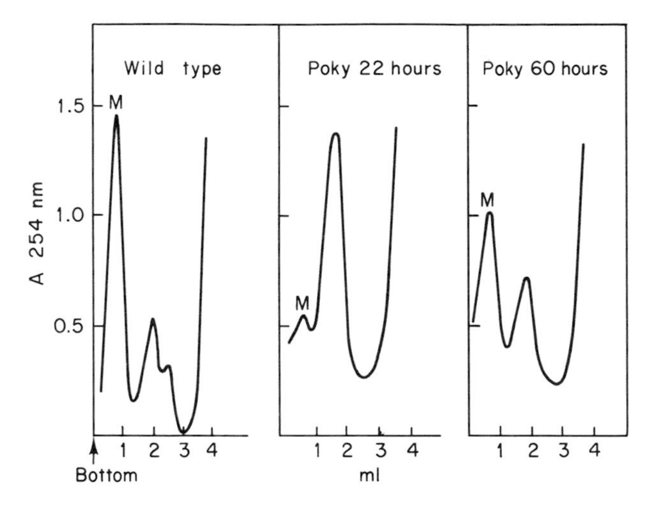

Fig. 5.13. Sedimentation profiles of mitochondrial lysates from *Neurospora.* The M peak corresponds to the monomeric form of mitochondrial ribosomes, the other peaks representing the large and small subunits. From (*29*).

STUDIES OF CYTOPLASMIC MUTANTS BY MICROINJECTION

Two other slow-growing strains of *Neurospora* have been described carrying mutations called *abnormal-1* (*abn-1*) and *abnormal-2* (*abn-2*) (*4, 7*). Both mutations were detected during routine subculture of laboratory stock cultures. Slow growth was found to be irreversible; faster rates could not be induced by subculture or by addition of special components to the medium.

The mutant strains contain higher than normal amounts of cytochrome c, indicating some alteration in the regulation of cytochrome synthesis, but respiratory rates are normal. Neither strain produces protoperithecia, but they do form conidia, and therefore can function as males but not as females in crosses.

When the mutants were used as the conidial parents in crosses with the wild type, all the progeny were wild type. Thus, further tests were necessary to establish the cytoplasmic and particulate nature of the mutations.

Heterokaryons were formed between strains carrying either *abn-1* or *abn-2* and strains of the opposite mating type with a wild-type cytoplasm carrying nuclear biochemical markers. In a few instances, slow-growing cultures, like the mutants, were obtained from the original heterokaryon and from the conidia produced by it. These conidia

TABLE 5.2

Inheritance of Mitochondrial Cytoplasmic Content and Ribosomal Monomer to Large Subunit Ratio[a]

	Cytochromes (nmoles/mg mitochondrial protein)			Ribosomal ratio
Strain	a	b	c + c_1	monomer/ large subunit
Wild type	0.40	0.84	1.07	(4.1–7.4)[b]
Poky (22 hour)	0.01	0.30	1.60	0.27
Poky (60 hour)	0.03	0.92	1.38	1.25
Wild type ♀ × *poky* ♂[c]				
6-2	0.49	0.99	1.42	5.74
6-4	0.57	0.89	1.03	5.90
6-6	0.49	0.76	1.44	4.34
6-8	0.59	0.79	1.27	4.11
Poky ♀ × wild type ♂				
early 2-1	0.01	0.44	1.97	0.78
2-3	0.01	0.25	2.14	0.69
2-5	0.01	0.30	1.61	0.27
2-7	0.01	0.44	1.70	0.25
late 2-1	0.17	1.09	2.25	2.25
2-3	0.14	1.05	1.22	1.75
2-5	0.09	1.28	2.20	0.94
2-7	0.01	1.74	2.54	0.47

[a] From (*29*), Tables 1 and 2.
[b] Although in wild type this ratio is generally between 4 and 7, it may be as low as 2.7.
[c] Reciprocal crosses were made and ascospores were isolated in order. Cultures derived from one member of each spore pair were studied. Since *poky* partially recovers during its growth cycle, early and late cultures of ascospore isolates with an equivalent mass to the parent *poky* strain at 20–22 hours and 60 hours of growth were examined.

contained nuclei derived from the other parent. Thus the heterokaryon test indicated the presence of a cytoplasmic determinant.

A more refined analysis was made possible by the development by J. F. Wilson of a technique for microinjection of cytoplasm from one mycelium into another (*38, 38a, 39*). In the studies with *abn-1* and *abn-2*, cytoplasm was injected from the mutant into one or two cell sections of wild-type mycelium, after which the injected cells were excised from the mycelium and grown separately on agar.

In a few successful experiments with *abn-2*, it was established that the artificial heterokaryons and the natural heterokaryons serving as controls behaved in very similar fashion, requiring several subcultures to establish the onset of abnormal symptoms. In one instance, transfer of *abn-2* cytoplasm was achieved without the transfer of any nuclei from the *abn-2* strain. Reciprocal experiments were also carried out in which normal cytoplasm was injected into mycelia of *abn-2*. Four out of seven such experiments led to the establishment of mycelia growing at wild-type rate.

The microinjection technique makes possible the injection of mitochondria from one mycelium into another and experiments of this type were attempted by Diakumakos *et al.* (*4*) with strains carrying *abn-1*. Their results provide evidence of changes in wild-type mycelia induced by injection of *abn-1* mitochondria which had been purified by banding in sucrose gradients. Three injected cultures showed a decrease or loss of cytochrome a activity as judged from oxidation–reduction difference spectra. These results suggest the transfer by injection of a genetic determinant affecting mitochondrial function.

Attempts were made to analyze the process of conversion of normal to abnormal cultures in greater detail. In unpublished studies which were briefly summarized in a review paper (*36*), the conversion process was followed during many successive conidial subcultures. As shown in Table 5.3, the shift from wild-type to *abn-1* type growth occurred slowly over a period of many conidial subcultures. Thus the mitochondrial microinjection procedure is at best qualitatively successful in transferring a determinant which competes successfully with the wild type, but the system does not lend itself to quantitative or kinetic analysis.

TABLE 5.3

Plating of Conidia from Serially Transferred Cultures[a,b]

Injected cultures, transfer number	Plating efficiency (%)	Growth character of conidial subcultures			
		Wild type	Intermediate	Very slow	Failed to grow
10	100	30	20	–	–
12	94	28	17	5	–
13	91	17	25	8	–
14	62	5	1	29	15
15	42	–	3	15	32

[a] From (*36*) Table 1.

[b] Conidia obtained from a hyphal segment injected with *abn-1* mitochondria were transferred serially in agar slants. From the transfer numbers indicated, small inocula were used to induce conidia formation. Conidia were plated on sorbose medium, and fifty colonies were selected for transfer to agar slants for determination of their growth characteristics.

These results are in line with the studies of cytohets discussed earlier in this chapter. Taken together all of these studies strongly indicate the existence of mitochondrial genes in *Neurospora* which are capable of mutating and competing successfully with the wild type for control of the phenotype. Superficially at least, the behavior of *Neurospora* cytohets resembles suppressiveness in yeast. Here too, replicative dominance may involve competition at the level of mitochondrial DNA. However, this possibility remains untested as yet in *Neurospora.*

STUDIES OF CYTOPLASMIC HEREDITY IN PODOSPORA

Two of the most intriguing of the unsolved problems in cytoplasmic genetics were described by Rizet *et al.* (*30*) during their studies of *Podospora*, an ascomycete related to *Neurospora.* Two phenomena, one called barrage and the other called senescence, were each shown to involve a cytoplasmic determinant. The two systems were intensively investigated by Rizet's students [barrage by Janine Beisson-Schecroun (*2*) and senescence by Denise Marcou (*22*)].

Barrage is a phenomenon of cellular incompatibility, regulated in *Podospora* by a nuclear gene with two alleles, *S* and *s*. When haploid mycelia of the two genotypes, *S* and *s*, meet, fusion of hyphal tips occurs, but the fused elements then disintegrate leaving a cleared space, called a barrage. Thus heterokaryons cannot be formed between the two genotypes. However, crosses can be made by sprinkling microconidia from either strain onto fruiting bodies (protoperithecia) of the other strain. In these crosses, most if not all of the cytoplasm comes from the female (protoperithecial) parent.

In crosses of $S \times s$, the ascospores of each resulting ascus segregate $2S:2s^*$. The symbol s^* refers to a new phenotype, characterized by the loss of incompatibility. The s^* strain does not form a barrage with either *S* or with *s*; it fuses freely with both. As shown in Fig. 5.14, s^* exhibits maternal inheritance in crosses with *s*. All the progeny of such crosses resemble the female (protoperithecial) parent whichever way the cross is made. On the contrary, in crosses of $S \times s^*$, the results are as in the initial crosses: segregation of $2S:2s^*$, regardless of the sexual polarity.

The problem posed by barrage is twofold: What is the genetic difference between *s* and s^*, and what is the molecular identity of the determinant responsible for this difference? Is s^* an altered form of the *s* gene; or is *s* unchanged in s^* strains, with s^* signifying a change in a cytoplasmic gene which interacts with the *S* locus in regulating cellular incompatibility?

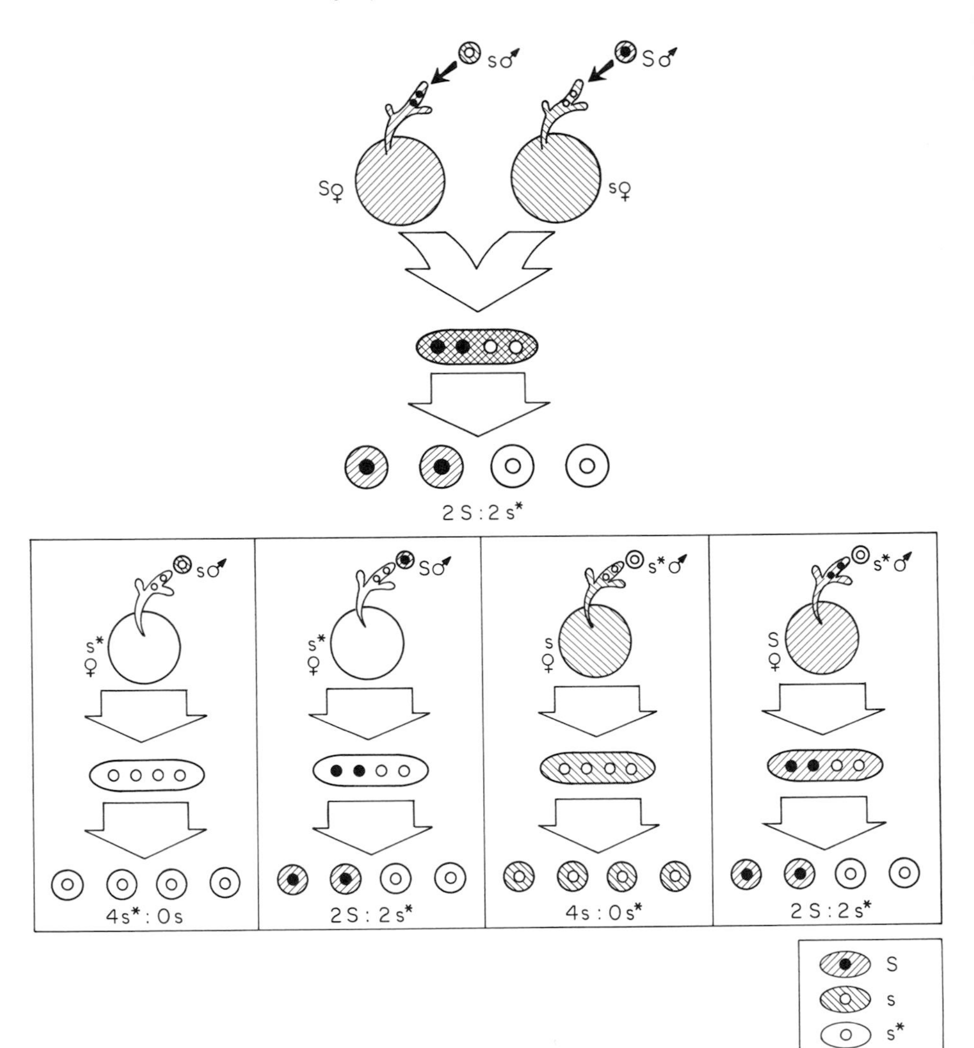

Fig. 5.14. Maternal inheritance of the barrage phenomenon in *Podospora.* In reciprocal crosses of *S* × *s*, the ascospores segregate 2*S*:2*s**. The parental *s* is replaced by a new phenotype, *s**, which has lost its incompatibility and does not form barrage with *S* or with *s*. The new phenotype shows maternal inheritance in crosses with *s* but not with *S*. See text for discussion. Based on (2).

To explore the relation of *s** to *s*, conditions were examined that are favorable for the conversion of *s* to *s** and conversely for the conversion of *s** back to *s*. Of prime importance was the finding that the *s* strain is infectious. In mycelial fusions between *s* and *s** strains, infection begins in as short a time as 15 minutes, always with the conversion of *s** back to *s*. The rate of this conversion was found to be at least ten times that of growth; it does not involve nuclear migration. The conversion is temperature dependent, being propagated almost twice as fast at 24°C as at 18°C, and substantially faster in young mycelia than in old ones, and similarly faster in mycelia growing on a rich medium than on a poor one.

These results suggest that the *s* mycelium contains a cytoplasmic factor which is not present in *s** mycelia, or is present in an altered form unable to compete with the original *s* factor. Experiments indicating competition between cytoplasmic factors from *S* and from *s* mycelia were carried out by grafting them both onto *s** mycelia, as shown in Fig. 5.15. The rate of conversion of *s** to *s* was found to be sharply decreased by the presence of cytoplasmic factors from *S*, although the *S* factor itself was not infectious in *s** mycelia.

Are these cytoplasmic factors just products of the nuclear *S* locus, or do they represent products of autonomous cytoplasmic genes? The hypothesis of genetic autonomy is strongly supported by the maternal pattern of inheritance of *s** × *s* crosses, and by the infectivity of the *s* factor.

This interpretation is further supported by evidence of discontinuities in the transmission of the *s* factor through microconidia. In *Podospora*, the microconidia, which act as males in crosses, may sometimes be induced to germinate asexually and produce a mycelium. Microconidia from *s* strains occasionally give rise to *s** mycelia. This observation is evidence that *s* is converted to *s** by the loss of an autonomous cytoplasmic gene. On the other hand, spontaneous reversion from *s** to *s* occurs with a rate estimate at about 10^{-7}, which is in the range of conventional mutation rates. The fact that *s** can revert spontaneously to *s* is evidence that the genetic difference between *s* and *s** is not an irreversible genetic loss.

Unfortunately, experiments with this system were restricted to the cellular level, and the infectious agent was never identified. These studies were carried out principally in the 1950's, before modern methods to examine cytoplasmic DNA's were available. The system might be a fruitful one to reexamine, particularly in the light of the ease and rapidity of infectious conversion of *s** mycelia to *s*. The system is also of interest in comparison with the phenomenon of senescence.

Senescence in fungi, i.e., the gradual loss of viability of cultures under continuous cultivation, appears to involve a cytoplasmic genetic component. In *Podospora*, different strains vary in longevity from 9 to 106 days of continuous growth on agar at 26°C. Elevated temperatures, continuous light, and frequent subculture all accelerate the appearance of senescence.

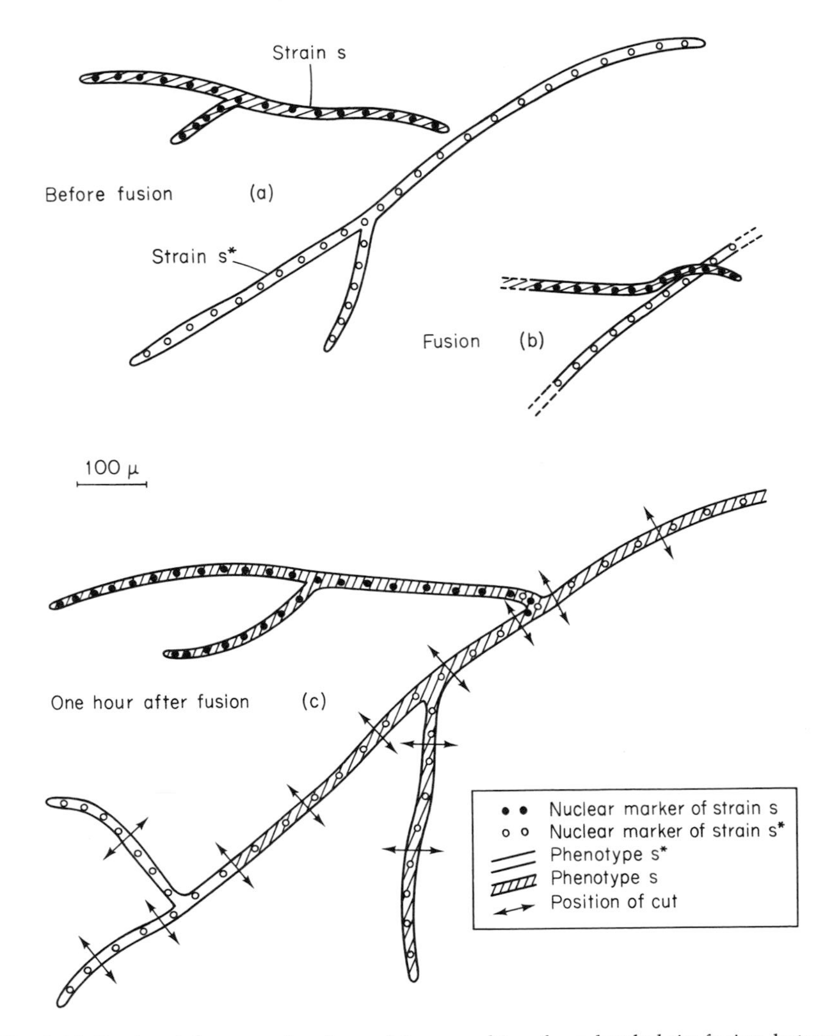

Fig. 5.15. Study of the reversion from s^* to s resulting from hyphal tip fusion between mycelia of the two types: s and s^*, each with different nuclear markers. Extent of transformation of s^* mycelium to s 1 hour after fusion is shown in (c) based upon sections (indicated by arrows) which were cut and subcultured to assay phenotype. From (2).

The inheritance of senescence was studied by means of reciprocal crosses between normal cultures and senescent ones. Fortunately, morphologically senescent cultures will still form fruiting bodies and microconidia, and consequently suitable crosses can be made. As shown in Fig. 5.16, senescence is sexually transmissible through the female parent but never through microconidia. However, senescent females give rise to both senescent and normal progeny. In some strains, entire peri-

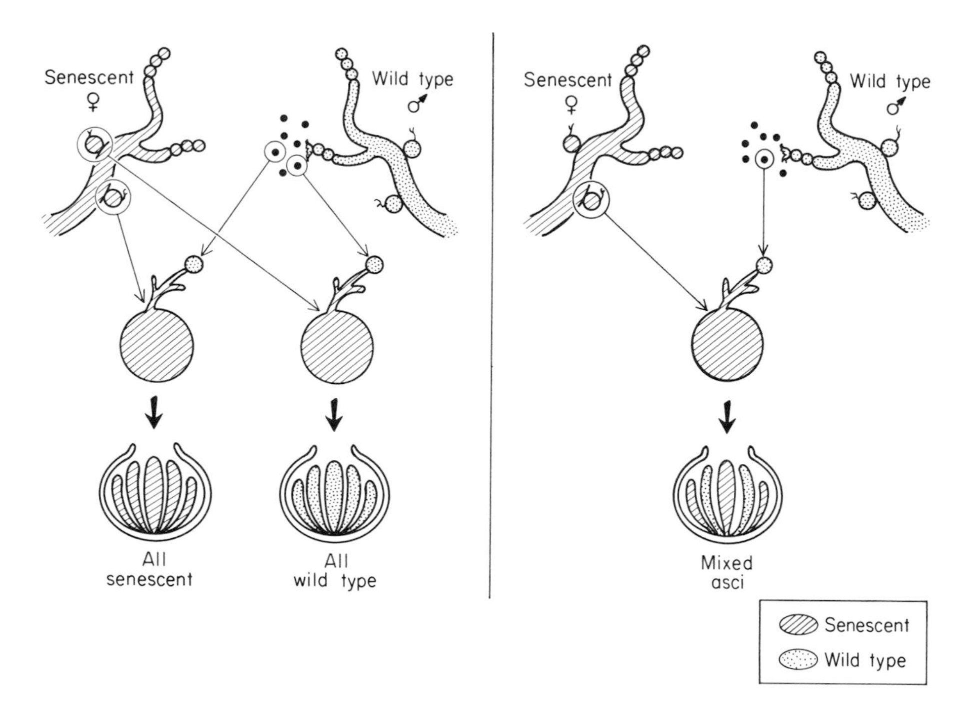

Fig. 5.16. Inheritance of senescence in *Podospora*. Protoperithecia from senescent mycelia fertilized by normal conidia give rise to two classes of perithecia: (left) homogeneous perithecia in which all ascospores of all asci are uniformly either senescent or normal; and (right) heterogeneous perithecia in which some asci contain all senescent ascospores and other asci contain all normal ascospores. Segregation within asci is very rare. Based on (22).

thecia from senescent females produce either all normal or all senescent progeny. In other strains, the perithecia are heterogeneous, containing both normal and senescent asci, each producing all normal or all senescent ascospores. Rarely is any segregation seen within asci. Thus, senescence shows maternal inheritance, but with discontinuities: senescent mycelia do not always give rise to senescent progeny.

Discontinuities are also seen in the vegetative transmission of senescence. Mycelia can be subcultured by fragmentation in a blender followed by growth of new mycelia from the fragments. Fragmentation of senescent mycelia leads to the emergence of some healthy hyphae as well as some senescent ones.

Senescence is infectious in mycelial grafts; senescent hyphae make healthy hyphae sick. An experiment demonstrating infectivity is shown in Fig. 5.17. The conversion is not accompanied by nuclear migration, but it is a slow process compared to the infectivity rates seen in the barrage system.

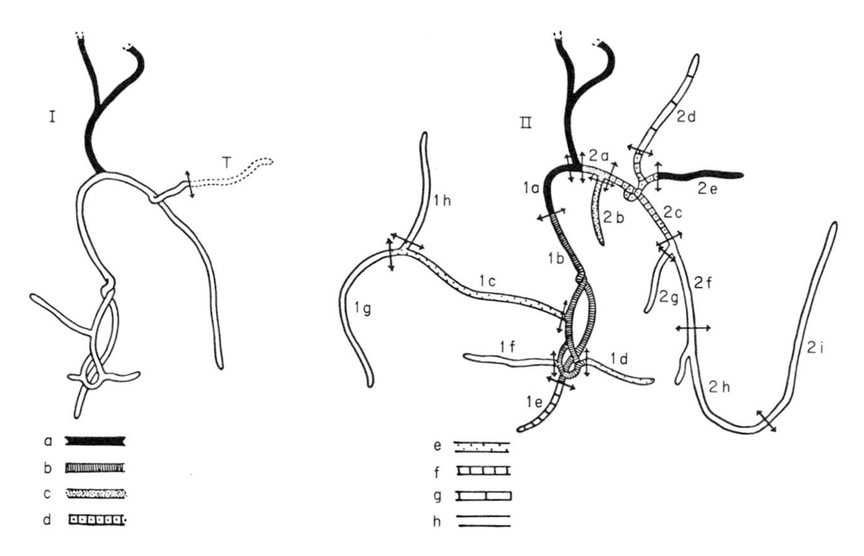

Fig. 5.17. Results of infection of a normal mycelial filament by a senescent filament following hyphal fusion. I. Condition immediately after fusion. II. Condition after 12 hours, when the filaments were fragmented with a micromanipulator. The phenotype of the cultures obtained from each of the fragments is indicated in the diagram. a, no further growth; b, grew only in a microdrop; c, grew less than 2 cm; d, grew 2–4 cm; e, grew 4–6 cm; f, grew 6–10 cm; g, grew 10–20 cm; h, normal longevity. From (*22*).

Some evidence of cytoplasmic genes influencing senescence has also been reported in other fungi, including *Aspergillus glaucus* (*34, 35*) and *Pestalozzia annulata* (*2b, 3*). The systems all have in common the dominance of the senescent condition in hyphal fusions or grafts, but only in *Podospora* has senescence been transmitted in crosses.

In *Podospora,* the strongest lines of evidence supporting the hypothesis of an autonomous cytoplasmic gene regulating senescence are (*a*) transmission of senescence in sexual crosses; (*b*) infectivity of senescence in hyphal grafts; and (*c*) discontinuities in transmission, indicating the particulate nature of the factor. Presumably, the senescent tissues contain a mixture of mutant genes and their normal counterpart. The mixture is observed in the emergence of normal as well as senescent fruiting bodies from senescent mycelia, and vegetatively in the growth of normal as well as senescent hyphae from fragments of senescent mycelia.

Taken together, the data support the hypothesis of a cytoplasmic gene responsible for senescence. The development of the senescent phenotype may be viewed as the result of mutation of a normal cytoplasmic gene to a new form, which then competes with the unmutated form, but the evidence in support of this view is all indirect. As in the case of barrage, so here too the problem is in need of investigation at the molecular level.

Rizet *et al.* (*30*) and Marcou (*22*) have pointed out formal similarities between the inheritance of senescence in *Podospora* and suppressiveness in yeast. Both phenomenona, as studied at the cellular level, seem to involve competition at the molecular level between wild-type and mutant cytoplasmic genomes. In *Podospora* however there is no evidence that mitochondrial DNA is the site of the cytoplasmic mutation that leads to senescence. Perhaps some other cytoplasmic DNA is involved. Microinjection of defined extracts from senescent into wild-type hypha might provide a fruitful approach.

STUDIES OF CYTOPLASMIC HEREDITY IN ASPERGILLUS

Following the popularity of *Neurospora* as an organism for genetic research, a number of investigators began to apply similar methods to the study of *Aspergillus* and the closely related *Penicillium*. *Penicillium* is a wholly asexual fungus in which genetic studies can only be pursued by heterokaryon formation. In *Aspergillus,* however, there are homothallic sexual strains in which classical genetic investigations can be carried out, as was done by Pontecorvo (*27*) and his students.

Homothallic strains can be maintained by hyphal tip transfer, by germination of asexual conidia, and by production of sexual ascospores. The asexual spores, or conidia, are haploid and uninucleate. Individual conidia give rise to homokaryotic haploid mycelia, which may become differentiated in localized regions by the fusion of two nuclei to form a homozygous diploid nucleus. The subsequent events resemble those occurring in heterothallic forms like *Neurospora.* Perithecia form, each perithicium grows to produce hundreds of asci, and within each ascus, meiosis and ascospore formation occurs.

If mycelia arising from genetically different conidia are brought together, fusion occurs with the formation of heterokaryotic mycelia containing two kinds of uninucleate haploid nuclei in a common cytoplasm. Three kinds of perithicia arise in such a heterokaryon: homozygous perithicia resulting from fusion of two nuclei from the same parent as well as a heterozygous diploid arising from fusion of two unlike parental nuclei. With the use of well-chosen biochemical mutations in each of the parental strains, it is possible to select the heterozygous perithicia. By this method asci coming from heterozygous diploids can be selected and used for classical genetic analysis.

In the 1950's Pontecorvo and his students developed the classical

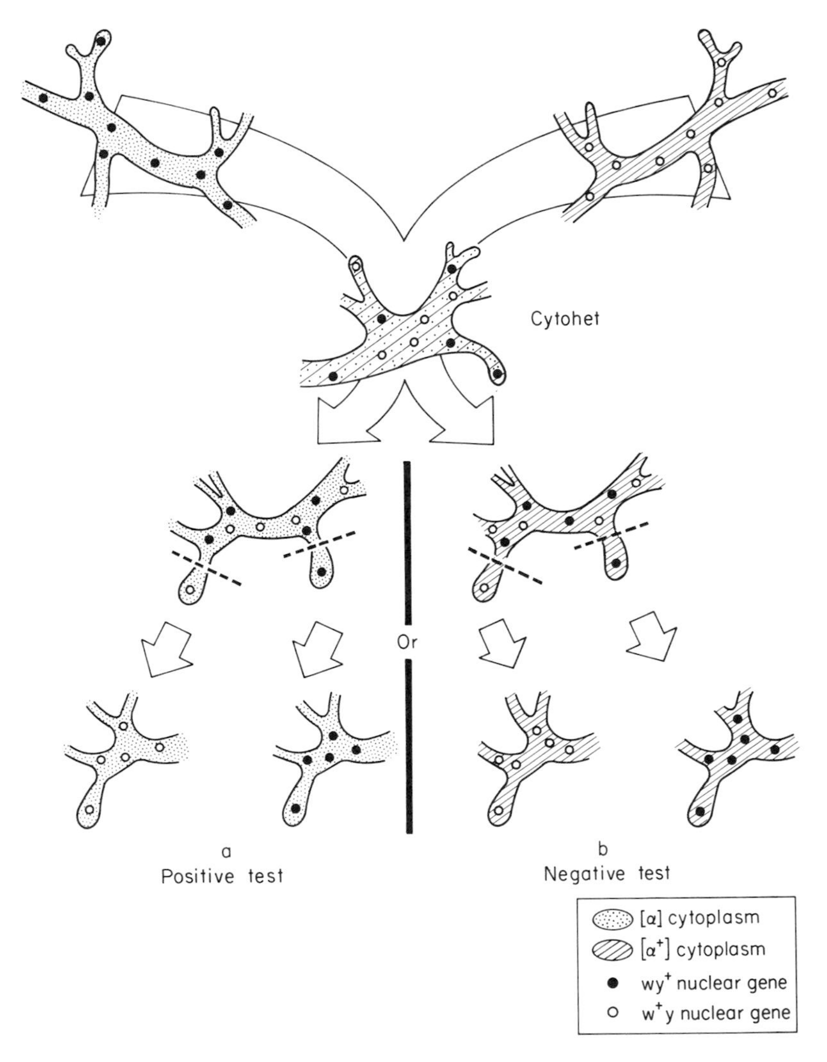

Fig. 5.18. The heterokaryon test for cytoplasmic inheritance. Two haploid mycelia: wy^+ $[\alpha]$ and w^+y $[\alpha^+]$ fuse to form a heterokaryotic mycelium. Then by hyphal tip isolation, each nuclear type is recovered independently. (a) Positive heterokaryon test: the mutant cytoplasmic phenotype $[\alpha]$ is associated with both nuclear types. (b) Negative heterokaryon test: the mutant phenotype is not associated with either nuclear type.

genetics of *Aspergillus* (*27*), with these methods, while a number of other investigators in Britain developed methods for studying cytoplasmic inheritance with the *Aspergillus* system (*19*).

Jinks (*17*) developed the heterokaryon test (Fig. 5.18) to provide evidence, in the absence of a sexual stage in the life cycle, of whether a given phenotype was determined by nuclear genes or not. In this test, heterokaryons are formed between mycelia which differ in the phenotype to be tested as well as in nuclear markers. After suitable mixing of nuclei and cytoplasm in the heterokaryon, the two nuclear types are extracted by hyphal tip isolation, and the phenotypes compared with the parental strains. This test is strong when the mutant cytoplasmic phenotype is dominant, or segregates independently of the nuclei and is perpetuated. When the mutant phenotype is lost, however, the test is weak; it does not distinguish between a cytoplasmic genetic basis and an epigenetic metabolic condition (a developmental state under the control of nuclear genes) of the kind discussed in Chapter 9. For this distinction one needs evidence of the particulate nature of the cytoplasmic factor by virtue of its discrete segregation at cell division.

Indeed, with the work of Jinks, we begin a consideration here and in the following chapter of evidence from many laboratories of phenotypes with obscure structural and molecular bases. Some of them may represent a cytoplasmic stabilization (i.e., epigenetic differentiation) of phenotypes governed by nuclear genes, while others may have their basis in genes of uniquely cytoplasmic identity. Methods for distinguishing these alternatives are not fully developed, and some of the data to be discussed is at present unclassifiable.

In 1954, Jinks (*13*) reported preliminary studies of somatic selection in fungi leading to the establishment of strains showing heritable differences which he attributed to cytoplasmic genetic factors. Starting with normal homokaryotic cultures he found that by continual subculture of asexual conidia, strains with impaired ability to form perithecia could be selected. This trend could be reversed by selecting ascospores from the remaining perithecia for subculture instead of conidia. Thus the changes were not irreversible.

When heterokaryons were formed between the selected poor conidial lines and wild type carrying genetically marked nuclei, the property of impaired spore formation was lost. Perithecia from such heterokaryons gave rise to ascospores with all the nuclear combinations, and in all cases, conidia formation was normal. Jinks concluded that the fungal differentiation for which he had selected must have been due to a cytoplasmic determinant since the phenotypes were lost in heterokaryon formation.

In a subsequent paper, Jinks (*14*) studied cytoplasmic variability in *Aspergillus glaucus* by selecting for differences not only in growth rate but also in germination rate, pigmentation, and sexuality. Selection was based on naturally occurring variation among single asexual spores, and a high correlation was noted between growth rate, pigmentation, perithecia formation, and rate of germination of ascospores. However, as in the preceding work, he found that all of the selected changes could be wiped out either by heterokaryon formation with a wild-type strain or by back selection toward the wild-type phenotype. Similar methods were used to study selection of adaptibility of *Aspergillus glaucus* to new environments (*15, 16*), and similar results were obtained.

A prime difficulty with these studies is the fact that the selected phenotypes are totally wiped out in the process of heterokaryon formation, and they never reappear. Thus it cannot be established whether the selected phenotypes are the result of genetic differences between the wild-type and selected strains, or whether what has been selected is an epigenetic condition, i.e., a stabilized metabolic state.

A similar difficulty was encountered by Mahoney and Wilkie in their investigation of the loss of perithecia-forming ability in *Aspergillus nidulans* (*20, 21*). They studied four wild-type strains, each of which characteristically gave rise to some asexual spores (conidia) in which the ability to form perithecia had been irreversibly lost. The percent of conidia behaving in this way varied from about 4% in strain *A* to about 83% in strain *D*.

The heterokaryon test was performed by making heterokaryons between stable, asexual variants isolated from each of the four strains and a wild-type normal green (*W*) biotin-requiring (*bi*) strain capable of producing normal perithecia. The resulting heterokaryons were fertile and gave rise to three types of perithecia: *W bi/W bi* and *w Bi/w Bi* homozygous diploids and *W bi/w Bi* the heterozygous diploid. All three classes of perithecia produced ascospores, and the mycelia from them all produced perithecia. Thus the mutant phenotype of sterility was lost.

Somewhat more compelling evidence in favor of a cytoplasmic genetic determinant influencing perithecia formation comes from the work of Subak-Sharpe with a mutant of *Aspergillus glaucus* (*34, 35*). He studied a strain which could exist in either the *A* form or the *B* form. The *B* form was the stable wild-type form; the *A* form was unstable, consisting of sectors of asexual mycelium no longer capable of perithecia formation. These sectors also differed from wild type in their growth rate, pigmentation, and abnormally high rate of conidia formation. When form *A* was subcultured with large inocula, it bred true, giving only *A*-type growth. However, very small inocula of form *A* were unstable, giving rise both to *A*- and to *B*-type growth in sectors. The nuclear genome was shown to have no effect on this sectoring process by the heterokaryon test.

Subak-Sharpe studied the process of conversion of *B* to *A* in great detail. He found that *A* (mutant) could infect mycelia of *B* (wild type) and spread in a sector-like pattern through the *B*-type mycelium. He proposed that *A* contains an infectious particle formally called x' which is a mutant form of x normally present in type *B*. Whenever x' is lost, *A* reverts to *B*. The amount of x' present could be varied greatly by the physiologically state of the culture—and, above all, x' was always lost during aging. Unfortunately, a detailed report of these studies was never published (*34, 35*) and the problem has not been pursued.

Further evidence for the existence of a particulate but unidentified system in fungi regulating growth and differentiation comes from work with a *red* mutant of *Aspergillus nidulans* (*1*). The mutant mycelium is characterized by the production of a red pigment excreted into the medium. The *red* regions produce protoperithecia but no viable sexual conidia and persistently segregate out wild-type sectors.

A heterokaryon was synthesized between a normal strain with a nuclear marker and the *red* strain. Spores from this heterokaryon produced colonies, some of which contained the nuclear markers from the normal strain in a mycelium that was phenotypically like the *red* mutant. This positive result of the heterokaryon test suggests the presence of a discrete genetic determinant responsible for the *red* phenotype located outside the nucleus.

However, the mutant *red* heterokaryotic mycelium was unstable and continued to produce wild-type sectors in a manner similar to the original *red* variant. The property of persistent segregation was examined further. Uninucleate, haploid spores produced mosaic colonies containing normal sectors and *red* mutant sectors. When hyphal transfers were carried out from the mutant sectors, they gave rise to two classes of mycelia, one called *r-normal* and the other *r-red*. The *r-normal* mycelia appeared to be normal unless they were grown very extensively, in which case they did give rise to a few *r-red* sectors. The *r-red* mycelial growth was similar to that of the strain initially isolated, consisting of sectors of apparently normal and of clearly abnormal growth, the abnormal regions resembling those described above.

Further cloning established that progeny producing a particular fraction of *r-red* offspring tended to transmit this quantitative property to the next generation. This result is reminiscent of the cloning experiments of Ephrussi and Grandchamp with suppressiveness in yeast (Chapter 4). Some evidence was obtained that nuclear genes of the wild-type partner influenced the quantitative results. Grindle (*9*) introduced the *red* cytoplasmic determinant into a number of different genetic strains by heterokaryon formation. The nuclear component was found to influence the relative survival of the *red* determinant. Some nuclei were neutral with respect to the relative growth rate of the mutant, other genotypes favored either the mutant or the wild type. Four nuclear genes were found that strongly favored the mutant and permitted the establishment of pure breeding *red* strains.

These results lead one to accept, at least *pro tem,* the conclusion that the *red* phenotype is determined by a particulate non-Mendelian genetic component. However, the component does not get transmitted through ascospores in sexual reproduction. This observation alone would seem to rule out mitochondrial DNA as the carrier of this genetic component. Clearly, additional information is needed to carry the analysis further.

An analogous study was done with the *minute* variant of *Aspergillus nidulans* (6), which is characterized by small colonies, few perithecia, and very little pigment production. Asexual spores carrying the *minute* determinant give rise to normal colonies and to *minutes,* which segregate persistently, much like the *red* mutant described above. *Minute* is not transmitted through sexual spores and can be maintained only by continual selection. Despite the absence of positive support for its genetic control by a non-Mendelian factor, its similarity in behavior to the *red* mutant suggests that its mode of inheritance may be the same.

CONCLUDING REMARKS

This chapter has summarized numerous lines of genetic evidence concerning cytoplasmic heredity in fungi. The most definitive studies are those with *Neurospora,* in which experimental evidence links the genetic changes with alterations in mitochondrial phenotypes. It is a strong inference, though not critically established, that the mutations affecting mitochondrial function—*poky, mi-3, mi-4, abn-1, abn-2,* and possibly *stp* and the *SG* strains—are the result of mutations in mitochondrial DNA.

The principal lines of genetic evidence are similar to those of other systems: 4:0 segregation in tetrads after meiosis and somatic segregation in cytoplasmic hybrids. In addition, the microinjection experiments implicate genes carried in mitochondria in the *abn-1* phenotype.

In *Podospora* and in *Aspergillus,* the evidence is extensive for the presence of cytoplasmic genes affecting a number of traits including sexual differentiation, incompatibility, and senescence. However, the relation of these phenotypes to specific organelles, cellular structures, or cytoplasmic DNA is unknown.

Can these phenomena be interpreted in terms of mitochondrial influences, or must one think in terms of other cytoplasmic genetic systems? Fungal senescence could be viewed as a mitochondrial disorder, since it involves cell growth as a whole. With respect to the other phenotypes, it is difficult to conceive of them as mitochondrial, since they involve only differentiated aspects of growth. Thus in the fungi we have encountered a number of cytoplasmic mutations which may not be mitochondrial. The lack of evidence of either a cytoplasmic structure or a cytoplasmic DNA associated with these genes has provoked a variety of hypotheses and speculations. Some of these proposals have been discussed in Chapter 1.

One of the most promising lines of work involves the recovery of respiratory-deficient strains of *Neurospora* that show non-Mendelian but biparental transmission in crosses. We may look forward to recombination analysis and mapping procedures with mutations such as these.

Suggested Review Articles

Ephrussi, B. "Nucleo-Cytoplasmic Relations in Micro-Oganisms." Clarendon, Oxford.

Esser, K., and Kuenen, R. (1967). "Genetics of Fungi." Springer-Verlag, Berlin and New York.

Fincham, J. R. S., and Day, P. R. (1965). "Fungal Genetics," 2nd Ed., Chap. 10. Blackwell, Oxford.

Jinks, J. L. (1963). Cytoplasmic inheritance in fungi. *In* "Methodology in Basic Genetics" (W. J. Burdette, ed.), pp. 325–354. Holden-Day, San Francisco, California.

Jinks, J. L. (1966). Mechanisms of inheritance. 4. Extranuclear inheritance. *In* "The Fungi" (G. C. Ainsworth and A. S. Sussman, eds.), Vol. 2, pp. 619–660. Academic Press, New York.

References

1. Arlett, C. F., Grindle, M., and J. L. Jinks (1962) The 'red' cytoplasmic variant of *Aspergillus nidulans. Heredity* **17,** 197–209.
2. Beisson-Schecroun, J. (1962). Incompatibilité cellulaire et interaction nucleo-cytoplasmique dans les phenomenes de "barrage" chez le *Podospora anserina. Ann. Genet.* **4,** 1.

2a. Bertrand, H., and Pittenger, T. H. (1969). Cytoplasmic mutants selected from continuously growing cultures of *Neurospora crassa. Genetics* **61,** 643–659.

2b. Chevaugeon, J., and Digbeu, S. (1960). Un second facteur cytoplasmique infectant chez le *Pestalozzia annulata. C. R. Acad. Sci. Ser. D.* **251,** 3043.

3. Chevaugeon, J., and Lefort, C. (1960). Sur l'apparition reguliere d'un "mutant" infectant chez un champignon du genre *Pestalozzia. C. R. Acad. Sci. Ser. D.* **250,** 2247.
4. Diacumakos, E. G., Garnjobst, L., and Tatum E. L. (1965). A cytoplasmic character in *Neurospora crassa.* The role of nuclei and mitochondria. *J. Cell Biol.* **26,** 427–443.
5. Edwards, D. L., and Woodward, D. O. (1969). An altered cytochrome oxidase in a cytoplasmic mutant of *Neurospora. FEBS Lett.* **4,** 193 (abstract).
6. Faulkner, B. M., and Arlett, C. F. (1964). The 'minute' cytoplasmic variant of *Aspergillus nidulans. Heredity* **19,** 63–73.
7. Garnjobst, L., Wilson, . F., and Tatum, E. L. (1965). Studies on a cytoplasmic character in *Neurospora crassa. J. Cell Biol.* **26,** 413–425.

8. Gowdridge, B. M. (1956). Heterocaryons between strains of *Neurospora crassa* with different cytoplasms. *Genetics* **41,** 780–789.
9. Grindle, M. (1964). Nucleo-cytoplasmic interactions in the 'red' cytoplasmic variant of *Aspergillus nidulans. Heredity* **19,** 75–95.
10. Hardesty, B. A., and Mitchell, H. K. (1963). Accumulation of free fatty acids in *poky,* a maternally inherited mutant of *Neurospora crassa. Arch. Biochem. Biophys.* **100,** 330–334.
10a. Haskins, F. A., Tissieres, A., Mitchell, H. K., and Mitchell, M. B. (1953). Cytochromes and the succinic acid oxidase system of *poky* strains of *Neurospora. J. Biol. Chem.* **200,** 819–826.
11. Howell, N., Zuiches, C. A., and Munkres, K. D. (1971). Mitochondrial biogenesis in *Neurospora crassa.* I. An ultrastructural and biochemical investigation of the effects of anaerobiosis and chloramphenicol inhibition. *J. Cell. Biol.* **50,** 721.
12. Infanger, A. M., and Srb, A. M. (1964). Nucleo-cytoplasmic relations in heterokaryons of *Neurospora sitophila.* Ph.D. Dissertation, Cornell University, Ithaca, New York.
13. Jinks, J. L. (1954). Somatic selection in fungi. *Nature (London)* **174,** 409–410.
14. Jinks, J. L. (1956). Naturally occurring cytoplasmic changes in fungi. *C. R. Trav. Lab. Carlsberg Ser. Physiol.* **26,** 183.
15. Jinks, J. L. (1959). Selection for adaptability to new environments in *Aspergillus glaucus. J. Gen. Microbiol.* **20,** 223–236.
16. Jinks, J. L. (1959). Lethal suppressive cytoplasms in aged clones of *Aspergillus glaucus. J. Gen Microbiol.* **21,** 397–409.
17. Jinks, J. L. (1963). Cytoplasmic inheritance in fungi. *In* "Methodology in Basic Genetics" (W. J. Burdette, ed.), pp. 325–354. Holden-Day, San Francisco, California.
18. Jinks, J. L. (1966). Mechanisms of inheritance. 4. Extranuclear inheritance. *In* "The Fungi" (G. C. Ainsworth and A. S. Sussman, eds.), Vol. 2, pp. 619–660. Academic Press, New York.
19. Jinks, J. L. (1964). "Extrachromosomal Inheritance." Prentice-Hall, Englewood Cliffs, New Jersey.
19a. Littlewood, R. K. (1970). Comparative studies on several cytoplasmic mutants of *Neurospora.* Ph.D. Thesis, Cornell University, Ithaca, New York.
19b. Littlewood, R. K., and Srb, A. M. (1969). Temperature interactions among four cytoplasmic mutants in *Neurospora crassa. Genetics* **61,** s37 (abstract).
19c. McDougall, K. J. (1964). Inheritance of induced and naturally occurring cytoplasmic alterations in *Neurospora crassa.* Ph.D. Thesis, Kansas State University. Univ. Micro Films, Ann Arbor, Michigan.
20. Mahoney, M., and Wilkie, D. (1958). An instance of cytoplasmic inheritance in *Aspergillus nidulans. Proc. Roy. Soc. Ser. B* **148,** 359–361.
21. Mahoney, M., and Wilkie, D. (1962). Nucleo-cytoplasmic control of perithecial formation in *Aspergillus nidulans. Proc. Roy. Soc. Ser. B* **156,** 524–532.
22. Marcou, D. (1961). Notion de longevité et nature cytoplasmique du determinant de la senescence chez quelques champignons. *Ann. Sci. Naturelles Bot.* **2** (Series 12), 653–764.
23. Mitchell, M. B., and Mitchell, H. K. (1952). A case of "maternal" inheritance in *Neurospora crassa. Proc. Nat. Acad. Sci. U. S.* **38,** 442.
24. Mitchell, M. B., and Mitchell, H. K. (1956). A nuclear gene suppressor of a cytoplasmically inherited character in *Neurospora crassa. J. Gen. Microbiol.* **14,** 84.
25. Mitchell, M. B., Mitchell, H. K., and Tissieres, A. (1953). Mendelian and non-Mendelian factors affecting the cytochrome system in *Neurospora crassa. Proc. Nat. Acad. Sci. U. S.* **39,** 606–613.

26. Pittenger, T. H. (1956). Synergism of two cytoplasmically inherited mutants in *Neurospora crassa. Proc. Nat. Acad. Sci. U. S.* **42,** 747–752.
27. Pontecorvo, G. (1958). "Trends in Genetic Analysis." Columbia Univ. Press, New York.
27a. Puhalla, J. E., and Srb, A. M. (1967). Heterokaryon studies of the cytoplasmic mutant SG in *Neurospora. Genet. Res.* **10,** 185–194.
28. Reich, E., and Luck, D. J. L. (1966). Replication and inheritance of mitochondrial DNA. *Proc. Nat Acad. Sci. U. S.* **55,** 1600–1608.
29. Rifkin, M. R., and Luck, D. J. L. (1971). Defective production of mitochondrial ribosomes in the *poky* mutant of *Neurospora crassa. Proc. Nat. Acad. Sci. U. S.* **68,** 287–290.
30. Rizet, G., Marcou, D., and Schecroun, J. (1958). Deux phenomenes d'heredité cytoplasmique chez l'ascomycete *P. anserina. Bull. Soc. Fr. Physiol. Veg.* **4** (No. 4), 136.
31. Silagi, S. (1965). Interactions between an extrachromosomal factor, *poky*, and nuclear genes in *Neurospora crassa. Genetics* **52,** 341–347.
32. Srb, A. M. (1963). Extrachromosomal factors in the genetic differentiation of *Neurospora. Symp. Soc. Exp. Biol.* **17,** 175–187.
33. Srb, A. M. (1966). Extrachromosomal heredity in fungi. *In* "Reproduction: Molecular, Subcellular and Cellular" (M. Locke, ed.), *Symp. Soc. Develop. Biol.*, Vol. 24, pp. 191–211. Academic Press, New York.
34. Subak-Sharpe, J., H. (1956). Heterokaryosis and extra-nuclear inheritance in a wild homothallic ascomycete. Ph.D. Thesis, University of Birmingham, England.
35. Subak-Sharpe, J. H. (1958). A closed system of cytoplasmic variation in *Aspergillus glaucus. Proc. Roy. Soc. Ser. B* **148,** 355–359.
36. Tatum, E. L., and Luck, D. J. L. (1967). Nuclear and cytoplasmic control of morphology in *Neurospora. Develop. Biol. Suppl.* **1,** 32–42.
37. Tissieres, A., and Mitchell, H. K. (1954). Cytochromes and respiratory activities in some slow growing strains of *Neurospora. J. Biol. Chem.* **208,** 41–249.
38. Wilson, J. F. (1961). Micrurgical techniques for *Neurospora. Amer. J. Bot.* **48,** 46–51.
38a. Wilson, J. F. (1963). Transplantation of nuclei in *Neurospora crassa. Amer. J. Bot.* **51,** 780–786.
39. Wilson, J. F., Garnjobst, L., and Tatum, E. L. (1961). Heterocaryon incompatibility in *Neurospora crassa:* micro-injection studies. *Amer. J. Bot.* **48,** 299.

6

Cytoplasmic Genes in Higher Plants

This chapter will consider some of the principal evidence for the existence and properties of cytoplasmic genes in higher plants. The most extensively studied examples are those affecting chloroplast development. The voluminous literature on this subject has recently been admirably summarized and discussed in "The Plastids" (*18*), thereby lightening the burden of information to be included in this chapter. In addition to plastid formation, however, cytoplasmic genes also affect many other developmental processes in plants, including pollen formation, sexual differentiation, and other aspects of morphogenesis. Except for pollen sterility (*7*), this subject matter has not been recently reviewed, and most of the work was done long ago. For some insight into these studies, the reader is referred to reviews by Rhoades (*32*), Smith (*36*), Michaelis (*20, 21*), Oehlkers (*26*), and Caspari (*3*).

The emphasis and focus of interest in this chapter is twofold: (*a*) to present a few exemplary studies which most clearly illustrate the methods and results of cytoplasmic genetic analysis with higher plants, and (*b*) to relate the findings wherever possible to studies with other organisms and, above all, to the behavior of organelle DNA's.

The recognition of cytoplasmic genes began in 1908 with two reports, by Carl Correns (*5*) and by Erwin Baur (*1*), each describing the non-Mendelian inheritance of a factor influencing chloroplast development.

In both reports the phenotypes involved leaf variegation, that is the presence of striped sectors of normal green and of mutant white leaf tissue. The patterns of inheritance in the two systems were different. In *Mirabilis,* the four o'clock flower, Correns found strict maternal inheritance, in which the pollen did not contribute genetically to the mutant phenotype. In contrast, Baur found in *Pelargonium* that both parents contributed to the variegation, but not with Mendelian ratios. In both *Mirabilis* and *Pelargonium,* many other genetic differences affecting chloroplast development and flower color showed Mendelian inheritance in the same studies. The two non-Mendelian systems are diagrammed in Fig. 6.1.

The identification of cytoplasmic genes is relatively straightforward in higher plants as seen in Fig. 6.1. The first criterion is the occurrence of differences in the outcome of reciprocal crosses. Even in plants like *Pelargonium,* which exhibit biparental inheritance, the results of reciprocal crosses are different and the ratios are variable and non-Mendelian. If the parents are not isogenic, the progeny should be backcrossed for several generations to the male parent, in order to insure that the nuclear genes of the female parent have been replaced by genes from the male.

The second criterion, similar to that in microorganisms, is somatic segregation of the two parental phenotypes during growth of the progeny. As in Mendelian heredity, segregation out of hybrids is evidence of the particulate nature of the elements involved (Chapter 3). Cytoplasmic mutations affecting chloroplast development characteristically give rise to green and white or green and yellow sectored plants, some of which incidentally are highly prized and beautiful ornamentals. Flower parts arising in mutant sectors can be utilized in test crosses to establish the genetic basis of the mutation. Similarly in cytoplasmic pollen sterility, sectors are often seen with normal and mutant male flower parts on the same plant giving rise to fertile and sterile pollen, respectively.

Historically, the two plants which have been the most exhaustively investigated with respect to cytoplasmic genetics are *Oenothera,* studied initially by Renner and his students (*27*), and more recently by Schotz (*34*) and by Stubbe (*38–41*); and *Epilobium* studied primarily by Michaelis and his students (*20, 21*).

The outstanding contribution of the literature of this period (1920–1960) has been to document the widespread occurrence of cytoplasmic genetic mutations throughout the plant kingdom, and to show the multiplicity of phenotypes affected by cytoplasmic genes. Some of the classical systems involving plastid abnormalities are listed

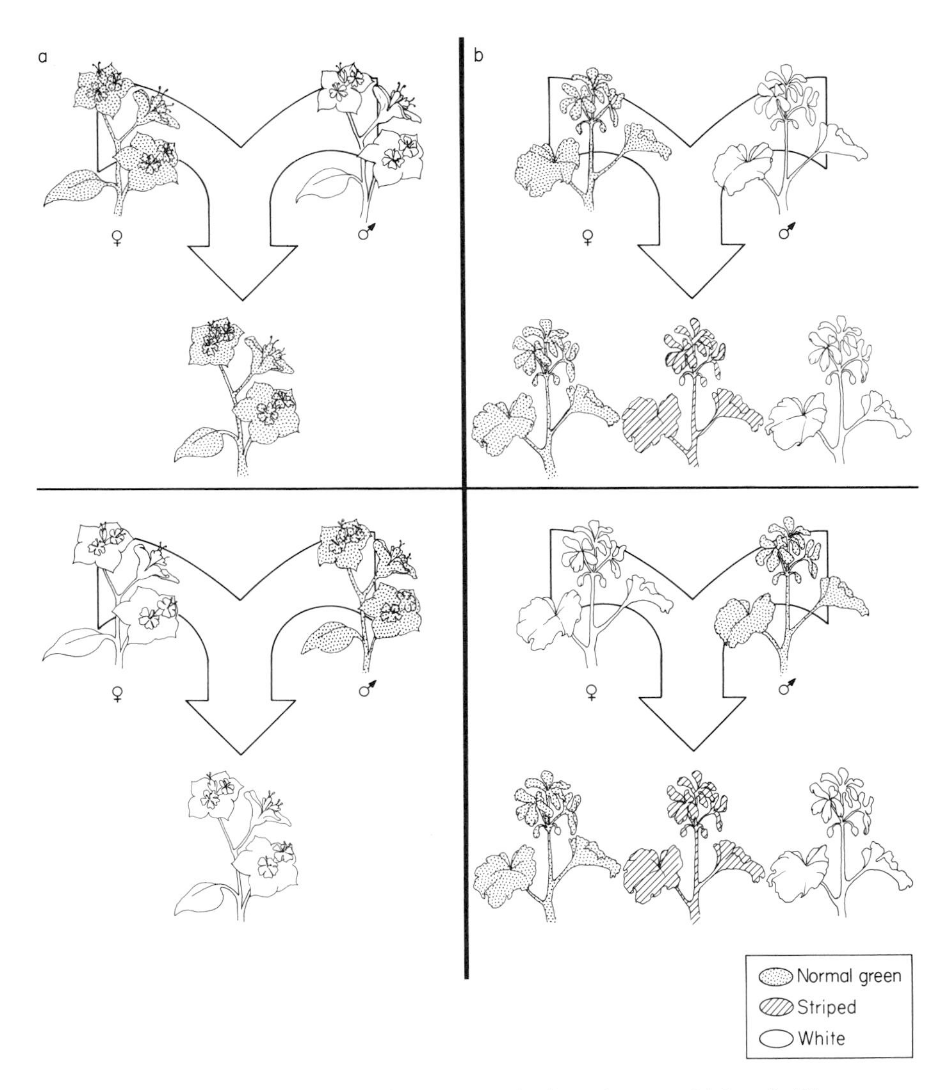

Fig. 6.1. Patterns of chloroplast inheritance in higher plants. a. *Maternal:* All progeny resemble the female parent; no transmission of chloroplast genes from the male parent. b. *Biparental:* Both female and male chloroplast genes are transmitted; ratios are non-Mendelian, variable, influenced by genotype and environment.

in Table 6.1. In addition to plastid abnormalities, cytoplasmic pollen sterility has been documented in more than thirty species of plants. The classical literature was reviewed by Edwardson (9), and the extensive studies carried out with the corn plant *Zea mays* were reviewed by Duvick (7). In most studies of plastid mutations and in some studies of pollen sterility, evidence of both non-Mendelian ratios and somatic segregation have been reported.

Effects of cytoplasmic genes upon other developmental phenotypes, including size, leaf shape, and differentiation of flower parts, have been described in a number of plants including *Oenothera* (*27*), *Epilobium* (*21*), *Funaria* (*3, 6a*), and *Cirsium* (*3, 6a*). The complexity of these phenotypes has made genetic analysis very difficult (e.g., absence of sectors), but the evidence they give of the pervasive influence of cytoplasmic genes is impressive.

In none of these studies have nucleic acids been implicated experimentally, nor has any evidence of recombination or of a linked set of genes been adduced. Thus, these studies do not bring us very far in modern terms. However, a reexamination of some of the data in the light of present knowledge leads to new experimental approaches which could be pursued with these extensively and systematically documented systems.

The absence of positive evidence concerning the identity or location of non-Mendelian genetic elements led to a controversy concerning the basis of chloroplast variegation: the plasmone versus the plastome. The term plasmone refers to the total cytoplasmic genome, whereas plastome refers to the chloroplast genome specifically. According to the plasmone hypothesis advocated most vigorously by Michaelis, non-Mendelian genes affecting chloroplast development are not located necessarily within the chloroplast, but may be anywhere outside the nucleus. Supporters of the opposing plastome hypothesis, including most investigators working with *Pelargonium* and *Oenothera,* have argued that the breeding results, as well as the detailed studies of the variegation pattern, can best be interpreted in terms of the plastids themselves as the carriers of these genetic determinants.

It seems likely that both views were formally correct in the sense that chloroplast development may be controlled not only by chloroplast DNA interacting with the nuclear genome, but also by other extranuclear genetic elements such as those of mitochondria. However, the particular contributions of each genetic system remain largely unidentified as yet.

In this chapter, we will first consider the evidence for the presence of cytoplasmic genes in the corn plant *Zea mays,* because the clearest examples of maternal inheritance are to be found here in the beautiful studies by Marcus Rhoades (*28–31*). The rigorous identification of cytoplasmic genes was greatly aided in these studies by the extensive knowledge and use of nuclear genes as controls in the investigation.

Second, we will consider an example of the biparental cytoplasmic

TABLE 6.1

Higher Plants with Maternal Inheritance of Plastid Mutations[a,b,c]

Plant		Demonstration mixed cells: Cytological observation	Demonstration mixed cells: Inferred by breeding: seedlings	Type of cross	Results of reciprocal crosses: Ia G × W / IIa G × V: G	V	W	Ib W × G / IIb V × G: G	V	W	Reference[d]
Dicotyledons											
Antirrhinum majus		–	G V W	IIa,b	G	–	–	4	–	12	*3*
		–	G V W	Ia,b	G	–	–	–	–	700	*42*
	P	–	–	Ia,b	7970	3	4	–	–	∞	*22*
		Yes	G V W	–	–	–	–	–	–	–	*31, 49, 50*
		Yes	G V W	–	–	–	–	–	–	–	*25*
Arabidopsis thaliana		Yes	G V W	I,b	–	–	–	–	–	323	*40*
Arabis albida	P	–	–	Ia,b	466	–	–	–	–	254	*15*
Aubrietia graeca	P	–	–	Ia,b	75	–	–	–	–	14	*15*
Aubrietia purpurea	P	–	–	Ia,b	92	–	–	–	–	25	*15*
Beta vulgaris		–	–	Ia,b	G	–	–	–	–	W	*34, 35, 44*
Capsicum annuum		–	G – W	Ia,b	277	–	–	–	1	21	*19*
Curcurbita maxima		–	G V W	Ia,b	2021	–	–	–	–	281	*27*
Epilobium hirsitum (numerous mutants)		Yes	G V W	Ia,b	∞	–	–	–	–	∞	*32*
Hydrangea hortensis	P	–	–	Ia,b	G	–	–	383G	–	216	*9*
								54Y	–	104	*9*
Lactuca sativa		–	–	Ia,b	G	–	–	–	–	W	*48*
Lycopersicum esculentum (*Solanum lycopersicum*)	P	–	–	Ia,b	G	–	–	–	–	2500	*43*
Mesembryanthemum cordifolium	P	–	–	Ia,b	21	–	–	–	–	12	*15*

Mimulus quinquevulnerus		–	G V W	IIa,b	G	–	–	G	V	W	*4, 5*
Mirabilis jalapa		–	G V W	Ia,b	94	–	–	–	3	12	*13, 14*
Nicotiana colossea	P	–	–	Ia,b	1200	–	–	–	–	1800	*26*
Nicotiana tabacum		Yes	G V W	IIa,b	500	–	–	31	131	–	*51*
	P	–	–	Ia,b	287	–	–	1	–	83	*6*
		Yes	G V W	Ia,b	322	–	–	–	–	184	*7*
		Yes	G V W	Ia,b	2235	–	–	–	–	1770	*47*
Petunia violacea		–	–	Ia,b	153	–	–	–	–	110	*45*
		–	G V W	IIa,b	442	–	–	195	11	861	*37*
Pharbitis nil		–	G V W	Ia,b	66	–	–	–	–	24	*33, 28*
Pisum sativum		–	G V W	IIa,b	12	–	–	2	7	4	*24*
Primula sinensis		Yes	G V W	Ia,b	70	–	–	–	–	11	*23*
Primula vulgaris		–	–	Ia,b	32	–	–	–	–	120	*8*
Stellaria media		Yes	G V W	Ia,b	100	–	–	–	–	49	*16–18, 21*
Trifolium pratense		–	G V W	IIa,b	83	–	–	35	25	11	*36*
Viola tricolor		–	G V –	IIa,b	22	–	–	4	6	–	*10, 11*
Monocotyledons											
Avena sativa		–	G V W	IIa,b	43	–	–	–	6	12	*1*
Avena sativa × sterilis		–	G V W	IIa,b	26	–	–	34	18	–	*30*
Chlorophytum comosum × celatum	P	–	G V W	Ia,b	438	6	1	1	11	143	*12*
Chlorophytum elatum	P	–	G V W	Ia,b	52	4	1	1	1	92	*37*
Hordeum vulgare		–	–	Ia,b	30	–	–	–	–	26	*41*
Hosta japonica	P	Yes	G V W	Ia,b	164	–	–	–	2	17	*52*
Sorghum vulgare		–	G V W	Ia,b	6	–	–	–	–	7	*29*
Triticum vulgare		–	G V W	IIa,b	G	–	–	G	V	W	*46*
		–	G V W	Ia,b	10	–	–	–	–	27	*38*
Zea mays		–	G V W	IIa,b	G	–	–	G	V	W	*2, 39*
		–	G V W	IIa,b	16675	–	–	60	48	137	*20, 53*

[a] From (*18*), Table VII.1.

[b] In each case at least a part of the typical sorting-out process was reported.

[c] Key: G = green, V = variegated, W = white, and Y = yellow germcells or seedlings; W is used in the general sense to signify any mutant plastid type. P = Breeding experiments which were made with stable periclinal chimeras. ∞ = Numerous seedlings.

Footnotes for Table 6.1 continued on page 186.

TABLE 6.1 Footnotes (Continued)

[d] Key to references:

1. Akerman, A. (1933). *Bot. Notis.*, 255–270.
2. Anderson, E. G. (1923). *Bot. Gaz. Chicago* **76,** 411–418.
3. Baur, E. (1910). *Z. Vererbungslehre* **3,** 34–98.
4. Brozek, A. (1923). *Stud. Pl. Physiol. Lab. Charles Univ.* **1,** 45–78.
5. Brozek, A. (1926). *J. Hered.* **17,** 113–129.
6. Burk, L. G., and Grosso, J. J. (1963). *J. Hered.* **54,** 23–25.
7. Burk, L. G., Stewart, R. N. and Dermen, H. (1964). *Amer. J. Bot.* **51,** 713–724.
8. Chattaway, M. M., and Snow, R. (1929). *J. Genet.* **21,** 81–83.
9. Chittenden, R. J. (1926). *J. Genet.* **16,** 43–61.
10. Clausen, J. (1927). *Hereditas* **9,** 245–256.
11. Clausen, J. (1930). *Hereditas* **13,** 342–356.
12. Collins, E. J. (1922). *J. Genet.* **12,** 1–17.
13. Correns, C. (1909). *Z. Vererbungslehre* **1,** 291–329.
14. Correns, C. (1909). *Z. Vererbungslehre* **2,** 331–340.
15. Correns, C. (1919). *S. B. Preuss. Akad. Wiss.*, 820–857.
16. Correns, C. (1922). *S. B. Preuss. Akad. Wiss.*, 460–486.
17. Correns, C. (1931). *S. B. Preuss. Akad. Wiss.*, 203–231.
18. Correns, C. (1931). *Z. Vererbungslehre* **59,** 275–280.
19. Dale, E. E. (1930). *Pap. Mich. Acad. Sci. Arts Lett.* **13,** 5–8.
20. Demerec, M. (1927). *Bot. Gaz.* **84,** 139–155.
21. Funaoka, S. (1924). *Biol. Zentralbl.* **44,** 343–384.
22. Gairdner, A. E., and Haldane, J. B. S. (1929). *J. Genet.* **21,** 315–325.
23. Gregory, R. P. (1915). *J. Genet.* **4,** 305–321.
24. de Haan, H. (1930). *Genetica* **12,** 321–440.
25. Hagemann, R. (1964). "Plasmatische Vererbung." Fischer, Jena.
26. Honing, J. A. (1927). *Genetica* **9,** 1–18.
27. Hutchins, A. E., and Youngner, V. B. (1952). *Proc. Amer. Soc. Hort. Sci.* **60,** 370–378.
28. Imai, Y. (1936). *Z. Vererbungslehre* **71,** 61–83.
29. Karper, R. E. (1934). *J. Hered.* **25,** 49–54.
30. Love, H. H., and Craig, W. T. (1936). *J. Amer. Soc. Agron.* **28,** 1005–1011.
31. Maly, R., and Wild, A. (1956). *Z. Vererbungslehre* **87,** 493–496.
32. Michaelis, P. (1954). *Advan. Genet.* **6,** 287–401.
33. Miyaki, K., and Imai, Y. (1935). *Bot. Gaz. Chicago* **96,** 571–574.
34. Munerati, O. (1928). *Proc. V. Int. Congr. Genet.* **2,** 1137–1142.
35. Munerati, O. (1942). *Zuechter* **14,** 214–215.
36. Nijdam, F. E. (1932). *Genetica* **14,** 161–278.
37. Pandey, K. K., and Blaydes, G. W. (1957) *Ohio J. Sci.* **57,** 135–147.
38. Pao, W. K., and Li, H. W. (1946). *J. Amer. Soc. Agron.* **38,** 90–94.
39. Randolph, L. F. (1922). *Bot. Gaz. Chicago* **73,** 337–375.
40. Robbelen, G. (1962). *Z. Vererbungslehre* **93,** 25–34.
41. Robertson, D. W. (1937). *Genetics* **22,** 104–113.
42. Scherz, W. (1927). *Z. Vererbungslehre* **45,** 1–40.
43. Schlosser, L. A. (1935). *Z. Vererbungslehre* **68,** 222–241.
44. Stehlik, V. (1921). *Z. Zuckerind. Cech. Repub.* **45,** 409–414.
45. Terao, H., and U, N. (1929). *Jap. J. Genet.* **4,** 86–89.
46. Umar, S. M. (1943). *Indian J. Genet.* **3,** 61–63.
47. von Wettstein, D., and Eriksson, G. (1965). *Genet. Today* **3,** 591–612.
48. Whitaker, T. W. (1944). *J. Hered.* **35,** 317–320.
49. Wild, A. (1958). *Planta* **50,** 379–387.
50. Wild, A. (1960). *Beitr. Biol. Pflanz.* **35,** 137–175.
51. Woods, M. W., and DuBuy, H. G. (1951). *J. Amer. Cancer Inst.* **11,** 1105–1151.
52. Yasui, K. (1929). *Cytologia* **1,** 192–215.
53. Zirkle, C. (1929). *Bot. Gaz. Chicago* **88,** 186–203.

inheritance originally described by Baur (Fig. 6.1b). Here the clearest evidence is to be found in the current work of Tilney-Bassett (*45*) who is reinvestigating and further developing Baur's classical studies with *Pelargonium.*

The studies with *Zea mays* and *Pelargonium* will illustrate some of the methods, results, and difficulties in cytoplasmic genetic analysis with higher plants. We will then turn briefly to the older, very extensive, elaborate, and complex studies with *Oenothera* and *Epilobium,* discussing a few selected experiments which highlight the principal observations, conclusions, and difficulties of those investigations.

CHLOROPLAST VARIEGATION AND POLLEN STERILITY IN ZEA MAYS

Rhoades investigated two phenotypes that showed the influence of cytoplasmic genes: chloroplast variegation and pollen sterility. Both of them exhibited strict maternal inheritance, of the sort first described by Correns and diagrammed in Fig. 6.1a. It is noteworthy that many different phenotypes in numerous plants, listed in Table 6.1, have shown this pattern of transmission, phenotypes including obscure morphogenetic alterations as well as blocks to chloroplast development and to pollen formation. The examples of plastid variegation and of pollen sterility are by far the most numerous, presumably because they are distinct and easy to follow.

Both pollen sterility and plastid variegation are of interest not only in the framework of basic science but also in applied science. Variegated plants are often prized as horticultural ornamentals. Cytoplasmic pollen sterility has been of extraordinary agricultural importance. Many crop plants including not only the corn plant *Zea mays,* but also beets, sugarcane, and onions are grown commercially from hybrid seeds. Since hybrid seed production involves controlled pollination, the use of male sterile plants has been of great practical value.

In hybrid corn production for example, it is essential to block self-pollination, a process which normally occurs readily since corn plants are self-fertile, containing both male (tassel) and female (ear) flower parts. The way in which cytoplasmic pollen sterility is utilized in hybrid seed production is diagrammed in Fig. 6.2.

In 1952, before the development of commercial hybrid seed lines carrying a cytoplasmic pollen sterility gene, it cost $10–$20 an acre to detassle corn plants. At present the cost of hand detassling is very high. Indeed it is amusing that while the existence of cytoplasmic genes was still being bitterly debated among geneticists, cytoplasmic pollen sterility was being used extensively by hybrid seed producers.

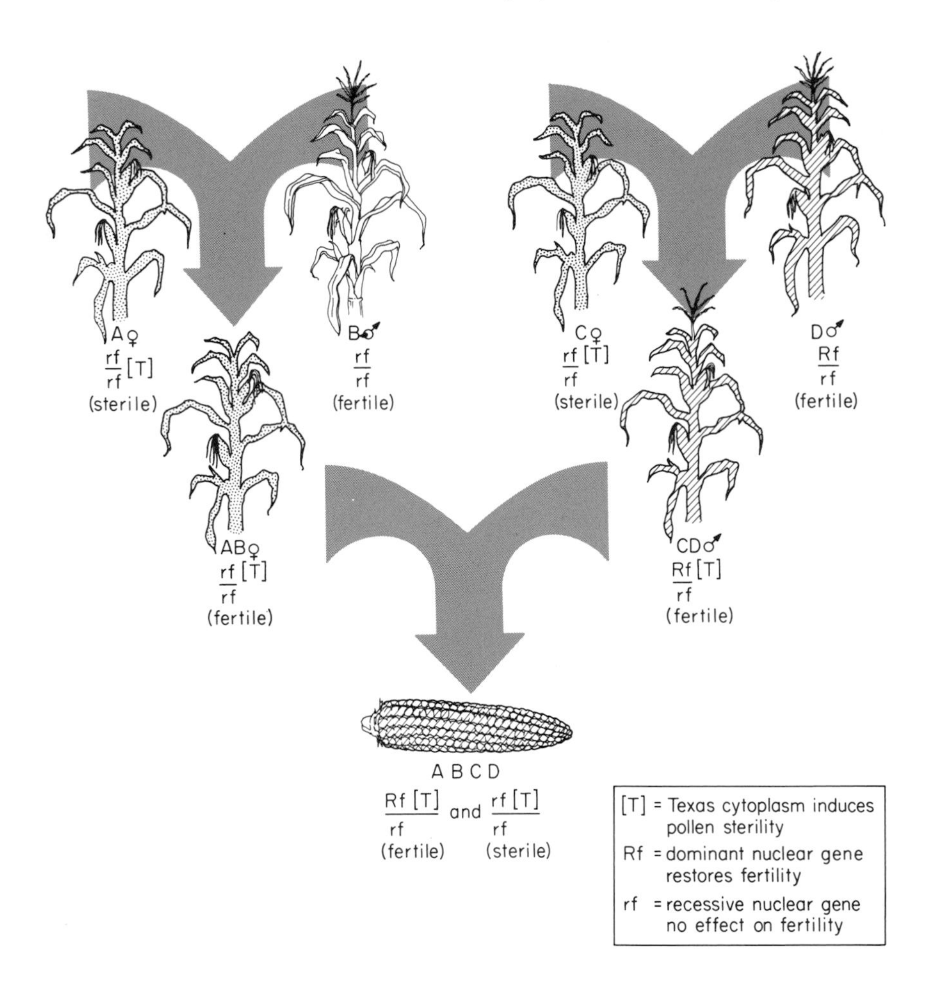

Fig. 6.2. Hybrid corn seed production. A, B, C, and D are four inbred lines. AB and CD are single cross hybrids produced the first year. Kernels from AB and CD produce plants for double cross made in second year. Kernels from double cross plants are used for hybrid corn production. *Rf* (dominant) and *rf* (recessive) are alleles of a nuclear gene that restores fertility in plants carrying [*T*] the cytoplasmic gene causing pollen sterility.

Within the last 2 years, the need for a better fundamental understanding of cytoplasmic inheritance in maize has been dramatized by the susceptibility of pollen-sterile lines of corn to two fungal diseases: yellow corn leaf blight and southern corn leaf blight. The Texas cytoplasm, which is commonly used as a source of pollen sterility, also transmits susceptibility to these infectious agents. This problem will be further discussed below (p. 371).

Inheritance of Iojap

Rhoades studied a particular kind of leaf variegation called *iojap* (*29, 30, 32*). He found that striped (variegated) plants used as the female parent gave rise to green, striped, and colorless seedlings with non-Mendelian ratios unaffected by the genotype of the male parent. In the reciprocal crosses, with a normal green female parent and a striped

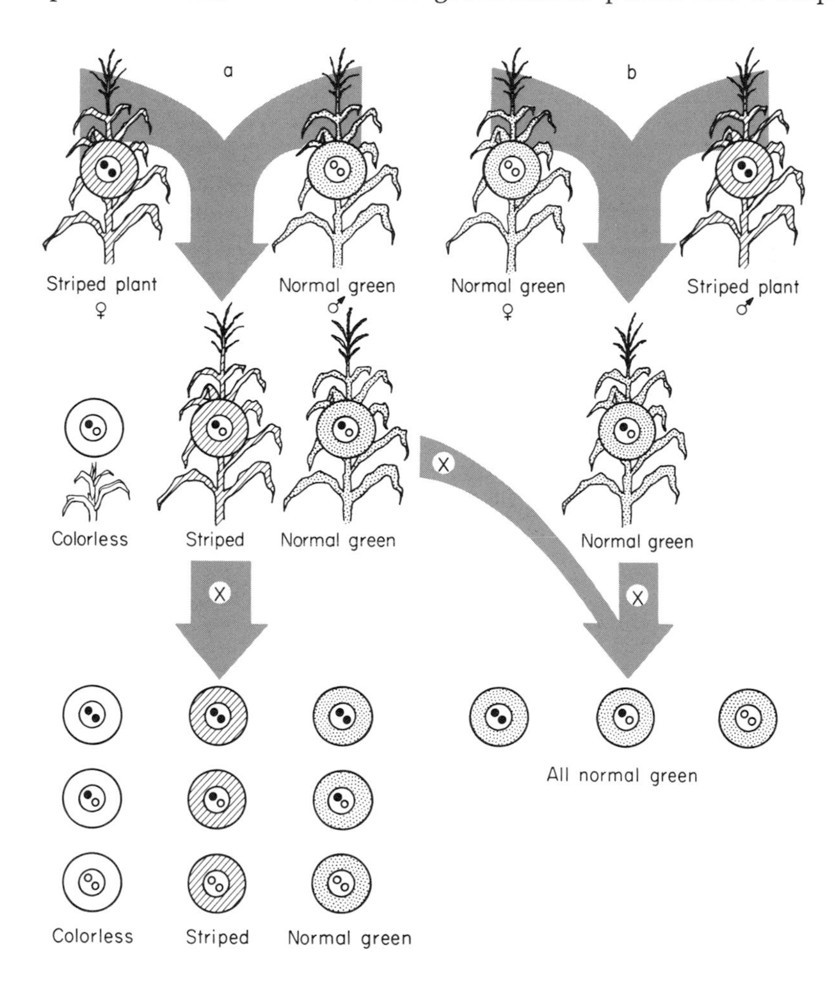

Fig. 6.3. Cytoplasmic inheritance of striping in higher plants. (a) Striped female parent × normal green male produces three kinds of seedlings in variable ratios: colorless, striped, and green. Crossing or self-pollinating striped plants again produces green, striped, and colorless F_2 regardless of nuclear gene segregation. (b) Normal green female × striped male produces only normal green progeny in F_1 and F_2 regardless of nuclear gene segregation. Code: ● = Ij; ○ = ij (iojap); ⊗ = self-pollination (see text for explanation).

male, the progeny and their descendants were all green, and the striped property was not transmitted. These results, shown in Fig. 6.3, provide one line of evidence that the striping factor is non-Mendelian.

A further test of the non-Mendelian nature of the chloroplast alteration was performed by a method originally developed by Correns to study somatic segregation in plants. The ears arising in a striped region were viewed as possible carriers of sectors in the germ line. To test this possibility the kernels were planted in rows according to their original location on the ear. An example of the results obtained with *iojap* are shown in Fig. 6.4. The sectorial pattern of the segregation of green and white seedlings is clearly evident.

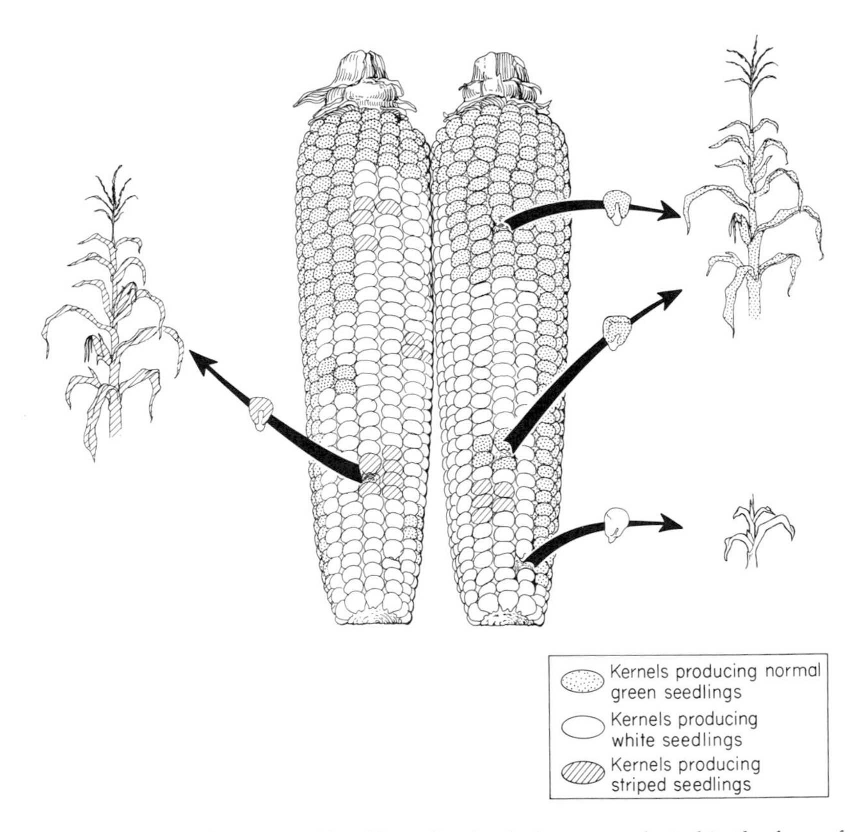

Fig. 6.4. Maize kernels from a self-pollinated striped plant were planted in the form of a map of the ear. A large sector was found, in which kernels gave rise to green, white, and striped seedlings. In the rest of the ear, all kernels gave rise to normal green seedlings. The sector constitutes evidence of somatic segregation of a genetic determinant for normal chloroplast development. After Anderson, E. G. (1923). *Bot. Gaz. (Chicago)* **76,** 411, Copyright 1923, by the University of Chicago.

Thus, the non-Mendelian nature of the plastid mutation is attested both by the strict pattern of maternal inheritance and by the evidence of somatic or clonal segregation in development of the ear. However, many questions remain unanswered. First, what is the origin of the variegation? Rhoades found that the variegation was initiated by a recessive Mendelian gene called *iojap* (*ij*). His evidence is diagrammed in Fig. 6.5. When green plants of the genotype *Ij ij* were crossed as female parent with males that were homozygous recessive (*ij ij*), the progeny included some striped plants. Half of the progeny were *Ij ij* and

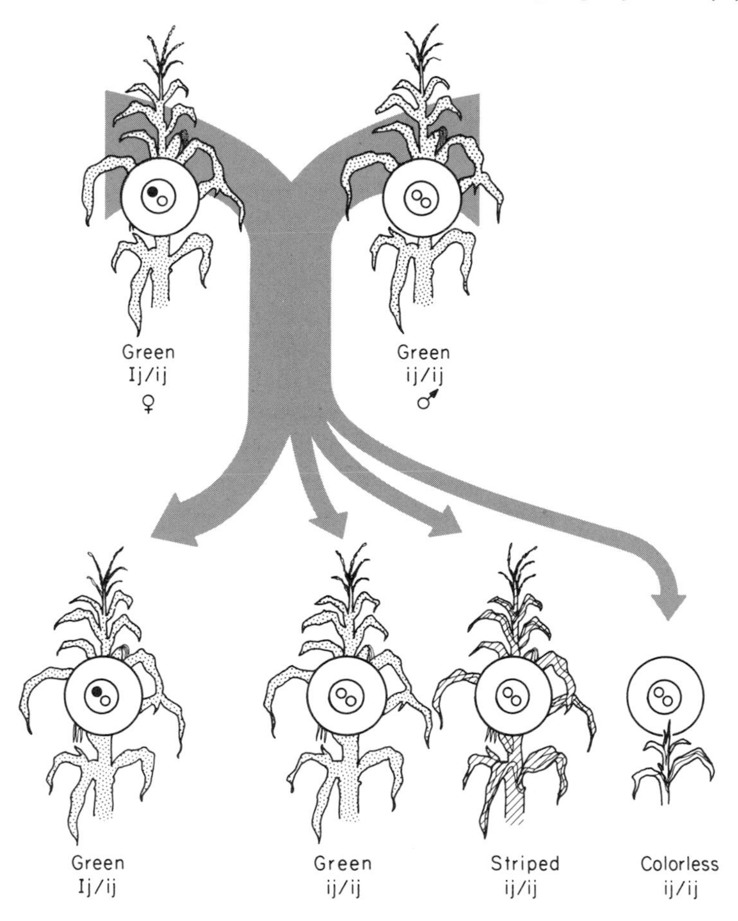

Fig. 6.5. Origin of *iojap* striped plants in maize. Cross of normal green female carrying *iojap* nuclear gene mutation (*Ij*/*ij* heterozygote) by normal green male carrying *iojap* (*ij*/*ij* homozygous recessive) genes produces two nuclear gene classes: heterozygotes which are all normal green and homozygous recessives which include green, striped, and colorless (lethal) seedings. Code: ● = *Ij*; ○ = *ij*(*iojap*).

these plants were all green, whereas among the *ij ij* plants, some were green, some striped, and some were colorless (and died as seedlings).

The striped plants used as female parent gave rise to green, striped, and white seedlings, regardless of the genotype of the male parent. Once the variegated condition had been established, the presence of the recessive *ij ij* genotype was not required. All further studies could be carried out with plants that were homozygous *Ij Ij*. Thus, permanent damage had been induced in chloroplast development by the *ij ij* genotype, but was not dependent on that genotype for its maintenance.

One may view the action of the *ij ij* genotype as mutagenic. Rhoades found that mutations to pollen sterility were also appearing in *ij ij* plants, supporting the idea that the plants produce some sort of a mutagen. In plants with both pollen sterility and plastid variegation induced by *iojap*, Rhoades found that the two mutations were transmitted independently. This evidence suggests that the two cytoplasmic genes are probably in different linkage groups, i.e., different DNA's.

In the light of present knowledge, we may assume that the mutations induced by the *iojap* plants are mutations in DNA—but which DNA? Since chloroplast development is affected in the leaf variegation, it seems likely that chloroplast DNA is the target, but Rhoades is careful to point out that this inference is not required by the data. The genetic data establish the particulate nature of the cytoplasmic determinant, but do not establish its location. Examples of nuclear gene-induced mutations of cytoplasmic genes leading to chloroplast variegation have also been reported in several other plants (*18*).

INHERITANCE OF POLLEN STERILITY

What about the genetic control of pollen sterility? Chloroplast DNA is probably not directly involved. Might pollen development be influenced by mutations in mitochondrial DNA? The fact that pollen-sterile plants are otherwise normal speaks against this view. However, let us consider in more detail the genetic and biochemical evidence about cytoplasmic pollen sterility. Some of the clearest examples of non-Mendelian heredity in higher plants involve the failure of pollen to develop normally. Instances have been reported in more than thirty genera of plants (*9*).

The first clearly analyzed example of cytoplasmic pollen sterility in *Zea mays* was described by Rhoades in 1933 (*28*). Since the phenotype of pollen sterility precluded the use of reciprocal crosses to analyze the pattern of inheritance, Rhoades crossed pollen-sterile plants as females with normal males for many generations to develop progeny that were homozygous for the genetic component of the male parent while still maintaining the cytoplasm of the female. With the availability of many markers on all of the maize chromosomes, it was possible for Rhoades to demonstrate the effectiveness of the repeated backcrosses in introducing the male genome into the backcross progeny. He found that the phenotype of pollen sterility could be maintained indefinitely in the backcross generations, thereby establishing its independence of the nuclear genome.

In subsequent years it was shown that some varieties of corn carried nuclear genes which allowed the sterile pollen to develop normally (*7*). These genes were called fertility restorers. Their interaction with cytoplasmic pollen sterility factors is shown in Fig. 6.2. Two different cytoplasmic pollen sterility factors can be distinguished in maize. These two, one called Texas [T] and the other USDA, have been distinguished by their different modes of interaction with the pollen fertility restorer genes, as well as by the morphology of the anthers. In the Texas type in completely sterile plants, the anthers are not exserted, whereas in the USDA type some or all of the anthers are exserted. Also, in conditions of partial sterility the Texas factor gives rise to twisted and misshapen anthers, whereas the morphology of partially sterile plants carrying the USDA factor is much more normal.

Studies are in progress on the biochemistry of pollen sterility in an attempt to identify the block to pollen development in maize and other plants. As a start, ninhydrin-positive components have been shown to accumulate in the anthers of sterile maize plants and excess asparagine and alanine have been detected (*7*).

These results do not yet shed much light on either the primary biochemical lesion or the location of the mutations responsible for the phenotype. It is interesting, however, that genes located on several chromosomes, two loci on chromosome 2 and others on chromosomes 3, 4, 7, 9, and 10, in maize, were shown to induce partial restoration of fertility in plants carrying the Texas sterility factor (*2, 7*). Thus genes throughout the nuclear genome interact with the cytoplasmic system. This arrangement is reminiscent of the mitochondrial-nuclear interactions found in the *petite* and *poky* mutations (Chapters 4 and 5).

Similar evidence of pollen sterility under the control of both cytoplasmic factors and fertility restoring genes has been described in the tobacco plant *Nicotiana.* Initially East (*8*) described a cytoplasmic factor responsible for pollen sterility. Subsequently eight different cytoplasmic male sterile factors have been described, each one causing different sorts of flower anomalies (*36*).

An extensive investigation of cytoplasmic pollen sterility in the genus

Solanum (potato) was carried out by Grun and his students (*13, 14*). They examined a wide range of species collected in Mexico and in South America where the genus presumably evolved. Many of the species are cross-fertile, making possible the kinds of interspecies crosses needed to test interactions of nuclear and cytoplasmic genes of diverse origin.

A number of different cytoplasmic factors were identified by their interactions with specific nuclear genes. For example, a cytoplasmic gene denoted as In^s conferred pollen sterility to plants when present in combination with any one of five nuclear *In* genes. In addition, two dominant restorer genes were identified. Either one could restore full fertility to plants which would otherwise be sterile because they carried In^s and a nuclear *In* gene. By studying the distribution of nuclear and cytoplasmic genes which interacted in pollen development, Grun *et al.* were able to show an independent role of the cytoplasmic factors in evolution within the genus. These studies complement those with *Oenothera,* to be discussed later in this chapter (p. 212), which also emphasize the role of cytoplasmic genes in evolution.

Fukasawa has recently reviewed evidence of cytoplasmic pollen sterility in wheat (*11*). In crosses of *Aegilops ovata* as female parent × *Triticum durum* (Emmer wheat) the progeny are pollen sterile. After twelve generations of backcrossing the female hybrids by *Triticum* male parents, the male sterility has remained unchanged and is accompanied throughout by poor vegetative growth. Some *Triticum* strains carry fertility-restoring genes. As in other systems, the restoration of pollen sterility is reversible—the cytoplasmic gene causing pollen sterility is unaltered in the pollen-fertile plants and sterility can be reestablished by crossing out the fertility-restoring genes.

Edwardson and Corbett (*10*) reported asexual transmission of cytoplasmic male sterility in *Petunia* by means of grafting. Initially the factor was transferred from stock to scion, using male-sterile plants as stock and fertile plants of a compatible genotype as scion. Subsequently the factor was transmitted through seed produced on male-sterile plants pollinated by the same compatible fertile line. The authors interpret their results as evidence that the factor is *viral* but do not consider the alternative possibility that a mutated cytoplasmic DNA of the male sterile host is the responsible agent.

An attempt to examine the distribution of pollen sterility determinants was made by Gabelman using maize plants which were partially sterile (*12*). He determined the percent of viable pollen per floret in five florets chosen at random from each of five tassels. He found that the percent of viable pollen per floret showed a discontinuous distribution, falling into a bimodal pattern. He interpreted the results as evidence of the particulate nature of a cytoplasmic factor, one copy of which was sufficient to cause sterility.

In summary, studies with *Zea mays* as well as other plants have provided clear evidence of particulate cytoplasmic genes operating on a defined nuclear genetic background. However, neither a method of inducing cytoplasmic mutations nor of mapping them has been developed. Further analysis of cytoplasmic inheritance in higher plants will require not only a source of more mutations, but above all a method to permit biparental inheritance and thereby to make possible genetic recombination and mapping. It has been widely assumed that the cause of maternal inheritance is the nontransmission of male cytoplasm into the fertilized egg. It may be that the mechanism of maternal inheritance in higher plants resembles that in *Chlamydomonas,* i.e., a restriction system in which the cytoplasmic DNA from only one parent is replicated. If so, maternal inheritance could be converted to biparental by blocking the restriction system—and then recombination and mapping might be possible. The practical importance to agriculture of recombination of cytoplasmic genes has been dramatized by the linkage of pollen sterility and disease susceptibility in maize.

A different approach to the general problem of cytoplasmic genetics in higher plants involves the investigation of those plants in which biparental transmission of cytoplasmic genes does occur, the prime example being *Pelargonium.*

BIPARENTAL INHERITANCE IN PELARGONIUM

The original studies of Baur, who discovered non-Mendelian biparental inheritance of plastid variegation in *Pelargonium,* were later confirmed by Roth (*33*) and by Imai (*17*). Recently, Tilney-Bassett, recognizing the fundamental importance of this discovery, undertook an elaborate and systematic breeding program to reexamine the phenomenon. His work will now be briefly summarized (*42–45*).

The strains used in these studies were horticultural varieties of *Pelargonium zonale,* most of them being periclinal chimeras of the so-called white-over-green type. In these chimeras there are three genetically distinct cell layers which can be thought of as a series of skins, wrapped one over the other. The germ line has been shown to originate from layer II. In the two varieties used extensively in this work, Flower of Spring (FS) and Dolly Varden (DV), layer II consists of mutant white cells. Both varieties also have isogenic lines which are wholly green. Consequently reciprocal isogenic crosses can be made between the green and chimeral strains of the same variety as well as between the

TABLE 6.2

Comparison between Seven Pairs of Reciprocal Crosses Showing the Gradation in the Transmission of Green and White Plastids from an Advantage through the Female to an Advantage through the Male Parent[a]

Crosses		Numbers of seedlings[b]			Percent variegated seedlings	Numbers of plastids[b]			Percent plastid distribution		Percent gradation in nonreciprocal plastid transmission
		G	V	W		G	W	Total	Female	Male	
Crystal Palace Gem[c]	G × W	67	4	–	5.7	138	4	142	97.2	2.8	Maternal advantage
× Dolly Varden	W × G	44	43	–	49.4	131	43	174	24.7	75.3	21.9
Crystal Palace Gem[c]	G × W	208	5	–	2.3	421	5	426	28.8	1.2	Maternal advantage
× J. C. Mapping	W × G	83	40	3	31.7	206	46	252	18.3	81.7	17.1
Dolly Varden	G × W	78	27	4	24.8	183	35	218	83.9	16.1	Paternal advantage
isogenic	W × G	36	13	–	26.5	85	13	98	13.3	86.7	2.8
Paul Crampel[c]	G × W	276	76	28	20.0	628	132	760	82.6	17.4	Paternal advantage
× F. of Spring	W × G	121	30	2	19.6	272	34	306	11.1	88.9	6.3
Paul Crampel[c]	G × W	13	11	11	32.4	35	33	68	51.5	48.5	Paternal advantage
× Miss B-Coutts	W × G	11	24	–	68.6	46	24	70	34.3	65.7	14.2
Paul Crampel[c]	G × W	108	139	83	42.1	355	305	660	53.8	46.2	Paternal advantage
× Dolly Varden	W × G	127	50	3	27.7	304	56	360	15.6	84.4	30.6
Flower of Spring	G × W	104	14	73	7.4	222	160	382	58.1	41.9	Paternal advantage
isogenic	W × G	115	27	2	18.8	257	31	288	10.8	89.2	31.1

[a] From (43), Table 4.
[b] G = green, V = variegated, W = white
[c] Green parent.

two varieties in all combinations. The splendid advantage of the chimera, of course, is that the green tissues nourish the white layer, making available for the breeder functional white flower parts, both male and female.

Some typical results of reciprocal crosses are given in Table 6.2. The occurrence of male transmission is evident from the sizeable number of white seedlings arising in crosses of G (♀) × W (♂), as well as the number of green seedlings arising in the reciprocal crosses, W (♀) × G (♂). Similar results are seen in crosses between different varieties, but the isogenic crosses are most impressive because they eliminate the possibility of differential nuclear effects.

In Table 6.2, the results of Tilney-Bassett are reported in terms of the number of seedlings recovered of each type—green, white, or variegated—and in terms of the percentage of green and white plastids. The plastid ratios were estimated from the proportions of seedling types by assuming that each zygote contained four functional copies and that the variegated seedlings arose from a mixed cell with either 3 + 1, 2 + 2, or 1 + 3 copies of the two types, depending on the ratio of green to white tissue.

Subsequently, a method of scoring embryos was developed which permitted examination of large numbers of progeny at early developmental stages. By scoring embryos it was possible to evaluate the role of selection against the white embryos which would have biased the seedling data. He found that the proportions of green, variegated, and white embryos did not change during development, and therefore that selection was not playing a role in the seedling types and plastid ratios seen. He was also able to rule out environmental influences on germination, and thereby to restrict the interpretation of the findings to wholly genetic mechanisms.

The results in Table 6.2 show that the ratios of G:V:W vary from one cross to another, and also differ markedly in reciprocal crosses, both isogenic and intervarietal. In all crosses an excess of green plastids was observed, but in some crosses, the excess of green plastids was higher in the G × W polarity (maternal advantage) and in other crosses, actually higher in the W × G polarity (paternal advantage).

The differences in ratios seen from one variety to another suggested a genotypic influence on the results, and raised the question whether that influence itself was maternal or not. To examine this question, the two chosen isogenic varieties, FS and DV, were compared in crosses with four other varieties (*45*). FS and DV were each used as the green female parent in crosses with all six varieties as white male parent; and then the reciprocal comparisons were made, with FS and DV as the white female parents in crosses with the six green strains as male parent.

The results of these twenty-four crosses are summarized in Table 6.3. The principal findings were: (*a*) the green female parent had a decisive role in determining progeny ratios, with little or no effect from the white male parent. With FS as female parent, the estimated percent of green plastids was consistently 50–56% in all six crosses, and with DV as female parent, the values were 77–93%. (*b*) In the reciprocal crosses between six different white female parents and FS and DV as green males, the influence of the white female parents was decisive, and the differences imposed by the green males, FS and DV, were minor, i.e., the frequencies of green progeny were in each instance a bit higher with DV as green male parent than with FS. (*c*) Each white female determined a characteristic G:W frequency, and the order, whether arranged by percent of green embryos or estimated percent of green plastids, was the same in crosses with FS and with DV as male parent. (*d*) The per-

TABLE 6.3

Percentages of Green, White, and Variegated Embryos and Estimated Percentage of Green Plastids in Twelve G × W Crosses and Reciprocal W × G Crosses[a]

Source of white plastids[c]	Green plastids: Flower of Spring				Green plastids: Dolly Varden			
	Embryos (%)[b]			Green plastids (%)	Embryos (%)[b]			Green plastids (%)
	G	V	W		G	V	W	
				(a) Green female × white male crosses				
FoS	54.8	2.0	43.2	56.3	76.0	17.7	6.3	89.1
FS	51.3	4.0	44.7	54.6	70.9	20.7	8.4	83.8
JCM	53.5	2.3	44.2	55.0	83.9	15.1	1.0	93.6
DV	43.5	20.4	36.1	54.5	57.6	33.1	9.3	77.2
LG	44.5	11.9	43.6	50.1	62.1	32.5	5.4	80.3
MBC	45.1	17.3	37.6	55.0	70.8	24.1	5.1	87.6
				(b) White female × green male crosses				
FoS	71.7	26.2	2.1	89.1	88.3	11.7	–	98.3
FS	69.0	27.2	3.8	85.2	79.9	18.9	1.2	95.5
JCM	56.9	32.9	10.2	77.3	67.9	29.9	2.2	88.1
DV	56.2	34.1	9.7	71.6	51.8	43.3	4.9	78.3
LG	23.2	58.6	18.2	51.3	25.4	67.2	7.4	62.6
MBC	4.9	73.8	21.3	34.6	0.3	84.8	14.9	42.0

[a] From (45), Table 2.

[b] G = Green, V = variegated, W = white.

[c] Abbreviations of *Pelargonium* cultivars (varieties): FoS, Foster's seedlings; FS, Flower of Spring; JCM, Mrs. J. C. Mappin; DV, Dolly Varden; LG, Lass o' Gowrie; MBC, Miss Burdett-Coutts.

centage of variegated progeny was consistently different from cross to cross, showing that the probability of variegation itself was under genetic control.

Thus the female parent, whether green or white, largely controls the G:W ratio. In the same study, it was discovered that the percentage of variegated embryos was regulated independently of the G:W ratio, and that this value could be altered greatly by the genotype of the female parent, and also influenced by the genotype of the male parent.

Let us now summarize the principal results which need to be taken into account in any interpretation of these findings.

1. Plastid phenotypes (and therefore chloroplast DNA's) from both male and female parents are transmitted to progeny in all crosses, i.e., *inheritance is biparental* as Baur (*1*) first reported. However, in contrast to Mendelian inheritance, some progeny are pure green like one parent, others are pure white like the other parent. Variegated progeny get some cytoplasmic genes from both parents.
2. Green plants are usually in excess whichever parent is green, although the actual G:W ratios are different in the reciprocal crosses.
3. Pure-type progeny (green or white) are usually more frequent than variegated ones, but the percent of variegated progeny differs from cross to cross. There is no simple relation between the G:W ratio and the % V.
4. The genotype of the female parent is the principal determinant of the G:W ratio and of the percent variegated. The male parent has some influence, though minor, on the final ratios.
5. Selection at the level of differential growth rates or survival of green versus white plastids seems to have been adequately ruled out in these experiments.
6. Studies of cell lineage in *Pelargonium* show that the zygote (i.e., fertilized egg) divides into a two-celled embryo, and only one of these cells, the "terminal" cell, gives rise to the entire plant.

We need a mechanism to account for the particular G:W ratios characteristic of each variety, for their control by the female parent, and for the separate control of the frequency of variegation.

Tilney-Bassett's data have ruled out most of the old-time favorite hypotheses, including random sorting-out of many copies, strong effects of environmental variables upon the G:W ratio, or selection against the white embryos. Since the majority of progeny are pure green or pure white, sorting-out must happen at the first zygotic division in most instances. Thus the zygote is acting like a genetic funnel—many gene copies come in but only a few go out to the terminal cell. Tilney-Bassett's proposed mechanism is that the first plastid that replicates effectively inhibits the replication of the rest; if two plastids happen to replicate more or less simultaneously, then variegation may result. An unexplained feature is the ability of the female parent to favor replication of one or the other type: the zygote must be able to tell green from white.

I have attempted to reevaluate the problem in terms of chloroplast DNA instead of green and white plastids. Although no direct evidence establishes chloroplast DNA as the carrier of the mutations blocking chloroplast development in *Pelargonium,* this interpretation is simplest and most consistent with present-day knowledge.

Reasoning from our recent findings with the *Chlamydomonas* system, two kinds of mechanisms may be at work in chloroplast heredity in *Pelargonium.* One is a mechanism by which only one of the two parental chloroplast DNA's is chosen for replication, which is in effect a kind of restriction system. The system might operate by enzymatic degradation as in bacterial modification-restriction, or by some other as yet unknown means. In *Pelargonium,* the fact that green plastids are usually favored over white regardless of parentage might result from alterations in the chloroplast DNA of the mutant, rendering it less likely to be replicated than the wild type. The idea of competition at the level of DNA replication has been postulated to explain suppressiveness in yeast (p. 141) and senescence in fungi (p. 167).

A second mechanism involves exchanges between parental DNA's, i.e., recombination or marker rescue, in which genes from one of the parental DNA's are transferred into the other one. Variegation has generally been interpreted as the result of a sorting-out process. The occurrence of variegated seedlings demonstrates that chloroplast genes from both parents have been transmitted simultaneously to the progeny. Having no additional markers, we cannot judge whether variegation involves the entire chloroplast genomes of both parents, or just one or a few genes. In the literature of this subject, recombination has not been invoked to explain seedling ratios or variegation, because there was no evidence of the occurrence of recombination. However the new knowledge that chloroplast and mitochondrial DNA's do recombine (as seen in *Chlamydomonas* and in yeast), leads one naturally to consider the role of recombination in plastid heredity in higher plants.

In the *Pelargonium* system, the genetic difference between green and white may be localized at a specific mutated site on chloroplast DNA. If so, white progeny could arise in these crosses by recombination of the mutant site into wild-type DNA. On this basis, the G:W ratios of seedlings might actually represent recombination frequencies. If the recombination is associated with degradation of the mutant DNA, then the G:W ratios might also involve the probability of marker rescue.

The *Pelargonium* system has been discussed in some detail because extensive data are available for evaluating a basic question in chloroplast heredity—the mechanism of variegation. Although the question has not been answered, some false leads have been ruled out, and the issue has been focused on the control of DNA replication, transmission and recombination.

CYTOLOGICAL IDENTIFICATION OF "MIXED" CELLS

All the investigators of chloroplast heredity have assumed that the green and white sectors seen in variegated plants result from some kind of a sorting-out process, i.e., somatic segregation. There the agreement ends. Somatic segregation of what? In the historic controversy between plastome and plasmone, it was thought that the identification of both mutant and wild-type plastids lying side by side in the same cell might solve the problem. If phenotypically distinct plastids could be found in the same cell, so-called mixed cells, then it could be argued that the genetic autonomy resided in the plastids themselves. On the other hand, if the green sectors contained cells with normal plastids and the white sectors contained cells with mutant plastids and no "mixed" cells could be found, then it could be argued that the determinant of plastid phenotype resided elsewhere in the cytoplasm. Consequently the quest for "mixed" cells became a central theme in chloroplast cytogenetics. In plants with striped leaves, "mixed" cells could sometimes be seen at the interface between the green and white sectors. However, "mixed" cells might easily have been the result of transient physiological interaction between the green and mutant sectors and would then represent "pseudo mixed" cells. In practice the distinction between mixed and pseudo mixed cells has been difficult to achieve, and consequently cytological evidence of mixed cells has not played a useful role in the elucidation of chloroplast genetics.

A particularly impressive example of mixed cells with two distinct plastid types was presented by Woods and Du Buy in their studies of the plant *Nepeta* (*46, 47*). Unfortunately their observations were never followed up with a thorough genetic analysis. Extensive studies of mixed cells with both light and electron microscopes have been carried out with the plant *Antirrhinum* (*16*), and mixed cells have been found throughout the tissue of variegated leaves. These observations are in line with genetic evidence that the wild type and mutant plastids carry unique genes conferring the observed morphological differences.

EPILOBIUM AND THE PLASMONE HYPOTHESIS

The willow-herb *Epilobium* was chosen by P. Michaelis for his intensive and extensive investigations of non-Mendelian inheritance carried out over the last 40-odd years, beginning with his studies on the cytology and embryonic development of *Epilobium* published in 1925 (*19*). In an extensive review summarizing his observations and conclusions in 1954, Michaelis stated, "We cannot distinguish manifestations of single plasmagene units, but are limited to recognition of the effect of all those hereditary components of the cell that do not show Mendelian segregation;" (*21*). He uses the term plasmone, originated by von Wettstein, to include all extranuclear hereditary elements of the cell (*21*). More recently, Michaelis has summarized his principal fundings and conclusions in a series of short papers (*22–25*).

In choosing an example to illustrate plasmone inheritance Michaelis has often described the following set of experiments illustrated with one of his diagrams in Fig. 6.6. The figure shows typical F_1 plants resulting from reciprocal crosses between two species of *Epilobium, E. hirsutum* I and *E. luteum*. The smaller plant on the right is pollen sterile, female × *E. hirsutum* male has fertile pollen. In Fig. 6.7 we see a similar cross between two geographic races of *E. hirsutum*. Michaelis carried out a series of backcrosses starting with the healthy F_1 hybrid from the orig-

Fig. 6.6. Typical F_1 plants resulting from reciprocal crosses between two species of *Epilobium, E. luteum* and *E. hirsutum*. The plant on the right, with stunted growth and sterile pollen, received its cytoplasm principally from the *E. hirsutum* parent, while the healthy plant on the left received its cytoplasm from the *E. luteum* parent. From (22).

Fig. 6.7. Typical F_1 plants from reciprocal crosses between two geographical races of *E. hirsutum* (*E. hirsutum* Jena and *E. hirsutum* Munchen). As in Fig. 6.6, the plants resulting from reciprocal crosses are very different. From (22).

inal *E. luteum* × *E. hirsutum* Jena cross. The normal appearance of the F_1 hybrid indicated that the *E. hirsutum* genes which came from the pollen were functioning adequately in the cytoplasm of the *E. luteum* female parent.

What would happen if all the *E. luteum* genes were replaced by *E. hirsutum* genes by means of repeated backcrosses with the *E. hirsutum* parent as male? Michaelis performed this celebrated series of backcrosses for twenty-five generations, testing the hybrids for their response in crosses with *E. luteum.* The question was whether totally replacing the *E. luteum* genome would produce a plant like the original homozygous *E. hirsutum* which, used as the female parent, gave poor F_1 progeny in crosses with *E. luteum* males. The answer was very clear. Even after twenty-five generations, crosses between the hybrid (*E. hir-*

sutum genes in a *E. luteum* cytoplasm) and *E. luteum* males gave normal progeny.

As Michaelis concluded, this experiment demonstrated the constancy and relative autonomy of the *E. luteum* cytoplasm in cells containing *E. hirsutum* genes.

Subsequently Michaelis undertook reciprocal crosses of a similar sort with a large assortment of different species and races in the genus *Epilobium*. He concluded, "This series of crosses leads to the result that reciprocal differences are not produced by the cytoplasm alone, but by the interaction of the cytoplasm with the nuclei. All the different plants possess the same cytoplasm as *Epilobium hirsutum* Jena but different nuclei. It can be said that the reciprocal differences are produced by difficulties in the interactions between nuclei and cytoplasm. In reciprocally equal hybrids the cytoplasm and the nuclei work together in a normal way."

It should be emphasized that the differences in reciprocal crosses which Michaelis observed included not only effects upon chloroplast development, but also more general morphogenetic effects such as stunted growth of plants, deformed flowers, and pollen sterility. In no case is the biochemical basis of the phenotypic change known. Poor growth may be attributed to poorly functioning chloroplasts, and by implication to mutations in chloroplast DNA. However, mutations affecting only flower formation and pollen development in plants that are otherwise normal can hardly be viewed as chloroplast mutations. Perhaps they are indications, as Michaelis has proposed, of mutations in some other extranuclear genetic system.

In other studies Michaelis attempted to analyze the pattern of variegation in leaf development. In particular, he attempted to evaluate the number of genetic copies per cell on the basis of a simple mathematical model of segregation. To develop a workable model it was necessary to make certain simplifying assumptions, in particular to assume that plastids, whether normal or mutant, multiply at the same rate and that plastids are distributed to daughter cells at cell division in a random and equal manner like green and white marbles in a black box. As might have been anticipated, his calculations indicated that one needed a system with a low number of units to account for the variegation pattern observed, a lower number than his plasmone hypothesis had envisaged.

Indeed, Michaelis tried to distinguish between genes present in many copies (plasmone genes) and those present in a few copies (plastome genes located in plastids) by analyzing segregation patterns in variegated seedlings. Applying this method to green-white variegation in striped seedlings, Michaelis found a result not too different from that in *Pelargonium*—that very few copies need be invoked to explain the data. Uncertainties in cell lineage and cell division rates make the quantitation imprecise. Nonetheless, it is a striking observation that far fewer DNA copies are necessary than would be expected on the basis of estimates of the number of plastids. This result is puzzling and reminiscent of a similar problem with mitochondrial DNA in yeast (cf. p. 120).

PLASTID AUTONOMY IN OENOTHERA

As many later investigators have emphasized, the pioneering studies of *Oenothera* by Renner and his students have provided the most extensive formal genetic evidence of chloroplast autonomy in all the higher plant literature. Here we will consider a few experiments excerpted from these voluminous investigations (*18, 27*).

The genus *Oenothera* was an excellent choice for studies of cytoplasmic genetics because of the peculiarities of the nuclear system. Virtually no recombination of the parental nuclear genomes occurs in F_1 hybrids coming from crosses between different races of *Oenothera*. Consequently the segregation of cytoplasmic genes can be examined on an almost constant nuclear background. The genomically constant hybrids were referred to as "true breeding complex-heterozygotes" and exploited experimentally by Renner long before the mechanism was unravelled in a series of beautiful cytogenetic studies by Cleland (*4*).

Oenothera is a diploid plant with seven pairs of chromosomes. Cleland showed that the chromosomes are organized into rings at meiosis as the result of a series of subterminal reciprocal translocations. As shown in Fig. 6.8, crossing-over between homologous regions at the ends of the chromosomes serves to hold the chromosomes together on the metaphase plate, and they segregate as a unit at anaphase. In some species, all fourteen chromosomes are included in a single ring, and segregate as shown in Fig. 6.9. In other species, one or more pairs of

Fig. 6.8. Diagram to show segregation of maternal and paternal complexes in first meiotic anaphase in *Oenothera*. The seven chromosomes from each parent are lined up at metaphase because of reciprocal translocation. The set shown here at anaphase will segregate without recombination at metaphase II, giving rise to gametes of identical genotype to the parental strains. From (*5*).

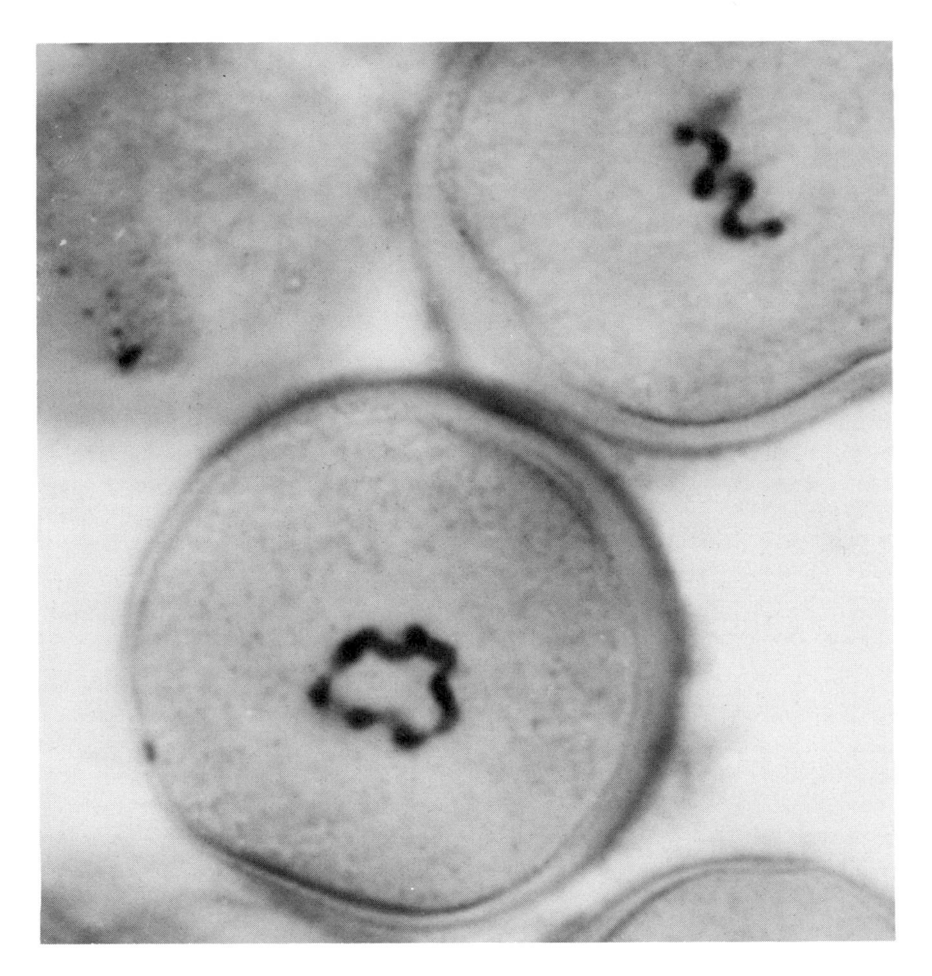

Fig. 6.9. Photomicrograph of first metaphase of meiosis in *Oenothera,* showing all 14 chromosomes in a circle on the metaphase plate. Above: side view; below: polar view. From (5).

chromosomes are not associated with the ring and segregate independently. The specific chromosome behavior of the species involved must be taken into account in the studies of cytoplasmic heredity.

In *Oenothera,* interspecies crosses are often fertile, but in some combinations the chloroplasts develop poorly. Renner used the level of plastid development to evaluate interactions between the nucleus and the plastids in sets of reciprocal interspecies crosses. These studies provided extensive evidence for what Renner interpreted as plastid autonomy.

One example often cited for its clarity involves reciprocal crosses between the two species *O. hookeri* and *O. lamarkiana*. The results are shown in Table 6.4. *Oenothera lamarkiana* has two chromosome complexes, called *gaudens* and *velans*, each representing a full haploid genome. The homozygous diploid is lethal, so the plant breeds true as a heterozygote (containing both *gaudens* and *velans* genomes). *Oenothera hookeri* is a homozygous diploid, so its haploid gametes are all identical, and are simply called *hookeri*.

As can be seen in Table 6.4, when *O. hookeri* was the female parent, both nuclear classes of progeny, *hookeri-gaudens* and *hookeri-valens*, were green. In the reciprocal cross, however, the *hookeri-gaudens* progeny were green, but the *hookeri-velans* progeny were yellow. Also, a few variegated plants appeared among the *hookeri-velans* progeny from both of the reciprocal crosses. Renner interpreted these results as evidence of plastid autonomy with the following reasoning. First, he assumed maternal inheritance of plastids, so that he considered the plastids themselves to be *hookeri* when *O. hookeri* was the female parent. On this basis, he reasoned that *hookeri* plastids were compatible with both nuclear genomes, but that *lamarkiana* plastids developed normally on one of the nuclear genetic backgrounds but not on the other. Since the nuclear genomes are identical, the *lamarkiana* plastids themselves must be different from the *hookeri* plastids.

Renner distinguished between plastids themselves as the carriers of this cytoplasmic difference in reciprocal crosses, as against some cytoplasmic factor outside of the plastids by studies of the variegated seedlings. He viewed the variegation as evidence of a small amount of male transmission of plastids through the pollen. For example in the cross with *O. hookeri* as female parent, he viewed the variegated progeny as the result of male transmission of a few *lamarkiana* plastids

TABLE 6.4

Plastid Autonomy in Oenothera

Cross	F_1 offspring
hookeri × *lamarckiana*	^{h}*hookeri-gaudens* (green)
	and
♀ Plastids *hookeri*	^{h}*hookeri-velans* (green + few variegated)
lamarckiana × *hookeri*	^{h}*hookeri-gaudens* (green)
	and
♀ Plastids *lamarckiana*	^{h}*hookeri-velans* (yellow + few variegated)

[a] From (*18*).

which are yellow in the *hookeri-velans* plants; and in the reciprocal cross, the variegation was attributed to male transmission of a few *hookeri* plastids.

Renner stressed the significance of hybrid variegation, and he and his students studied this phenomenon extensively. Hybrid variegation is the mutant response of one plastid type but not the other, interacting with a particular nuclear genome when both plastid types are present in the same plant. The permanence of the plastid autonomy was shown by repeated generations of back crosses in which the plastids maintained their mutant phenotype after as many as twenty backcross generations but could be restored to normality by being crossed into a compatible nuclear genome. These experiments established the genetic autonomy of plastids by demonstrating the clear separation of plastid phenotype and genotype. They also demonstrated the importance of coordination between nuclear and plastid genomes in chloroplast development.

The weakness of Renner's interpretation lies in the absence of any positive evidence that plastid transmission, rather than the transmission of some other cytoplasmic component, lies at the basis of variegation. Renner's brilliant analysis of hybrid variegation in the *O. hookeri* × *O. lamarkiana* (and many other) F_1 hybrids, surely indicates somatic segregation of some genetic determinant. What is its identity? Chloroplast DNA is by far the best candidate, primarily because of its existence and the growing body of knowledge about its functions.

At the genetic level, further evidence about the *Oenothera* plastids comes from studies by Schotz and Stubbe. An intensive investigation of reciprocal crosses between normal green and mutant white gametes was performed by Schotz (*34*). In this system, the mutant plastids were unable to develop normally on any nuclear background. Since the white seedlings were lethal, Schotz used a variegated plant that produced a continual supply of white flower parts. The plant was identified as *O. hookeri*, having the *hookeri* nucleus, but the mutant plastids came from *O. biennis*, as a result of a previous cross (*O. biennis* × *O. hookeri*). This plant was utilized as the source of white gametes in reciprocal crosses, G × W and W × G, with eleven different species as green parent. Subsequently, another set of G × W crosses were made with the same eleven green female parents and a different male containing mutant white plastids from *O. lamarkiana*.

The principal results shown in Table 6.5 demonstrate a number of points.

1. G × W crosses produced green and variegated progeny, but no white seedlings, and W × G crosses produced white and variegated

TABLE 6.5

Percentage of Variegated Offspring from Crosses between a Series of Oenothera Species, used as a Variable Source of Green Plastids, and Two Mutant Plants, used as a Constant Source of White Plastids[a]

Source of plastids		No. offspring[b]			Percent		No. offspring[b]			Percent
Female	1st male	G	V	W	Variegated	2nd male	G	V	W	Variegated
Green × white crosses										
O. hookeri	× *biennis*	797	–	–	–	*lamarckiana*	380	65	–	14.7
O. lamarckiana	× *biennis*	555	14	–	2.5	*lamarckiana*	376	140	–	27.2
O. bauri	× *biennis*	256	26	–	9.3	*lamarckiana*	119	91	–	43.3
O. rubricaulis	× *biennis*	255	84	–	24.8	*lamarckiana*	90	56	–	38.4
O. suaveolens	× *biennis*	264	71	–	21.1	*lamarckiana*	105	117	–	52.6
O. biennis	× *biennis*	118	68	–	36.4	*lamarckiana*	77	64	–	45.4
O. syrticola	× *biennis*	169	108	–	38.9	*lamarckiana*	72	134	–	65.0
O. parviflora	× *biennis*	219	158	–	41.9	*lamarckiana*	64	83	–	56.4
O. rubricuspis	× *biennis*	183	108	–	37.1	*lamarckiana*	27	57	–	67.8
O. ammophila	× *biennis*	63	51	–	44.6	*lamarckiana*	–	—	–	–
O. atrovirens	× *biennis*	148	344	–	69.9	*lamarckiana*	27	51	–	65.3
White × green crosses										
O. biennis	× *O. hookeri*	–	12	76	13.5					
O. biennis	× *O. lamarckiana*	–	124	68	68.2					
O. biennis	× *O. bauri*	–	54	159	25.5					
O. biennis	× *O. rubricaulis*	–	34	55	38.2					
O. biennis	× *O. suaveolens*	–	8	52	13.3					
O. biennis	× *O. biennis*	–	11	45	20.5					
O. biennis	× *O. syrticola*	–	–	150	–					
O. biennis	× *O. parviflora*	–	1	65	1.5					
O. biennis	× *O. rubricuspis*	–	–	–	–					
O. biennis	× *O. ammophila*	–	–	204	–					
O. biennis	× *O. atrovirens*	–	–	207	–					

[a] Adapted from (*18*) Table VIII.3.
[b] G = Green, V = variegated, W = white.

seedlings but no greens. This result is clearly different from that obtained with an analogous set of crosses in *Pelargonium* (cf. p. 199), in which green, white, and variegated seedlings were recovered from each G × W cross.

2. The percent of variegated progeny recovered (% V) differed in each cross, and the species could be arranged according to increasing % V. The arrangement of green species by increasing % V was the same in both sets of G × W crosses, one set with the white male carrying plastids from *O. biennis* and in the other set from *O. lamarkiana*. However, the actual frequencies of variegated progeny in the two sets were quite different. The fact that the order of eleven species was the same in both sets demonstrated a role of the female parent, while the difference between the two sets indicated an effect of the male, presumably of the male plastids themselves.

3. In the reciprocal crosses with *O. biennis* as white female parent, the % V also varied from cross to cross, indicating an influence of the female parent, and the order was roughly that seen in the reciprocal crosses. In general, the probability of white plastids appearing in the progeny increased from cross to cross as in the G × W crosses, but the actual frequencies were very different.

Schotz interpreted these data as evidence of differential multiplication rates of plastids from the various species; he concluded that the nuclear genome contributed only a secondary influence on the outcome. His conclusion was based primarily on differences in % V of crosses with the same nuclear complement but different plastids from the green parent. He ruled out the possibility that different females contribute different numbers of plastids by actually counting the numbers in egg cells of different species and finding them very similar.

Schotz's experiment (Table 6.5) has been discussed in detail because it has played an important role in popularizing the view, first proposed by Renner, that different plastid types multiply at different rates on a constant genetic background. Schotz assumed that the percent of variegated progeny recovered in these crosses was a measure of the relative multiplication rates of different plastid types. Thus, the hypothesis of different multiplication rates depends upon the validity of this assumption.

In the light of present day knowledge, Schotz's results can also be evaluated at the level of chloroplast DNA. In the hybrid crosses of Table 6.5, the maternal chloroplast DNA was always transmitted to the progeny, and some fraction of them received a paternal contribution as well; this latter group were variegated. We may then ask what molecular mechanism might be responsible for the frequency of % V seen in each cross, and for the different frequencies seen in the different crosses. This question has already been discussed in reference to the similar experiments performed with *Pelargonium* (Table 6.2 and p. 197). Two mechanisms were proposed: competition at the level of DNA replication, and introduction of genes from the paternal DNA into the maternal chloroplast genome by recombination. Either or both of these mechanisms could be operative in *Oenothera;* there are no data on the subject.

An interesting difference between the results with *Oenothera* and those with *Pelargonium* is the recovery of green, white, and variegated progeny in the reciprocal G × W and W × G crosses in *Pelargonium,* but only of maternal and variegated progeny in *Oenothera.* This difference may reflect the different fates of paternal chloroplast DNA's during the fertilization process in the two species. While the overall pattern of inheritance seems to be species specific, the quantitative probabilities of success of the male and female chloroplast genomes seem to be regulated primarily by the female parent.

Recently, Schotz (*35*) has further examined the effects of hybrid variegation upon chloroplast development in the electron microscope. Specifically he examined the development of *lamarkiana* plastids in plants with the *velans-hookeri* nuclear genome (from *O. hookeri* × *O. lamarkiana* crosses). He described four distinguishable steps in plastid development: (*a*) detachment of invaginations from the inner membrane of the plastid envelope; (*b*) growth of lamellar membranes as two-dimensional sheets; (*c*) increase in the number of lamellar membranes; and (*d*) partial fusion of these membranes into their final, stacked configuration. Blockage of chloroplast development was seen at each of these stages in the variegated leaves of *O. hookeri* × *O. lamarkiana* progeny. Schotz speaks of the "relative intensity of disharmony" between genome and plastome and suggests that the "disharmony" may vary in intensity at different locations in the plant and at different times in development.

In the language of molecular genetics, "disharmony" may be read as failures of regulatory interactions between nuclear and plastid genomes. As yet little is known of regulation in plastid development at the molecular level (cf. Chapter 8). Consequently, Schotz's cytological observations cannot yet be interpreted in biochemical terms. Higher plants do not lend themselves to the kind of biochemical and biophysical studies necessary to test the molecular hypotheses advanced here.

As fundamental aspects of regulation are worked out in simpler systems, especially in the sexual green algae, it should become possible to apply the knowledge to the *Oenothera* mutants, complementing the extensive genetic knowledge which is available.

THE ROLE OF PLASTID GENES IN EVOLUTION IN THE GENUS OENOTHERA

The proposal of Schotz that plastids from different species of *Oenothera* might multiply at different rates in response to particular nuclear genomes was adopted by Stubbe to help in the interpretation of the role of the plastid genome in evolution in the genus. However, Stubbe's exhaustive investigations of compatibility between nuclear and plastid genomes can also be interpreted at the DNA level. The findings show a striking interaction between nuclear and chloroplast genomes during evolution.

Renner had concluded from his studies of interspecies crosses that different plastid types existed in the genus, as judged by their ability to develop normally on different genetic backgrounds. Renner emphasized that the inability of plastids to become green in some hybrid combinations was not the result of mutation, since their ability to develop normally could be restored by a suitable subsequent cross into a compatible background.

Stubbe studied more than 400 combinations of nuclear and plastid genomes in order to evaluate the compatibility of the plastid types with the nuclear complexes. The results of these extensive studies are summarized in Figs. 6.10, 6.11, and 6.12 (*38–41*).

Stubbe's classification of the nuclear complexes agrees well with the extensive taxonomic studies of Cleland and of Renner. As Stubbe points out, the relationships in Figs. 6.11 indicate only a limited compatibility between nuclear and plastid genomes. Nonetheless even within that limited range of compatibilities, not all of the possible combinations have been found in nature. Figure 6.12 shows the distribution of naturally occurring species according to the same classification scheme.

Stubbe suggests that the missing combinations may have existed at an earlier time, but they have been superceded by faster multiplying plastid types. This proposal is based on assigning relative multiplication rates to the plastid types according to the findings of Schotz. Whether selection operated at the level of chloroplast multiplication or of DNA replication, or by some other mechanism, the data are in agreement with the concept that the plastid genome has played an independent role in evolution in the genus *Oenothera*.

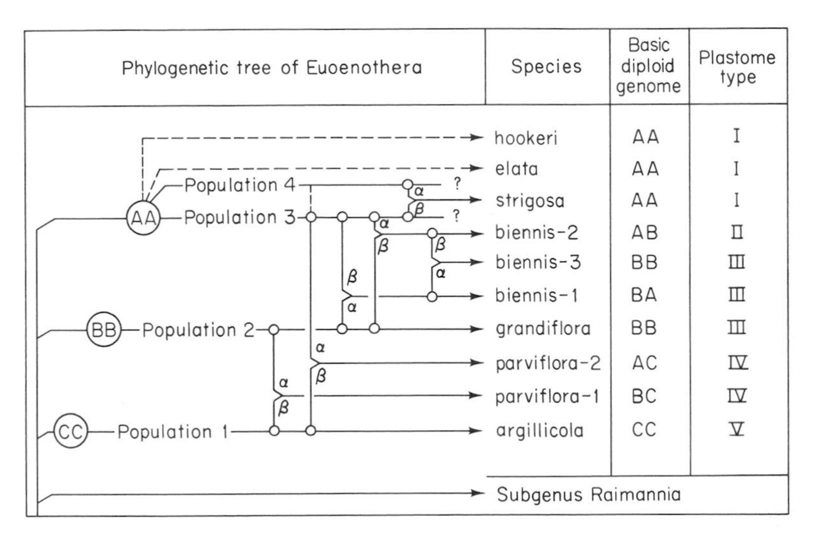

Fig. 6.10. Phylogenetic tree of the subgenus *Euoenothera*, showing the relationship of the basic nuclear genotypes and plastid types of species found in nature. α and β refer to parental genomes. From (*40*).

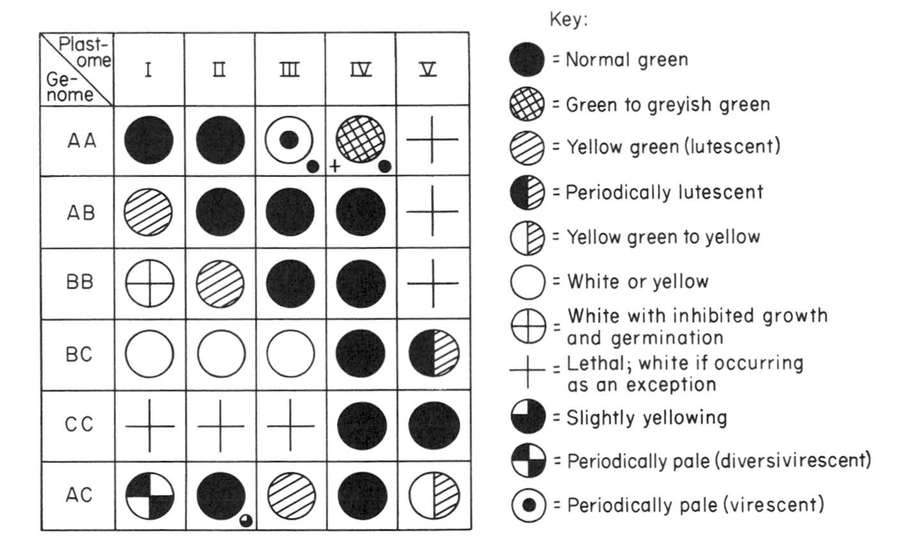

Fig. 6.11. Compatibility relations between different genotypes and plastid types. From (*40*).

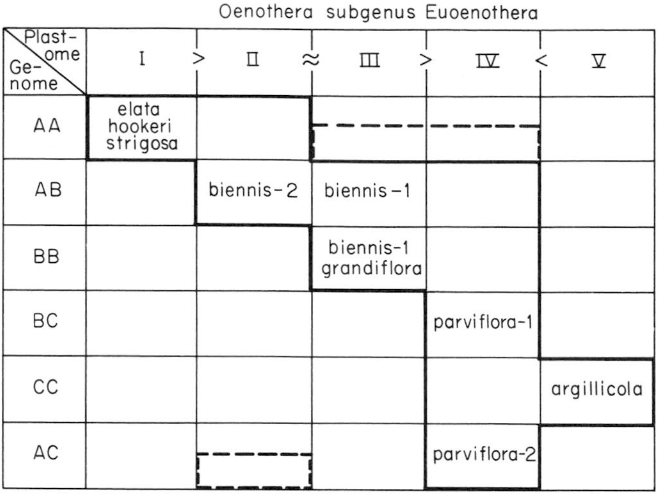

Oenothera subgenus Euoenothera

Plastome: I > II ≈ III > IV < V

Plastome / Genome	I	II	III	IV	V
AA	elata hookeri strigosa				
AB		biennis-2	biennis-1		
BB			biennis-1 grandiflora		
BC				parviflora-1	
CC					argillicola
AC				parviflora-2	

Fig. 6.12. Distribution of the natural species among the compatible combinations (framed squares). The signs between the plastomes I–V refer to the relative strength (multiplication rate) of the plastids. Dotted lines mark position of missing combinations which may have existed in the past. From (*41*).

CONCLUDING REMARKS

This chapter has summarized some of the leading investigations of cytoplasmic genes in higher plants. These studies have relied principally on evidence of non-Mendelian ratios, differences in progeny from reciprocal crosses, and somatic segregation for the identification of cytoplasmic genes. No evidence of recombination, linkage, or association with any cytoplasmic DNA's have been adduced in any of these systems. Nonetheless, all the reported evidence is consistent with the view that these genes like those in microbial systems are located in cytoplasmic DNA's.

The phenotypes involved in cytoplasmically inherited mutations include chloroplast development (seen as entire white seedlings or white striped sectors in green plants), pollen development (seen as pollen sterility), and a host of other morphogenetic abnormalities. Although these phenotypes are poorly defined, they indicate a wide ranging influence of cytoplasmic genes on plant development. Cytoplasmic genes have also been implicated in plant evolution, not only on the basis of exhaustive studies in the genus *Oenothera,* but also in the genus *Solanum,* the domestic potato. In *Oenothera,* Stubbe has proposed that the chloroplast genome played a role in evolution by virtue of different multiplication rates in different nuclear backgrounds. In *Solanum,* Grun has suggested that cytoplasmic pollen sterility played an important role in speciation by blocking various interspecies crosses, and Stinson has provided evidence of an analogous process in *Oenothera* (*37*).

Cytoplasmic genes have been of great importance in both agriculture and in horticulture. Cytoplasmically inherited pollen sterility is effectively utilized in hybrid seed production; and plants of many species with white stripes or sectors resulting from mutations of chloroplast genes are prized ornamentals. Currently, the susceptibility of hybrid corn to southern corn leaf blight has been found linked to cytoplasmic pollen sterility. In the absence of recombination techniques, plant geneticists have found it necessary to discard the enormously valuable sterility factor in order to gain blight resistance. The present crisis in hybrid corn production vividly highlights the need for the application of new developments in cytoplasmic genetics to agricultural crops.

This chapter has focused primarily on studies that shed light on the molecular basis and behavior of cytoplasmic genes in replication and transmission. In plants like *Pelargonium* and *Oenothera* in which biparental transmission of cytogenes does occur, efforts to establish a recombination system might well succeed. In plants like *Zea mays* which show strict maternal inheritance, the mechanism of exclusion of the male cytogenes is unknown. If no cytogenes enter the pollen, then the possibilities for recombination are nil. If, however, exclusion operates at a later stage in fertilization, by an enzymatic mechanism analogous to that in *Chlamydomonas,* then it might be possible to rescue the male genome.

Biparental inheritance, with its potential for recombination, might be engineered by the selection of suitable mutations. The development of methods for the detection and analysis of recombination of cytogenes could open a new era in higher plant genetics—in basic science and in applications to agriculture.

Suggested Review Articles

Duvick, D. N. (1965). Cytoplasmic pollen sterility in corn. *Advan. Genet.* **13,** 1–56.

Hagemann, R. (1964). "Plasmatische Vererbung." Fischer, Jena.

Kirk, J. T. O., and Tilney-Bassett, R. A. E. (1967). "The Plastids," Chaps. 7, 8, and 9. Freeman, London.

Michaelis, P. (1954). Cytoplasmic inheritance in *Epilobium* and its theoretical significance. *Advan. Genet.* **6,** 287–401.

Rhoades, M. M. (1955). Interaction of genic and non-genic hereditary units and the physiology of non-genic inheritance. *In* "Encyclopedia of Plant Physiology" (W. Ruhland, ed.), Vol. 1, Springer-Verlag, Berlin and New York.

References

1. Baur, E. (1909). Das Wesen und die Erblichkeitsverhaltnisse der "Varietates albomarginatae hort" von *Pelargonium zonale. Z. Vererbungslehre* **1,** 330–351.
2. Becket, J. B. (1966). Inheritance of partial male fertility in maize in the presence of Texas sterile cytoplasm. *Crop Sci.* **6,** 183–184.
3. Caspari, E. (1948). Cytoplasmic inheritance. *Advan. Genet.* **2,** 1–66.
4. Cleland, R. E. (1962). The cytogenetics of *Oenothera. Advan. Genet.* **11,** 147–237.
5. Correns, C. (1909). Vererbungsuersuche mit blass (gelb) grunen und buntblattrigen Sippen bei *Mirabilis jalapa, Urtica pilulifera* und *Lunaria annua. Z. Vererbungslehre* **1,** 291–329.
6. Correns, C. (1909). Zur Kenntnis der Rolle von Kern und Plasma bei der Vererbung. *Z. Vererbungslehre* **2,** 331–340.

6a. Correns, C. (1937). "Nicht Mendelnde Vererbung" (F. von Wettstein, ed.). Borntraeger, Berlin.

7. Duvick, D. N. (1965). Cytoplasmic pollen sterility in corn. *Advan. Genet.* **13,** 1–56.
8. East, E. M. (1934). The nucleus-plasma problem. *Amer. Natur.* **68,** 289–303, 402–439.
9. Edwardson, J. R. (1956). Cytoplasmic male-sterility. *Bot. Rev.* **22,** 696–738.
10. Edwardson, J. R., and Corbett, M. K. (1961). Asexual transmission of cytoplasmic male sterility. *Proc. Nat. Acad. Sci. U. S.* **47,** 390.
11. Fukasawa, H. (1967). Constancy of cytoplasmic property during successive backcrosses. *Amer. Natur.* **101,** 41–46.
12. Gabelman, W. H. (1949). Reproduction and distribution of the cytoplasmic factor for male sterility in maize. *Proc. Nat. Acad. Sci. U. S.* **35,** 634–639.
13. Grun, P., and Aubertin, M. (1965). Evolutionary pathways of cytoplasmic male sterility in *Solanum. Genetics* **51,** 399–409.
14. Grun, P., Aubertin, M., and Radlow, A. (1962). Multiple differentiation of plasmones of diploid species of *Solanum. Genetics* **47,** 1321.
15. Hagemann, R. (1964). "Plasmatische Vererbung." Fischer, Jena.
16. Hagemann, R. (1964). Advances in the field of plastid inheritance in higher plants. *Genet. Today* **3,** 613–625.
17. Imai, Y. (1936). Geno- and plasmotypes of variegated pelargoniums. *J. Genet.* **33,** 169–195.
18. Kirk, J. T. O., and Tilney-Bassett, R. A. E. (1967). "The Plastids." Freeman, London.
19. Michaelis, P. (1925). Zur cytologie und embryoentwicklung von *Epilobium. Ber. Deut. Bot. Ges.* **43,** 61–67.
20. Michaelis, P. (1951). Interactions between genes and cytoplasm in *Epilobium. Cold Spring Harbor Symp. Quant. Biol.* **16,** 121–130.
21. Michaelis, P. (1954). Cytoplasmic inheritance in *Epilobium* and its theoretical significance. *Advan. Genet.* **6,** 287–401.
22. Michaelis, P. (1965). Cytoplasmic inheritance in *Epilobium* (a survey). *The Nucleus* **8,** 83–92.
23. Michaelis, P. (1965). II. The occurrence of plasmon-differences in the genus *Epilobium* and the interactions between cytoplasm and nuclear genes (a historic survey). *The Nucleus* **8,** 93–108.
24. Michaelis, P. (1966). I. The proof of cytoplasmic inheritance in *Epilobium* (a historical survey as example for the necessary proceeding). *The Nucleus* **9,** 1–16.
25. Michaelis, P. (1966). III. The sum of plasmon constituents and the plasmon analysis. *The Nucleus* **9,** 103–118.

26. Oehlkers, F. (1964). Cytoplasmic inheritance in *Streptocarpus. Advan. Genet.* **12,** 329.
27. Renner, O. (1936). Zur kenntnis der nichtmendelnden Buntheit der Laubblatter. *Flora Jena* N. F. **30,** 218–290.
28. Rhoades, M. M. (1933). The cytoplasmic inheritance of male sterility in *Zea mays. J. Genet.* **27,** 71–93.
29. Rhoades, M. M. (1943). Genic induction of an inherited cytoplasmic difference. *Proc. Nat. Acad. Sci. U. S.* **29,** 327–329.
30. Rhoades, M. M. (1946). Plastid mutations. *Cold Spring Harbor Symp. Quant. Biol.* **11,** 202–207.
31. Rhoades, M. M. (1950). Gene induced mutation of a heritable cytoplasmic factor producing male sterility in maize. *Proc. Nat. Acad. Sci. U. S.* **36,** 634–635.
32. Rhoades, M. M. (1955). Interaction of genic and non-genic hereditary units and the physiology of non-genic inheritance. *In* "Encyclopedia of Plant Physiology," (W. Ruhland, ed.), Vol. 1, pp. 19–57. Springer, Berlin.
33. Roth, L. (1927). Untersuchungen uber die periklinal bunten Rassen von *Pelargonium zonale. Z. Vererbungslehre.* **45,** 125–259.
34. Schotz, F. (1954). Uber Plastidenkonkurrenz bei *Oenothera. Planta (Berlin)* **43,** 182–240.
35. Schotz, F. (1970). Effects of the disharmony between genome and plastome on the differentiation of the thylakoid system in *Oenothera. In* "Control of Organelle Development" (P. L. Miller, ed.), *Soc. Expt. Biol. Symp.,* Vol. 24, pp. 39–54. Cambridge Univ. Press, London.
36. Smith, H. H. (1968). Cytoplasmic inheritance in *Nicotiana. Advan. Genet.* **14,** 1–54.
37. Stinson, H. T., Jr. (1960). Extranuclear barriers to interspecies hybridization between *Oenothera hookeri* and *Oenothera argillicola. Genetics* **45,** 819.
38. Stubbe, W. (1959). Genetische Analyse des Zusammenwirkens von Genom und Plastom bei *Oenothera. Z. Vererbungslehre* **90,** 288–298.
39. Stubbe, W. (1960). Untersuchungen zur genetischen Analyse des Plastoms von *Oenothera. Z. Bot.* **48,** 191–218.
40. Stubbe, W. (1963). Die Rolle des Plastoms in der Evolution der Oenotheren. *Ber. Deut. Bot. Ges.* **76,** 154–167.
41. Stubbe, W. (1964). The role of the plastome in evolution of the genus *Oenothera. Genetica* **35,** 28–33.
42. Tilney-Bassett, R. A. E. (1963). Genetics and plastid physiology in *Pelargonium. Heredity* **18,** 485–504.
43. Tilney-Bassett, R. A. E. (1965). Genetics and plastid physiology in *Pelargonium.* II. *Heredity* **20,** 451–466.
44. Tilney-Bassett, R. A. E. (1970). Genetics and plastid physiology in *Pelargonium.* III. Effect of cultivar and plastids on fertilisation and embryo survival. *Heredity* **25,** 89–103.
45. Tilney-Bassett, R. A. E. (1970). The control of plastid inheritance in *Pelargonium, Genet. Res.* **16,** 49–61.
46. Woods, M. W., and DuBuy, H. G. (1951). The action of mutant chondriogenes and viruses on plant cells with special reference to the plastids. *Amer. J. Bot.* **38,** 419–434.
47. Woods, M. W., and DuBuy, H. G. (1951). Hereditary and pathogenic nature of mutant mitochondria in *Nepeta. J. Amer. Cancer Inst.* **11,** 1105–1151.

Patterns of Transmission of Cytoplasmic Genes: A Summary

Cytoplasmic genes are identified by the same fundamental method that Mendel used, by the segregation of parental alleles out of hybrids which arose from a controlled cross of the parental strains. In nuclear systems, allelic segregation is determined by the behavior of chromosomes and the spindle apparatus. In cytoplasmic systems, the rules of segregation are different from the nuclear ones. It has been a major goal in studies of cytoplasmic genes to establish the rules of their segregation and to identify the structures responsible for the observed regularities.

In the cytoplasmic systems which have been surveyed in the preceding four chapters, the identification of cytoplasmic genes has rested primarily upon formal genetic criteria: non-Mendelian ratios, differences in the progeny of reciprocal crosses, somatic segregation during vegetative growth, and infectivity in grafts and hyphal fusion. All of these criteria depend upon two underlying properties of cytoplasmic systems: (*a*) preferential transmission to progeny of alleles from one of the parental strains; and (*b*) rapid conversion of heterozygotes to homozygotes, often by somatic segregation.

Preferential transmission and homozygosis are achieved at different stages of the life cycle and probably by different mechanisms in different organisms. Nonetheless, the pervasive occurrence of preferential transmission seen for example as maternal inheritance in *Chlamydomonas* and in most higher plants, and seen as "mitochondrial sex" in yeast, suggests that it is a pattern of fundamental importance in cytoplasmic heredity.

How is preferential transmission achieved? We will attempt in this summary to compare and relate the behavior of cytoplasmic genes in the several organisms considered in Chapters 3–6. The table presents a summary of the principal features of the systems discussed in these four chapters.

Two terms have been introduced and put to continual use in these chapters: *cytohet* and *replicative dominance*. Cytohet is an abbreviated form of cytoplasmic heterozygote, the hybrid condition that constitutes the starting point for genetic analysis. *Replicative dominance* refers to the emergence from cytohets of one of the two parental alleles as sole or preferential product, with no implication as to mechanism.

MUTAGENESIS

Most of the cytoplasmic mutations identified in higher plants and in the fungi are of spontaneous origin and present in wild-type populations. The systematic use of UV irradiation and of chemical mutagens to induce cytoplasmic mutations has been pursued mainly with the *petites* of yeast (Table 4.2). The results cannot be freely generalized to other systems owing to the immensely high spontaneous rate of mutation to the cytoplasmic *petite* condition. Acridine dyes which are such potent inducers of the *petite* mutations in yeast are also effective in *Neurospora* and *Aspergillus*. Streptomycin, which is a highly effective mutagen for chloroplast genes in *Chlamydomonas*, is also effective in blocking chloroplast development in higher plants and in *Euglena*. A systematic investigation of the mutagenic response of cytoplasmic genes to conventional mutagens as well as to radiation and chemical hazards in the environment has not been carried out with any organism and is very much needed.

PHENOTYPES

Among the cellular systems affected by mutations of cytoplasmic genes, chloroplasts and mitochondria head the list. The second half of this book will take up in detail the known functions of cytoplasmic genes, especially in the biogenesis of these organelles. Much less well known and virtually unexplored, are the effects of cytoplasmic genes on other phenotypes, such as the differentiation of fruiting bodies, other aspects of morphogenesis, fungal incompatibility, and senescence (discussed in Chapter 5). These phenotypes, may not be determined by chloroplast or mitochondrial genes, and if not, the cellular localization of their genomes is quite unknown.

IDENTIFICATION

The identification of cytoplasmic genes has rested primarily upon the segregation of parental alleles in cytoplasmic heterozygotes. In cytoplasmic systems, cytohets may be formed either in sexual crosses or vegetative cell fusions, e.g., heterokaryon formation in fungi. Also cytohets for single genes may arise following mutation of one allele in cells which are diploid or polyploid for their cytoplasmic genome.

Cytoplasmic systems differ from nuclear ones in the mode of distribution of alleles at cell division. In nuclear systems, the segregation and recombination which occurs in meiosis but not in mitosis is governed by the behavior of nuclear chromosomes. In cytoplasmic systems, the patterns of segregation and recombination reflect the behavior in mitosis and meiosis of the cytoplasmic DNA's. Thus the differences in transmission which serve to distinguish nuclear and cytoplasmic genes also inform us about differences in the modes of replication, recombination and distribution of nuclear and cytoplasmic DNA's.

PREFERENTIAL TRANSMISSION

The salient features of cytogene segregation in the various systems discussed in this book have been summarized in the accompanying table. Are there any unifying generalizations which can be inferred from these observations? The principal generalization has already been stated: *cytohets tend to become homozygous as rapidly as possible, usually with the preferential segregation of one of the parental types.*

In order to compare the cytoplasmic transmission systems of different organisms, we must first consider differences in the life cycles which influence transmission patterns. In yeast and in higher plants, fertilization is followed by vegetative multiplication of diploid cells. If the zygote is a cytohet, then segregation of cytogenes will occur during vegetative growth. In this respect, segregation in plant seedlings and in zygote clones of yeast are directly comparable. In both systems, meiosis

Preferential Transmission of Cytogenes in Meiosis and Mitosis

Meiosis	Mitosis
Chlamydomonas A. *Maternal.* 4:0 segregation in tetrads Mechanism: Exclusion of paternal cytogenes initially present in zygote. B. *Biparental.* No segregation in meiosis Mechanism: Inhibition of paternal exclusion process. Replication and transmission of both maternal and paternal cytogenes to all zoospores. No preferential transmission.	*Chlamydomonas* 1. Segregation occurs in vegetative clones of cytohet zoospores, with segregation ratios approximately 1:1. 2. Segregation patterns: (a) 1:1 by reciprocal exchange (b) average 1:1 by nonreciprocal conversion events. 3. Segregation mechanism: recombination of DNA molecules at 4-strand stage. 4. No preferential transmission or replicative dominance seen in mitosis.
Yeast A. *Uniparental.* 4:0 segregation in tetrads Mechanism: Sporulating diploids are cytoplasmic homozygotes before sporulation. B. *Biparental.* Irregular segregation in tetrads Mechanism: Sporulating diploids are cytohets and segregation in meiosis is irregular. Segregation mechanism unknown.	Yeast 1. Segregation in vegetative zygote clones not 1:1. 2. Segregation patterns: (*a*) preferential transmission (polarity) determined by mitochondrial sex factor. (*b*) replicative dominance of some mitochondrial DNA's (suppressiveness). 3. Segregation mechanisms unknown.
Fungi A. *Uniparental.* 4:0 segregation in tetrads One parental genome excluded in ascus. Mechanism unknown. B. *Biparental.* No segregation in meiosis. Mechanism unknown.	Fungi 1. Segregation in mycelial growth, no quantitative data. 2. Evidence of replicative dominance in *Neurospora* and in *Podospora.* 3. Segregation mechanisms unknown.
Higher plants A. *Maternal.* Paternal genome excluded, mechanism unknown. B. *Biparental.* Both maternal and paternal cytogenes transmitted through zygote, usually with maternal preference. Mechanism unknown.	Higher plants 1. Segregation of cytohets from biparental zygotes. 2. Preferential segregation with various ratios. 3. Segregation mechanisms unknown.

occurs at a time well separated from fertilization, after vegetative growth of diploids, and after adequate opportunities for somatic segregation. Consequently, the cells in which meiosis occurs are usually homozygous for all cytogenes; and therefore uniparental transmission occurs in meiosis because only one parental type is present. In these systems, segregation and recombination are seen primarily (or entirely) in diploids just after fertilization.

In *Chlamydomonas* and in fungi, however, meiosis occurs soon after fertilization, so that the maternal ratios seen in tetrad analysis reflect events which occurred in zygotes between the time of fertilization and tetrad formation. In *Chlamydomonas* the paternal genome is excluded by a precise biochemical mechanism discussed in Chapter 3; and in the fungi it seems likely that an analogous mechanism is involved. In *Neurospora,* 4:0 segregations have been reported for most of the cytogenes discussed in Chapter 5, and for mitochondrial DNA itself. In all these instances, the pattern was "maternal" involving transmission of cytogenes and mitochondrial DNA from the protoperithecal parent. However, one instance of paternal transmission has been reported involving *ac-7.* If the exclusion mechanisms in *Neurospora* are similar to those in *Chlamydomonas,* some instances of biparental transmission should be found.

Let us now consider the behavior of cytohets in meiosis, as summarized in the table. In yeast, the usual condition is that the cells undergoing meiosis are homozygous for their cytogene complement, and therefore no segregation is seen at meiosis. In higher plants, meiotic segregation of cytoplasmic genes has not been studied on its own, although it would be technically possible in plants in which the tetrad of pollen grains coming from a single meiosis stay together. At present one cannot judge whether any of the complexities of somatic segregation patterns in plants following biparental transmission can be attributed to meiotic segregation in either male or female cell lines.

The life cycle of *Chlamydomonas* is very favorable for examining cytogene segregation and recombination in meiosis. We find that the progeny of biparental zygotes segregate on the average 1:1 for each of the cytogene markers present, indicating an equal contribution from each parent. Thus, we infer that the DNA's from the two parents, in the absence of exclusion, receive equal treatment in the zygote: both are replicated and transmitted to each zoospore.

As a consequence of these differences in the life cycle, the patterns of transmission of cytogenes may appear quite different in various organisms. The clearest differences so far noted are between *Chlamydomonas* and yeast. In the former, tetrad analysis provides the first evidence of cytoplasmic inheritance, and segregation and recombination are seen only in the exceptional cytohet progeny. In yeast, zygotes are initially cytohet, and segregation and recombination occur in the first few mitotic doublings of diploids, whereas tetrads provide a much later check on the transmission pattern.

Let us return now to the two fundamental properties of cytoplasmic systems which we have been discussing, *preferential transmission* and *somatic segregation*. Of the two, the process of somatic segregation is the simpler one, determined by and reflecting the distribution of the corresponding cytoplasmic DNA at cell division. Thus the behavior of cytohets provides genetic evidence about the behavior of the DNA's which carry the appropriate marker genes.

The analysis of preferential transmission is more complex since a variety of different molecular mechanisms can influence this result. In higher plants, for example, maternal inheritance may be brought about by exclusion of cytogenes from the pollen, or exclusion prior to or during fertilization, or exclusion in the zygote after fertilization by a mechanism like host restriction, or exclusion in the first few divisions of the zygote. Whatever the specific mechanism, preferential transmission is a pervasive feature of cytoplasmic inheritance. In view of this pervasiveness, it seems evident that it must be the result of powerful selective pressures operating in evolution. Why is this phenomenon of selective value? One possibility is that preferential transmission operates to restrict recombination.

The selective value of restricted recombination has been discussed by many geneticists, going back to R. A. Fisher "The Genetical Theory of Natural Selection." Clarendon Press, Oxford, 1930. Lewontin (*Proc. Nat. Acad. Sci. U. S.* **68,** 984; 1971) has provided new mathematical evidence that in polymorphic populations, highly restricted recombination maximizes fitness at the population level.

The evidence that elaborate mechanisms are present in natural populations, restricting the recombination of cytogenes, is particularly interesting because it further emphasizes the importance to the organism of these cytoplasmic genomes.

CYTOPLASMIC GENES AND ORGANELLE BIOGENESIS

Introductory Remarks

The next two chapters will consider the role of cytoplasmic genes in the biogenesis of two organelle systems—mitochondria and chloroplasts. Ever since the early 1900's, cytologists have stressed the hereditary continuity of mitochondria and chloroplasts: that these organelles appear to arise only from preexisting organelles of the same kind. The modern counterpart of these observations is the evidence that there are unique genomes associated with these organelles. Indeed, the correlations between the behavior of organelle DNA's and of cytoplasmic genes, discussed in previous chapters, leave little doubt that chloroplasts and mitochondria contain DNA-based genetic systems with unique functions.

The full analysis of organelle biogenesis is a problem of very broad scope, involving the contributions not only of cytoplasmic genes but also of many nuclear genes and of the complex regulatory interactions between the genomes about which virtually nothing is yet known. The aim in these chapters is a relatively modest one: to concentrate on what is known about the role of the organelle genomes per se. A rationale for this approach is "divide and conquer." Since the organelle genomes are relatively small and circumscribed, perhaps they provide key functions in the overall scheme, and by focusing on them we may gain special insights into the biogenetic process.

In these chapters we will first examine the molecular composition and structural organization of the organelles, especially the membranes with their specific and characteristic lipids and proteins. Membrane assembly is a process of central importance in subcellular organization, and chloroplasts and mitochondria are providing particularly favorable material for studying the process. Membranes offer a special challenge to genetic theory because they can vary enormously in composition and stoichiometry while maintaining specificity. Whether DNA plays a direct role in this kind of specificity is a very important question just starting to be studied at the molecular level. Some evidence on membrane formation will be considered in the next two chapters and further discussed in Chapter 9 in comparison with other systems, especially morphogenesis of phage T4 and cortical inheritance in *Paramecium.*

Most of the evidence to be presented in the next two chapters on the role of organelle DNA's is based on the application of well-developed methodologies of molecular biology: DNA–RNA hybridization for the identification of RNA transcripts of organelle DNA's and incorporation studies with radioisotope-labeled substrates to follow the synthesis of nuclear acids and proteins in isolated organelles and in intact cells.

These methods have been successful in establishing that ribosomal RNA's and most if not all tRNA's of chloroplasts and mitochondria are transcribed from the corresponding DNA's. The methods have been far less successful in identifying specific proteins either coded by organelle genes or synthesized within the organelles. Considerable knowledge has been amassed concerning biosynthetic processes in organelles. However, at this writing not a single protein has been identified unambiguously as the translational product of an organelle DNA, and very few proteins have been identified as synthesized within either organelle. Despite these lacks, potentialities for accomplishing both of these objectives are present, and both organelle systems appear poised on the threshold of new advances. In particular, suitable mutant strains carrying specific organelle gene mutations are just becoming available for investigation and may be expected to provide specific identification of some organelle-coded proteins as well as invaluable aids for biochemical dissection of biosynthetic processes. These and other possibilities for the near future will be discussed in the concluding remarks at the end of this section.

7

The Role of Mitochondrial Genes in Mitochondrial Biogenesis

Biochemists studied the pathways of oxidative energy metabolism for many years before mitochondria were recognized as unique structures in which these enzymatic activities were localized. Isolation of the chondriome, as mitochondria were collectively called, was first achieved by Albert Claude in the 1940's. Since that time, our understanding of mitochondrial function has been strongly influenced by evidence from structural studies of this organelle system (*54*).

Within the past few years, a third approach has begun to play an important role in mitochondrial investigations: the analysis of organelle formation or biogenesis. The dramatic discovery of DNA and of a protein-synthesizing system in mitochondria has opened the way for a new and much more powerful investigation of mitochondrial function. Utilization of the tools of genetic analysis and molecular biology as well as those of biochemistry and electron microscopy has promoted our understanding of this complex organelle.

This third approach—analysis of organelle formation—has just begun. We cannot yet assess the role of the mitochondrial genome in the biogenesis and function of the organelle, but certain key observations have clarified the picture. In the first place, it has become apparent that most of the proteins found in mitochondria are coded by nuclear genes and synthesized in the cytoplasm. Mitochondrial DNA is large enough to code for only some twenty to thirty proteins in animal cells, and perhaps five times as many in eukaryotic microorganisms.

Furthermore, it seems likely that all or most of these mitochondria-coded proteins are membrane-bound components. In examining the literature on mitochondrial structure, it turns out that surprisingly little is yet known about the proteins that comprise the mitochondrial membranes, or about the organization of the proteins, lipids, and other components in the membranes. We do know that the central functions of electron transport and oxidative phosphorylation are membrane bound, as are transport systems for the many substrates, cofactors, and ions which move across the mitochondrial membranes.

The problem of mitochondrial biogenesis is intimately concerned with membrane formation and with the functions of mitochondrial DNA, RNA's, and ribosomes. Organelle membranes are not stoichiometrically constant in composition; new components can be inserted into preexisting membranes, and components can be removed. Thus membranes are in a dynamic state; and genes, both nuclear and cytoplasmic, are continually at work in the synthesis of mitochondrial components and the regulation of their activity.

Both nuclear and mitochondrial genomes contribute to mitochondrial biogenesis. Ever since the discovery of mitochondrial DNA, investigators have been puzzled and fascinated by the question of what special function is served by this second genetic system. Why are not all organelle components coded by nuclear genes? What unique properties of mitochondrial coded proteins require that they be transcribed and translated within the organelle?

At this writing, these questions remain largely unanswered. What is important is that these new questions have been raised and that a new approach can be brought to bear upon their solution. A new molecular genetics is emerging in which the kinds of analysis which have been successful in elucidating biosynthetic pathways are being applied to the vastly more complex problem of organelle biogenesis.

STRUCTURAL ORGANIZATION (94, 95)

As seen in electron microscope studies of thin-sectioned cells, mitochondria consist of two concentric membrane-bound compartments (Plate XII). The outer membrane is smooth and continuous.

The inner membrane is also continuous, but is thrown into many folds, called cristae, which are platelike or fingerlike, depending upon the species, and which vary in extensiveness, not only with the species but also with the tissue and its nutritional state. The surface area of the cristae is roughly proportional to the respiratory activity of the mitochondria. The compartment between the inner and outer membranes is called the intermembrane space, and the region within the inner membrane is called the matrix.

The inner and outer membranes exhibit no structural differences when seen in sectioned preparations in the electron microscope. Both are about 50–70 Å in diameter in osmium-fixed material, and show the typical tripartite structure (two dense lines with the less dense core) of the so-called unit membrane. However, their physical properties and composition are very different (*29, 75, 93*). The outer membrane is less dense than the inner membrane and swells irreversibly in hypotonic phosphate, leading to breakage and detachment. The swelling of the outer membrane thus provides a means to separate outer from inner membrane preparation (*67*). Another separation method involves differential solubility in digitonin (*83*). Some of the characteristic properties and components of the membranes are listed in Tables 7.1 and 7.2. A diagrammatic representation of a mitochondrion is shown in Fig. 7.1.

The relation between the outer mitochondrial membrane and the microsomal fraction, i.e., the rough-surfaced membranes of the endoplasmic reticulum, is not entirely clear (*29, 93*). Judging from its composition and behavior, some investigators consider the outer membrane to be a part of the endoplasmic reticulum. Some differences between them have been reported. In particular, a number of enzymes

TABLE 7.1

Comparison of Inner and Outer Mitochondrial Membranes[a]

Property	Inner membrane	Outer membrane
Thickness (Å)	50–70	50–70
Shape	Folded	Distended
Surface	Smooth	Smooth
Density	1.21	1.13
Osmotic behavior	Reversible folding, unfolding	Irreversible stretching, leading to rupture
Permeability	Highly specific	Nonspecific–up to molecular wt. ~10,000
Phospholipid/protein (w/w)	0.301	0.878
Cardiolipin	21.5	3.2
Phosphatidyl inosital	4.2	13.5
Cholesterol (μg/mg protein)	5.06	30.1
Ubiquinone	Present	Absent

[a] Data from (*29*).

TABLE 7.2

***Localization of Enzymes in Liver Mitochondria*[a]**

Outer Membrane	Intermembrane space
"Rotenone-insensitive" NADH-cytochrome c reductase (NADH-cytochrome b_5 reductase; cytochrome b_5) Monoamine oxidase Kynurenine hydroxylase ATP-dependent fatty acyl-CoA synthetase Glycerolphosphate-acyl transferase Lysophosphatidate-acyl transferase Lysolecithin-acyl transferase Cholinephosphotransferase Phosphatidate phosphatase Phospholipase A_{11} Nucleoside diphosphokinase Fatty acid elongation system Xylitol dehydrogenase (NAD-specfic?)	Adenylate kinase Nucleoside diphosphokinase Nucleoside monophosphokinase Xylitol dehydrogenase (NAD-specific?)
Inner Membrane	**Matrix**
Respiratory chain (cytochromes b, c_1, c, a, a_3; succinate dehydrogenase; succinate-cytochrome c reductase; succinate oxidase; "rotenone-sensitive" NADH-cytochrome c reductase; NADH oxidase; choline-cytochrome c reductase; cytochrome c oxidase; respiratory chain-linked phosphorylation) β-Hydroxybutyrate denydrogenase Ferrochelatase δ-Aminolevulinic acid synthetase? Carnitine palmityl-transferase Fatty acid oxidation system? Fatty acid elongation system Xylitol dehydrogenase (NADP-specific?)	Malate dehydrogenase Isocitrate dehydrogenase (NADP-specific) Isocitrate dehydrogenase (NAD-specific) Glutamate dehydrogenase α-Ketoglutarate dehydrogenase (lipoyl dehydrogenase) Citrate synthetase Aconitase Fumarase Pyruvate carboxylase Phosphopyruvate carboxylase Aspartate aminotransferase Ornithine-carbamoyl transferase Fatty acyl-CoA synthetase(s) Fatty acid oxidation systems? (β-hydroxybutyryl-CoA dehydrogenase) Xylitol dehydrogenase (NADP-specific?)

[a] Data from (29).

with similar but not identical activities have been described: NADH-cytochrome b reductase, cytochrome b_5, ATP-dependent fatty acyl CoA synthetase, glyceraldehyde phosphate acyl CoA transferase, and phospholipase.

It would be useful to know whether fusions occur between the outer mitochondrial membranes and the endoplasmic reticulum. If so, mi-

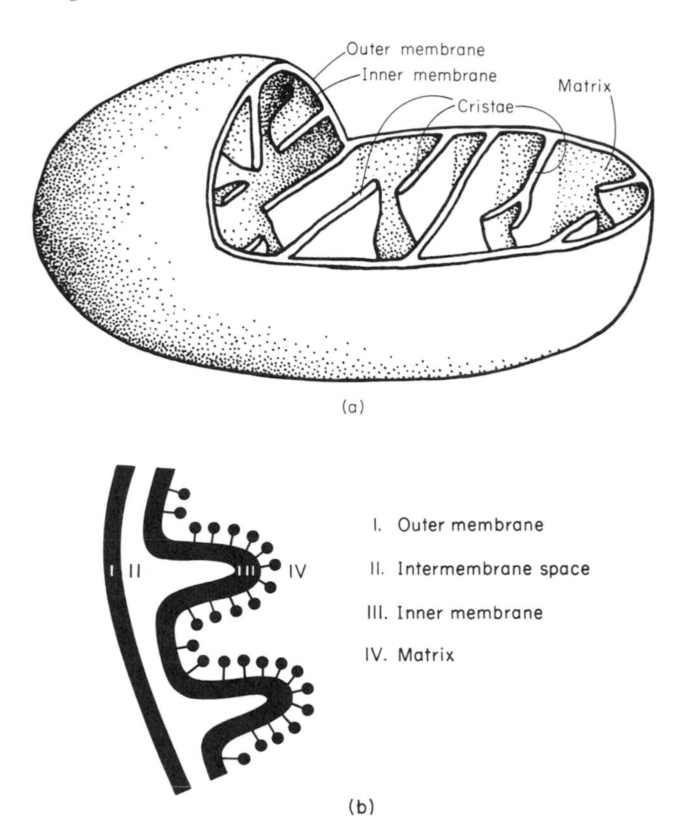

Fig. 7.1. (a) Drawing of a mitochondrion showing three-dimensional arrangement of membranes. From (*98a*). (b) Definition of mitochondrial compartments. I, Outer membrane; II, intermembrane space; III, inner membrane; IV, matrix. From (*29*).

tochondria would have easy access to proteins and other components synthesized in the cytoplasm but destined to function inside the mitochondrial system, by the fusion of elements of the endoplasmic reticulum with the outer mitochondrial membrane. Presumably, fusions do not occur between the inner and outer mitochondrial membranes. If fusions did occur, the many differences between the membranes listed in Tables 7.1 and 7.2 would not exist.

The inner mitochondrial membrane is exceedingly complex, with many discrete but coupled functions and components (*75*). In addition to the proteins of electron transport and oxidative phosphorylation, some steps in various biosynthetic pathways and numerous transport systems are localized there. The list in Table 7.2 is far from complete; new enzymes and new activities are still being found.

Most of the evidence about molecular organization has come from studies of submitochondrial fractions, examined for activity and composition by biochemical methods correlated with structural studies by electron microscopy. New technical developments in electron microscopy have contributed uniquely to the analysis.

The most dramatic feature of the inner membrane, both structurally and functionally, is the presence of spherical particles, about 80 Å in diameter, projecting on short stalks into the mitochondrial matrix. These particles were discovered by Fernández-Morán with the use of the (then) new method of negative staining (*31*). Subsequently, the development of freeze-etch methods which reveal surface structure with great precision, have focused attention on the arrangement of globular proteins in the membrane faces (*63*). Kagawa and Racker (*44*) have shown that these particles consist of a protein of about 284,000 molecular weight, called F_1 which has ATPase activity and which, according to Racker, is involved in oxidative phosphorylation at each of the three sites shown in Fig. 7.2. The ATPase activity of F_1 bound to the mitochondrial membrane is cold-stable and sensitive to the antibiotic oligomycin. In contrast, the purified F_1 protein is cold-labile and exhibits an ATPase activity that is resistant to oligomycin. Inner membrane preparations can be stripped of their particles, thereby losing ATPase activity. Purified F_1 can then be added back to the stripped membranes with the reappearance of particles on the membranes and the restitution of ATPase activity.

Kagawa and Racker examined a preparation of inner membranes called the CF_0 fraction, shown in Plate VIII, from which the cytochromes and most of the lipids have been extracted. This fraction can be combined with F_1; the resulting complex has no ATPase activity, but in the electron microscope, as shown in Plate IX, particles can be seen attached by stalks just as in intact membranes. When phospholipid is added to this complex, oligomycin-sensitive ATPase activity is restored, and the complex appears as shown in Plate X.

This enzymatic change is accompanied by a dramatic change in

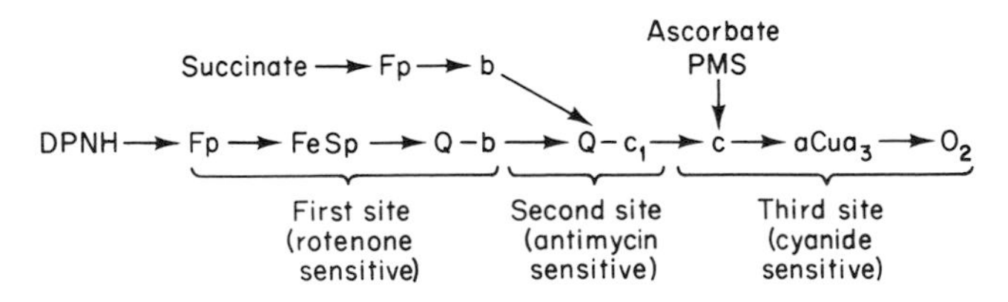

Fig. 7.2. The oxidation chain in mitochondria. Fp, flavoprotein; FeSp, nonheme iron protein; a_1, a_3,c, c_1 cytochromes. After (*75*).

morphology as well, as seen in the electron micrographs. CF_0 preparations consist of thin flakes with very little structure. When lipids are added, smooth membrane-bound vesicles appear, and in section the typical unit membrane profile can be seen. Thus the phospholipids appear to play a very important part in structural organization as well as enzymatic activity. As shown so beautifully in this experiment, the disposition of the enzymes in the membrane surface, regulated in part by phospholipids, is crucial to enzymatic activity.

The term "allotopic" was introduced to describe the activity of an enzyme which depends upon its association with a membrane (*75*). The sensitivity of F_1 to oligomycin when it is membrane bound, and its insensitivity when in solution, is an example of an allotopic response. However, this useful concept which concerns the catalytic role of an enzyme does not fully cover the experimental situation, since F_1 for instance has been shown also to play a structural role. Let us consider a few examples that provide some insight into the complex interactions of mitochondrial membrane components.

F_1 can be treated with iodine so that all of its ATPase activity is lost, but in this condition it still retains some phosphorylation-stimulating activity. For example, submitochondrial particles which have been partially depleted of ATPase activity, can be markedly stimulated in phosphorylating activity by addition of F_1 even in its iodine-treated form.

Schatz *et al.* (*82*) performed experiments in which they found that submitochondrial particles from beef deficient in F_1 can be activated with an F_1 preparation from yeast. However, in these "hybrid" particles, antibodies to yeast F_1 inhibit ATPase activity, but not phosphorylation, whereas the beef antibody inhibits phosphorylation. In interpreting their results, the authors concluded that the yeast enzyme could function structurally in phosphorylation in the beef particles, and that this function was not inhibited by antibody, whereas the catalytic function, which yeast ATPase could not perform in the beef particle, was highly susceptible to inhibition by the antibody. More direct evidence for a structural role for F_1 has come from studies of an F_1 requirement for activity of a reaction not involving ATPase (the succinate-driven trans-hydrogenation of TPN by DPNH).

The fact that F_1 attaches only to the matrix side of the inner membrane suggests a topological sidedness which is supported by other lines of evidence as well. For example, succinate and NADH have been shown to interact with the electron transport system only from the matrix side, whereas cytochrome c appears to attach on the outside of the inner membrane.

Recently, Crane *et al.* (*15*) have provided evidence from freeze-etch electron micrographs of detergent-extracted membranes that the inner mitochondrial membrane is in fact double, with 70 Å being the peak-to-peak distance. Their schematic diagram is shown in Fig. 7.3.

In addition to the electron transport system and its coupled phosphorylation machinery, the inner membrane also contains the fatty acid elongation system and the so-called transporters or carriers of Krebs cycle intermediates, cations, ATP, ADP, P_i, and other small molecules involved in mitochondrial growth and metabolism. Of particular relevance to us is the likelihood that many of these enzymes are subject to catabolite repression or to regulation by other mechanisms, leading to their addition to or removal from preexisting membranes.

There is a controversy regarding the existence of structural protein (*17*)—defined as a special class of proteins which have a major structural role as the proteinaceous backbone of the mitochondrial membrane. Do all membrane proteins, like F_1, have both enzymatic and structural roles? Is there a non-enzymatic component holding the active molecules in register? Above all, is there a single protein species functioning in a structural capacity? The principal reason for this controversy has been the difficulty in identifying many of the proteins found in the mitochondrial membranes. Most membrane proteins are insoluble under physiological conditions and are difficult to disag-

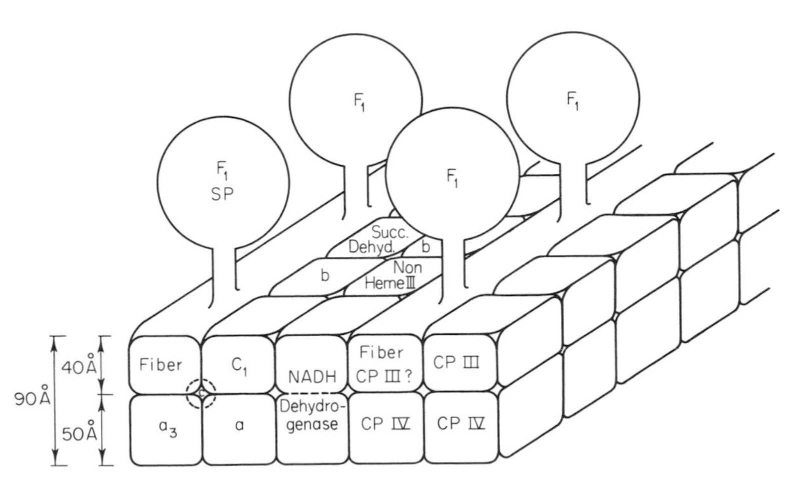

Fig. 7.3. Diagram of binary structure in mitochondrial cristae membranes, showing proposed arrangement of apoproteins of cytochrome (a, a_3, b, c_1) and other components of electron transport chain. F_1 = ATPase; SP = structural protein; CP = core protein. From (*15*).

gregate, requiring fairly drastic treatment to solubilize and separate. Powerful new methods of gel electrophoresis have been developed to separate and display membrane proteins, but enzymatic activities are lost in the process. Consequently, except for the studies just discussed, few components of the mitochondrial membrane have been isolated as yet. Many of the enzymes listed in Table 7.2 have been identified solely on the basis of activity.

What is known about the lipids which constitute some 30–40% of the total mass of the mitochondrial membranes? The lipid composition of mitochondria from different organisms and from different organs of animal cells differ in detail, but all contain three major classes of phospholipids which in sum constitute about 90% of the total lipid *(13)*. The three classes are: phosphatidyl choline (PC), phosphatidyl ethanolamine (PE), and cardiolipin (CL). Each of the phospholipids is esterified with a number of fatty acids, both saturated and unsaturated. An example of the distribution in rat liver and beef heart is given in Table 7.3.

The lipid composition of mitochondrial membranes has been shown to vary considerably with medium and growing conditions. An extreme example is the comparison of mitochondrial membranes from aerobic and anaerobically grown yeast. Indeed until recently there was disagreement about whether mitochondrial membranes existed at all in anaerobic yeast, so great was their compositional alteration.

TABLE 7.3

Phospholipids of Rat Liver and Beef Heart Mitochondria[a,b]

		Cardiolipin fraction			Phosphatidyl-ethanolamine fraction			Phosphatidyl-choline fraction		
Fatty acid		Rat liver		Beef heart	Rat liver		Beef heart	Rat liver		Beef heart
Palmitic	16:0	–	4		17.5	20		13	20	24
Stearic	18:0				30.4	19.5	38	22	19	5
Oleic	18:1	10	13	5	5	10	4	14	12	19
Linoleic	18:2	84	74	84	4	16	15	20	20	37
Linolenic	18:3			6						4
Arachidonic	20:4				21	21	33	14	19.5	4
Docosahexaenoic	22:6				14	9		8	3.4	

[a] From *(13)*.
[b] All figures expressed as area percentages, and only those components present to the extent of 4% included. All data rounded off to nearest whole number.

Bloomfield and Bloch (*8*) showed that in anaerobically grown yeast supplemented with Tween 80 and ergosterol, the principal fatty acids are monounsaturated C_{16} and C_{18} compounds, whereas the same cells grown anaerobically without Tween 80 as a source of long-chain fatty acids have primarily saturated C_{10} and C_{12} components. It is now known that enzymes of fatty acid desaturation have an O_2 requirement; thus anaerobic cells need an exogenous supply of long-chain fatty acids as well as ergosterol (*8*).

Recently, Schatz *et al.* (*16, 69, 72, 73*) have reexamined the chemical composition and electron microscopic evidence on mitochondria from strictly anaerobically grown yeast in comparison with the same strain grown aerobically. They found that structures which they called "promitochondria" were present in the anaerobically grown cells. These structures were greatly altered in their lipid content, as others had previously noted, containing mostly saturated short-chain fatty acids. However all major types of phospholipids were present, albeit in different amounts from their characteristic concentrations in mitochondria of aerobically grown cells. Their results are shown in Tables 7.4 and 7.5.

Confusion about the presence of mitochondria in anaerobic yeast probably resulted from the low content of unsaturated lipids which led to inadequate staining for electron microscopy. A more fundamental question is how much variation in composition is consistent with the maintenance of functional activity. In the reconstitution experiments of Kagawa and Racker, it may be recalled (p. 232) that ATPase activity did not reappear until the phospholipid fraction was added back to the preparation. The specific functions of individual phospholipids in mitochondrial activity are still largely unknown. However, their essential role is attested by their universal presence in membranes, and enzymatic nonfunction in their absence.

Further evidence of the compositional flexibility of mitochondrial membranes has come from studies of mitochondrial biogenesis (*56*). In their studies of anaerobically grown yeast, Criddle and Schatz (*16*) found that promitochondria contain the entire oligomycin-sensitive ATPase (F_1) complex indistinguishable from their aerobic counterparts. Furthermore the amount per cell of mitochondria and of promitochondria was about the same in aerobic and anaerobically grown yeast. With the use of the powerful freeze-etch technique, which is relatively insensitive to changes in chemical composition, Plattner and Schatz (*72*) showed the presence in promitochondria of cristae, as well as inner and outer membranes. Thus, anaerobic growth does not inhibit the synthesis of mitochondrial inner membranes but greatly modifies their composition.

TABLE 7.4

***Phospholipids of Aerobic and Anaerobic Yeast*[a]**

Particle preparation	% of Total phospholipid phosphorus						
	Cardiolipin	Phosphatidyl-ethanolamine	Phos-phatidyl-serine	Phos-phatidyl-inositol	Lysophosphatidyl-ethanolamine + lysophosphatidyl-serine	Lyso-phos-phatidyl-choline	Phos-phatidyl-choline
Mitochondria (from cells grown in presence of lipids)	10.9	30.6	4.2	8.1	4.5	3.2	38.5
Promitochondria (from cells grown in presence of lipids)	6.2	19.3	10.0	12.6	4.1	0.3	47.5
Promitochondria (from cells grown without added lipids)	8.9	17.9	3.9	26.0	8.2	0.8	34.3

[a] From (*69*).

TABLE 7.5

Fatty Acid Composition of Mitochondria from Aerobic and Anaerobic Yeast[a]

Particle preparation	Wt % of total fatty acids							
	C_{10}	C_{12}	C_{14}	C_{16}	$C_{16.1}$	C_{18}	$C_{18.1}$	$C_{20.1}$
Fatty acid composition of total phospholipids								
Mitochondria (from cells grown in presence of added lipids)	tr	tr	0.6	17.9	43.7	3.6	34.2	tr
Promitochondria (from cells grown in presence of added lipids)	tr	tr	4.5	20.5	6.5	3.9	61.5	3.1
Promitochondria (from cells grown without added lipids)	14.3	8.9	10.4	33.7	12.0	13.7	7.0	tr
Fatty acid composition of neutral lipids								
Mitochondria (from cells grown in presence of lipids)	tr	tr	1.2	16.1	43.1	3.6	36.0	tr
Promitochondria (from cells grown in presence of lipids)	tr	tr	4.9	18.0	2.1	10.3	63.0	1.7
Promitochondria (from cells grown without added lipids)	8.5	10.5	14.4	34.0	7.7	17.5	6.4	1.0

[a] From (*69*).

Adaptation to oxygen then would be expected to involve the addition of electron transport components into undifferentiated preexisting mitochondrial membranes. Thus the preexisting membranes found in anaerobic cells may be considered to represent the structural backbone or shell of the aerobic membrane.

Another view of the compositional flexibility of mitochondrial membranes comes from studies by Luck (*59, 60*) who was investigating the origin and growth of mitochondrial membranes. Using a choline-requiring mutant of *Neurospora,* Luck was able to follow the growth of mitochondria in *Neurospora* mycelia by following the uptake of choline, a constituent of lecithin, into growing mitochondrial membranes. Cultures prelabeled with ^{3}H-choline were transferred to unlabeled medium and permitted to grow for three doublings of mycelial mass during which the distribution of label was measured by quantitative radioautography. The results showed that throughout three doubling cycles the label was uniformly distributed among the mitochondria.

Subsequently Luck found that in the choline-requiring mutant the chemical composition of mitochondria could be varied substantially by changing the available level of choline in the medium. Mitochondria produced in cultures growing with different levels of choline were found to differ sufficiently in their densities so that they could be distinguished in a sucrose gradient. The density differences resulted from differences in the phospholipid-to-protein ratio.

These findings made it possible to examine the growth of mitochondria by following changes in density during the switch-over from a low choline to a high choline medium. As shown in Fig. 7.4, over a period of 90 minutes after a shift from low to high choline, there was a gradual shift towards lower density of the entire mitochondrial population. Fractions were collected from the gradient and assayed for the distribution of cytochrome oxidase and of total protein. The assays demonstrated not only that the bands were mitochondrial, but also that the density shift involved the entire population.

These experiments provide dramatic evidence that in exponentially growing cultures of *Neurospora,* mitochondria grow by intercalation of new components into preexisting membranes and that this process occurs all over the mitochondrial surface in a manner which is random at the level of these observations. These experiments do not indicate whether the incorporation of new components is random at the molecular level, and whether there is a molecular architecture into which these components are integrated. The possibility of secondary randomization

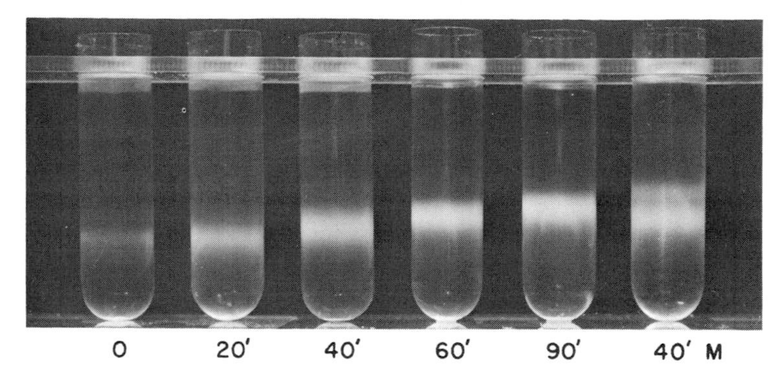

Fig. 7.4. *Neurospora* mitochondria with various lipid-to-protein ratios. Photograph of centrifuge tubes after isopycnic centrifugation of mitochondria from a choline-requiring mutant of *Neurospora* at various times after a shift from low to high choline growth medium. Tube 40′M contains a mixture of mitochondria from a culture grown for 15 hours in high choline (upper band) with mitochondria from 40-minute culture to show resolving power of the method. From *(60)*.

of phospholipids after their initial incorporation into the membrane also needs further consideration.

What is known about the spatial relations in the mitochondrial membrane of all these component proteins and of the lipids which constitute some 30–40% of the total mass? Two extreme models have been widely discussed: the Davson-Danielli bimolecular leaflet (*95, 96*) and the subunit model (*37*).

The Davson-Danielli model postulates a continuous bimolecular lipid layer with the hydrophobic fatty acid side chains lined up perpendicular to the membrane surface and the more hydrophilic heads of the phospholipid molecules in association with the proteins which coat them on both surfaces like a sandwich. Among the observations inconsistent with this model are (*a*) evidence of the importance of hydrophobic interactions between protein and lipid; (*b*) the ease of attacking phospholipids enzymatically; (*c*) the removal of lipids without changing the electron microscope image; and (*d*) the requirement for special proteins, like carrier molecules to traverse the thickness of the membrane. To accommodate these and other specific objections, the model has been considerably modified (*95, 96*). The lipid bilayer concept has been modified to allow hydrophobic interactions; greater fluidity in the lipid region and occasional interruptions of the lipid phase have been proposed in the form of protein "cores" for transport activities.

From the biological viewpoint, the greatest weakness of the model is its nonspecificity. What characterizes differences among membranes? What determines the asymmetrical nature of membranes (i.e., inside vs. outside), and above all, where does the specificity, regulation, and possibility for growth reside? This model which seems quite satisfying to some investigators because of its simple physical properties, is much too general for the biologist who wishes to know about precisely those features with which the model is not concerned.

At the other extreme is the subunit model, which proposes a rigid regularity in the design of membranes by virtue of their assembly from identical repeating subunits (*37*). It may be argued intuitively that at some level there must be repeating subunits or subassemblies in membranes, since multienzyme complexes with their conformational specificities are present in many copies. Here again, however, the model is too rigid, not providing for the remarkable compositional variability which we have seen to be a pervasive feature of mitochondrial membrane organization.

A new property of mitochondrial membranes, contractility, or the ability to undergo remarkable conformational changes, has recently been described, and new models of membrane organization have been proposed to meet this requirement (*29*). However, conformational flexibility is a very different requirement from the stoichiometric flexibility we have been discussing. It seems likely that much more understanding will be necessary before a comprehensive model of molecular organization of mitochondrial membranes will be forthcoming.

In summary, the principal findings and concepts in this section concern membrane structure in relation to function. Mitochondrial membranes appear as typical tripartite unit membranes in electron micrographs of sectioned cells, but are found in biochemical investigations to contain a complex assemblage of proteins and lipids in ordered arrays. The outer membrane resembles in its enzymatic and lipid composition the endoplasmic reticulum with which it is in close proximity. The inner membrane is unique in composition, containing the multicomponent assemblies for electron transport, oxidative phosphorylation, and fatty acid elongation as well as the protein carriers which regulate the transport of cations and small molecules across the inner membrane.

The activity of enzymes and of multienzyme complexes is dependent upon their proper integration into the membrane structure, and phospholipids play a key role in determining the active configurations. Membranes recognized by the ATPase-containing spheres on their surface are present in anaerobically grown yeast cells in which many of the lipids and enzymes characteristic of the inner membrane are absent. The adaptation to oxygen involves the intercalation of newly synthesized components into these preexisting membranes.

It seems likely therefore that under usual growing conditions mitochondrial membranes increase, differentiate, and dedifferentiate by the intercalation of new molecules into old membranes. The problem of biogenesis, then, turns on the origin and regulatory control of these intercalating molecules. We need therefore to examine the genetic and synthetic abilities of the mitochondrial system.

MITOCHONDRIAL RNA'S AND RIBOSOMES

The search for DNA-dependent RNA synthesis in mitochondria and the characterization of mitochondrial RNA's was sparked by the discovery of mitochondrial DNA and by preliminary evidence of mitochondrial protein synthesis. Is mitochondrial DNA transcribed? If so, what are its transcription products?

Mitochondrial RNA Polymerase

The first evidence for the existence of a mitochondrial RNA polymerase came from studies with isolated mitochondria from *Neurospora* (*61*), rat liver (*67a*), and yeast (*103*). The incorporation of radioactive nucleotide triphosphates into polyribonucleosides of high molecular weight was examined and found to be dependent on the four nucleotide triphosphates and sensitive to actinomycin.

Recently the first mitochondrial RNA polymerase to be highly purified has been reported by Küntzel (*52*). The enzyme, extracted from *Neurospora* mitochondria, has a molecular weight of 64,000 estimated from its position on polyacrylamide gels. *In vitro* incorporation studies showed that the enzyme is sensitive to rifampicin, an inhibitor of the initiation of transcription, and is resistant to α-amanitin. In both respects it resembles the RNA polymerases of bacteria and differs from those which have been characterized from the nuclei of eukaryotic cells. It is noteworthy that the mitochondrial RNA polymerase is much smaller than the bacterial and animal nuclear enzymes which are over 150,000 daltons and composed of several subunits.

Biochemical studies of transcription have been dramatically abetted by new techniques for visualizing the molecules in the electron microscope. A beautiful example of this coordinated approach is shown in Plate XI. A circular double-stranded molecule of mitochondrial DNA from a HeLa cell is shown with globular molecules of RNA polymerase and many single stranded "bushes" of newly transcribed RNA attached. The newly transcribed RNA appears as "bushes" because the single stranded molecules collapse during preparation for electron microscopy (*1, 4*).

Classes of Mitochondrial RNA

Three major discrete size classes of RNA's have been found in mitochondria, and several lines of evidence have established them as ribosomal and transfer RNA's. In addition, in sucrose gradients, heterogeneous RNA's have been seen which may include messenger RNA's of mitochondrial origin (*4, 20, 33, 34, 92*).

Identification of ribosomal and transfer RNA's has been accomplished in a number of laboratories by (*a*) extraction of radioisotope labeled RNA's from purified mitochondria, (*b*) separation of the various size classes on sucrose gradients and on polyacrylamide gels, and (*c*) identification of each size class as a mitochondrial transcript by hybridization with the homologous mitochondrial DNA. The tRNA's have been further identified by their activity as amino acyl acceptors.

Mitochondrial ribosomes have been so difficult to extract and purify that the ribosomal RNA's have been characterized primarily after extraction from intact mitochondria rather than from the ribosomes per se. However, the identification of ribosomal RNA's has been established ultimately by isolation mitochondrial ribosomes (*3, 64, 98*).

In all organisms studied so far, the mitochondrial ribosomal RNA's have been shown to differ from their cytoplasmic counterparts in sedimentation rates in sucrose gradients, and in electrophoretic mobility on gels.

Mitochondrial ribosomal RNA's have been described from a number of eukaryotic microorganisms including *Neurospora* (*51, 77*), *Aspergillus* (*22*), yeast (*30, 55, 64, 104*), and *Tetrahymena* (*14*), and from a number of animal tissues including rat liver, tissue cultured cells of mouse, hamster (*20*), man (i.e., HeLa cells) (*4*) and oocytes of the amphibian *Xenopus* (*19*). Striking differences have been reported between the mitochondrial ribosomes and ribosomal RNA's of eukaryotic microorganisms and those of animal cells, as shown in Tables 7.6, 7.7, and 7.8.

TABLE 7.6

Sedimentation Values of Ribosomal RNA's

	Mitochondria		Cytoplasm		
Organism	Large	Small	Large	Small	References
Yeast	25	15	26	18	*64*
	23	16	28	18	*104*
Neurospora	23	16	26	16	*49*
Xenopus	18	13	28	18	*98*
HeLa cells	16	12	28	18	*3*

TABLE 7.7

Base Composition of Mitochondrial RNA's

Organism	A	G	C	U
Yeast[a]				
Cytoplasmic ribosomes	26.4	28.2	17.0	28.4
Mitochondrial ribosomes				
15 S	36.1	17.7	12.5	34.5
25 S	38.4	16.5	12.2	32.8
Cytoplasmic 4 S	18.8	30.5	24.2	26.5
Mitochondrial 4 S	26.5	23.0	15.6	35.1
Neurospora[b]				
Cytoplasmic ribosomes	25.3	27.7	21.6	25.4
Mitochondrial ribosomes				
16 S	31.8	20.4	16.0	31.7
23 S	33.9	19.1	15.0	31.9

[a] Morimoto *et al.* (*64*).
[b] Rifkin *et al.* (*77*).

TABLE 7.8

Sedimentation Values of Mitochondrial Ribosomes and Their Subunits

Organism	Monosome	Subunits		References
		Large	Small	
Yeast	74 ± 1	53–58	35–40	*10*
	80	60	40	*64*
Neurospora	73	50	37	*49*
Rat Liver	50–55			*68*
Xenopus	60	43	32	*98*
Man (HeLa cells)	60	45	35	3

Some disagreement persists concerning the precise sedimentation values of mitochondrial ribosomes and their RNA's (*10, 22*). Regardless of the precise values, however, certain generalities are now well established. (*a*) Mitochondrial ribosomes are smaller than their cytoplasmic counterparts. (*b*) Mitochondrial ribosomal RNA's are correspondingly smaller than their cytoplasmic counterparts. (*c*) The physical properties of mitochondrial ribosomal RNA's show a number of unique features which further distinguish them from the cytoplasmic ones.

Ribosomal RNA's. Considerable difficulties were encountered in the characterization of mitochondrial ribosomal RNA's, owing to their relatively small size and unusual physical properties compared with their cytoplasmic counterparts. The special physical properties of mitochondrial ribosomal RNA's were first noted on the basis of discrepancies between estimates of molecular weight from sedimentation rates in gradients and from electrophoretic mobility on gels.

This problem has recently been intensively investigated by Littauer and his collaborators (*22*), in a series of studies comparing the ribosomal RNA's of *Aspergillus, Neurospora,* and *Trichoderma* with those of *E. coli.* These investigators conclude on the basis of base composition, thermal denaturation patterns, and circular dichroism measurements, as well as sedimentation rates in gradients and electrophoretic mobility in various buffers in polyacrylamide gels, that the mitochondrial ribosomal RNA's differ structurally from the homologous cytoplasmic RNA's of the same organisms, and from the bacterial ribosomal RNA's as well.

The structural differences, seen as greater thermal instability, and increased sensitivity to small changes in ionic concentrations, appear to result from a looser molecular configuration of the mitochondrial ribosomal RNA's than of the cytoplasmic ones. The differences are partly attributable to base composition of the mitochondrial RNA's, which are low in GC content, and partly to other features which influence their secondary structure in solution. How mitochondrial and cytoplasmic RNA's may differ in the intact ribosome is still unknown.

As a consequence of their physical properties, it has not been possible to establish the molecular weights of these molecules accurately from their positions in gradients and gels. However, the same properties which make physical characterization difficult, have helped in distinguishing and separating the mitochondrial ribosomal RNA's from the corresponding cytoplasmic ones. The best estimates of molecular weight of ribosomal RNA's come from measurements on electron micrographs of molecules prepared under strongly denaturing conditions to destroy their secondary structure.

The mitochondrial ribosomes and ribosomal RNA's of animal cells are particularly intriguing because of their small size. When first described, the "miniribosomes" of animal cells with sedimentation constants of 50–60 S were identified with the large subunit rather than with the intact ribosomes. Similarly, the RNA's with S values of 12 S and 16 S in sucrose gradients were suspected to be degradation products of larger molecules. The recently achieved acceptance of the "mini" values has resulted from the concordance of evidence from several laboratories and several experimental systems (Tables 7.6, 7.7, and 7.8).

Transfer RNA's. Barnett and co-workers (*5, 6, 21, 27, 28*) have investigated the transfer RNA's and corresponding amino acid synthetases in *Neurospora* comparing the components present in isolated washed mitochondria with those present in whole cell extracts. They found that *Neurospora* mitochondria contained transfer RNA's for all of the amino acids. In a subsequent study, Epler and Barnett (*28*) reported that mitochondrial leucine transfer RNA showed different coding properties from those of cytoplasmic leucine transfer RNA. The mitochondrial transfer RNA was found to respond only to a UC copolymer in the binding reaction, whereas the cytoplasmic leucine transfer RNA responded to (UC), (UG), and also to polyuridylic acid.

In the rat liver system, Buck and Nass (*11*) found that a particular leucine transfer RNA was present only in mitochondria, and could be acylated only by the corresponding mitochondrial synthetase; similarly a tyrosine transfer RNA of mitochondrial origin was charged only by the mitochondrial synthetase.

The presence of *N*-formylmethionyl-tRNA has been demonstrated in mitochondria of animal cells (*35*), *Neurospora* (*28*), rat liver (*91*), and yeast (*91*). This finding is very important in relation to the evidence that the protein-synthesizing apparatus of mitochondria is bacteria-like in many of its properties. Since *N*-formylmethionyl-tRNA regulates the initiation of translation of polypeptide chains in bacterial systems but not in the cytoplasm of eukaryotic cells, its presence in mitochondria provides another means for the independent regulation of mitochondrial and cytoplasmic protein synthesis in these cells.

Hybridization Studies

Are mitochondrial RNA's transcribed from mitochondrial DNA? This question has been investigated by hybridization of mitochondrial DNA with radioisotope-labeled rRNA's and tRNA's purified in sucrose gradients by several investigators (*4, 19, 39, 66, 104, 105*). For example, Dawid (*19*) examined mitochondrial DNA–RNA hybridization with nucleic acids isolated from oocytes of *Xenopus,* a superior source in which the mitochondrial nucleic acids are far more abundant than the nuclear ones (*18*). He showed that the large and small ribosomal RNA's from mitochondria hybridized with mitochondrial DNA and that the hybrids were not competed out by cytoplasmic RNA's. Similarly, mitochondrial RNA's did not hybridize with nuclear DNA beyond the extent of nonspecific hybridization with *E. coli* DNA. He found that the two ribosomal RNA's hybridized with mitochondrial DNA to the extent of about 12% and the 4 S RNA's to about 3%. These values correspond to one cistron for each rRNA and about twelve cistrons for tRNA's.

Attardi *et al.* (*4*) identified the 16 S and 12 S ribosomal RNA's of HeLa cell mitochondria in sucrose gradients as shown in Figs. 7.5 and 7.6. Labeling of the contaminating cytoplasmic 18 S and 28 S components was preferentially inhibited by actinomycin D and, reciprocally, labeling of the 16 S and 12 S components was inhibited by ethidium bromide.

HeLa cell mitochondrial DNA, like that of many other animal cells, can be separated into light and heavy strands because of base compositional differences. Attardi *et al.* (*4*) used the strand-separated DNA in hybridization studies to find out how much of each of the strands was transcribed. They found that virtually all of the heavy strand could be hybridized with mitochondrial RNA synthesized *in vivo,* but that only a small fraction of the light strand was similarly hybridized. From the extent of hybridization of the DNA with purified fractions of 16 S, 12 S, and 4 S tRNA's, it was estimated that the heavy strand of DNA contains one copy each of the 16 S and 12 S components, and probably eight

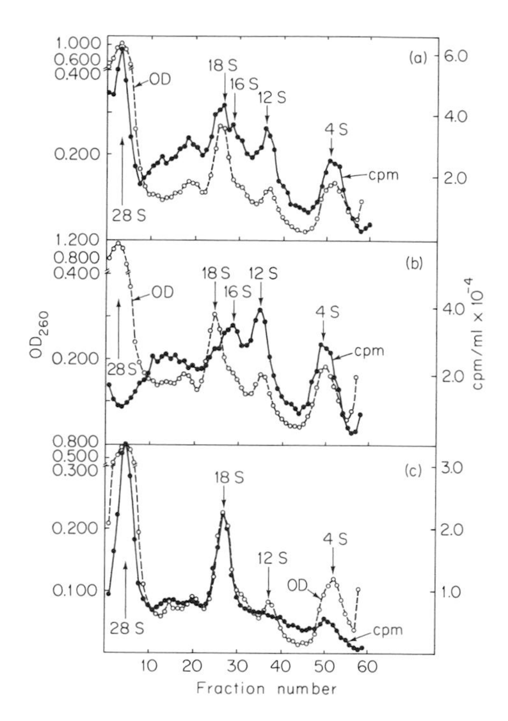

Fig. 7.5. Sedimentation pattern of RNA components sedimenting slower than 28 S RNA from the mitochondrial fraction of HeLa cells exposed to 5-^{3}H-uridine for 90 minutes in (a) the absence of either drug; (b) the presence of 0.04 μg/ml actinomycin D; (c) the presence of 1 μg/ml ethidium bromide. Graphs show RNA components separated in a 15–30% (w/w) sucrose gradient in SDS buffer (0.5% SDS, 0.1 *M* NaCl, 0.01 *M* TRIS at pH 7.0, 0.001 *M* EDTA) in the SW 25.3 Spinco rotor at 25,000 rpm for 28 hours at 20°C. From (*4*).

copies of 4 S tRNA's. The light strand hybridized with 4 S RNA's to an extent corresponding to three tRNA's.

With the use of newly developed, delicate techniques for visualizing nucleic acid molecules in the electron microscope, a tentative mapping of ribosomal and tRNA's on the mitochondrial DNA was achieved (*1*). In this procedure, double-stranded regions appear as linear rods, whereas single-stranded regions, which are random coils, appear as "bushes." The heavy strand of mitochondrial DNA was hybridized with 12 S and 16 S RNA's and then examined in the electron microscope to look for bushes. Linear regions corresponding to the length of the 12 S and 16 S RNA's were seen with bushes on both sides, but no bushes were seen between the 12 S and 16 S hybridized linear regions. Thus it appears that the two ribosomal cistrons lie side by side on the

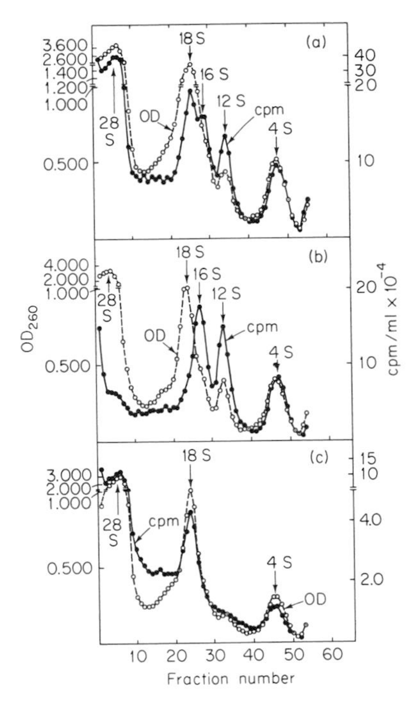

Fig. 7.6. Sedimentation pattern of RNA components sedimenting slower than 28 S RNA from the mitochondrial fraction of HeLa cells exposed to 5-^{3}H-uridine for 4 hours in (a) the absence of either drug; (b) the presence of 0.04 μg/ml actinomycin D; (c) the presence of 1 μg/ml ethidium bromide. Graphs show RNA components separated in a 15–30% (w/w) sucrose gradient in SDS buffer and 0.03 *M* EDTA, and centrifuged in an SW 27 Spinco rotor at 25,000 rpm for 25 hours at 20°C. From (4).

heavy strand. A similar study with 4 S RNA indicated that the eight tRNA cistrons do not lie side by side but are scattered around the DNA.

How much of mitochondrial DNA is transcribed as ribosomal and transfer RNA's? The available estimates are summarized in Table 7.9. In order to make this calculation, it is necessary to know the genomic size of the mitochondrial DNA. The clearest evidence on this point comes from animal cell studies, in which the small closed circles have been repeatedly reported as about 5 μ in contour length, corresponding to a molecular weight of about 1×10^7 daltons. No evidence of reiteration has been seen in reannealing studies, which give a genomic size corresponding to the molecular weight. On this basis we may assume the value of 10^7 daltons as a fair approximation of the genomic size (double-stranded) of mitochondrial DNA from animal cells. In yeast, on the basis of similar evidence from electron micrographs and from rean-

TABLE 7.9

Transcription of Mitochondrial DNA's

Organism	Genomic size of mitochondrial DNA[a]	Amount hybridized[a]		Remaining coding potential[a]	References
		4 S RNA	ribosomal RNA		
HeLa cells	15,000	900	3600	10,500	*4*
Xenopus	15,000	900	3600	10,500	*19*
Yeast	75,000	1600[b]	3600	70,000	*64*

[a] In nucleotide pairs.
[b] Estimated on basis of one cistron for each of twenty tRNA's.

nealing kinetics, we assume the value of 5×10^7 daltons (*9*). Renaturation studies with mitochondrial DNA from *Tetrahymena* have given a genomic size close to that of the reported molecular size of 17.6 μ or about 3.5×10^7 daltons (*9*). In *Neurospora* the genomic size has been estimated as $6–7 \times 10^7$ daltons (*105*).

The coding potentials of mitochondrial DNA's given in Table 7.9 are based on the best estimates available for the percent of the genome transcribed as ribosomal and transfer RNA's. At present these RNA's are the only transcription products of the mitochondrial DNA which have been identified. Thus even in animal cell mitochondria, the major fraction of the DNA is of unknown function. Attardi and his colleagues (*1, 4*) have provided the first direct evidence that most if not all of HeLa cell mitochondrial DNA is transcribed. Indeed, the evidence that some tRNA's are transcribed from the light strand raises the possibility that both strands may be transcribed in some regions.

As yet no one has shown that both strands of mitochondrial DNA are transcribed in the same region. If this occurred it would of course increase the information content of the DNA as much as twofold. Evidence on this point may be expected to come from further hybridization studies, and from measurements of self-annealing of RNA transcription products.

An independent line of evidence that mitochondrial ribosomal RNA's and 4 S RNA's are transcribed from mitochondrial DNA is based on the fact that their transcription can be blocked preferentially by ethidium bromide, the acridine dye which has been shown to bind preferentially to mitochondrial DNA. The selective inhibition of mitochondrial ribosomal RNA synthesis was shown with intact HeLa cells (*112*) and the inhibitory effect of the drug on 4 S RNA synthesis was demonstrated with isolated mitochondria (*111*).

What are the functions of the unidentified RNA's transcribed from mitochondrial DNA? In HeLa cells and presumably in other animal cell mitochondria, these RNA's could code for about 3500 amino acids (assuming that only one strand of DNA is coding), corresponding to about twenty to thirty proteins. The mitochondrial DNA of yeast and *Neurospora* is so much larger than that of animal cells, that five times as many proteins could be encoded.

Despite the availability of this coding potential, as yet not a single protein has been identified unambiguously as a gene product of mitochondrial DNA. In the rest of this chapter we will examine the evidence concerning the presence of ribosomes in mitochondria, the occurrence of mitochondria protein synthesis, and the various indirect lines of evidence on the origin of mitochondrial proteins.

Mitochondrial Ribosomes

The isolation and characterization of mitochondrial ribosomes has been difficult to accomplish. Nonetheless, investigators pursued the elusive ribosomal particles, spurred on by the presence of tRNA's and ribosomal-type RNA's in isolated mitochondria. By now, mitochondrial ribosomes have been characterized from eukaryotic microorganisms and from animal cells (*3, 51, 64, 68, 74, 77, 98*) and their activity in *in vitro* amino acid incorporation has been demonstrated in several systems (*3, 64, 74, 98*).

The proteins from mitochondrial ribosomes have been examined by fractionation on carboxymethyl cellulose columns and compared with those of the corresponding cytoplasmic ribosomes in *Neurospora* (*50*). The differences between the proteins of the mitochondrial and cytoplasmic ribosomes are shown dramatically by the noncorrespondence of patterns in a double labeling experiment. Figures 7.7a and b show the results of cochromatography of proteins from the large (a) and small (b) subunits, in which the mitochondrial proteins were labeled with ^{3}H-labeled lysine and the cytoplasmic proteins with ^{14}C-labeled lysine. From these patterns, it appears that virtually all of the proteins are different. Similar conclusions have been reported for yeast (*64*).

The most striking characteristic of mitochondrial ribosomes is their size (Table 7.8). Those from animal cells are "miniribosomes," no larger than the large subunit of the cytoplasmic ribosomes, while those from eukaryotic microorganisms are larger than the animal cell particles, but smaller than their cytoplasmic counterparts. Mitochondrial ribosomes resemble bacterial ones in their sensitivity to low magnesium ions, to chloramphenicol, and to many other inhibitors of bacterial protein synthesis; and in their resistance to cycloheximide, a classical inhibitor of

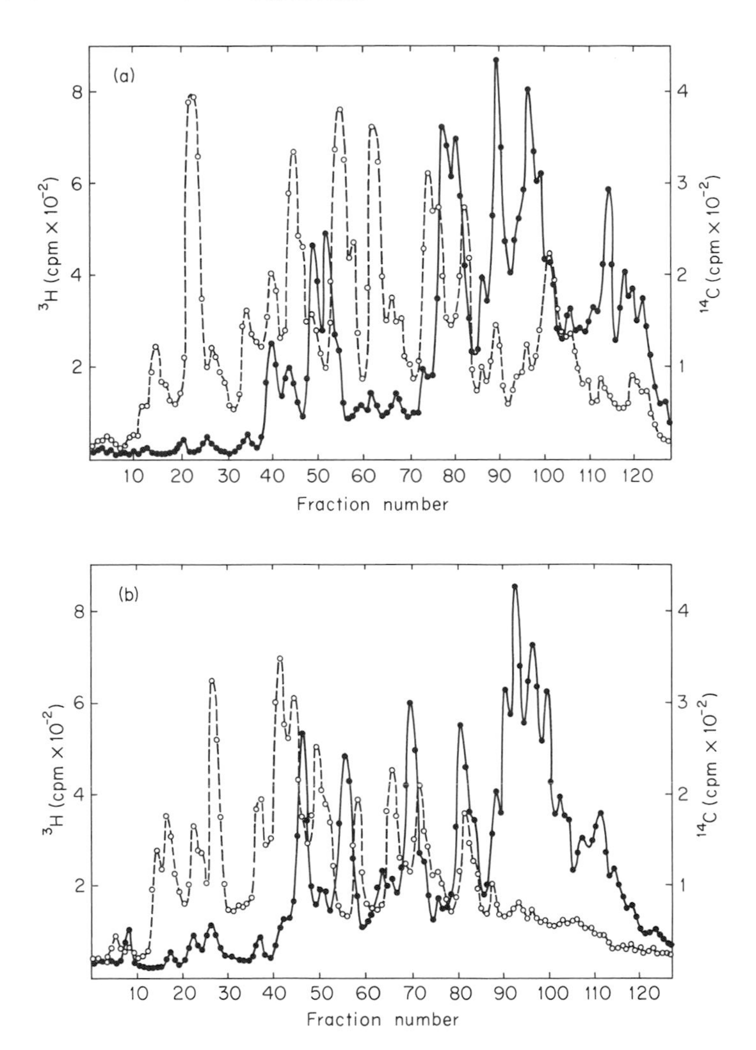

Fig. 7.7. (a) Cochromatography of ^{3}H-lysine-labeled proteins from mitochondrial 50 S subunits (○---○) with ^{14}C-lysine-labeled proteins from cytoplasmic 60 S subunits (●—●). (b) Cochromatography of ^{3}H-lysine-labeled proteins from mitochondrial 37 S subunits (○---○) with ^{14}C-lysine-labeled proteins from cytoplasmic 37 S subunits (●—●). From (*50*).

protein synthesis on cytoplasmic but not on bacterial ribosomes (*45; 56*).

Another property of mitochondrial ribosomes which distinguishes them as a class is their mode of control of protein synthesis. Initiation requires formylation of a special f-met-tRNA as in bacterial systems (*91*); and elongation factors have been isolated from *Neurospora* (*36*) and from yeast (*64, 70*). 5 S RNA has not yet been detected in mitochondrial ribosomes (*57*), but neither its presence or absence has been established. If it is smaller than the cytoplasmic 5 S RNA, it may be difficult to separate from the 4 S RNA's (*10*).

Now that mitochondrial ribosomes have been unambiguously identified and characterized, a few fundamental questions can be raised about their structure and functions, as well as their relationship to cytoplasmic and to bacterial ribosomes. First, let us consider their structural features. Despite their remarkable range in size, all mitochondrial ribosomes have numerous properties in common: their rRNA's are transcribed from mitochondrial DNA, and their proteins differ from those of their cytoplasmic counterparts. The rRNA's have unique physical properties which differ from those of cytoplasmic and bacterial ribosomes. The signals to which mitochondrial ribosomes respond in protein synthesis resemble the bacterial system: the use of f-met-tRNA for initiation, the interchangeability of initiation and of peptide-bond elongation factors with *E. coli,* the sensitivity to the same classes of inhibitors of protein synthesis, such as chloramphenicol and erythromycin, and insensitivity to cycloheximide.

What is the functional significance of the size range displayed by mitochondrial ribosomes? As yet no differerences in protein-synthesizing ability have been demonstrated between mitochondrial ribosomes and those of the cell-sap. If a 55–60 S particle could functionally replace an 80 S ribosome, then presumably no 80 S ribosomes would be found. The fact that 80 S ribosomes do abound in the cytoplasm of eukaryotes suggests some superiority, as yet unknown, of these large particles.

The bacterial origin of mitochondrial ribosomes has been debated, but no clarifying concepts or facts have yet emerged. Functionally, mitochondrial ribosomes resemble bacterial ones more than they resemble their cytoplamic counterparts. One clear observation is that mitochondrial ribosomes as a class respond to different signals than do their cytoplasmic counterparts. This dichotomy should provide the key to understanding their unique functions, and ultimately should also shed light on the significance of their unique structures.

As a pure speculation, one might equate ribosomal size with functional complexity. On this basis, cytoplasmic ribosomes would have functional capacities beyond those of all the others. The bacterial ribosomes would be ranked with mitochondrial ribosomes of eukaryotic microorganisms like yeast and *Neurospora,* and those of animal cells would be functionally stripped to the bare essentials. In considering the protein-synthesizing functions of mitochondria in the rest of this chapter, it will be useful to keep in mind the distinction between the ribosomes of these size classes.

ORIGIN OF MITOCHONDRIAL PROTEINS

The evidence that mitochondrial DNA's have coding potential for a number of proteins, minimally about twenty in animal cells and many more in mitochondria from eukaryotic microorganisms, leads us now to examine the origin of mitochondrial proteins. Which ones are synthesized in mitochondria? Which ones are coded by mitochondrial DNA?

One might suppose that any protein synthesized on mitochondrial ribosomes would be the product of a mitochondrial messenger RNA and consequently that mitochondrial DNA would contain a cistron coding for that protein. On this simple view, one could establish which proteins are coded on mitochondrial DNA by showing that they are synthesized in the organelle. This approach has been used by several investigators, and most of the evidence to be presented in this section concerns the translational origin of mitochondrial proteins.

However, we must also consider the possibility that messenger RNA's of nuclear origin could be translated on mitochondrial ribosomes. If so, the evidence that a particular protein was synthesized in the organelle would not be critical evidence that it was coded by mitochondrial DNA. As yet the only evidence on RNA transport across mitochondrial membranes *in vivo* is a preliminary report of RNA's which hybridize with mitochondrial DNA and are found associated with nonmitochondrial cytoplasmic membranes (*2*).

Thus, the possibility of messenger RNA transport in or out of mitochondria is primarily of heuristic value at this time, but must be kept in mind as the complex tangle of nuclear–cytoplasmic–organellar interactions is unraveled.

A number of different experimental approaches are currently being utilized in attempts to track down the origin of mitochondrial proteins. Two general lines of experimentation can be distinguished. (*a*) *In vitro* studies of amino acid incorporation into isolated mitochondria. (*b*) Studies of protein synthesis *in vivo* followed by fractionation to separate mitochondrial from other cell components. (*c*) Comparison of proteins present in mitochondrial mutants with those of wild-type cells.

We will consider the *in vitro* studies first and then turn to the more complex findings with various *in vivo* systems.

Protein Synthesis in Isolated Mitochondria

The study of protein synthesis in isolated mitochondria might be envisaged as the most direct approach to identifying the proteins coded by mitochondrial DNA. One might hope that intact mitochondria would incorporate amino acids into all the proteins synthesized in the organelle *in vivo* and that the entire process from DNA via messenger RNA to protein might be charted with *in vitro* experiments.

However, protein synthesis in a multicomponent organelle is far more complex than similar studies with isolated ribosomes both in experimental design and in the interpretation of findings. There are large numbers of proteins, some in the matrix and some in the membranes, some present in high concentrations and other very low, some of them varying in amount with the metabolic state of the cell (i.e., inducible or repressible), and many of them regulated in synthesis or activity by other components of the organelle. Thus, the kind of information that can be obtained from *in vitro* experiments, while very important, is circumscribed. Nonetheless, *in vitro* experiments do provide a *minimal* estimate of the amount of protein synthesis occurring in mitochondria as well as a means of identifying specific proteins synthesized there.

In examining mitochondrial protein synthesis, one essential question is whether the mitochondrial ribosomes themselves are capable of amino acid incorporating activity in an organelle-free system. Successful demonstrations of the activity of purified ribosomes from mitochondria of yeast (*64*), of chick liver (*74*), and of *Xenopus* (*98*) have been reported. In addition to the ribosomes, some of the supernatant factors required for initiation and for elongation have been identified (*48, 64, 70, 91*).

Further requirements for amino acid incorporating activity of isolated mitochondria are much more complex than those for the ribosomes per se. Conditions have been described by a number of investigators, working principally with animal mitochondria (*46, 101, 107*). To retain amino acid incorporating activity mitochondria must be intact and capable of a good rate of oxidative phosphorylation. A high level of O_2 availability is essential; the concentration of an oxidizable substrate is critical, too high a concentration being inhibitory. There is a sharp pH optimum at 7.2; Mg^{2+}, AMP, and inorganic phosphate ions are required; and the incorporation rate is not decreased by treatment with ribonuclease (evidence of intactness of the organelle).

Incorporation of amino acids into protein in the *in vitro* system depends not only upon the structural integrity of the organelle, but also on the host tissue and its metabolic state. For example, using locust flight muscle, Kleinow *et al.* (*46*) found that mitochondria from growing tissues were much more active than those from nongrowing muscle as shown in Fig. 7.8 and were 100 times as active as those from rat liver. With rat liver mitochondria, pretreatment of the rats with thyroid hormone increased the incorporation rates of the isolated organelles.

Given an actively incorporating preparation, where do the labeled

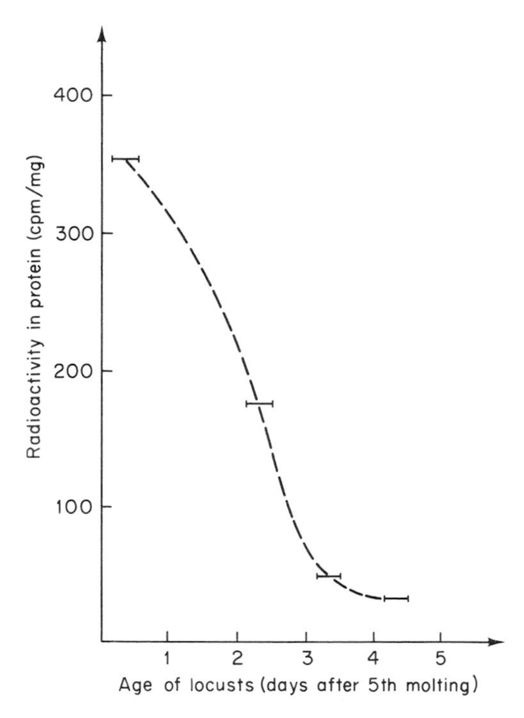

Fig. 7.8. Rate of *in vitro* incorporation of ^{14}C-isoleucine into mitochondria isolated from developing flight muscle of the insect *Locusta migratoria* at different stages of development. From (*46*).

amino acids go? Investigators agree that virtually all of the label is taken up into membrane proteins and not detectably into soluble proteins of the mitochondrial matrix such as malate dehydrogenase or catalase (*41, 41a, 45, 46, 84, 107*). For example, Kleinow *et al.* (*46*) compared the incorporation of labeled amino acids into mitochondrial proteins of locust flight muscle in both *in vivo* and *in vitro* studies, as shown in Fig. 7.9. They found incorporation mainly into a few proteins of the insoluble fraction.

Neupert *et al.* (*67*) found that inner membrane preparations of rat liver mitochondria took up over 95% of the total amino acid label incorporated, and that the outer membrane took up almost none. They distinguished the inner and outer membranes by the method of preparation and by the use of enzymatic markers: succinate cytochrome c reductase marked the inner membrane and monoamine oxidase marked the outer membrane. In separate experiments they used radioactive arginine, isoleucine, phenylalanine, valine and leucine. With each amino acid the labeling pattern was essentially the same: the outer

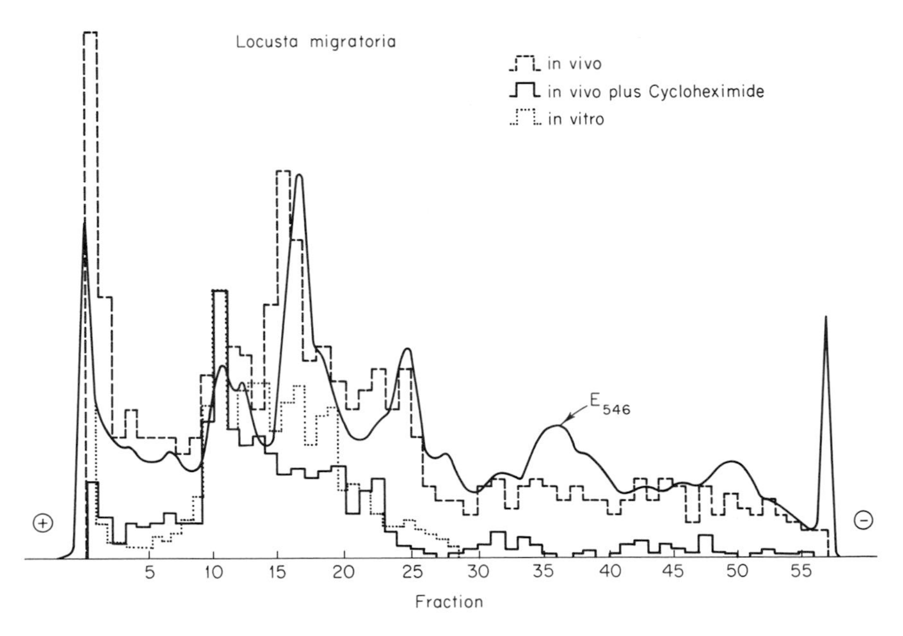

Fig. 7.9. Densitogram of amido black stained insoluble mitochondrial protein after gel electrophoresis (smooth line); distribution of incorporated radioactivity (dashed line) *in vivo* 1 hour; same plus 50 μg cycloheximide per locust (solid line); *in vitro* incorporation by isolated mitochondria (dotted line). From (*46*).

membrane took up less than 5% as much label as the inner membrane.

Haldar *et al.* (*40*) described conditions under which amino acid incorporation continued for some time into isolated rat liver mitochondria which were then fractionated. None of the label was incorporated into cytochrome c or malic dehydrogenase, but rather it was concentrated in the insoluble membrane fraction. With *in vivo* labeling experiments in ascites tumor cells they found some incorporation of label into mitochondria too early to represent translocation from the cytoplasm. They treated the cells with ribonuclease and a slight osmotic shock to get the ribonuclease into the cytoplasm and found that microsomal protein synthesis was inhibited much more than mitochondrial.

Protein fractionation experiments indicated that the label was incorporated primarily into a fraction corresponding to the "structural protein" of Criddle (p. 234). Similar observations have been reported from other laboratories (e.g., *108*), but because of the difficulty in characterizing these proteins more satisfactorily, this approach to the problem has not been enthusiastically pursued. At present, several investigators have estimated that about 5% of the total protein or about 10–15% of the membrane-bound mitochondrial protein is synthesized *in situ,* as judged from incorporation studies with isolated mitochondria. Of course, this estimate is minimal, since any required cofactors washed out during isolation of the mitochondria would result in decreased incorporation.

A further criticism of the *in vitro* studies is based on the observation (*101*) that the labeled polypeptides are unstable and continuously degraded to acid-soluble form. This degradation could lead to an underestimate of the actual protein-synthesizing activity. Another possible source of error in the direction of underestimation is the possibility that labeled products of the *in vitro* system may leak out of the organelle before the reaction is terminated.

At the time when the first successful studies of amino acid incorporation into isolated mitochondria were being carried out, one surprising and consistent finding was reported. No amino acids were being incorporated into cytochrome c. Indeed unraveling the biogenesis of cytochrome c has been particularly instructive. Cytochrome c is not synthesized within mitochondria; it is synthesized on the cytoplasmic microsomes and then transferred into mitochondria where it is bound into the inner membrane as part of the electron transport chain.

Kadenbach (*43*) showed that prelabeled microsomes containing cytochrome c which was synthesized *in vivo* can transfer cytochrome c to mitochondria *in vitro,* but if cytochrome c is purified and then added to the mitochondrial system, no transfer occurs. Recently he has reported evidence that the heme group of cytochrome c is attached after the apoprotein has been transferred into mitochondria from the endoplasmic reticulum where it was synthesized.

Work and his colleagues (*107*) found that mild osmotic shock plus ribonuclease treatment would inhibit protein synthesis in the microsomal fraction, but not in mitochondria of mammalian cells, in tissue culture. Under these conditions, cytochrome c transfer continues, so that after 10 minutes all the cellular cytochrome c has accumulated in the mitochondria. The authors suggest that transport occurs while the protein is being synthesized.

The fact that cytochrome c is synthesized on cytoplasmic ribosomes is consistent with the genetic evidence that the structural gene for cytochrome c in yeast is nuclear (*86*). With the accumulating evidence that many mitochondrial proteins are coded by nuclear genes and are synthesized in the cytoplasm, the mechanism of transfer of proteins across organelle barriers and the regulation of their synthesis, transfer, and insertion have become central questions for experimental investigation.

In summary, studies of protein synthesis with isolated mitochondria have provided a minimal value of some 5% of total mitochondrial protein, or about 15% of membrane-bound protein, capable of incorporating amino acids under *in vitro* conditions. Separation of these proteins by gel electrophoresis has shown that they are largely if not entirely components of the inner mitochondrial membrane; none of them have been identified per se. Although most of these studies were carried out with rat liver mitochondria, similar results have been reported with mitochondria from *Neurospora* and yeast.

These experiments also do not tell us definitely whether any or all of these proteins are coded by mitochondria DNA. To answer that question we need to identify the individual genes that code for these proteins or at least to know the transcriptional origin of the messenger RNA's from which these proteins were translated. Since the *in vitro* studies have not answered our questions, we must turn to other methods. One approach has been to identify proteins associated with the respiratory activity of mitochondria, and to ask where that group of proteins are synthesized and coded. Studies of this kind have been carried out primarily with yeast, which is a facultative anaerobe. In this organism, the synthesis of some proteins associated with respiration can be blocked by growth in the absence of O_2, by growth on high concentrations of glucose, by growth in the presence of certain antibiotics, and by mutation to respiratory inability. All these methods have been examined and the results will now be discussed.

Mitochondrial Proteins of Yeast

In order to identify proteins synthesized in mitochondria in intact cells, one needs to distinguish them from proteins of cytoplasmic origin. The first and still the most extensive evidence comes from biochemical studies of mitochondrial *petite* (*rho*$^-$) mutants of yeast. More recently, the results with *petite* strains have been compared with those of wild-type yeast grown anaerobically, or under conditions of catabolite (glucose) repression, or in the presence of various inhibitors of protein synthesis. These studies have given a minimal estimate of the amount of protein which may be of mitochondrial origin, quite in line with the results of the *in vitro* studies just discussed.

The single most striking conclusion to be drawn from all the studies, *in vitro* and *in vivo,* is that most mitochondrial proteins are of cytoplasmic origin. Consequently, the problem of establishing which proteins are synthesized in mitochondria, and which are coded by mitochondrial DNA is exceedingly difficult: one must identify individual proteins which comprise a small minority of the total population. A further problem is posed by the fact that the proteins which take up label are mostly if not all components of the inner mitochondrial membrane. Most of these proteins have not been identified, characterized, or related to the many enzyme activities associated with the inner membrane. The first requirement of a study of biogenesis is the ability to recognize the components of the system, but this requirement has not yet been met with mitochondrial membranes.

The experimental problem is further complicated by the fact that some proteins must be present in order that other may be properly bound. For example, a protein known as the oligomycin-sensitivity conferring factor must be synthesized before the F_1-ATPase, previously present, exhibits any oligomycin sensitivity (*75, 80*). Another example is succinate dehydrogenase. Conflicting reports have appeared from different laboratories about whether or not the enzyme is present in mitochondria of anaerobically grown cells. It now appears that the extent of its activity is strongly influenced by the membrane lipid composition which in turn is strongly influenced by the growth medium (*56, 81*).

Because of these problems as well as the complexities of mitochondrial biogenesis, with interlocking components of cytoplasmic and mitochondrial origin, only a few proteins have tentatively been identified as products of mitochondrial protein synthesis or of the mitochondrial genome. On the other hand, many mitochondrial proteins have been classified with some assurance as cytoplasmic in origin. As an aid to discussion, Tables 7.10 and 7.11 have been drawn up, listing the principal mitochondrial proteins tentatively considered as synthesized in mitochondria or in the cell sap.

To evaluate the evidence on which these tables are based, it will be necessary first to consider briefly some of the effects of anaerobiosis, glucose repression, antibiotics, and the *petite* mutation upon mitochondrial structure and function. Most of the proteins listed in Tables 7.10 and 7.11 have shown similar response in more than one of the experimental systems; and this concordance of results has supported the inferences which have been drawn.

Structural and Functional Consequences of the petite Mutation and Catabolite Repression. As discussed in Chapters 2 and 4, mitochondrial *petite* mutants of yeast comprise a heterogeneous set of strains in which more or less extensive alterations have occurred in the mitochondrial DNA. Consequently, different *petite* strains retain more or less of the mitochondrial genome. However, at the level of respiratory activity, no differences have been seen. Rather, all *petites* have shown very similar if

TABLE 7.10

Proteins Associated with the Mitochondrial Genome[a]

	Evidence of association with mitochondrial genome				
		Inhibited by CAP or ERY[c]			
Protein or factor	Missing in *rho*$^{-}$ [b]	wt.	anaerobic → aerobic	glucose repression → derepression	References
cyt a + a_3 (spectral)	●	●	●	●	
cyt b (spectral)	●	●	●	●	
cyt c_1 (spectral)	●	●	●	●	
cyt c oxidase activity	●	●	●	●	
(NADH) cyt c reductase activity (antimycin-sensitive)	●	●	●	●	
Succinate dehydrogenase activity	● (reduced)	● (partial)	● (partial)		
Oligomycin sensitivity	●		●		*80*
P_i-ATP exchange activity	●	●	●		*38*
Erythromycin resistance	● (some strains)				
Chloramphenicol resistance	● (some strains)				
Paramomycin resistance	● (some strains)				
Protein necessary for ethidium bromide induction of *rho*$^{-}$	●	●			*71*
Proteins needed for integration of other proteins into inner membrane				●	
Mitochondrial regulator(s) of catabolite repression				●	

[a] See general references *40, 41, 41a, 56, 81* except where indicated.
[b] Evidence of coding by mitochondrial DNA.
[c] Evidence of synthesis on mitochondrial ribosomes. CAP, chloramphenicol; ERY, erythromycin.

TABLE 7.11

Mitochondrial Proteins Probably Not Associated with the Mitochondrial Genome

Protein or factor	Evidence of nonassociation: Present in *rho⁻*	Evidence of nonassociation: Inhibited by CHI and not by CAP	References
Most ribosomal proteins	NT[a]	●	*50*
F_1-ATPase	●	●	*16*
Most soluble matrix proteins including Krebs cycle enzymes	●	●	*41, 41a, 56, 81*
Cytochrome c	●	●	*43*
Mitochondrial DNA polymerase (or repair enzyme)	●	●	*102a*
Mitochondrial RNA polymerase	● (reduced)	●	*102a*
Many mitochondrial membrane proteins	NT	●	*41, 41a, 56, 81*

[a] NT = not tested.

not identical phenotypes. The description to be given here of the basic *petite* phenotype comes primarily from studies by Ephrussi and collaborators (*24, 25, 87, 88, 90*) working with a neutral *petite* strain.

Petite mutants were identified initially by their inability to grow on a nonfermentable carbon source and by the small colonies which they formed on glucose agar. Under anaerobic conditions, the wild type and the *petite* mutant grew at about the same rate and their growth properties were indistinguishable. Under aerobic conditions however, even in the presence of glucose, normal yeast cells grew faster than did *petite* and produced about four times as much cell mass.

Growth studies showed that the *petite* mutants had virtually no terminal respiratory system; in the presence of oxygen their respiratory rate was only about 8% of wild type and this residual respiration was not sensitive to KCN. The absence of respiratory activity in the *petite* mutants was shown to result from the absence of the electron transport chain of terminal respiration. Spectral studies showed that *petite* mutants had neither cytochrome b nor cytochrome $a + a_3$, while cytochrome c was higher than in normal cells (*24*). The *petite* mutant, possessing neither cytochrome oxidase nor succinic cytochrome c reductase activity, cannot oxidize or reduce its cytochrome c, via the usual pathway. A number of the dehydrogenases and enzymes of the Krebs cycle which are soluble mitochondrial enzymes are present in both wild-type and *petite* strains.

To compare the *petite* phenotype with that of wild-type cells grown on glucose, we must first distinguish between glucose repression and the Pasteur effect. The Pasteur effect, classically described as the inhibition of fermentation by molecular oxygen, is shown best with nongrowing cultures in glucose under aerobic conditions. The glucose is oxidized to CO_2 and water, very little ethanol is formed, respiration is fast, and the fermentation rate is slow. *Petites* do not show a classic Pasteur effect: in the presence of O_2, the fermentation rate is unaltered.

The results are very different, however, if the yeast cells are growing at the time that oxygen is introduced. A second process counters the Pasteur effect, a process now known as glucose (more generally, catabolite) repression. In aerobic growing cultures, glucose blocks the synthesis of the electron transport chain required for terminal respiration and the uptake of oxygen. So long as glucose is present, it is converted to ethanol by aerobic fermentation. When the glucose has been used up, glucose repression is overcome, the terminal respiratory chain appears, and oxygen is taken up. Glucose repression operates at the level of enzyme formation, not of function. In this respect, it is quite different from the Pasteur effect.

Effects of glucose repression upon wild-type yeast were first clearly shown by Ephrussi *et al.* (*26*) in studies of biochemical and cytological changes occurring during the aerobic cycle in wild-type yeast. The respiration rate falls considerably during exponential growth on glucose, reaching its original value only after the glucose has been exhausted. Aerobic fermentation shows essentially the inverse responses, being fastest during exponential growth. These metabolic variations have their counterpart in changes in the cytochrome spectrum. During exponential growth, the cytochromes decrease so that they can hardly be seen except at the temperature of liquid nitrogen. Then they reappear as terminal respiration rises and the rate of fermentation falls.

The *petite* mutant also responds to glucose repression to the best of its ability. Since cytochromes a and b are missing, one can observe glucose repression by examining the spectrum of cytochrome c. During aerobic growth on glucose the intensity of the α band of cytochrome c is faint, very much as in normal yeast. When the glucose is used up, cytochrome c synthesis proceeds and its concentration may increase as much as fourfold. Changes in the cytochrome c content of mitochondria of the *petite* mutant unable to use the enzyme mimic the changes induced in functional mitochondria by glucose and O_2. Apparently then certain regulatory mechanisms involved in mitochondrial biogenesis still function in some *petite* strains. Whether or not they function in the most extreme *petites*, such as those without mitochondrial DNA (cf. Chapter 2), needs to be reinvestigated.

Thus, glucose-repressed wild-type cells resemble *petites* in the virtual absence of cytochromes and of terminal respiratory activity. When glucose is used up, both wild type and *petite* strains escape from catabolite repression. The wild type develops electron transport activity while the mutant shows an increase in cytochrome c, the only component of the electron transport chain known to be coded in the nucleus and synthesized in the cytoplasm.

The structural changes accompanying these biochemical transformations in mitochondrial activity were first described by Yotsuyanagi (*109, 110*). In 1962 two detailed papers were published, comparing the fine structure of yeast mitochondria of wild-type cells and of *petites*. Yotsuyanagi was the first investigator to demonstrate the presence of mitochondrial membranes in *petites*.

He began with wild-type yeast cells in stationary phase in which the respiratory capacity was very high and the ability to ferment was very low. The mitochondria were small and dispersed around the periphery of the cell. During the course of growth on glucose, a progressive decrease in the number of mitochondrial profiles was noted, paralleling the fall in respiratory capacity and increase in fermentation. When the glucose was used up and respiratory activity appeared, a progressive increase in the number of mitochondrial profiles was seen. After cell division stopped, the mitochondria broke up into many small elements with much the same morphology as was seen in the inoculum.

The mitochondria of glucose-repressed cells as seen in electron micrographs, were poorly developed structurally, compared to those of respiring cells. During the transition stage, between logarithmic and stationary growth, the mitochondria changed spectacularly, developing into large structures with many cristae. Subsequently if cells were kept in the stationary phase, their respiration rate diminished and the mitochondrial structures began to break down.

Yotsuyanagi compared wild-type cells with mitochondrial *petites*, with nuclear *petites*, and with double mutants containing both nuclear and mitochondrial mutations. Again, it was important to consider the stage in the growth cycle during which observations were made. It was noted that, as with normal yeasts, the mutant mitochondria were best developed just at the end of the growth cycle. Thus morphological evidence is consistent with biochemical evidence that *petites* undergo glucose repression and derepression.

During exponential growth, the mitochondria of normal and *petite* strains show few differences. But as the growth rate slows down, the differences increase. The normal cells develop mitochondria with regularly disposed cristae and fairly homogenous size and shape, whereas the mutants develop a highly polymorphic set of mitochondria in which the internal development appears anarchic.

In similar studies of three nuclear *petite* strains, p_5, p_6, and p_7, quite a different picture was observed. (Mutants carrying p_5 lack cytochrome a; those with p_6 and p_7 lack both cytochromes a and b.) During logarithmic growth no differences were discernible between the normal yeast and the nuclear *petites.* As cells went into the stationary stage, few changes occurred in the mitochondria of p_5 or p_7. Neither their number nor their size changed very much. They behaved as if their development were arrested at the end of the log phase of growth. In p_6 the number of mitochondria and the degree of cristae development was much less than in the wild type. At the end of the log phase, p_6 did show normal mitochondrial development, in contrast to p_5 and p_7. The double mutants resembled the mitochondrial and not the nuclear *petites.* Even the p_6 double mutant formed multilamellar structures never present in the p_6 single mutant from which it could thereby be clearly distinguished. Of special importance was the demonstration that *petites,* despite their respiratory incapacity, still contained mitochondria quite similar in structure to those of glucose-repressed wild-type cells.

Mitochondria of Anaerobic Yeast. These findings were all but forgotten during the subsequent controversy over the presence or absence of mitochondria in anaerobically grown cells. The mitochondria of anaerobic yeast, called "promitochondria," have recently been described by Schatz (*81*) (cf. p. 236). Promitochondria seen in electron micrographs of frozen-etched preparations, resemble normal mitochondria in possessing both outer and inner membranes, the latter with typical infolding or cristae. The lipid composition of these promitochondria depends on the growth medium, but in all media, differ considerably from that of normal mitochondria from aerobic cells (Tables 7.4 and 7.5).

However, promitochondria differ most strikingly from aerobic mitochondria in the absence of respiratory pigments and the electron transport chain. Thus, the physiological switch from anaerobic to aerobic growth provides another experimental situation in which to study the synthesis and assembly of the missing electron transport proteins.

The differences in mitochondrial enzyme activities of anaerobic and aerobic yeast were first extensively studied by Ephrussi and Slonimski (*24, 25, 89*). More recently, the problem has been reexamined by Criddle and Schatz (*16*). The cytochromes of terminal respiration are totally missing or inactive in anaerobic yeast, but the activity of succinic dehydrogenase may be negligible or as high as 50% of the aerobic control, depending on the medium. The particular lipid composition influences succinic dehydrogenase activity markedly, while having no apparent effect on the other electron transport enzymes. Promitochondria do contain the oligomycin-sensitive F_1-ATPase although the amount is somewhat lower than the aerobic level. Their results indicate major similarities in the protein composition of mitochondria from aerobic and anaerobic yeast. Recently, Groot *et al.* (*38*) have shown that promitochondria are able to carry out the P_i-ATP exchange reaction which is a partial reaction of oxidative phosphorylation. Thus, this system is not coupled to terminal respiration. The activity of the system leads one to suggest that promitochondrial membranes may be more functional than had previously been suspected. When cells are grown anaerobically in the presence of erythromycin, an antibiotic which blocks protein synthesis within mitochondria, this exchange activity is lost. Thus, some mitochondrial protein synthesis does normally occur in the anaerobic promitochondria.

We may summarize this section as follows. Anaerobically grown yeast shares with glucose repressed wild type and *petite* strains, the absence of respiratory activity. The physiological switch from anaerobic to aerobic growth, like that from glucose repressed to derepressed, provides a good experimental situation in which to study the synthesis and assembly of electron transport (and perhaps additional) proteins into preexisting mitochondrial membranes.

Effects of Antibiotics which Inhibit Protein Synthesis. The effects on amino acid incorporation into isolated yeast mitochondria of a series of antibiotics including chloramphenicol and cycloheximide were first reported by Wintersberger (102a) and subsequently by Lamb *et al.* (*53*). Cycloheximide was shown to have no effect on the mitochondrial incorporating system although it is a potent inhibitor of cytoplasmic protein synthesis on 80 S ribosomes of yeast (as well as on the 80 S ribosomes of algae and higher plant and animal cells). On the other hand, chloramphenicol effectively blocked incorporation into mitochondrial proteins in *in vitro* studies; and in *in vivo* studies, chloramphenicol was effective in blocking appearance of respiratory activity during adaptation of anaerobically grown cells to molecular oxygen. The mitochondrial protein-synthesizing system was also blocked by erythromycin and related antibiotics, and by the aminoglycosides such as streptomycin and neomycin. In these responses, mitochondrial protein synthesis in yeast resembles the bacterial system which also is blocked by chloramphenicol, erythromycin, and the aminoglycoside antibiotics but not by cycloheximide. Mitochondria of animal cells have been reported to differ from those of yeast in being more resistant to the commonly used inhibitors of mitochondrial protein synthesis with the exception of chloramphenicol (*56*).

At the cellular level, chloramphenicol and erythromycin were found to block incorporation of ^{14}C-leucine into mitochondrial proteins to the extent of 13% in derepressed cells and 4% in glucose repressed cells (*45*). In the same series of short-time labeling experiments, cycloheximide inhibited cytoplasmic protein synthesis completely and mitochondrial protein synthesis 85–90%. Similar results reported by several investigators (e. g., *41, 41a*) led to the view noted above that most mitochondrial proteins are of cytoplasmic origin. In none of these studies, however, have the individual proteins responsible for the radioisotope labeling of mitochondrial protein fraction been identified.

Mahler and co-workers (*41, 41a*) examined mitochondrial protein synthesis during derepression of yeast which had been grown with 1% glucose as carbon source. Looking at the effects of inhibitors upon incorporation and upon derepression, they found that derepression could be inhibited by either chloramphenicol or by cycloheximide. Thus, this process required proteins synthesized by both systems. During the derepression process itself, chloramphenicol inhibited cellular protein synthesis by 22%, and cycloheximide inhibited it by 91%. Together the two inhibitors decreased it 94%. Kinetic studies of the rate of labeling of different mitochondrial fractions suggested that most proteins were of cytoplasmic origin but that a particular membrane protein fraction was probably synthesized *in situ*. However, the individual proteins in this fraction were not identified further.

In a further study of the synthesis of particular proteins during derepression, Henson *et al.* (*41, 41a*) followed the activities of the mitochondrial enzymes cytochrome oxidase. NADH-cytochrome c reductase, NADH-dehydrogenase and L-malate dehydrogenase, as well as four cytoplasmic enzymes: cytoplasmic malate dehydrogenase, NADPH:cytochrome c reductase, and cytoplasmic forms of NADH:cytochrome c reductase (insensitive to antimycin A) and NADH dehydrogenase. They found that all the cytoplasmic enzymes were inhibited to various extents by chloramphenicol during derepression, and that the mitochondrial enzymes were also inhibited to varying extents. They did not show coordinated derepression.

The complexities of their data led them to propose that all soluble proteins of the mitochondrial matrix as well as some of the proteins of the inner membrane are coded by nuclear genes and synthesized in the cytoplasm, and that during derepression these proteins are individually transported from the endoplasmic reticulum into mitochondria; and furthermore, that integration of these components into the mitochondrial membranes requires the concomitant synthesis of products (unidentified) of the chloramphenicol-sensitive protein synthesizing system of the mitochondria. Finally they proposed that the functioning mitochondria produce substances which regulate the continuing inflow of cytoplasmic components.

Antibiotics have been used with some success in conjunction with other methods to examine the origin of a few specific mitochondrial proteins. As shown in Table 7.10, the mitochondrial origin of the oligomycin-sensitivity conferring factor has been inferred from its absence in *rho*$^-$ mutants, from the blocking of its synthesis with chloramphenicol, and from the existence of oligomycin-resistant mitochondrial mutants. Similarly, the P_i-ATP exchange activity of mitochondria has been shown to be absent in a *rho*$^-$ strain, and its synthesis can be blocked by erythromycin.

In principle, the most satisfactory method to demonstrate that a protein is coded by mitochondrial DNA is to identify mitochondrial mutations which alter the protein. A start in this direction has been made with the identification of mitochondrial mutations to antibiotic resistance. As discussed in Chapter 4, erythromycin-resistant, chloramphenicol-resistant and paromomycin-resistant strains have been described. No protein alterations correlated with these mutations have yet been reported, but work along these lines is in progress in several laboratories. Bunn *et al.* (*12*) described another class of mitochondrial mutants in which resistance arose simultaneously to a number of different antibiotics: mikamycin, chloramphenicol, lincomycin, carbomycin, and tetracycline. *In vitro* studies showed that mitochondrial protein synthesis was sensitive to inhibition by these antibiotics, whereas in the same experiments, mitochondria from an erythromycin-resistant strain were resistant to the drug. They propose that multiple drug resistance may result from changes in the mitochondrial membrane, whereas the erythromycin resistance may reflect an altered mitochondrial ribosomal protein.

The origin of mitochondrial ribosomal proteins has been examined most extensively by Küntzel (*50*) and by Sebald *et al.* (*84*) comparing radioisotope labeling patterns of untreated controls with those from cycloheximide and chloramphenicol treated cultures of *Neurospora*. Küntzel's results are shown in Table 7.12. The cytoplasmic and mitochondrial supernatant fractions include membrane proteins which were solubilized during preparation of the ribosomes. These results show clearly that the overall labeling pattern of the ribosomal proteins followed that of the cytoplasmic ribosomes and cytoplasmic superna-

TABLE 7.12

Effect of CHI and CAP on In Vivo Incorporation of Amino Acids into Mitochondrial and Cytoplasmic Proteins[a]

Fraction		cpm/mg Protein	% Inhibition
In vivo incorporation of ^{3}H-lysine into mitochondrial and cytoplasmic proteins in presence and absence of cycloheximide (CHI)			
Cytoplasmic ribosomes	−CHI	67,430	
Cytoplasmic ribosomes	+CHI	2,090	96.9
Mitochondrial ribosomes	−CHI	69,200	
Mitochondrial ribosomes	+CHI	1,940	97.2
Cytoplasmic supernatant	−CHI	56,820	
Cytoplasmic supernatant	+CHI	1,250	97.8
Mitochondrial supernatant	−CHI	54,900	
Mitochondrial supernatant	+CHI	5,920	89.2
In vivo incorporation of ^{3}H-lysine into mitochondrial and cytoplasmic proteins in presence and absence of chloramphenicol (CAP)			
Cytoplasmic ribosomes	−CAP	91,500	
Cytoplasmic ribosomes	+CAP	89,000	2.8
Mitochondrial ribosomes	−CAP	85,820	
Mitochondrial ribosomes	+CAP	83,790	2.2
Cytoplasmic supernatant	−CAP	76,430	
Cytoplasmic supernatant	+CAP	74,250	3.1
Mitochondrial supernatant	−CAP	79,890	
Mitochondrial supernatant	+CAP	65,450	18.0

[a] From (*50*).

tant fractions, and differed from that of the mitochondrial supernatant. The ribosomal proteins were not examined individually for their labeling specificity in this study, so it remains possible that a few of the proteins did not follow the same labeling pattern as the majority.

Küntzel has interpreted these data to mean that most if not all mitochondrial ribosomal proteins are synthesized in the cytoplasm, and most other investigators have accepted this interpretation. However, the evidence is indirect, and depends upon how faithfully *in vivo* inhibition of proteins synthesis follows the *in vitro* pattern. Cycloheximide could shut down synthesis of the mitochondrial ribosomal proteins by blocking synthesis of a required cofactor or inducer. Chloramphenicol could be ineffective in blocking mitochondrial synthesis of ribosomal proteins because a higher concentration is required than for blocking other synthetic processes. Thus, supporting evidence from different kinds of experimentation is required to establish Küntzel's interpretation.

One protein that may be partly synthesized in mitochondria is cytochrome oxidase. Spectrally, the enzyme is known as cytochrome $a + a_3$; the apoenzyme has not yet been fully characterized. Based on inhibitor studies, Henson *et al.* (*41, 41a*) concluded that cytochrome oxidase was probably synthesized in mitochondria. No enzyme activity has been found in *rho*$^-$ strains, but Kraml and Mahler (*47*) found cross-reacting material to cytochrome oxidase antiserum in mitochondria from a *rho*$^-$ strain. Tuppy and Birkmayer (*99*) recently reported that *petite* mitochondria do contain the apo-protein of cytochrome oxidase, but other investigators have questioned their results (*56, 81*). Edwards and Woodward (*23*) found an altered cytochrome oxidase in *poky* mutants of *Neurospora,* as judged by a shift in its spectral properties, but not by studies of the apoprotein. Thus it is evident that further characterization of the apoprotein is needed before the question of its origin can be resolved.

A new activity, presumably enzymatic, has recently been associated with wild-type mitochondria: the ability to respond to ethidium bromide by alterations in mitochondrial DNA leading to the *rho*$^-$ condition (*71*). Studies of the process of mutagenesis by ethidium bromide have shown that chloramphenicol-grown yeast cells are resistant to this effect of ethidium bromide, and also that the chloramphenicol-blocked component, called compound X, is temperature-sensitive. Heat shock also makes the cells resistant to the induction of *rho*$^-$ mutations by ethidium bromide. It seems likely that compound X is a DNA-repair enzyme, responsible for carrying out the alterations induced by ethidium bromide. The fact that this activity is blocked by growth in the presence of chloramphenicol, and also that *rho*$^-$ cells are resistant to further alterations of ethidium bromide treatment, together suggest that compound X is of mitochondrial origin.

SUMMARY

This chapter began with a description of normal mitochondria of wild-type cells: the distribution of enzymes in the matrix, outer and inner membranes, the lipids and their natural variability in different organisms and under different growing conditions, the evidence of structural continuity in mitochondrial growth, and the evidence that mitochondrial membranes are multicomponent complex lipoprotein structures without a fixed stoichiometry. The structural continuity between membranes of promitochondria from anaerobically grown cells and normal mitochondria of respiring cells represents a striking demonstration of the flexibility of mitochondrial membranes in incorporating new components into preformed structures. The central problem of mitochondrial biogenesis then turns on the mechanisms for assembly of new components into old shells.

The presence of DNA, RNA polymerase and a protein-synthesizing apparatus in mitochondria contributes to the control of biogenesis, by providing unique products of the mitochondrial genome to be used in mitochondrial growth. We saw that mitochondria contain ribosomes which differ from those of the cytoplasm in containing unique RNA and proteins and in responding to different inhibitors. Mitochondrial proteins synthesis differs from that of the cytoplasm also in utilizing different transfer RNA's and synthetases, as well as different initiation and elongation factors. This wealth of differences provides numerous steps at which the organelle system can be regulated independently of cytoplasmic protein synthesis.

Given this splendid system for mitochondrial protein synthesis, how is it utilized? With this question we come almost to the end of our solid understanding of mitochondrial biogenesis. The rest of this chapter has summarized a vast body of experiments which provide some indirect evidence and many clues, but few hard facts to illuminate the mechanisms by which mitochondrial membranes are assembled.

The studies of *in vitro* protein synthesis by isolated mitochondria show that they are capable of protein synthesis directed by polynucleotides present within the organelles. The rate of incorporation in the *in vitro* system is low, but may be limited by extraneous factors such as the rate of transport of the labeled amino acids across the mitochondrial membranes.

More important than the rate, the extent of labeling in the *in vitro* system is low, representing some 5% of the total mitochondrial protein. Labeled amino acids are incorporated almost entirely into proteins of the inner membrane which have not been individually identified.

In vivo studies of incorporation of labeled amino acids into mitochondrial proteins in the presence of cycloheximide or chloramphenicol confirm the *in vitro* results in implicating only a small fraction of membrane-bound proteins as products of chloramphenicol-sensitive protein synthesis. Similarly by this method all or most of the mitochondrial ribosomal proteins appear to be of cytoplasmic origin. The conclusions from inhibitor studies have been confirmed by other methods, especially the kinetics of labeling during glucose derepression. Although several enzyme activites are missing in *rho*$^-$ mutants, the patterns of membrane protein components seen by gel electrophoresis do not differ much from the wild type.

Thus the principal finding from both *in vitro* and *in vivo* studies is that most mitochondrial proteins are of cytoplasmic origin, and are present in both respiring and nonrespiring yeast. Some investigators have estimated the fraction of mitochondrial proteins actually synthesized within the organelle at about 5%. The value is certainly minimal. Whether 5% is a fair estimate of the biological situation is not clear at this time.

One reason why investigators have not challenged the 5% value is the small size of the mitochondrial genome in animal cells: 1×10^7 daltons or about 15,000 nucleotide pairs. About 15% of these pairs are transcribed as ribosomal and transfer RNA's, leaving enough nucleotide pairs to code for only about twenty to thirty proteins. This calculation leads one to anticipate that only a few mitochondrial proteins will be of mitochondrial origin.

However, most studies of mitochondrial protein synthesis have been done with yeast or *Neurospora* in which the mitochondrial DNA is about five times the size of that from animal cells, and may therefore code for 100 or more proteins. No differences have yet been noted in the extent of protein synthesis with mitochondria from microorganisms and from animal cells. Indeed, the biological significance of the size differences in mitochondrial DNA, as in miniribosomes, is totally unknown at this time.

These considerations further expose our ignorance. In terms of selection pressures and evolution, one must assume that the smaller size of mitochondrial macromolecules in animal cells presages a contracted function over that in microorganisms.

The biogenesis problem is intimately bound up with the identification of products of the mitochondrial genome. These mitochondrial gene products may be anticipated to play a central role in the assembly of inner membrane components. Two lines of investigation seem of the most central importance: development of new techniques for the identification of membrane-bound proteins, and identification of all mitochondrial gene products. With the recognition that mutations to *rho*$^-$ are principally deletions, it is now possible to utilize both *rho*$^-$ *petites* and point mutations in mitochondrial DNA as source material for identifying individual mitochondrial gene products. From the genetic point of view, acquisition of new mutations in mitochondrial DNA should provide a comprehensive source of material for further investigation of mitochondrial biogenesis.

Suggested Review Articles

Boardman, N. K., Linnane, A. W., and Smillie, R. M., eds. (1970). "Autonomy and Biogenesis of Mitochondria and Chloroplasts," *Aust. Acad. Sci. Symp.* North-Holland Publ., Amsterdam.

Kroon, A. M. (1969). DNA and RNA from mitochondria and chloroplasts (biochemistry). *In* "Handbook of Molecular Cytology" (A. Lima-De-Faria, ed.). North-Holland, Publ., Amsterdam.

Linnane, A. W., and Haslam, J. M. (1970). The biogenesis of yeast mitochondria. *In* "Current Topics in Cellular Regulation" (B.L. Horecker and E. R. Stadtman, eds.), Vol. 2 pp. 102–172. Academic Press, New York.

Miller, P. L., ed. (1970). "Control of Organelle Development," *Symp. Soc. Exp. Biol.*, Vol. 24. Cambridge Univ. Press, London and New York.

Racker, E., ed. (1970). "Membranes of Mitochondria and Chloroplasts." Van Nostrand-Reinhold, New York.

References

1. Aloni, Y., and Attardi, G. (1972). *J. Mol. Biol.* (submitted for publication).
2. Attardi, B., and Attardi, G. (1967). A membrane-associated RNA of cytoplasmic origin in HeLa cells. *Proc. Nat. Acad. Sci. U. S.* **58,** 1051.
3. Attardi, G., and Ojala, D. (1971). Mitochondrial miniribosomes in HeLa cells. *Nature New Biol.* **229,** 133.
4. Attardi, G., Aloni, Y., Attardi, B., Ojala, D., Pica-Mattoccia, L., Robberson, D. L., and Storrie, B. (1970). Transcription of Mitochondrial DNA in HeLa cells. *Cold Spring Harbor Symp. Quant. Biol.* **35,** 599.
5. Barnett, W. E., and Brown, D. H. (1967), Mitochondrial transfer ribonucleic acids. *Proc. Nat Acad. Sci. U. S.* **57,** 452.
6. Barnett, W. E., Brown, D. H., and Epler, J. L. (1967). Mitochondrial specific aminoacyl-RNA synthetases. *Proc. Nat. Acad. Sci. U. S.* **57,** 1775.
7. Beck, J. C., Parker, J. H., Balcavage, W. X., and Mattoon, J. R. (1970). Mendelian genes affecting development and functions of yeast mitochondria. *In* "Autonomy and Biogenesis of Mitochondria and Chloroplasts" (N. K. Boardman, A. W. Linnane, and R. M. Smillie, eds.), *Aust. Acad. Sci. Symp.*, pp. 194–204. North-Holland Publ., Amsterdam.
8. Bloomfield, D. K., and K. Bloch (1960). The formation of Δ^9-unsaturated fatty acids. *J. Biol. Chem.* **235,** 337.
9. Borst, P. (1970). Size, structure and information content of mitochondrial DNA. *In* "Autonomy and Biogenesis of Mitochondria and Chloroplasts". (N. K. Boardman, A. W. Linnane, and R. M. Smillie, eds.), *Aust. Acad. Sci. Symp.* (pp. 260–266. North-Holland Publ., Amsterdam.
10. Borst, P., and Grivell, L. A. (1971), Mitochondrial ribosomes. *FEBS Lett.* **13,** 73.
11. Buck, C. A., and Nass, M. M. K. (1969). Studies on mitochondrial tRNA from animal cells. I. A comparison of mitochondrial and cytoplasmic tRNA and aminoacyl-tRNA synthetases. *J. Mol. Biol.* **41,** 67.
12. Bunn, C. L., Mitchell, C. H., Lukins, H. B., and Linnane, A. W. (1970). Biogenesis of mitochondria. XVIII. A new class of cytoplasmically determined antibiotic resistant mutants in *Saccharomyces cerevisiae*. *Proc. Nat. Acad. Sci. U. S.* **67,** 1233.
13. Chapman, D., and Leslie, R. B. (1970). Structure and function of phospholipids in membranes. *In* "Membranes of Mitochondria and Chloroplasts" (R. Racker, ed.), pp. 91–126. Van Nostrand-Reinhold, New York.

14. Chi, J. C. H., and Suyama, Y. (1970). Comparative studies on mitochondrial and cytoplasmic ribosomes of *Tetrahymena pyriformis*. *J. Mol. Biol.* **53,** 1531.
15. Crane, F. L., Arntzen, C. J., Hall, J. D., and Ruzicka, F. J. (1970). Binary membranes in mitochondria and chloroplasts. *In* "Autonomy and Biogenesis of Mitochondria and Chloroplasts" (N. K. Boardman, A. W. Linnane, and R. M. Smillie, eds.), *Aust. Acad. Sci. Symp.*, pp. 53–69. North-Holland Publ., Amsterdam.
16. Criddle, R. S., and Schatz, G. (1969). Promitochondria of anaerobically grown yeast. I. Isolation and biochemical properties. *Biochemistry* **8,** 322.
17. Criddle, R. S., Bock, R. M., Green, D. E., and Tisdale, H. (1962). Physical characteristics of proteins of the electron transfer system and interpretation of the structure of the mitochondrion. *Biochemistry* **1,** 827.
18. Dawid, I. B. (1966). Evidence for the mitochondrial origin of frog egg cytoplasmic DNA. *Proc. Nat. Acad. Sci. U. S.* **56,** 269.
19. Dawid, I. B. (1970). The nature of mitochondrial RNA in oocytes of *Xenopus laevis* and its relation to mitochondrial DNA. *In* "Control of Organelle Development" (P. L. Miller, ed.), *Symp. Soc. Exp. Biol.*, Vol. 24, pp. 227–246. Cambridge Univ. Press, London and New York.
20. Dubin, D. T., and Montenecourt, B. S. (1970). Mitochondrial RNA from cultured animal cells. Distinctive high molecular weight and 4 S species. *J. Mol. Biol.* **48,** 279.
21. Dure, S. L., Epler, J. L., and Barnett, W. E. (1967). Sedimentation properties of mitochondrial and cytoplasmic ribosomal RNA's from *Neurspora*. *Proc. Nat. Acad. Sci. U. S.* **58,** 1883.
22. Edelman, M., Verma, I. M., Herzog, R., Galun, E., and Littauer, U. Z. (1971). Physicochemical properties of mitochondrial ribosomal RNA from fungi. *Eur. J. Biochem.* **19,** 372.
23. Edwards, D. L., and Woodward, D. O. (1969). An altered cytochrome oxidase in a cytoplasmic mutant of *Neurospora*. *FEBS Lett.* **4,** 193.
24. Ephrussi, B. (1952). The interplay of heredity and environment in the synthesis of respiratory enzymes in yeast. *Harvey Lect. Ser. XLVI,* **1950–1951,** p. 45.
25. Ephrussi, B., and Slonimski, P. P. (1950). La synthèse adaptative des cytochromes chez la levure de boulangerie. *Biochim. Biophys. Acta* **6,** 256.
26. Ephrussi, B., Slonimski, P. P., Yotsuyanagi, Y., and Tavlitski, J. (1956). Variations physiologiques et cytologiques de la levure au cours du cycle de la croissance aerobie. *C. R. Trav. Lab. Carlsberg, Ser. Physiol.* **26,** 87.
27. Epler, J. L. (1969). The mitochondrial and cytoplasmic transfer ribonucleic acids of *Neurospora crassa*. *Biochemistry* **8,** 2285.
28. Epler, J. L., Shugart, L. R., Barnett, W. E. (1970), *N*-Formylmethionyl transfer ribonucleic acid in mitochondria from *Neurospora*. *Biochemistry* **9,** 3575.
29. Ernster, L., and Kuylenstierna, B. (1970). Outer membrane of mitochondrka. *In* "Membranes of Mitochondria and Chloroplasts" (E. Racker, ed.), pp. 172–212. Van Nostrand-Reinhold New York.
30. Fauman, M., Rabinowitz, M., and Getz, G. S. (1969). Base composition and sedimentation properties of mitochondrial RNA of *Saccharomyces cerevisiae*. *Biochim. Biophys. Acta* **182,** 355.
31. Fernandez-Moran, H., Oda, T., Blair, P. V., and Green, D. E. (1964). A macromolecular repeating unit of mitochondrial structure and function. *J. Cell Biol.* **22,** 63.
32. Flavell, R. B. (1971). Mitochondrion as a multi-functional organelle. *Nature* (*London*). **230,** 504.
33. Fukahara, H. (1967). Informational role of mitochondrial DNA studied by hybridization with different classes of RNA in yeast. *Proc. Nat. Acad. Sci. U. S.* **58,** 1065.

34. Fukahara, H. (1970). Transcriptional origin of RNA in a mitochondrial fraction of yeast and its bearing on the problem of sequence homology between mitochondrial and nuclear DNA. *Mol. Gen. Genet.* **107,** 58.
35. Galper, J. B, and Darnell, J. E. (1969). The presence of *N*-formylmethionyl-tRNA in HeLa cell mitochondria. *Biochem. Biophys. Res. Commun.* **34,** 205.
36. Grandi, M., and Küntzel, H. (1970). Mitochondrial peptide chain elongation factors from *Neurospora. FEBS Lett.* **10,** 25.
37. Green, D. E., and Perdue, J. F. (1966). Membranes as expressions of repeating units. *Proc. Nat. Acad. Sci. U. S.* **55,** 1294.
38. Groot, G. S. P., Kovac, L., and Schatz, G. (1971). Promitochondria of anaerobically grown yeast. V. Energy transfer in the absence of an electron transfer chain. *Proc. Nat. Acad. Sci. U. S.* **58,** 308.
39. Halbreich, A., and Rabinowitz, M. (1971). Isolation of *S. cerevisiae* mitochondrial formyltetrahydrofolic acid: Methionyl-tRNA transformylase and the hybridization of mitochondrial fMET-tRNA with mitochondrial DNA. *Proc. Nat. Acad. Sci. U. S.* **68,** 294.
40. Haldar, D., Freeman, K., and Work, T. S. (1966). Biogenesis of mitochondria. *Nature (London)* **211,** 9.
41. Henson, C. P., Weber, C. N., and Mahler, H. R. (1968). Formation of yeast mitochondria. I. Kinetics of amino acid incorporation. *Biochemistry* **7,** 4431.
41a. Henson, C. P., Weber, C. N., Mahler, H. R. (1968). Formation of yeast mitochondria. II. Effects of antibiotics on enzyme activity during derepression. *Biochemistry* **7,** 4445.
42. Jakob, H. (1965). Complementation entre mutants à déficience respiratoire de *Saccharomyces cerevisiae:* Etablissment et régulation de la respiration dans les zygotes et dans leur proche déscendance. *Genetics* **52,** 75.
43. Kadenbach, B. (1971). Biosynthesis of mitochondrial cytochromes. *In* "Autonomy and Biogenesis of Mitochondria and Chloroplasts" (N. K. Boardman, A. W. Linnane and R. M. Smillie, eds.) *Aust. Acad. Sci. Symp.,* pp. 360–371. North-Holland Publ., Amsterdam.
44. Kagawa, Y., Racker E., (1966). Partial resolution of the enzymes catalyzing oxidative phosphorylation. VIII. Properties of a factor conferring oligomycin sensitivity on mitochondrial adenosine triphosphatase; IX. Reconstruction of oligomycin-sensitive adenosine triphosphatase; X. Correlation of morphology and function in submitochondrial particles. *J. Biol. Chem.* **241,** 2461, 2467, 2475.
45. Kellerman, G. M., Griffiths, D. E. Hansky, J. E., Lamb, A. J., and Linnane A. W. (1971). The protein synthetic capacity of yeast mitochondria and the role of the mitochondrial genome in the economy of the cell. *In* "Autonomy and Biogenesis of Mitochondria and Chloroplasts" (N. K. Boardman, A. W. Linnane, and R. M. Smillie, eds.) *Aust. Acad. Sci. Symp.,* pp. 346–359. North-Holland Publ. Amsterdam.
46. Kleinow, W., Sebald, W., Neupert, W., and Bucher, T. (1971). Formation of mitochondria of *Locusta migratoria* flight muscle. *In* "Autonomy and Biogenesis of Mitochondria and Chloroplasts" (N. K. Boardman, A. W. Linnane, and R. M. Smillie, eds.), *Aust. Acad. Sci. Symp.,* pp. 140–151. North-Holland Publ., Amsterdam.
47. Kraml, J. R., and Mahler H. R. (1967). Biochemical correlates of respiratory deficiency. VIII. Precipitating anti-serum against cytochrome oxidase of yeast and its use in the study of respiratory deficiency. *Immunochemistry* **4,** 213.
48. Küntzel, H. (1969). Specificity of mitochondrial and cytoplasmic ribosomes from *Neurospora crassa* in poly-U dependent cell free systems. *FEBS Lett.* **4,** 140.
49. Küntzel, H. (1969). Mitochondrial and cytoplasmic ribosomes from *Neurospora crassa* characterization of their subunits. *J. Mol. Biol.* **40,** 315.

50. Küntzel, H. (1969). Proteins of mitochondrial and cytoplasmic ribosomes from *Neurospora crassa. Nature (London)* **222,** 142.
51. Küntzel, H., and Noll, H. (1967). Mitochondrial and cytoplasmic polysomes from *Neurospora crassa. Nature (London)* **215,** 1340.
52. Küntzel, H., and Schafer, K. P. (1971). Mitochondrial RNA polymerase from *Neurospora crassa. Nature New Biol.* **231,** 265.
53. Lamb, A. J., Clark-Walker, G. D., and Linnane, A. W. (1968). The biogenesis of mitochondria and the differentiation of mitochondrial and cytoplasmic proteins synthesizing systems *in vitro* by antibiotics. *Biochim. Biophys. Acta* **161,** 415.
54. Lehninger, A. L. (1970), "Biochemistry: The Molecular Basis of Cell Structure and Function." Worth, New York.
55. Leon, S. A., and Mahler, H. R. (1968). Isolation and properties of mitochondrial RNA from yeast. *Arch. Biochem. Biophys.* **126,** 305.
56. Linnane, A. W., and Haslam, J. M. (1970). Biogenesis of yeast mitochondria. *In* "Current Topics in Cellular Regulation" (B. L. Horecker and E. R. Stadtman, eds.), Vol. 2, pp. 102–172. Academic Press, New York.
57. Lizardi, P., and Luck, D. J. L. (1971). Mitochondrial ribosomes of *Neurospora crassa:* absence of a 5 S RNA component. *Nature New Biol.* **229,** 140.
58. Loening, U. E., and Ingle, J. (1967). Diversity of RNA components in green plant tissues. *Nature (London)* **215,** 363.
59. Luck, D. J. L. (1965). Formation of mitochondria in *Neurospora crassa. Amer. Natur.* **99,** 241.
60. Luck, D. J. L. (1965). Formation of mitochondria in *Neurospora.* A study based on mitochondrial density changes. *J. Cell Biol.* **24,** 461.
61. Luck, D. J. L., and Reich, E. (1964). DNA in mitochondria of *Neurospora crassa. Proc. Nat. Acad. Sci. U. S.* **52,** 931.
62. Malkin, L. I. (1970). Amino acid incorporation by isolated rat liver mitochondria during liver regeneration. *Proc. Nat. Acad. Sci. U. S.* **67,** 1695.
63. Moor, H., and Mühlethaler, K. (1963). Fine structure in frozen-etched yeast cells. *J. Cell Biol.* **17,** 609.
64. Morimoto, H., Scragg, A. H., Nekhorocheff, J., Villa, V., and Halvorson, H. O. (1970). Comparison of the protein synthesizing systems from mitochondria and cytoplasmic of yeast. *In* "Autonomy and Biogenesis of Mitochondria and Chloroplasts," (N. K. Boardman, A. W. Linnane, and R. M. Smillie, eds.), *Aust. Acad. Sci, Symp.,* pp. 282–292. North-Holland Publ., Amsterdam.
65. Munkres, K. D., Swank, R. T., and Sheir, G. S. (1971). Monomers of *Neurospora* structural protein. *In* "Autonomy and Biogenesis of Mitochondria and Chloroplasts" (N. K. Boardman, A. W. Linnane, and R. M. Smillie, eds.), *Aust. Acad. Sci, Symp.,* pp. 152–161. North-Holland Publ., Amsterdam.
66. Nass, M. M. K., and Buck, C. A. (1969). Comparative hybridization of mitochondrial and cytoplasmic aminoacyl transfer RNA with mitochondrial DNA from rat liver. *Proc. Nat. Acad. Sci. U. S.* **62,** 506.
67. Neupert, W., Brdiczka, D., and Sebald, W. (1968). Incorporation of amino acids into the outer and inner membranes of isolated rat liver mitochondria. *In* "Biochemical Aspects of the Biogenesis of Mitochondria" (E. C. Slater, J. M. Tager, S. Papa, and E. Quagliariello, eds.), pp. 395–408. Adriatica Editrice, Bari.
67a. Neupert, W., Sebald, W., Schwab, A. J., Pfaller, A., and Bücher, T. (1969). Incorporation *in vivo* of ^{14}C-labelled amino acids into proteins of mitochondrial ribosomes from *Neurospora crassa* sensitive to cycloheximide and insensitive to chloramphenicol. *Eur. J. Biochem* **10,** 589.

68. O'Brien, T. W., and Kalf, G. F. (1967). Ribosomes from rat liver mitochondria. I. Isolation procedure and contamination studies. *J. Biol. Chem.* **242,** 2172.
69. Paltauf, F., and Schatz, G. (1969). Promitochondria of anaerobically grown yeast. II. Lipid composition. *Biochemistry* **8,** 335.
70. Parisi, B., and Cella, R. (1971). Origin of the ribosome specific fractors responsible for peptide chain elongation in yeast. *FEBS Lett.* **14,** 209.
71. Perlman, P. S., and Mahler, H. R. (1971). Molecular consequences of ethidium bromide mutagenesis. *Nature New Biol.* **231,** 12.
72. Plattner, H., and Schatz, G. (1969). Promitochondria of anaerobically grown yeast. III. Morphology. *Biochemistry* **8,** 339.
73. Plattner, H., Salpeter, M. M., Saltzgaber, J., and Schatz, G. (1970). Promitochondria of anaerobically grown yeast. IV. Conversion into respiring mitochondria. *Proc. Nat. Acad. Sci. U. S.* **66,** 1252.
74. Rabbitts, T. H., and Work, T. S. (1971). The mitochondrial ribosome and ribosomal RNA of the chick. *FEBS Lett.* **14,** 214.
75. Racker, E. (1970). Function and structure of the inner membrane of mitochondria and chloroplasts. *In* "Membranes of Mitochondria and Chloroplasts" (E. Racker, ed.), pp. 127–171. Van Nostrand-Reinhold, New York.
76. Reilly, C., and Sherman, F. (1965). Glucose repression of cytochrome a synthesis on cytochrome-deficient mutants of yeast. *Biochim. Biophys. Acta* **95,** 640.
77. Rifkin, M. R., Wood D. D., and Luck, D. J. L. (1967). Ribosomal RNA and ribosomes from mitochondria of *Neurospora crassa. Proc. Nat. Acad. Sci. U. S.* **58,** 1025.
78. Roodyn, D. B. (1966). Factors affecting the incorporation of amino acids into protein by isolated mitochondria. *In* "Regulation of Metabolic Process in Mitochondria" (J. M. Tager, S. Papa, E. Quagliariello, and E. C. Slater, eds.), *Biochim. Biophys. Acta Library,* Vol. 7, pp. 562–563. Elsevier, Amsterdam.
79. Roodyn, D. B., and Wilkie, D. (1968). "The Biogenesis of Mitochondria." Methuen, London.
80. Schatz, G. (1968). Impaired binding of mitochondrial adenosine triphosphatase in the cytoplasmic "petite" mutant of *Saccharomyces cerevisiae. J. Biol. Chem.* **243,** 2192.
81. Schatz, G. (1970). Biogenesis of mitochondria. *In* "Membranes of Mitochondria and Chloroplasts" (E. Racker, ed.), pp. 251–299. Van Nostrand-Reinhold, New York.
82. Schatz, G., Penefsky, H. S., Racker, E. (1967). Partial resolution of the enzymes catalyzing oxidative phosphorylation. XIV. Interaction of purified mitochondrial adenosine triphosphatase from bakers' yeast with submitochondrial particles from beef heart. *J. Biol. Chem* **242,** 2552.
83. Schnaitman, C., and Greenawalt, J. W. (1968). Enzymatic properties of the inner and outer membranes of rat liver mitochondria. *J. Cell Biol.* **38,** 158.
84. Sebald, W., Berkenayer, G. D., Schwab, A. J., and Weiss, H. (1971). Incorporation of amino acids into electrophoretic and chromatographic fractions of mitochondrial membrane proteins. *In* "Autonomy and Biogenesis of Mitochondria and Chloroplasts" (N. K. Boardman, A. Linnane, and R. M. Smillie, eds.), *Aust. Acad. Sci. Symp.,* pp. 339–345. North-Holland Publ., Amsterdam.
85. Sherman, F., and Slonimski, P. P. (1964). Respiration-deficient mutants of yeast. II. Biochemistry. *Biochim. Biophys. Acta* **90,** 1.
86. Sherman, F., Stewart, J. W., Parker, J. H., Putterman, G. J., Agrawal, B. B. L., and Marfoliash, E. (1970). The relationship of gene structure and protein structure of iso-I-cytochrome c from yeast. *In* "Control of Organelle Development" (P. L. Miller, ed.), *Symp. Soc. Exp. Biol.,* Vol. 24, pp. 85–108. Cambridge Univ. Press, London and New York.

87. Slonimski, P. P. (1949). Action de l'acriflavine sur les levures. IV. Mode d'utilisation du glucose par les mutants "petite colonie." *Ann. Inst. Pasteur (Paris)* **76,** 510.
88. Slonimski, P. P. (1949). Action de l'acriflavine sur les levures. VII. Sur I'activité catalytique du cytochrome c des mutants "petite colonie" de la levure. *Ann. Inst. Pasteur (Paris)* **77,** 774.
89. Slonimski, P. P. (1956). Adaptation respiratoire: dévelopment du système hémoproteique induit par l'oxygène. *In* "Proceedings of the Third International Congress of Biochemistry, Brussels, 1955" (C. Lieberg, ed.), pp. 242–252. Academic Press, New York.
90. Slonimski, P. P., and Ephrussi, B. (1949). Action de l'acriflavine sur les levures. V. Le systeme des cytochromes des mutants "petite colonie." *Ann. Inst. Pasteur (Paris)* **77,** 47.
91. Smith, A. E., and Marcker K. A. (1968). *N*-Formylmethionyl transfer RNA in mitochondria from yeast and rat liver. *J. Mol. Biol.* **35,** 241.
92. Smith, D. J., and Mahler, H. R. (1968). RNA synthesis in yeast mitochondria: A derepression activity. *Nature (London)* **218,** 1226.
93. Smoly, J. M., Kuylenstierna, B., and Ernster, L. (1970). Topological and functional organization of the mitochondrion. *Proc. Nat. Acad. Sci. U. S.* **66,** 125.
94. Stoeckenius, W. (1966). Structural organization of the mitochondrion. *In* "Principles of Biomolecular Organization" (G. E. W. Wolstenholme, ed.), *CIBA Found. Symp.*, pp. 418–445. Little, Brown, Boston, Massachusetts.
95. Stoeckenius, W. (1970). Electron microscopy of mitochondrial and model membranes. *In* "Membranes of Mitochondria and Chloroplasts" (E. Racker, ed.). Van Nostrand-Reinhold, New York.
96. Stoeckenius, W., and Engelman, D. M. (1969). Current models for the structure of biological membranes. *J. Cell Biol.* **42,** 613.
97. Suyama, Y, and Eyer, J. (1967). Leucyl tRNA and leucyl tRNA synthetase in mitochondria of *Tetrahymena pyriformis. Biochem. Biophys. Res. Commun.* **28,** 746.
98. Swanson, R. F., and Dawid, I. B. (1970). The mitochondrial ribosome of *Xenopus laevis. Proc. Nat. Acad. Sci. U. S.* **66,** 117.
98a. Tandler, B., and Hoppel, C. L. (1972). "Mitochondria" (Series on the Ultrastructure of Cells and Organisms, edited by M. Locke). Academic Press, New York.
99. Tuppy, H., and Berkmayer, D. G. (1969). Cytochrome oxidase apoprotein in "petite" mutant yeast mitochondria. Reconstitution of cytochrome oxidase by combining apoprotein with cytohemin. *Eur. J. Biochem.* **8,** 237.
100. Watson, K., Haslem, J. M., Veitch, B., Linnane, A. W. (1971). Mitochondrial precursors in anaerobically grown yeast: evidence for their conversion into functional mitochondria, during respiratory adatiation. *In* "Autonomy and Biogenesis of Mitochondria and Chloroplasts" (N. K. Boardman, A. W. Linnane, and R. M. Smillie, eds.), *Aust. Acad. Sci. Symp.*, pp. 162–174. North-Holland Publ., Amsterdam.
101. Wheeldon, L. W., and Lehninger, A. L. (1966). Energy linked synthesis and decay of membrane proteins in isolated rat liver mitochondria. *Biochemistry* **5,** 3533.
102. Wilkie, D. (1970). Analysis of mitochondrial drug resistance in *Saccharomyces cerevisiae. In* "Control of Organelle Development" (P. L. Miller, ed.), *Symp. Soc. Exp. Biol.*, Vol. 24, pp. 71–84. Cambridge Univ. Press, London and New York.
102a. Wintersberger, E. (1965). Proteinsythese in isolierten Hefemitochondrien. *Biochem.* **341,** 409.
103. Winterberger, E., and Tuppy, H. (1965). DNA-abhangige RNA-synthese in isolierten Hefe-mitochondrien. *Biochem. Z.* **341,** 399.

104. Wintersberger, E., and Viehhauser, G. (1968). Function of mitochondrial DNA in yeast. *Nature (London)* **220,** 699.
105. Wood, D. D., and Luck, D. J. L. (1969). Hybridization of mitochondrial ribosomal RNA. *J. Mol. Biol.* **41,** 211.
106. Woodward, D. O., Edwards, D. L., and Flavell, R. B. (1970). Nucleocytoplasmic interactions in the control of mitochondrial structure and function in *Neurospora. In* "Control of Organelle Development" (P. L. Miller, ed.), *Symp. Soc. Exp. Biol.,* Vol. 24, pp. 55–70. Cambridge Univ. Press, London and New York.
107. Work, T. S., Coote, G. L., and Ashwell, M. (1968). Biogenesis of mitochondria, *Fed. Proc. Fed. Amer. Soc. Exp. Biol.* **27,** 1174.
108. Yang, S., and Criddle, R. S. (1969). Identification of a major membrane fraction as a product of synthesis by isolated yeast mitochondria. *Biochem. Biophys. Res. Commun.* **35,** 429.
109. Yotsuyanagi, Y. (1962). Études sur le chondriome de la levure. I. Variation de l'ultrastructure du chondriome au cours du cycle de la croissance aerobie. *J. Ultrastruct. Res.* **7,** 121.
110. Yotsuyanagi, Y. (1962). Études sur le chondriome de la levure. II. Chondriomes des mutants à déficience respiratoire. *J. Ultrastruct Res.* **7,** 141.
111. Zylber, E., and Penman, S. (1969). Mitochondria associated 4 S RNA synthesis inhibition by ethidium bromide. *J. Mol. Biol* **46,** 201.
112. Zylber, E., Vesco, C., and Penman, S. (1969). Selective inhibition of the synthesis of mitochondria-associated RNA by ethidium bromide. *J. Mol. Biol.* **44,** 195.

8

The Role of Cytoplasmic Genes in the Biogenesis of Chloroplasts

The chloroplast has long been regarded by many investigators as a cell within a cell. This simplistic view of the chloroplast as an autonomous organelle was further strengthened (briefly) by the discovery of chloroplast DNA and ribosomes, providing evidence of genetic as well as functional autonomy. Nonetheless, this view has now been shattered by the growing body of evidence revealing the chloroplast as a tightly integrated compartment within the overall economy of the cell. Indeed, the new lines of evidence are shifting the focus of thinking and research from the concept of semiautonomy to one of interdependence and interaction.

The question of autonomy arose more persistently with chloroplasts than with mitochondria. After all, chloroplasts carry out the activities of photosynthesis and photophosphorylation very much on their own, supplying the rest of the cell with carbohydrates and with ATP. Thus, unlike mitochondria, they are not dependent on the rest of the cell for their substrates. Furthermore, the amount of chloroplast DNA is quantitatively the equivalent of a bacterial genome, enough to code for at least a thousand proteins, and is vastly more than the amount of mitochondrial DNA.

Nonetheless, the concept of autonomy is fundamentally misleading because nuclear genes provide information for the synthesis of very many chloroplast components, and the chemical composition, growth rate, and functional efficiency of the chloroplast are regulated by both the nuclear and chloroplast genomes. Thus, the concept of regulatory interaction provides a more appropriate framework than autonomy within which to view the process of chloroplast biogenesis, just as we have already seen in the mitochondrial system.

This chapter is organized parallel to Chapter 7. The first section will consider the molecular composition and structure of chloroplast membranes: What are the principal components? How are the lipids, the proteins, and the photosynthetic reaction systems organized in the membranes? Are there experimental clues to the nature of the assembly process? As with mitochondria, the discovery of DNA and a protein synthesizing machinery in chloroplasts has opened the door to new lines of investigation.

Accordingly, the second section of the chapter will discuss chloroplast RNA's, ribosomes, and the molecular apparatus for protein synthesis. The third section will consider the evidence on the pivotal question of which chloroplast proteins are synthesized within the organelle. This question has been explored with isolated chloroplasts and with the use of inhibitors and mutants in intact cells. As in the mitochondrial system, the conclusions adduced from these studies are still tentative. Furthermore, not a single protein has yet been identified directly as a product of the chloroplast genome. Nonetheless, a wealth of indirect evidence on protein synthesis and lamellar membrane assembly has come from studies of compositional changes during the greening process, i.e., the switch from dark to light growth conditions, during which chlorophyll synthesis and lamellar formation go hand in hand at an accelerated rate. The principal difficulty in all these studies is the complexity of the system. A further complication in the chloroplast problem is the presence of mitochondria; their role in chloroplast biogenesis remains as yet totally undefined. Thus, one is faced with a three-ring circus: nuclear, mitochondrial, and chloroplast genomes interacting in the smooth propulsion of growth. How can we dissect the system? Are there special control points to identify?

MOLECULAR COMPOSITION AND STRUCTURE OF CHLOROPLASTS

General Structure of the Chloroplast (15, 38, 39, 68, 77, 79, 94, 98, 141, 144)

As seen in electron micrographs (Plates XII, XIII, XIV, XV, and XVI), the chloroplasts of algae and higher plants all have the same fundamental features of structural organization. The photosynthetic apparatus is located in lamellar membranes that are arranged in closed flattened vesicles called *discs* or *thylakoids*. Because the membranes consist of flattened vesicles, they always have a clearly defined inner and outer surface. The inner surface encloses the *intralamellar space* and the outer surface is bathed in the *matrix* or *stroma* of the chloroplast. The entire complex is surrounded by an envelope consisting of two continuous smooth membranes.

The outer chloroplast membrane appears in electron micrographs as a tripartite unit membrane, resembling the outer mitochondrial membrane and the smooth membranes of the endoplasmic reticulum with which it sometimes fuses (*98*). The outer membrane is distinctly different in appearance from the inner membrane of the envelope, and there is a somewhat variable distance between the two as seen in cross sections in electron micrographs. In its width and fuzziness, the inner chloroplast membrane resembles the lamellar membranes which it surrounds and to which it is sometimes attached. Superficially, great differences are seen in the arrangement of lamellar discs in the chloroplast as one surveys the green, brown, and red algae, and the higher plants (*68, 75, 141, 142, 144*). The discs may be embedded singly in the matrix of the chloroplast, fused in parallel arrays, or stacked into bundles called *grana*. A schematic representation of a higher plant chloroplast is shown in Figs. 8.1 and 8.2.

During the later stages of chloroplast development, especially in higher plants, chlorophyll is deposited preferentially in the region between adjoining stacked discs in the grana; in mature systems this may be the region of major photosynthetic activity. Weier and Benson (*144*) have called it the partition region and have located all the chlorophyll there. Weier *et al.* (*145*) have also proposed that the occasional un-

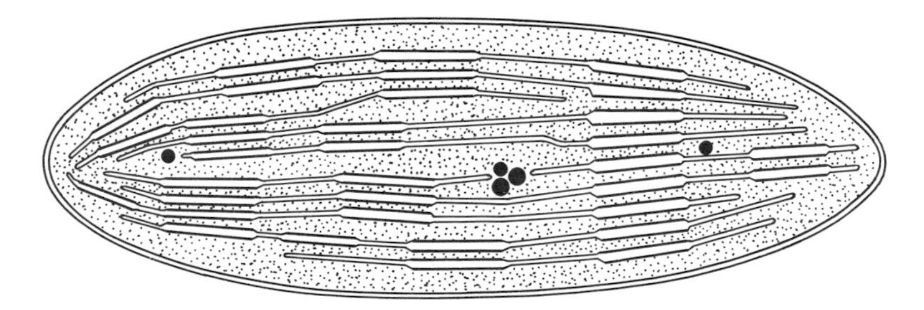

Fig. 8.1. Arrangement of lamellar discs in a higher plant chloroplast. In this schematic diagram, discs are shown singly or in stacks called grana. Small grana with few discs are found mainly in young chloroplasts; as the leaves mature, larger stacks are found. Lamellar membranes in the grana regions, called grana lamellae, are denser, containing more chlorophyll than the stroma lamellae located between the grana. From von Wettstein (*141*).

stacked discs, which connect separate grana and which they have called *frets,* provide channels for transport of photosynthetic intermediates between grana and also between discs within the grana. Their view of this arrangement is shown in Fig. 8.2. Frets are generally referred to as *stroma lamellae,* i.e., lamellar discs located free in the stroma (or matrix).

The proteins and pigments of photosynthetic electron transport are located in the lamellar membranes, as are the components of the photophosphorylation system. The matrix in which these membranes are embedded contains the soluble enzymes of CO_2 fixation and the Calvin cycle, as well as DNA and the RNA's, and ribosomes of the protein-synthesizing apparatus. The contents of the intralamellar spaces have not been identified as such, but as we shall see later, various proposals have been made for different photosynthetic activities on the inner and outer faces of the membranes themselves.

Fig. 8.2. Three-dimensional model of grana with interconnecting fretwork system in the higher plant chloroplast. From Weier *et al.* (*145*).

Chemical Composition of the Chloroplast

Reliable estimates of the total composition of the chloroplast have depended upon the development of methods to keep the isolated organelles intact during their isolation in order not to lose the soluble components by leakage. Two general methods have been used. One approach has been the application of the Behrens procedure (*126*): grinding lyophilized cells and separating chloroplasts by their density in organic solvents that do not extract water-soluble proteins. Enzymes which do not lose activity under these conditions have been definitely localized in the chloroplast (*118, 119*).

Using a different approach, Wildman and co-workers (*147*) developed a water-soluble medium in which chloroplasts still retaining their water-soluble enzymes could be recovered. They described two protein fractions from the chloroplast matrix, as seen in the ultracentrifuge: Fraction I, representing about 30% of the protein, and distributed as a homogeneous 18 S peak; and Fraction II, a much broader band at about 3–4 S. Fraction I was shown to contain carboxydismutase [ribulose-1, 5-diphosphate (RuDP) carboxylase], the enzyme that catalyzes the uptake of CO_2 into RuDP to form a 6-carbon intermediate which is subsequently split to give two molecules of 3-phosphoglyceric acid. Fraction II contains many of the enzymes of the photosynthetic carbon cycle.

The enzymes which have so far been located in the chloroplast by one means or another are listed in Table 8.1. In general, the enzymes of CO_2 fixation and carbohydrate and nucleotide metabolism are soluble enzymes located in the matrix, while the enzymes of photosynthetic electron transfer and photophosphorylation are membrane-bound. Some of the enzymes of Table 8.1 have not been fully identified with either the membrane or the soluble phase.

The components of the photosynthetic reactions systems I and II and their postulated positions in the overall scheme of photosynthesis are shown in Table 8.2. The arrangement shown follows the "Z scheme" first proposed by Hill and Bendall (*50*). This representation is now accepted by most investigators, although many questions are not yet resolved. Cyclic phosphorylation is known to be associated with photosystem I, but its precise position is not established, nor is that of the coupling factors for noncyclic photophosphorylation. Most important, some components of the electron transport and coupled phosphorylation systems have not yet been identified.

The best available estimates of molecular composition of the known chloroplast components are summarized in Table 8.3. Some 70% of the dry weight of the chloroplast is protein, and within this fraction, some-

TABLE 8.1

Enzymes Found in Chloroplasts[a]

Function	Enzyme	References
CO_2 Fixation: Calvin cycle	RuDP-carboxylase	*46, 86, 118*
	3-Phosphoglycerate kinase	*46, 86, 118*
	NADP-glyceraldehyde-3-P dehydrogenase	*46, 86, 118*
	Triosephosphate isomerase	*46, 86, 118*
	FDP aldolase	*46, 86, 118*
	Transketolase	*46, 86, 118*
	Xylulose-5-P epimerase	*46, 86, 118*
	Ribose-5-P isomerase	*46, 86, 118*
	Ribulose-5-P kinase	*46, 86, 118*
	Sedoheptulose-1,7-diphosphatase	*46, 86, 118*
	Alkaline FDPase	*46, 86, 118*
	FDPase (ferredoxin-activated)	*18a*
	Carbonic anhydrase	*32*
C_4-Dicarboxylic pathway	Pyruvate-P_i-dikinase	*44*
	Phosphoenolpyruvate carboxylase	*117*
Photosynthetic electron transfer and phosphorylation	Cytochrome f	*27*
	Cytochrome b_{559}	*72, 8*
	Cytochrome b_{562}	*49*
	Ferredoxin	*103*
	Plastocyanin	*60*
	Rubimedin	*47*
	Ferredoxin-NADP-reductase (Pyridine nucleotide transhydrogenase)	*63*
	Coupling factor (ATPase)	*74*
Protein-binding?	Structural protein	*26*
Porphyrin synthesis	5-Aminolevulinate dehydratase	*20*
	Ferrochelatase	*58*
	Chlorophyllase	*131*
Lipid metabolism	Fatty acid synthetase	*127*
	Stearoyl-acyl-carrier protein desaturase	*81*
	Galactolipid hydrolases	*105*
	Mevalonate kinase	*93*
	Carotenoid oxidase	*35*
Carbohydrate metabolism	Phosphorylase	*126*
	UDPG: F-6-P glucosyltransferase	*6*
	UDPG: fructose glucosyltransferase	*6*
	UDPG pyrophosphorylase	*6*
	Phosphoglucomutase	*6*
	Glucose-P isomerase	*6*
	Sucrose-6-phosphatase	*6*
Nucleic acid and nucleotide metabolism	DNA polymerase	*124*
	RNA polymerase	*111, 136, 64, 65*
	Ribonuclease	*46, 135a*

TABLE 8.1 (*Continued*)

Function	Enzyme	References
	Aminoacyl-sRNA synthetase	*13, 48*
	Adenosine-5′-triphosphatase	*3*
	Adenylate kinase	*118*
	Pyrimidine nucleotide synthesis	*59*
Nitrite reduction	Nitrite reductase	*92*
Sulfate reduction	Adenylylsulfate kinase	*28*
	Sulfate adenylyl-transferase	*28*
Other enzymes	Inorganic pyrophosphatase	*119a*
	Glutamate–oxalacetate transaminase	*45*
	Oxalate oxidase	*80*
	Acid phosphatase	*88*

[a] From Smillie and Scott (*119*).

what more than half is water-soluble, the rest being lamellar proteins. Among the water-soluble proteins, the Fraction I protein, with RuDP-carboxylase activity, accounts for $\frac{1}{4}$ to $\frac{1}{2}$ of the total weight. This protein, of molecular weight about 500,000 daltons, is composed of two kinds of subunits, one with a molecular weight of about 20,000 and the other about 10,000. The amino acid composition of the subunits is shown in Table 8.4 (*62, 147*). It is noteworthy that the amino acid composition of the large subunit is identical (within 3% error) in the three plants, whereas there are many differences in the composition of the smaller subunit. Kawashima and Wildman (*62*) have speculated that the small protein is coded by nuclear DNA and subject to species diversification, while the large protein, coded by chloroplast DNA, has remained constant. This situation, of one enzyme present in enormous excess over all others, strikingly distinguishes chloroplasts from mitochondria, in which the many soluble proteins are present in modest quantities.

Similarly, there is evidence (*137, 138*) that the chloroplast lamellar proteins include two which bind chlorophyll and which together comprise some 75% of the total lamellar protein. Here too the presence of major membrane protein components differs from the situation in mitochondria, where it now appears that there is no single major protein (cf. Chapter 7). Thornber *et al.* (*137, 138*) have isolated and characterized the two chlorophyll-binding proteins present in spinach, as well as in other higher plants and in green algae. The blue-green algae and photosynthetic green bacteria contain a chlorophyll–protein complex very similar to one of those found in higher plants, the chlorophyll a–protein complex. The protein of this complex has been investigated more exten-

TABLE 8.2

Components of Photosystems PS I and PS II of Green Plant Photosynthesis[a]

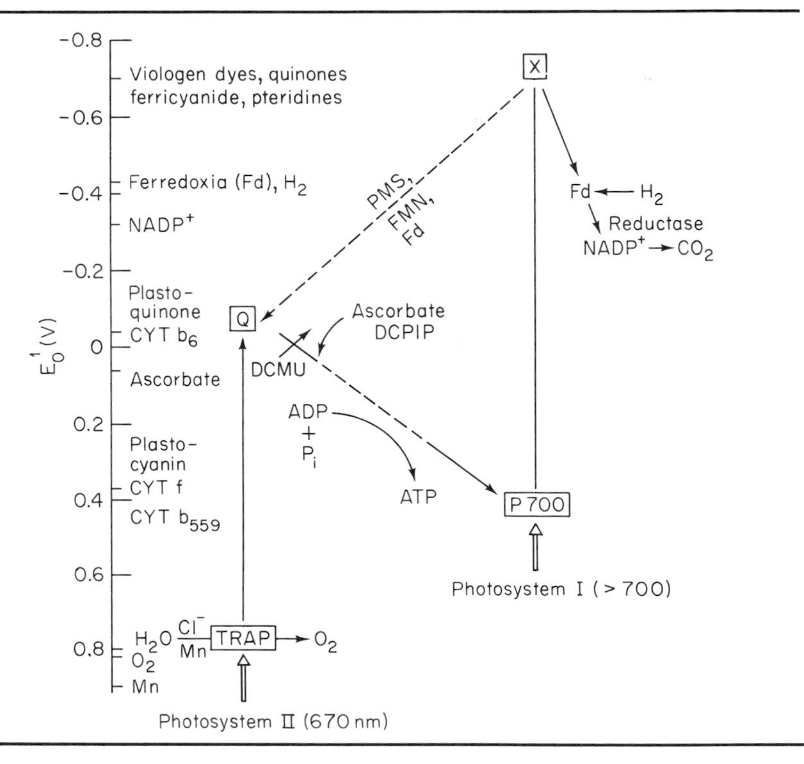

[a] The positions of the three cytochromes, plastocyanin, and plastoquinone are not established but probably are located in the bridge between PS II and PS I. The vertical scale indicates voltage charges associated with electron transport. Mn = manganese, required for PS II activity. TRAP, Q, and X are unidentified components, P700 is a special state of chlorophyll. Electrons flow from positive to negative. PMS, FMN, Fd can donate electrons to Q from X thus short-circuiting the flow to NADP. From Zelitch (*148*).

sively in the photosynthetic bacterium, *Chloropseudomonas ethylicum,* than in plants, but all forms have certain characteristics in common: very similar amino acid composition (Table 8.5), a molecular weight of 150,000 to 160,000 for the tetramer, and the presence of five molecules of chlorophyll bound noncovalently within each of the four subunit. Preparations from blue-green algae and from spinach both contain one molecule of P700, the chlorophyll a form associated with photosystem I, per about 70 chlorophyll molecules. If correct, this value indicates that not every tetramer (with 20 chlorophyll molecules) contains a photosynthetic reaction center.

TABLE 8.3

Overall Chemical Composition of Spinach Chloroplasts[a]

Component	Percentage of dry weight of chloroplasts	
	Values for chloroplasts isolated in water	Values corrected for loss of soluble proteins
Total protein:	50	69
Water-insoluble protein	50	31
Water-soluble protein	0	38
Total lipid:	34	21
Chlorophyll	8.0	5.0
Carotenoids	1.1	0.7
RNA	–	1.0–7.5
DNA	–	0.02–0.1
Carbohydrate (starch, etc.)	Variable	

[a] From Kirk and Tilney-Bassett (*68*).

TABLE 8.4

Amino Acid Composition of Subunits Obtained from Different Species of Fraction I Protein[a]

	Larger fragment				Smaller fragment			
Plant source	Spinach	Spinach	Tobacco	Spinach beet	Spinach	Spinach	Tobacco	Spinach beet
Dissociation and separation	SDS G-100	SDS G-100	SDS G-100	Urea G-200	SDS G-100	SDS G-100	SDS G-100	Urea G-200
Phenylalanine	1.00[b]	1.00	1.00	1.00	1.00	1.00	1.00	1.00
Lysine	1.18	1.04	1.07	1.15	1.21	*1.19*	*1.77*	1.44
Histidine	0.67	0.66	0.62	0.62	0.45	*0.53*	0.42	*0.21*
Arginine	1.44	1.42	1.44	1.26	*0.99*	0.96	0.96	*0.58*
Aspartate	2.18	2.35	2.27	2.00	2.14	2.31	*2.88*	*1.44*
Threonine	1.75	1.63	1.45	1.41	1.19	1.29	*1.60*	*0.77*
Serine	0.81	0.87	0.86	0.93	*0.77*	0.78	*1.51*	0.91
Glutamate	2.20	2.42	2.52	2.30	2.26	2.77	*4.14*	*2.17*
Proline	1.13	1.17	1.13	1.06	1.56	1.56	1.64	1.43
Glycine	2.10	2.43	2.44	2.22	*1.15*	1.28	*2.33*	1.58
Alanine	2.16	2.21	2.23	2.13	*0.86*	0.98	*1.63*	1.09
Valine	—	1.52	1.49	1.57	—	1.15	1.36	1.40
Methionine	0.42	0.35	0.36	0.33	0.46	0.43	0.41	0.30
Isoleucine	0.88	0.69	0.76	0.88	0.56	*0.47*	*0.84*	0.70
Leucine	2.09	2.19	2.16	1.91	1.63	1.81	*2.10*	*1.54*
Tyrosine	0.92	0.91	0.85	0.76	1.57	1.43	*1.92*	*0.99*

[a] From Wildman (*147*).
[b] These numbers are calculated as relative molar ratios to phenylalanine. Italicized numbers indicate differences greater than normal error of method.

TABLE 8.5

***Amino Acid Composition of Chlorophyll a Binding Protein of Beta vulgaris*[a]**

	Moles % of amino acids
Polar residues (total):	38
Acidic residues	
Aspartic	9
Glutamic	8
Basic residues	
Lysine	2
Histidine	5
Arginine	4
Hydroxyl residues	
Threonine	5
Serine	5
Nonpolar residues (total):	63
Aliphatic residues	
Glycine	11
Alanine	9
Valine	6
Leucine	12
Isoleucine	7
Proline	5
Methionine	2
Cysteic	1
Aromatic residues	
Tryosine	3
Phenylalanine	7
Tryptophan	1

[a] Recalculated from Thornber *et al.* (*138*).

Green algae and higher plants, in contrast to all other photosynthesizers, contain a second photosystem and a second chlorophyll binding protein. This complex contains chlorophyll a and b in equal amounts and exhibits photosystem II activity. The protein is only 30,000 daltons and probably binds one molecule each of chlorophylls a and b. Its amino acid composition differs from that of the complex I protein (*137*).

Other investigators have fractionated lamellar membranes by following photosynthetic activity, rather than chlorophyll per se. The most popular procedures use digitonin or the nonionic detergent Triton X-100 to solubilize the lamellae (*1, 2, 8, 140*). After detergent treatment, the photosystem II activity is found mainly in large fragments sedimenting at 10,000 *g*, and photosystem I activity is principally in much smaller fragments which sediment at 144,000 *g*. Both of these fractions are membranous as seen in the electron microscope, and both contain many proteins. The protein banding patterns of the two fractions seen in gel electrophoresis studies are complex but distinctly different. The specific identification of protein components in the gel patterns is just beginning.

The proteins listed in Table 8.2 as members of the photosystems are presumably present in the reaction center preparations. In addition, a coupling factor has been isolated from spinach chloroplasts which is necessary for photophosphorylation (*74, 87*). The factor, called CF_1, has many properties similar to mitochondrial F_1 (cf. p. 232), including its location in globular "knobs" on the lamellar surface. The amino acid compositions are similar, as shown in Table 8.6. The two proteins are immunologically different (antibodies prepared against one do not cross react with the other) neither are they functionally interchangeable in stimulating photophosphorylation in spinach chloroplast preparations.

TABLE 8.6

Amino Acid Composition of Chloroplast Coupling Factor CF_1 Compared with Mitochondrial F_1[a]

	Residues per half cystine	
Amino acid	F_1	CF_1
Lysine	15	11
Histidine	4	2
Arginine	14	15
Aspartic acid	19	18
Threonine	14	19
Serine	15	20
Glutamic acid	28	36
Proline	10	10
Glycine	22	22
Alanine	25	25
Half cystine	1	1
Valine	18	20
Methionine	5	6
Isoleucine	15	19
Leucine	21	27
Tyrosine	7	7
Phenylalanine	7	7

[a] From Racker (*87*).

Another lamellar component is the protein to which protochlorophyll is attached, the so-called protochlorophyllide holochrome. This protein plays an essential role in the photoreduction of protochlorophyll to chlorophyll. In classic studies using dark-grown bean seedlings, Smith (*121*) isolated a protein fraction containing protochlorophyllide which was transformable *in vitro* to chlorophyllide by reduction in the light. Recently, Schopfer and Siegelman (*107*) purified this holochrome fraction and obtained molecular aggregates of 300,000 MW, containing one protochlorophyll, and of 550,000 MW, containing two protochlorophylls, each exhibiting about 75% transformation of protochlorophyllide to chlorophyll upon illumination. No carotenoids or other visible pigments were present in these preparations.

Beyond this preliminary information about a few lamellar proteins, little is known to supplement the list given in Table 8.1. Most of the enzymes of the photosynthetic apparatus have not been isolated and characterized in detail as proteins with the exception of ferredoxin (*68*). Procedures for the separation and identification of lamellar proteins are just beginning to be developed (*33, 51, 91*). Before discussing the available evidence on the organization of proteins and other components in the lamellar membranes, we will first briefly consider the composition of the lipid fraction, which includes the chlorophylls, carotenoids, and the nonpigmented lipids.

The lipids comprise about half of the dry weight of the lamellae, the rest being protein. The major classes of lipids and their approximate relative molecular concentrations are listed in Table 8.7. Structural for-

TABLE 8.7

***Major Classes of Chloroplast Lipids in Spinach*[a]**

	% by weight of total lipid
Chlorophyll a and b	20.8
Carotenoids	2.8
Quinones, tocopherols, vitamin K	3.3
Phospholipids	9.1
(50% phosphotidyl glycerol)	
Glycolipids	44.3
Monogalacytosyl diglyceride	26.8
Digalactosyl diglyceride	13.4
Sulfolipid	4.1
Sterols	2.2
Total	82.5
Unidentified lipids	17.5

[a] From Kirk and Tilney-Bassett (*68*).

mulas of phosphatidyl glycerol (the major phospholipid of lamellae), major glycolipids, unique sulfolipid, and of chlorophyll are shown in Fig. 8.3.

Green algae and higher plants contain chlorophylls a and b with characteristic absorption spectra that have been extensively investigated (*68*). With the development of spectrophotometric devices for measuring spectra in intact cells and leaf tissues, it became possible to study the states of chlorophyll in the intact organism (*116*).

Chlorophyll b absorbing at 650 nm presents no identification problem, but other chlorophyll absorption peaks are seen at 673, 683, and 700 nm, and during greening of etiolated seedlings peaks are seen at 668 and 678 nm as well. In *Euglena* and *Ochromonas* there is also a distinct peak at 695 nm. All these peaks are *in vivo* manifestations of a single chemical species of chlorophyll a. The peak at 700 nm represents P700, the initial photoreceptor pigment of photosystem I (cf. Table 8.2). It is still not established to what extent these differences in absorption maxima result primarily from aggregated states of chlorophyll or to what extent chlorophyll–protein interactions also contribute to the spectral shifts. In reviewing this problem recently, Butler (*19*) concluded that chlorophyll 673 is in the monomeric disaggregated form, and, as such, is a component of photosystem II and that the 683 peak represents an aggregated form as it occurs in mature lamellar membranes and is a component of photosystem I. Boardman *et al.* (*11a*) have followed the spectral shifts in greening seedlings as a means of monitoring membrane changes during lamellar formation.

Sometime shortly after the conversion of protochlorophyllide to chlorophyllide, the phytyl side chain is added. This reaction is thought not to alter the absorption spectrum of chlorophyll, but it probably is of considerable importance in the development of lamellar structure. As Butler points out (*19*), "The lipophyllic phytyl tail on the chlorophyll molecule would allow the lipid-dissolved carotene molecule to come into close proximity with the chromophoric porphyrin head of the molecule. The efficiency of energy transfer increases as the chlorophyll and carotene are brought together in the developing lamellar structures." Carotene absorption can be detected by fluorescence measurements, and the time course of phytylization has been correlated with the appearance of energy transfer from carotene to chlorophyll.

The role of carotenoids in green plant chloroplasts is still not firmly established. There is considerable evidence that in the absence of carotenoids chlorophyll acts as photodynamic dye, killing cells which are exposed to light. If carotenoids are an essential part of the photosynthetic process, they should be found associated with both photosystems. In a study of fractions from spinach and from three algae, *Anabena, Por-*

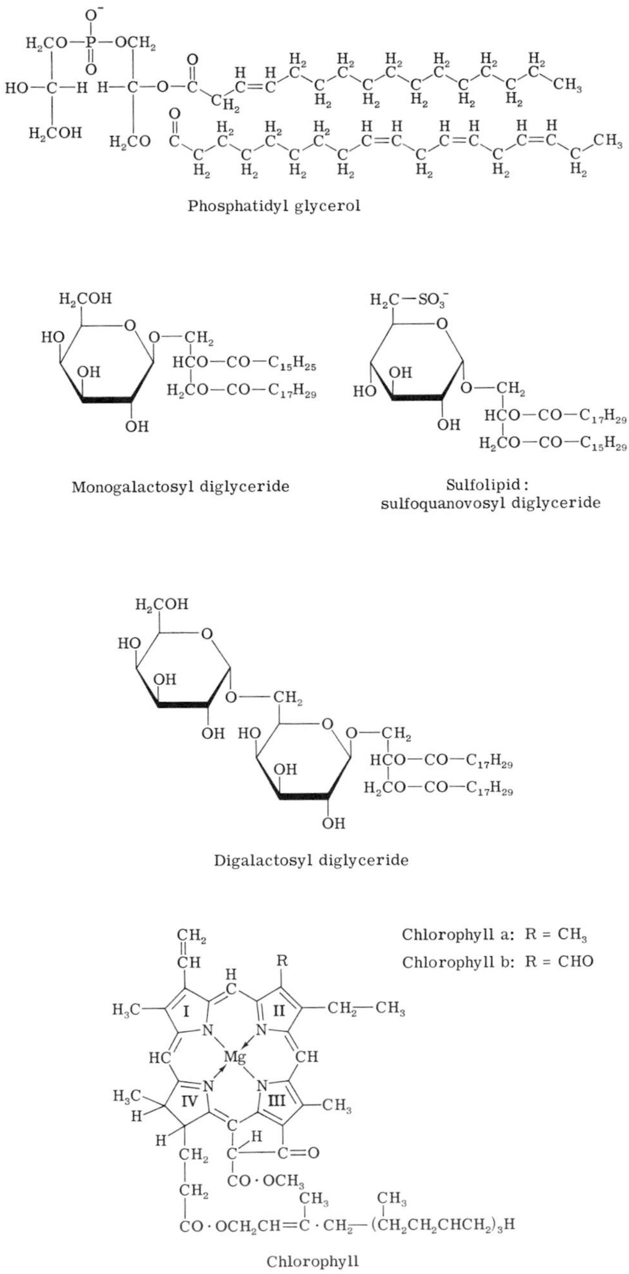

Phosphatidyl glycerol

Monogalactosyl diglyceride

Sulfolipid: sulfoquanovosyl diglyceride

Digalactosyl diglyceride

Chlorophyll a: R = CH_3
Chlorophyll b: R = CHO

Chlorophyll

Fig. 8.3. Structural formulas of phosphatidyl glycerol, major glycolipids, a unique sulfolipid, and chlorophyll.

phyra, and *Phaeodactylum* (a diatom), Ogawa *et al.* (*82*) found β-carotene preferentially associated with photosystem I and certain xanthophylls characteristically associated with photosystem II. This result is in agreement with a recently proposed mechanism involving reduction of NADP in the light by carotene through ferredoxin, a component of photosystem I. There are a number of experimental findings which implicate the xanthophylls in oxygen evolution, a function of photosystem II.

The colorless lipids of the chloroplast are highly characteristic compounds, for the most part found nowhere else in plant or animal cells (*4*). In particular, the galactolipids and sulfolipid shown in Fig. 8.3 are unique to chloroplast lamellae. The monogalactosyl diglyceride and the sulfolipid are each esterified with two different fatty acids, one being a C_{16} polyunsaturated fatty acid and the other a C_{18} component. The digalactosyl diglycerides, on the other hand, are esterified mainly with the C_{18} unsaturated fatty acid, dilinolenic acid. During lamellar biogenesis in *Chlorella*, the ratio of saturated to unsaturated fatty acids decreases and the ratio of mono- to digalactosyl diglycerides increases (*5, 7*). To some extent, different fatty acid constituents are found characteristically in particular plants.

In a comparative study of lipids in light-grown and dark-grown *Euglena gracilis*, Erwin and Bloch (*31*) showed that the lipid composition of the same organism grown as photoautotroph and as heterotroph was strikingly different. The light-grown cells contained galactolipids with their esterified polyunsaturated fatty acids, α-linolenic and hexadecatetraenoic acids; the phospholipids were represented principally by phosphatidylserine. In the heterotrophic dark-grown cells, the principal fatty acids were saturated; there were no galactolipids, but rather those phospholipids that are characteristic of animal mitochondria, e.g., phosphatidylcholine. Thus the differences in lipid composition of light-grown and dark-grown cells correlated closely with the switch from chloroplast to mitochondrial energetics. These experiments show in an elegant manner that the lipid composition is organelle-specific, reflecting the relative proportion of chloroplast to mitochondrial development in the same organism.

Recently Bloch, *et al.* (*7*) have suggested that differences in the typical fatty acids of chloroplast and mitochondria might be generalized as shown in Fig. 8.4 in terms of the saturation of the methyl end of the molecule. In the plant acids there are only two saturated terminal carbons, making the distal portion of the molecule relatively rigid, whereas in the animal fatty acids there are five saturated carbons giving the molecule more flexibility and greater suitability for hydrophobic interaction.

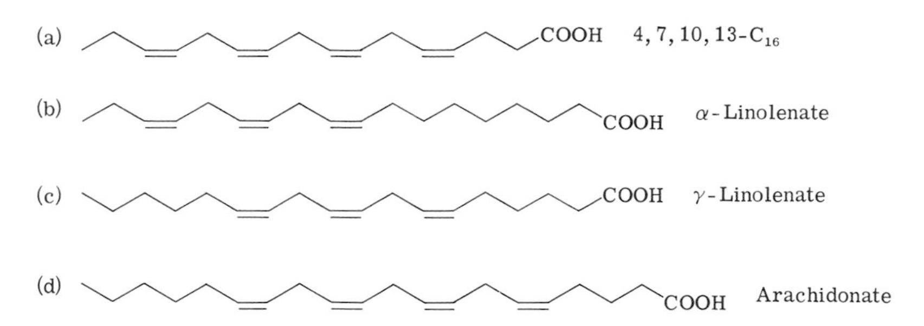

Fig. 8.4. Backbone structures of the major fatty acids of chloroplasts and mitochondria. As discussed in the text, the chloroplast fatty acids (a) and (b) are maximally extended toward the methyl end of the molecule, making the distal part of the molecule more rigid than in the animal fatty acids (c) and (d), in which the five terminal carbons are saturated, allowing more molecular flexibility. From Bloch *et al.* (*7*).

How are the major components arranged in the lamellar membranes? Studies of the molecular orientation of proteins and lipids in chloroplast membranes began in the 1930's with physical measurements of double refraction and dichroism. The results suggested the presence of a layered structure with lipid molecules oriented perpendicular to the lamellar plane. This formulation fit very well with the 1934 Danielli and Davson model of a generalized membrane, consisting of a lipid bimolecular leaflet coated on both sides by protein (cf. Chapter 7). For over thirty years this model provided the framework for most investigations and speculations on the organization of biological membranes, including those of chloroplast lamellae. As previously discussed (p. 240), the model is so general that it is hard to refute without detailed evidence, some of which is just now becoming available.

New lines of evidence, especially from studies of lipid–protein interactions (*5*), freeze-etch electron microscopy (*15, 77, 78, 85*), and biochemical characterization of chloroplast fragments (*2, 25, 104*), are beginning to provide a detailed and specific picture of molecular organization of chloroplast lamellae (*67*).

Benson, a foremost authority on chloroplast lipids and their role in lamellar membrane organization, has stressed the importance of hydrophobic interactions between lipids and proteins in membrane stabilization. Benson writes (*5*): "The concept (of hydrophobic interaction) defined by Kauzman (*61*) is based upon observed entropy changes as a hydrophobic group or molecule is transferred from water to a hydrophobic medium. The entropy derives largely from the enforced organization of water by the hydrocarbon and destruction of this organization when the hydrocarbon is removed from the aqueous environment. As hydrocarbon groups approach each other, clusters of oriented water

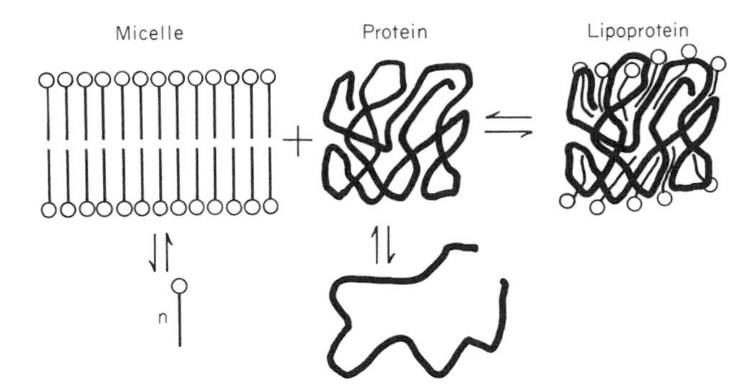

Fig. 8.5. Hydrophobic association of amphipathic lipids and membrane protein. Equilibria expressed by vertical arrows are associated with large changes in water entropy. Membrane lipoprotein is represented by the diagram at the right. From Benson *et al.* (5).

must disorganize and form liquid water. This 'melting' of ice structure involves a considerable entropy increase, apparently enough to drive the equilibrium far toward micelle or folded protein structure. As a consequence amphipathic lipids form micelles and (amphipathic) proteins coil to produce hydrophilic molecules with their hydrophobic groups concentrated in the interior." This process is shown schematically in Fig. 8.5.

The term "amphipathic" describes molecules which have spatially separate hydrophilic and hydrophobic regions. The amphipathic nature of chlorophyll is dramatized in Fig. 8.6. The hydrophobic phytol side chains and hydrophilic porphyrin ring are contrasted in this drawing of a scale model of the molecule.

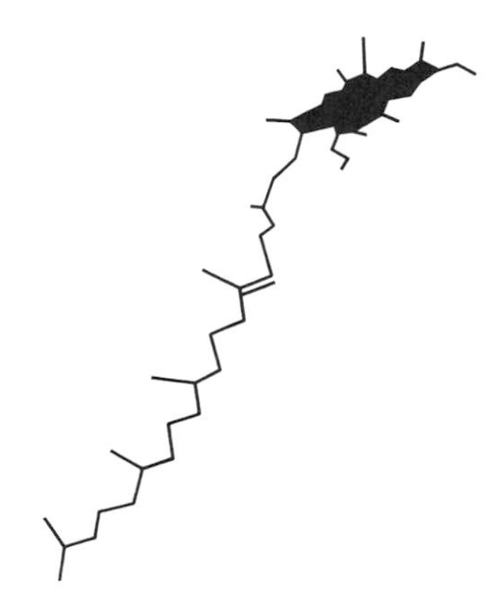

Fig. 8.6. Chlorophyll, an amphipathic lipid. From Benson *et al.* (5).

A recent molecular model of the chloroplast lamella developed by Ji (*57c*) and discussed by Benson (*5*) is shown in Fig. 8.7. The continuous phase is provided by lipoprotein aggregates. In this model the chlorophyll molecules as well as the phospholipids are shown associated with small globular protein molecules along the membrane surfaces. Larger protein aggregates are shown embedded more centrally within the membrane. This model is concerned primarily with the position of the lipids in stabilizing membrane structure.

Most other investigators have been principally concerned with the identification and localization of the major proteins and of the reaction centers of photosystems I and II. A technical advance of key importance in these investigations has been the development of freeze-etch electron microscopy (*79*). This method, already discussed in Chapter 7, has been used extensively in recent studies of chloroplast membranes and has provided new insights into the arrangement of the major protein components and the reaction centers.

In this method, isolated chloroplasts are suspended in glycerol to displace some of the water, then rapidly frozen at −150°C, and sectioned at −100°C. At this temperature, sectioning causes fractures, and the fracture faces provide surface views of the membranes. The ex-

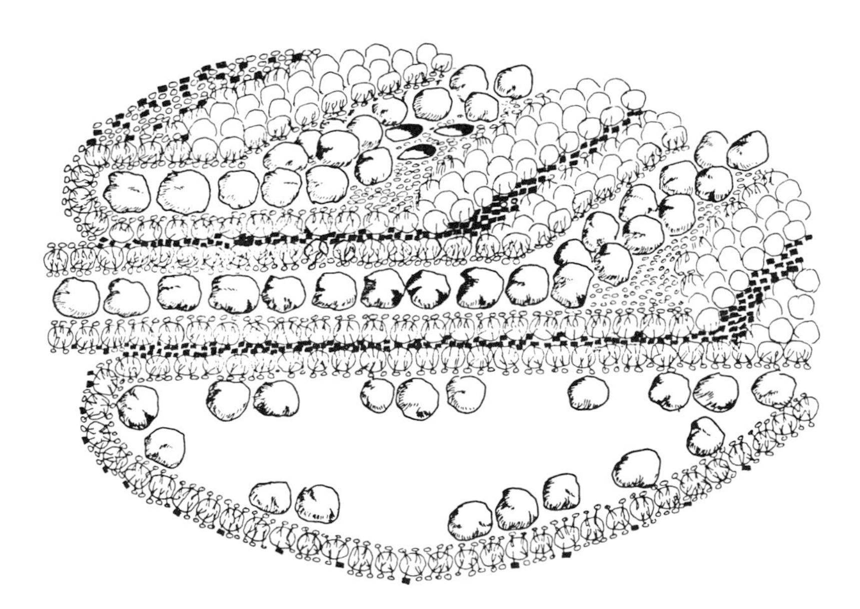

Fig. 8.7. Molecular model for lamellar membrane system based on the lipoprotein monolayer. Amphipathic membrane lipids and chlorophylls are associated hydrophobically with membrane proteins. From Ji (*57c*).

posed surfaces are layered with a thin platinum–carbon film deposited by evaporation, and this replica is observed in the electron microscope.

Mühlethaler first described spinach chloroplasts prepared by the freeze-etch technique. He found the fracture faces dotted with large and small particles partly embedded in the matrix (*78*). By negative staining the larger particles were shown to consist of aggregates, usually four, of the smaller particles. He proposed that the membranes consist of a continuous lipid bilayer, in which the globular proteins are partially embedded on the surface. He interpreted the fracture faces as the surfaces of the lamellae.

Branton and Park (*16*) published similar micrographs, which they interpreted very differently from Mühlethaler. They too postulated a continuous lipid phase, but proposed that the globular proteins seen on the fracture faces were actually located inside the membrane rather than on the surface. Thus, they proposed that the freeze-etch process had cleaved the membrane internally, peeling it open and revealing internal structure. Mühlethaler has recently come to the same conclusion on the basis of new studies in his laboratory (*67*). Indeed, many independent lines of evidence have now converged to support this interpretation, not only for chloroplast lamellae but for other membranes as well. For example, in recent experiments, the red cell membrane was labeled with ferritin (*86a*) and with fibrous actin (*139*), and then prepared by the freeze-etch process. The labels were not found on the fracture faces exposed by the freezing process, but only on the outer membrane surfaces revealed by deep etching.

Branton and Park (*16*) first demonstrated that two sizes of particles are differently distributed on the opposite fracture faces of the chloroplast lamellae. They found 175 Å globules predominantly on one face and 110 Å globules predominantly on the other. Some confusion was initially engendered by the presence of the knobs of CF_1, the photophosphorylation coupling factor, and by the aggregates of RuDP-carboxylase. These components, identified by Howell and Moudrianakis (*54, 55*), have been shown to lie on the outer surface of the lamellae facing the matrix where all the soluble components are located. They can be removed from the membranes without loss of activity of either photosystem I or II.

A comprehensive view of lamellar organization has now been proposed by Crane and his students (*2, 25*) on the basis of a combination of biochemical fractionation of isolated chloroplasts and electron microscopy of the intact chloroplasts and of the fractions. Their proposed model, shown in Fig. 8.8, describes *grana* lamellae of higher plant chloroplasts, but not *stroma* lamellae (cf. Fig. 8.2) shown by Sane *et al.* (*104*) to contain only photosystem I activity.

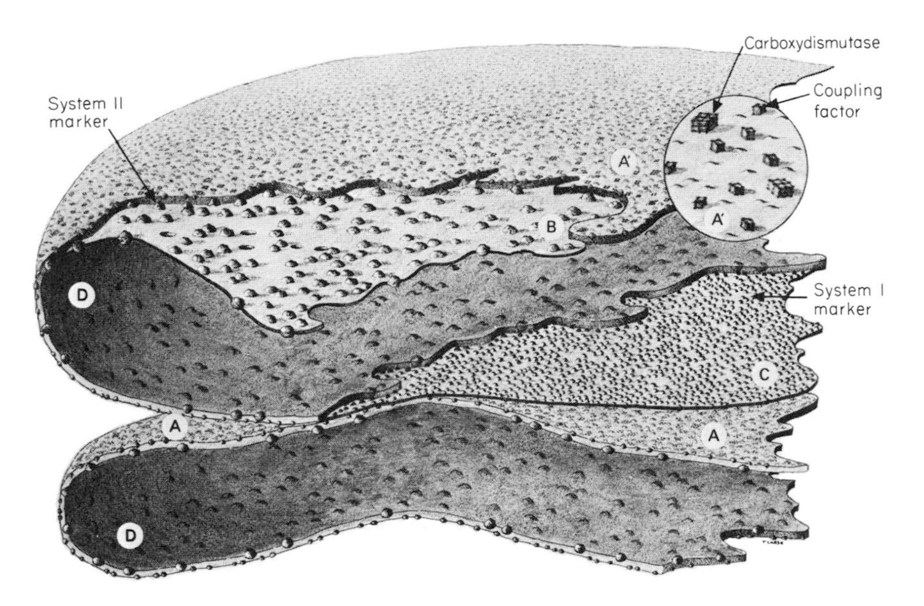

Fig. 8.8. A schematic representation of two chloroplast thylakoids showing a binary membrane structure. Face *B* revealed by freeze-etching contains the large 175 Å particles. The 10,000 *g* digitonin fraction is enriched in these particles and in photosystem II activity. Face *C* contains the 110 Å particles seen by freeze-etching. The 144,000 *g* digitonin fraction is characterized by these particles and by having high photosystem I activity. The external surface of the thylakoid membrane is the site of attachment of carboxydismutase and the coupling factor protein. From Arntzen *et al.* (*2*).

Their studies of intact chloroplasts agree with Branton and Park in distinguishing 175 Å globules on one fracture face and 110 Å globules on the other. They fractionated the membranes with digitonin and prepared separate fractions retaining the activity of photosystems I and II. These fractions were then examined in the electron microscope. In freeze-etch preparations, the photosystem II membrane fragments showed ridges made up of globules 175 Å in diameter, partly embedded in the continuous phase. The photosystem I fraction showed ridges of the 110 Å globules. Thus, the digitonin treatment had apparently fractionated the large and small globules in the same way as did the freeze-etch process.

This conclusion was supported by the thickness of the fractions. In sectioned material, they found that the photosystem I material was about 40–50 Å thick, as was the photosystem II material, while the unfractionated lamellae were about 80 Å. On this basis, they suggested that the thinner fragments could have arisen by splitting of the original membrane with some overlap or interdigitation of one layer into the other. The fact that the freeze-etch preparations were thicker than the chemically fixed and sectioned material was attributed to extraction of some components during chemical fixation.

These findings support the view of the lamellar membrane as a bipartite structure, with photosystem I localized toward the matrix and photosystem II localized toward the intralamellar space and both systems embedded in the lipid interior of the membrane. In its general conception, this view is comparable to that proposed by Crane *et al.* (*25*) for the inner mitochondrial membrane. Since, functionally, the electron transport and coupled phosphorylating activities of the two membrane systems are so similar, it seems likely a priori that significant structural similarities should be found. Whether the bipartite nature of these membranes represents *the* structural correlate of electron transport or rather represents some more general feature of membrane organization remains to be investigated.

Although the interpretation of lamellar organization shown in Fig. 8.8 seems in good agreement with much of the available data, there are some discrepancies which have recently been discussed in detail by Kirk (*67*). A foremost problem concerns reconciling this view with the x-ray diffraction data of Kreutz (*69*). The details are too complex for discussion here. In essence, Kreutz has derived a cross-sectional structure for the lamellar membrane based upon the distribution of electron densities in both intact and dried chloroplasts by small angle x-ray diffraction studies. He proposes a water layer about 10 Å thick in the center of the membrane, with three successive layers of different densities on either side, corresponding, from the inside to the surface, to lipids, chlorophyll, and protein. By arguing against some of the postulated values, Kirk has attempted to bring these results into line with those of Branton and Park (*15, 16, 85*) and Crane *et al.* (*25*).

Sane *et al.* (*104*) have challenged the bipartite model of Crane *et al.* by questioning whether the digitonin procedure provides the same fractionation as the freeze-etch cleavage. They prepared chloroplast fragments by mechanical breakage with the French press and recovered a size fraction containing only 110 Å globules. They identified this fraction in electron micrographs as stroma lamellae (cf. Fig. 8.2) and suggested that the digitonin procedure had also separated stroma from grana lamellae.

Finally, there is an interesting problem concerning the so-called "partition" region between the grana lamellae. As cells mature, this region becomes more and more electron-dense, and appears to accumulate increasing amounts of chlorophyll. Weier and Benson (*144*) proposed that the region is fully hydrophobic. This view seems particularly unlikely in view of the remarkable studies of Izawa and Good (*57b*) showing that the stacking of grana is reversible by changing the salt concentration of the medium. The importance of stacking in photosynthesis has been stressed by Homann and Schmid (*50a*) who described a tobacco mutant, containing only unstacked lamellar discs, which had only photosystem I but no photosystem II activity. Similarly, the bundle sheath chloroplasts (located in a single layer of cells surrounding the vascular bundles) of maize and related plants have been shown to contain only unstacked lamellar discs and to exhibit only photosystem I activity (*120a*). On the other hand, Goodenough *et al.* (*37a*) have described a nuclear gene mutation in the alga *Chlamydomonas* in which stacking is decreased, without any correlated loss of photosystem II activity.

A new insight into the reversibility of the stacking of grana has come from studies of sequential changes in lamellar organization during the cell cycle in synchronous cultures of *Chlamydomonas.* As shown in Plates XIII and XIV and discussed below, Palade has found a diurnal cycle of stacking and unstacking associated with biosynthesis of new components and growth of the membranes.

CHLOROPLAST RIBOSOMES AND RNA'S

In contrast to the difficulties encountered in isolating and characterizing mitochondrial ribosomes, the identification of chloroplast ribosomes has been relatively unambiguous and devoid of controversy. In particular, the ribosomes of higher plant chloroplasts are quite stable and amenable to characterization in sucrose gradients.

Chloroplast Ribosomes

The presence of a distinct class of chloroplast ribosomes was first reported by Lyttleton in 1962 (*73*). Subsequently Clark *et al.* (*23, 24*) showed that in the Chinese cabbage, *Brassica,* the cytoplasm contained 83 S ribosomes and that the chloroplast ribosomes were 68 S and constituted between 20 and 35% of the total ribosome population. They also demonstrated polysomes in the chloroplast fraction, indicating the presence of messenger RNA. Boardman *et al.* (*9*) isolated chloroplast ribosomes from tobacco leaves and showed that they were 70 S in contrast to the 80 S ribosomes of the cytoplasm. Subsequently, Chen and Wildman (*22*) described the isolation of chloroplast polyribosomes from tobacco. Boardman (*7a*) followed the changes in ribosome composition during the greening of dark-grown bean seedlings. He found that 70 S ribosomes were already present in the etiolated leaves of dark-grown seedlings in approximately the same proportion as in the leaves of light-grown plants, i.e., about 40%.

The uniformity of size of higher plant chloroplast ribosomes shown in Table 8.8 became evident as many reports, including the very careful measurements reported by Stutz and Noll (*129*), all agreed. In the higher plants, chloroplast ribosomes are all in the 67–70 S class, in contrast to the corresponding cytoplasmic ribosomes which are in the 80 S class. More difficulty was encountered with algal ribosomes. In *Euglena* the chloroplast ribosomes were first isolated and characterized by Brawerman and Eisenstadt (*18*), but they were unable to stabilize the cytoplasmic ribosomes. Only in 1969, using an elaborate extraction procedure, were Rawson and Stutz (*90*) finally able to establish the sedimentation values of both classes of ribosomes and their respective RNA's in *Euglena*, as shown in Table 8.8.

The chloroplast ribosomes of *Chlamydomonas* are much more fragile than those of higher plants or of *Euglena*, requiring 25 m*M* Mg^{2+} for stabilization. Extensive efforts to characterize them were eventually successful in demonstrating their presence as a 69 S particle in compari-

TABLE 8.8

Chloroplast and Cytoplasmic Ribosomes and rRNA's of Algae and Higher Plants

	Chloroplast			Cytoplasmic			
Organism	Ribosomes	Subunits	rRNA's	Ribosomes	Subunits	rRNA's	References
Chlamydomonas	68–70 S	53 and 28 S	23 and 16 S	80	60 and 40 S	25 and 18 S	*52, 96*
Euglena	70 S	50 and 30 S	23 and 16 S	87 S	67 and 45 S	26 and 21 S	*90*
Acetabularia			23 and 16 S			25 and 18 S	*110*
Radish, tobacco, bean, spinach, pea	68–70 S	47–50 S and 30–33 S	23 and 16 S	79–83 S	55–60 S and 40 S	25 and 16 S	*10, 24, 71, 73, 129*

son with 83 S cytoplasmic ribosomes (*96*). Subsequently, the diameters of the two classes of ribosomes were measured in electron micrographs, and those in the chloroplast were found to be demonstrably smaller (*84*). In further work, the subunits of both classes were also compared and were assigned values of 33 S and 28 S for the chloroplast ribosomal subunits and 43 S and 30 S for the cytoplasmic ones: under osmotic conditions in which rat liver ribosomal subunits were 47 S and 32 S (*52*).

Indirect evidence that some components of chloroplast ribosomes are encoded in chloroplast DNA comes from the antibiotic-resistant phenotypes of some mutants which map in the cytoplasmic linkage group of *Chlamydomonas* (Chapter 3). The phenotypes include streptomycin resistance, streptomycin dependence, and conditional streptomycin dependence, as well as resistance to spectinomycin, erythromycin, carbamycin, oleandomycin, spiramycin, and cleocin. These antibiotics have all been shown to block protein synthesis in bacteria at the ribosome level, and mutations to resistance toward streptomycin, spectinomycin, and erythromycin have each been shown to result from altered ribosomal proteins (*146*). It seems likely that the same will be true in the case of these mutations in *Chlamydomonas*, which are now under investigation.

Recently, Gillham *et al.* (*36*) reported that in strains of *Chlamydomonas* containing some but not other cytoplasmic mutations to streptomycin resistance, the chloroplast ribosomes were 66 S rather than 70 S. These values were established by comparison with cytoplasmic 80 S ribosomes from the same preparations in preparative sucrose gradients. A similar difference in S value was also seen in the large subunit but not in the small subunit from these 66 S ribosomes. These preliminary results are suggestive of a chloroplast ribosomal alteration resulting from mutation of a cytoplasmic gene, but the molecular basis of the 66 S value has not yet been ascertained. The same aberrant S value, 66 S, was also found for chloroplast ribosomes in a strain carrying the nuclear gene mutation, *ac-20* (*14*).

Chloroplast and cytoplasmic ribosomal proteins from several higher plants have been examined by gel electrophoresis (*40, 40a*). The patterns given by the proteins from the 70 S chloroplast and 80 S cytoplasmic ribosomes were found to differ in many bands, but it was not possible to determine whether any proteins of the two sets were identical. Similarly, the patterns shown by the proteins extracted from the large and small subunits of the 70 S ribosomes were found to be distinctly different. The authors also compared patterns of 80 S and 70 S ribosomal proteins from several plants and found a greater homology among the 70 S chloroplast proteins than among the 80 S cytoplasmic components.

More recently, a powerful method of two-dimensional gel electrophoresis of ribosomal proteins has been developed in the laboratory of Wittman. The published results have shown remarkable resolution of proteins from the ribosomal subunits of *E. coli* (*59a*). The method, now being applied to chloroplast ribosomes, will facilitate detailed comparisons of ribosomal proteins from mutants and from different compartments of the same cell, as well as from different species.

Ribosomal RNA's of Chloroplasts

The ribosomal RNA's from chloroplast ribosomes were first distinguished from those of cytoplasmic ribosomes by Stutz and Noll (*129*) on the basis of sedimentation rates in sucrose gradients. Subsequently, a powerful gel electrophoresis method was developed by Loening and Ingle (*71*) which can give excellent discrimination between RNA molecules of similar but not identical size. Using this method, Loening and Ingle (*71*) showed that the major RNA components in radish seedlings were 25 S, 23 S, 18 S and 16 S, and that the 18 S and 25 S RNA's came from the cytoplasmic ribosomes, while the other two were of chloroplast ribosomal origin. In further studies, the ribosomal RNA's of many green and nongreen plant tissues were examined (*57, 57a*). An example of the results is given in Fig. 8.9. A principal difficulty encountered in plant material is the ease with which the 23 S component is preferentially degraded during extraction of the RNA's. The extent of this degradation varies from one plant species to another, and also varies with the age and physiological state of the plant, but in most materials is appreciable. Presumably, this preferential susceptibility reflects some special feature of the structure of the 23 S molecule, but no investigation of the molecular basis of this phenomenon has yet been reported.

Evidence that chloroplast ribosomal RNA's are transcribed on chloroplast DNA's comes entirely from DNA–RNA hybridization studies with tobacco (*136*) and several other higher plants (*57, 57a*) and *Euglena* (*119, 130*). In all studies, the occurrence of hybridization, so-called homologous hybridization, between chloroplast DNA and the corresponding ribosomal RNA has been demonstrated clearly; and very low amounts of heterologous hybridization between cytoplasmic ribosomal RNA and chloroplast DNA have been reported. However, a significant amount of heterologous hybridization between nuclear DNA and chloroplast ribosomal RNA has been found in higher plants and in *Euglena*.

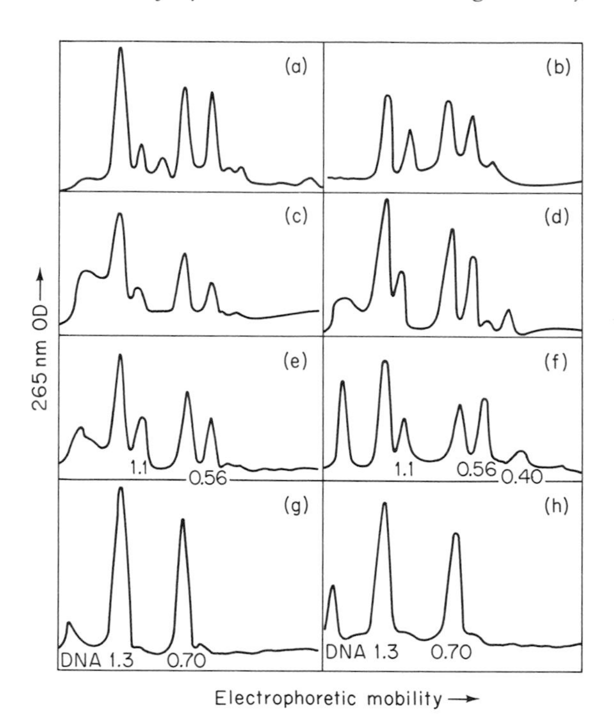

Fig. 8.9. Comparison of nucleic acids from green and nongreen tissues. (a) French-bean leaf, (b) lettuce leaf, (c) cocklebur leaf, (d) corn leaf, (e) pea leaf, (f) barley leaf, (g) pea root tips, (h) artichoke tuber. Total nucleic acid was prepared by a phenol–detergent method and fractionated by gel electrophoresis. All fractionations are in the prescence of EDTA unless otherwise stated. RNA components are referred to as their molecular weight in millions. From Ingle *et al.* (*57*).

To interpret this finding, Tewari and Wildman proposed the existence of two classes of chloroplast ribosomes, one class having RNA's of nuclear origin (*136*). Ingle *et al.* (*57, 57a*) examined the hybrids formed from carefully purified ribosomal RNA's and DNA's of a number of higher plants. They found that cytoplasmic ribosomal RNA's hybridized principally with DNA having a buoyant density of about 1.705–6 gm/cm^3 in CsCl. Having obtained the same result with four different plants, they concluded that the 1.705–6 gm/cm^3 DNA represented nuclear cistrons diverging from the buoyant density of the main nuclear peak of 1.695 gm/cm^3.

They then compared the hybridization of chloroplast and cytoplasmic ribosomal RNA's with this nuclear DNA component and found about the same extent of hybridization of the two. In a comparable experiment with chloroplast DNA, they found that the chloroplast RNA was much more effective than cytoplasmic RNA in hybridizing with chloroplast DNA. These results confirm and extend the earlier report by Tewari and Wildman (*136*). However, Ingle *et al.* (*57a*) then compared the specificity of the cytoplasmic RNA–nuclear DNA hybrids by comparing interspecies hybrids of four plants: wheat, swisschard, onion, and artichoke, as shown in Table 8.9. The lack of species discrimination seen in this comparison led Ingle *et al.* to conclude that more rigorous tests of the stability of the hybrids were required before any conclusions could be drawn concerning the nuclear origin of chloroplast ribosomal RNA's.

Stutz and Rawson (*128, 130*) have probably resolved the problem in *Euglena* by an experiment showing that the heterologous hybrids are less stable than the homologous ones, as judged by their low T_m (denaturation temperature) and broad melting profile. The instability is evidence of poor and incomplete hybridization, which could result from pairing of similar but nonidentical sequences. In general, ribosomal cistrons have remained more constant in evolution than the bulk DNA of the same organisms, so it would not be surprising if the ribosomal cistrons for cytoplasmic and chloroplast RNA's in the same organism were fairly similar. Thus it is likely, though not yet fully settled, that chloroplast ribosomal RNA's are entirely transcribed on chloroplast DNA and that the heterologous hybrids of chloroplast RNA with nuclear DNA are artifacts.

They (*130*) took the study of chloroplast ribosomal cistrons one step further by examining the strand specificity of hybridization. They found that heat-denatured *Euglena* chloroplast DNA gives a double band in CsCl with buoyant densities of 1.704 gm/cm^3 and 1.696 gm/cm^3.

TABLE 8.9

Comparative Specificity of Hybridization between Total DNA and Cytoplasmic Ribosomal RNA's of Four Plant Species[a]

	Hybridization (% total DNA)				
^{32}P ribosomal RNA [cytoplasmic]	Wheat DNA	Swiss chard DNA	Onion DNA	Artichoke DNA	Cucumber DNA
Wheat	**0.069**	0.22	0.076	0.035	0.56
Swiss chard	0.073	**0.30**	0.078	0.043	0.75
Onion	0.083	0.29	**0.089**	0.048	0.69
Artichoke	0.069	0.21	0.081	**0.042**	0.83

[a] From Ingle *et al.* (*57a*).

This observation provided the basis for a separation procedure which yielded purified light and heavy single-stranded DNA's. In hybridization studies, Stutz (*128*) found that the heavy strand bound approximately nine times more chloroplast ribosomal RNA than did the light strand. This result is similar to that of Attardi *et al.* (cf. p. 246) with mitochondrial DNA–RNA hybridization.

The extent of hybridization between chloroplast DNA and the homologous chloroplast ribosomal RNA has been measured in tobacco (*136*) and in *Euglena* (*130*). In both systems, the plateau values for the fraction of total chloroplast DNA hybridized with chloroplast ribosomal RNA's at saturation were about 2% (doubled-stranded value). In both organisms, the amount of DNA per chloroplast is about 0.5×10^{-14} gm or 3×10^{9} daltons (cf. p. 30). The genomic size estimated by reannealing kinetics is about $1.2–1.5 \times 10^{8}$ daltons (cf. p. 29), and on this basis the amount of reiteration is about twentyfold. Hybridization values indicate 20–30 cistrons corresponding to each ribosomal RNA per chloroplast or one cistron for each of the reiterated sequences. The similarity in values for two such disparate organisms as tobacco and *Euglena* is remarkable. The presence of a DNA component with similar genomic size in *Chlamydomonas* (cf. p. 30) suggests that this component ($1–2 \times 10^{8}$ daltons) may be a fundamental genetic complex of some sort, perhaps coding for a set of photosynthetic and/or biosynthetic enzymes.

Evidence for the existence of specific tRNA's associated with chloroplasts of higher plants has been presented by Dure and Merrick (*29*). Chloroplasts have been shown to contain a 5 S ribosomal RNA (*30, 135*), whereas this component has not yet been identified in any mitochondrial system.

Tewari and Wildman (*136*) reported hybridization of tRNA's from tobacco chloroplasts with tobacco chloroplast DNA, as shown in Fig. 8.10. The plateau value varied from 0.4 to 0.7% in different experiments, corresponding to $4.4–7.9 \times 10^{5}$ daltons in approximately 10^{8} daltons of chloroplast DNA, the size of the repeating genomic unit identified by reannealing kinetics. On this basis, each genomic unit would contain 25–30 tRNA cistrons as well as one cistron each for the ribosomal RNA's, as previously noted.

Origin of Chloroplast RNA's

The ability of isolated chloroplasts to synthesize RNA has been demonstrated by several investigators (*64, 65, 110b, 112, 113, 119, 120, 124*). In all these studies, incorporation was dependent upon the presence of all four nucleoside triphosphates and was inhibited by DNase, RNase,

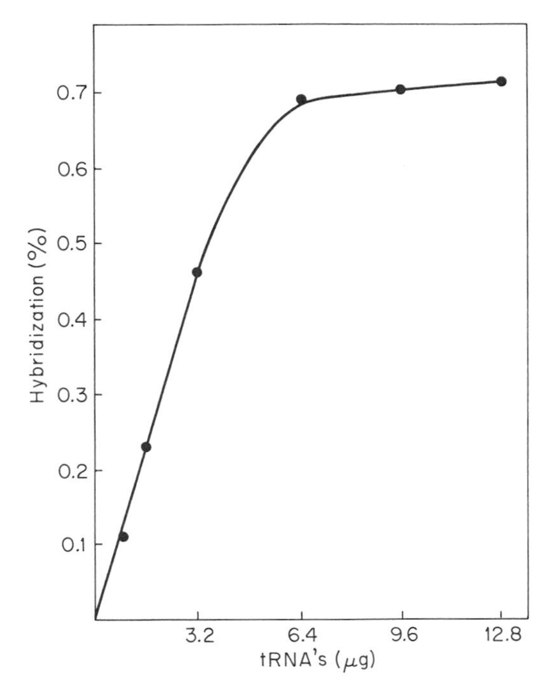

Fig. 8.10. Hybridization of tRNA's. Twenty μg of chloroplast DNA hybridized with increasing concentrations of chloroplast tRNA's (activity 2012 cpm/μg). From Tewari and Wildman (*136*).

and by pretreatment with actinomycin D. However, the product was not examined by hybridization in any of these studies. Consequently, the origin of the RNA synthesized as a template product of chloroplast DNA was not established.

Spencer and Whitfeld (*124*) characterized the product of short-term *in vitro* RNA synthesis by isolated spinach chloroplasts in sucrose gradients and found the product to be very polydisperse, ranging in size from 5 S to more than 25 S, with a peak around 12 S. They concluded that this material was probably a population of messenger RNA molecules because, of the size range and speed of labeling, and because long-term labeling *in vivo* yielded radioactivity primarily in ribosomal and transfer RNA's. Subsequently, the *in vitro* studies were repeated (*125*) using polyacrylamide gel fractionation to characterize the RNA's more precisely than had been possible with sucrose gradients. Essentially the same results were obtained. They concluded: "The product of RNA synthesis by isolated chloroplasts comprises a heterogeneous population of molecules ranging in size up to about 2×10^6 daltons molecular weight."

These results provide preliminary evidence of messenger RNA synthesis in chloroplasts. An indirect approach utilizes the rifamycin antibiotics which inhibit initiation of RNA transcription in bacterial systems by combining with RNA polymerase. Spencer *et al.* (*125*) undertook extensive studies with isolated chloroplasts and with *in vivo* studies to look for evidence of an inhibitory effect of rifamycin on chloroplast RNA synthesis. They found no effect whatsoever, and concluded that the RNA polymerase of higher plant chloroplasts is not sensitive to rifamycin.

A different conclusion was reached by Surzycki (*132*) in studies with *Chlamydomonas*. He reported inhibitory effects of rifampin [3-(4-methylpiperazinyliminomethyl)rifamycin SV] on adenine uptake into whole cells, and upon uptake of ^{32}P label of ^{32}P-ATP into isolated chloroplasts. The drug at a concentration of 250 μg/ml (*132*) or 300 μg/ml (*134*) inhibited incorporation of labeled adenine into whole cell RNA by 44% within an hour, compared with controls. Which cellular RNA's are inhibited? It is stated (*132*) that chloroplast ribosomal RNA's are not labeled after 4 hours growth in the presence of the drug, whereas the cytoplasmic ribosomal RNA's are labeled in this interval, but no data have been presented on this point. A radioisotope-labeled peak attributed to 5 S chloroplast RNA is absent after 4 hours with the drug, whereas the corresponding 5 S cytoplasmic species is present. Further evidence of the effect of the drug on transcription of chloroplast RNA's comes from *in vitro* studies utilizing a ^{32}P label in ATP. The very low incorporation rates obtained in the controls were inhibited by preincubating chloroplasts with 50 μg/ml of rifampin. The product of the reaction was characterized as high molecular weight RNA by its elution from a MAK chromatographic column together with carrier RNA.

Further evidence on the effect of rifampin on *Chlamydomonas* comes from long-term growth of cells in the presence of 250 μg/ml rifampin with acetate as a carbon source for four to five doublings. At the end of growth, the cells contained little or no chloroplast ribosomal RNA's. In electron micrographs of these cells, few chloroplast ribosomes were found. Thus, growth in the presence of the drug resulted in blockage of formation of chloroplast ribosomal RNA and ribosomes. The structural alterations in the chloroplasts of these cells resemble those seen in acetate-grown cells carrying the nuclear gene mutation *ac-20* (*37*). This mutation, as well as others affecting chloroplast structure and function, will be discussed in the next section, following consideration of the effects of various other antibiotics upon growth of *Chlamydomonas*.

A specific point concerning Surzycki's results needs to be discussed here. He has concluded that rifamycin blocks transcription in intact cells of all chloroplast DNA cistrons because (a) he found about 40% inhibition of adenine incorporation, approximately corresponding to the fraction of cellular RNA attributed to the chloroplast; and (b) virtually complete inhibition was seen with isolated chloroplasts. Neither of these results provide adequate evidence that the drug blocks transcription of messenger RNA's. Thus it cannot be concluded from these data, as Surzycki (*132*) and Surzycki *et al.* (*134*) have done, that cells grown with rifampicin would be blocked in the transcription of cistrons coding for DNA and RNA polymerases, were they located in chloroplast DNA. One of the classic weaknesses of inhibitor studies, as already discussed in Chapter 7, is the inability to judge whether all or only a part of the syntheses are blocked by a given concentration of the inhibitor. The complexity of RNA polymerases, the fact that some but not others are blocked by rifamycin, and the possibility that multiple forms of the enzyme are present in the chloroplast further complicate the interpretation of these findings. The results do show that 250–300 μg/ml of the drug can block the formation of ribosomal RNA and ribosomes after several days of growth in acetate medium. Whether this effect is direct or indirect remains to be demonstrated.

Let us consider one further system in which the RNA synthesizing capacity of the chloroplast has been examined. In principle, *Acetabularia* is an ideal organism for studying RNA synthesis independent of nuclear DNA (*109, 110*). One can excise the nucleus of *Acetabularia,* and the enucleated cell will continue to grow for some weeks. Unfortunately, contamination with bacteria and other algae has been a persistent problem, casting doubt on the meaning of the results, especially those involving radioisotopic labeling. Recently, methods have been developed for growing sterile cultures of *Acetabularia,* but as yet the organisms grow very slowly under these conditions.

Shephard (*114, 115*) used autoradioaugraphy to examine the uptake of radioisotope label into RNA and into protein using sterile enucleated plants. He showed that there was very little difference in the incorporation of label into RNA in enucleated and in control (nucleated) plants. The same was true with incorporation into protein.

Schweiger *et al.* (*110a*) described the incorporation of ^{14}C-uracil into ribosomal RNA of chloroplasts using bacteria-free (but not uni-algal) cultures of *Acetabularia.* They demonstrated the presence in enucleated *Acetabularia* of RNA synthesizing systems that were responsible for the biosynthesis of both ribosomal and transfer RNA's. In this paper they say "previous claims of RNA synthesis in enucleated cells as far as they are based on the incorporation of labeled precursors have suffered from the presence of contaminating microorganisms in the *Acetabularia* cultures and from lack of adequate identification of the incorporation product." The future use of *Acetabularia* for studies of macromolecular synthesis and regulation will depend upon developing a medium that supports normal growth under sterile conditions.

In summary, the studies of chloroplast ribosomes and RNA's to date have indicated the presence of machinery for the transcription of chloroplast DNA, as well as machinery for protein synthesis. Thus, chloroplast DNA is probably transcribed to produce chloroplast ribosomal and transfer RNA's, and messenger RNA's as well. The evidence is more fragmentary than in the mitochondrial systems described in Chapter 7, but no contrary evidence has been encountered. Chloroplast ribosomes are remarkably uniform in monosome size, size of subunits, and size of their respective RNA's in the algae and the higher plants. The chloroplast genome of *Chlamydomonas* resembles a bacterial genome in that it confers sensitivity (and by mutation, resistance) to numerous antibiotics, including streptomycin, spectinomycin, erythromycin, and their derivatives. Whether these mutations in fact result in alterations in chloroplast ribosomes has not been established, but the phenotypes, especially streptomycin dependence and conditional streptomycin dependence, provide strong indirect evidence to support this view.

Given the presence of a protein synthesizing apparatus in chloroplasts, let us turn our attention to the products of this apparatus. Which proteins are synthesized in the chloroplast and which of them are coded for by chloroplast DNA?

ORIGIN OF CHLOROPLAST PROTEINS

Identifying the genetic origins and sites of synthesis of chloroplast proteins comprises a complex puzzle. The problem has been approached by methods similar to those discussed in Chapter 7 for mitochondria: attempts to identify the proteins synthesized *in vitro* in isolated organelles and attempts to dissect the *in vivo* process with antibiotics, with physiological shifts in growing conditions, and most recently with cytoplasmic gene mutations.

The study of isolated chloroplasts per se, as we shall see below, has been unrewarding, particularly because of the very low rates of incorporation of radioisotopes into chloroplast proteins. Studies of intact cells, using antibiotics together with physiological shifts from growth in dark to growth in light or variations during the diurnal light–dark cycle have provided some indirect evidence of proteins synthesized either on cytoplasmic or on chloroplast ribosomes.

The use of mutants, potentially the most powerful method to identify the origin of specific proteins and to dissect the process of biogenesis, is just beginning to yield positive results. All of the nuclear mutations affecting chloroplast function that have been examined so far have shown complex pleiotropic phenotypes in which the primary effect of the mutation has not been identified. Mutations in chloroplast DNA, which are now being identified and mapped (Chapter 3), include phenotypes of antibiotic resistance and temperature sensitivity, both very promising as a source of mutationally altered proteins. However, as in the mitochondrial system, a key technical problem is still the development of adequate fractionation procedures for the separation and characterization of membrane proteins.

Protein Synthesis in Isolated Chloroplasts

Most reports of protein synthesis in isolated chloroplasts of higher plants published before 1965 are suspect on the grounds of bacterial contamination (*119*). Subsequently, thanks in part to the public airing given this problem by App and Jagendorf (*1a*), far greater attention has been paid to maintaining sterility, no mean problem with chloroplasts isolated from leaves. By 1967, protein synthesis had been successfully demonstrated in chloroplast preparations from a number of higher plants, as discussed by Smillie and Scott (*119*). However, neither intactness nor purity (i.e., absence of other cell fractions) were claimed for these preparations, and their activities were uniformly very low.

Among the best current studies are those of Spencer *et al.* (125) following procedures they had previously developed (*34, 122, 123*). Using spinach chloroplasts isolated from very young greenhouse grown seedlings, they have routinely obtained incorporation rates of 50–100 pmoles of radioactive amino acids incorporated per milligram of chlorophyll in 10–30 minutes. The radioactivity has been found distributed in the soluble fraction, the ribosomes, and the lamellar fraction. In the ribosome fraction, none of the radioactivity was associated with ribosomal proteins themselves, but rather with other proteins initially attached, but later dissociated, during preparation of ribosomal proteins for fractionation on polyacrylamide gels.

In studies of the soluble fraction, two peaks were tentatively identified, one probably representing Fraction I protein and the other probably being ferredoxin. The authors are cautious in their conclusions, owing to the low radioisotope incorporation rates obtained. However, it should be noted that almost as much incorporation was seen into the lamellar fraction as into the soluble (supernate) fraction. Thus, in the preliminary studies, there is already a hint that both soluble and membrane-bound proteins may be synthesized in the chloroplast, in contrast to the situation in mitochondria in which only membrane-bound components incorporate amino acids in the isolated organelle.

The authors conclude, rather pessimistically, that the method they have been developing, i.e., fractionation of proteins after amino acid incorporation into an isolated chloroplast preparation, is not sufficiently powerful to identify individual proteins. Their principal reasons for pessimism are low incorporating activity and a large number of different proteins being synthesized simultaneously. They favor the future development of reconstituted systems in which some highly active but nonspecific components (e.g., *E. coli* ribosomes) are used to synthesize identifiable chloroplast proteins from a chloroplast DNA template.

Two of the proteins present in rather high amount in the chloroplast are the carboxylase of Fraction I protein and the ATP-coupling fraction, CF_1. Some preliminary evidence of amino acid incorporation into these proteins has been reported (*76, 89*). Further evidence of the origin of chloroplast proteins comes from studies of intact cells, using inhibitors or mutations in an effort to dissect the system. Let us turn then to the far more indirect but extensive studies with intact cells.

Origin of Chloroplast Proteins: Evidence from Studies of Intact Cells

Our focus in this chapter is not upon the process of chloroplast biogenesis as a whole, but rather upon the chloroplast proteins, their sites of origin, and biosynthesis. Very little direct information is available on this question, and, consequently, we shall have to examine some of the indirect evidence, especially for clues. In the study of chloroplast biogenesis, a favorite approach has been to examine the transition from growth in the dark to growth in the light, the so-called greening process. This experimental approach is analogous to the studies of mitochondrial changes occurring in the transition from anaerobic to aerobic growth, and similarly provides an opportunity to observe and inhibit the synthesis of particular components.

Until very recently, the compositional changes studied during greening were confined to correlations between the appearance of various pigments and other lipids and the emergence of photosynthetic activity. In the future, with the development of improved methods for protein separation, the use of well-defined greening conditions should be of increasing value in the identification of specific proteins appearing during biogenesis. Currently, preliminary studies of protein changes during greening have been reported with two organisms:

Chlamydomonas and *Euglena*. For convenience in comparing the systems, we will consider them separately.

Studies of the Greening Process in *Chlamydomonas*. Wild-type *Chlamydomonas* synthesize chlorophyll in the dark, and the morphology of light-grown and dark-grown chloroplasts is very similar (*97, 98*). The greening process has been studied with a yellow strain, *y-1*, which loses its ability to form chlorophyll and lamellar membranes when grown in the dark. (The designation *y-1* has been used both for the mutant gene itself and for strains carrying this gene.) The mutant *does* contain protochlorophyll (*95*), and consequently the mutant block is probably located in the dark conversion of protochlorophyll to chlorophyll.

In our initial electron microscope studies of the yellow mutant, we (*97*) found that the dark-grown cells contained little or no chlorophyll and only a few disorganized vesicles in place of the highly structure stacks of lamellar discs seen in the normal green cells. The resulting differences in morphology between the wild-type and the dark-grown mutant are shown in Plates III and IV. The dark-grown cell contains a recognizable organelle, the leucoplast, which occupies the same volume as does the chloroplast of the light-grown cells. In both light-grown and dark-grown cells, the plastid contains pyrenoid, eyespot, DNA, ribosomes, and starch-synthesizing capacity. The plastids, whether green or yellow, appear to be essential to the life of *Chlamydomonas*. No plastid-free mutants have ever been obtained, and even *Polytoma*, a colorless relative, contains a large leucoplast in which starch is synthesized and stored.

When dark-grown mutant cells are returned to light, normal lamellae develop in parallel with the appearance of chlorophyll. Indeed, the 1:1 correspondence between the presence of chlorophyll and of lamellar membranes seen in the yellow mutant is typical not only of *Chlamydomonas* but of virtually all green algae and higher plants, with rare exceptions (*12, 94, 143*). This correspondence provides one line of evidence that the light-induced synthesis of chlorophyll is a key step in the regulation of lamellar membrane formation.

Studies of a pale-green mutant of *Chlamydomonas* (mutant #95) correlating photosynthetic activity (*21*), pigment composition (*102*), and structure (*94*), provided further evidence on the relationship between chlorophyll content, lamellar membrane formation, and photosynthetic activity in chloroplast development. The pale-green mutant contained about 10% of the wild-type content of chlorophyll a, and a similar fraction of wild-type lamellar membranes. Photosynthetic activity was above normal on a chlorophyll basis despite the virtual absence of chlorophyll b and of carotenoids. Because of its low total chlorophyll content, the mutant provided material for the first direct observation of the oxidation of cytochrome f upon illumination of intact cells. The very low carotenoid content demonstrated that those pigments were not essential for lamellar formation, and consequently that functional lamellae of mutant and wild-type cells could contain vastly different chlorophyll-carotenoid ratios. These studies, carried out in the 1950's, foreshadowed the much more extensive studies with the yellow mutant (*28a, 33, 36a, 51, 53, 83, 84*) and with a number of green nonphotosynthetic mutants (*70*) that have been carried out subsequently.

The yellow mutant has been utilized in a series of investigations describing the structural, compositional, and enzymatic changes in the chloroplast that occur during greening of dark-grown cells. Ohad *et al.* (*83, 84*) examined the degreening process by following the changes occurring in light-grown cells transferred to the dark. The cells were then returned to the light, and greening was followed over the 6–8 hour period required for synthesis of most constituents of the chloroplast, restoring the cells to the initial light-grown condition.

Some degreening data are shown in Fig. 8.11. The amount of chlorophyll per cell fell as calculated for dilution at cell division, with no new synthesis and no destruction. The amount of lamellar material, estimated from electron micrographs, fell at a slightly slower rate, indicating a small amount of dark synthesis. The Hill reaction, an index of photosynthetic activity, fell more slowly at first, and then more precipitously than the membranes or chlorophyll itself, suggesting a need for more structural integrity than remained after four doublings in the dark. Cytochrome f, involved in photosynthetic electron transport and weakly bound to the membranes, was apparently synthesized in the dark since its activity fell very slowly, leveling off at about 5% of the initial value after several doublings. The other photosynthetic enzymes measured, ribulose-1,5-diphosphate carboxylase, glyceraldehyde-3-phosphate dehydrogenase, fructose-1,6-diphosphatase, and ferredoxin, remained essentially constant. The mitochondrial enzyme, succinic dehydrogenase, increased three-fold and then remained constant. The photoreduction of NADP fell rapidly, paralleling the decrease in chlorophyll for two doublings; it could not be detected at lower concentrations in subsequently doublings.

Dark-grown mutant cells continue to synthesize carotenoids at about the same rate as in the light. They also synthesize the typical lipids of the chloroplasts (the galactolipids, sulfolipid, and phosphatidylglycerol) at rates that are roughly half that of the light-grown cells as estimated by incorporation of ^{14}C-labeled acetate. Thus, the dark-grown mutants contain a partially differentiated chloroplast lacking only chlorophyll and the lamellar membranes.

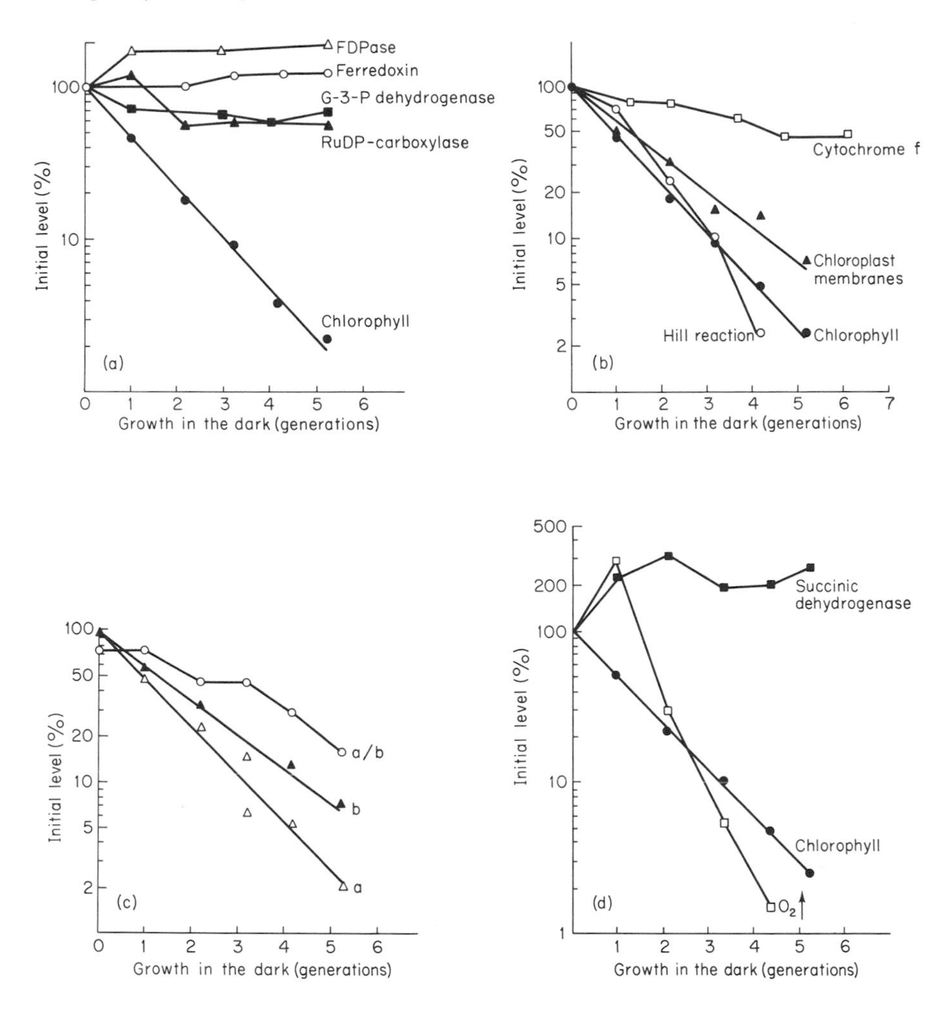

Fig. 8.11. Concentration of chlorophyll, chloroplast membranes, and various enzymes in mutant cells as a function of number of generations of growth in the dark in a semicontinuous culture apparatus. The maximal cell concentration allowed before dilution was 1.8×10^6 cells/ml. The generation time was 22 hours. The initial levels (100%) were as follows: (a) chlorophyll 34 μg/10^7 cells; FDPase, pH 8, 33 μmole P_i/30 min/10^7 cells; RuDP-carboxylase, 50.3 nmoles CO_2-fixed/10 min/10^7 cells; ferredoxin, 42 nmoles NADP-reduced/min/10^9 cells; G-3-P dehydrogenase, 0.072 μmole NADP reduced/min 10^7 cells. (b) Hill reaction, 0.088 μmole DC1 reduced/min/10^7 cells; cytochrome f, 2×10^{-3} OD/10^7 cells; chloroplast membranes index, 1.9 intersections/cm^2 for a total of 2217 membrane profiles counted. (c) Grana fusion lines index, 0.68/cm^2; ratio of grana fusion lines index to total chloroplast membrane index (%). a, Number of grana per chloroplast unit area; b, number of total chloroplast membrane profiles per chloroplast unit area. (d) Succinic dehydrogenase, 1.5 μmoles/10 min/10^8 cells; oxygen evolution, 4.2 μmoles/10 min/10^8 cells. From Ohad *et al.* (*84*).

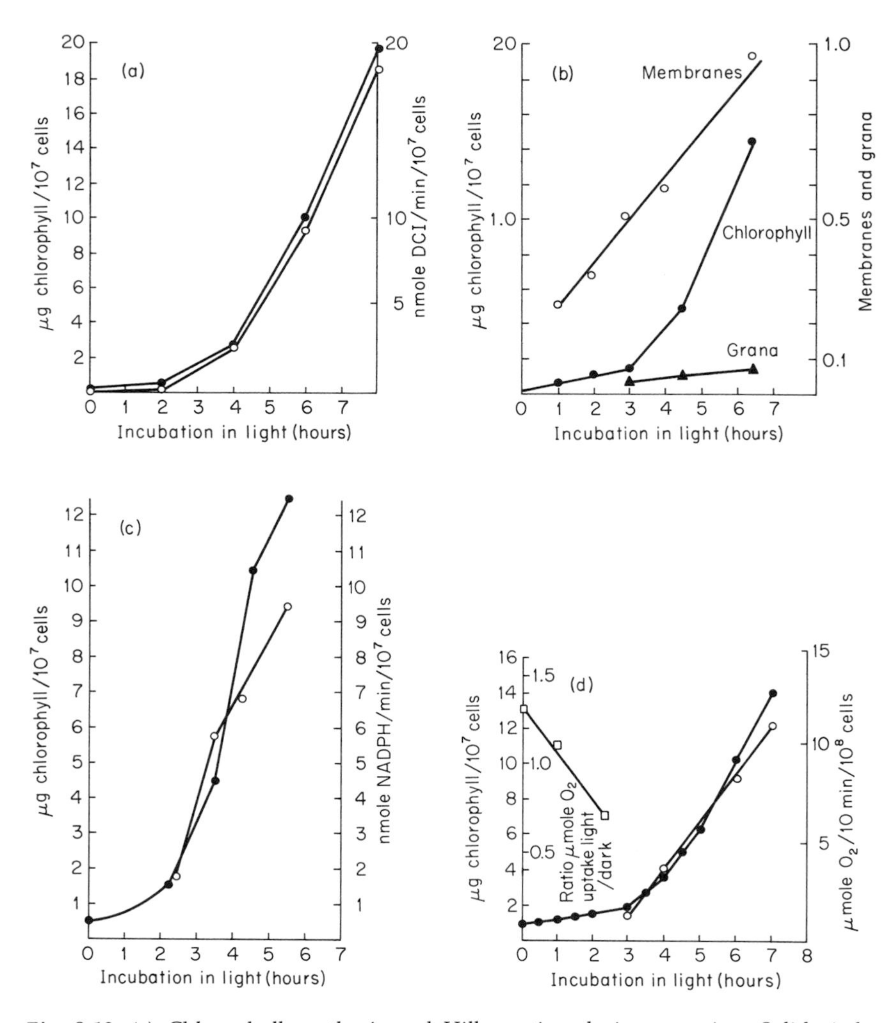

Fig. 8.12. (a) Chlorophyll synthesis and Hill reaction during greening. Solid circles, chlorophyll; open circles, Hill reaction. The cells were grown in the dark in batch culture for 6 days and exposed to light at a final concentration of 6.5×10^7 cells/ml. (b) Synthesis of chlorophyll and increase in amount of chloroplast membranes. Solid circles, chlorophyll; open circles, membrane index; triangles, grana index. Cells grown in batch-type culture for 6 days in dark resuspended in fresh medium at a final concentration of 3.3×10^6 cells/ml, and exposed to light. Samples were taken at different times for chlorophyll measurements and electron microscopy. (c) Photosynthetic pyridine nucleotide reductase activity during the greening process. Solid circles, chlorophyll; open circles, enzyme activity. Cells grown in dark for six generations in batch culture, resuspended in fresh medium at a final concentration of 1.1×10^7 cells/ml, and exposed to light. (d) Respiration (left) and oxygen evolution (right) during the greening process. Solid circles, chlorophyll; open circles, oxygen evolution; open squares, ratio O_2 uptake, light/dark. Cells were taken at different times (left) from the incubation mixture, washed, and resuspended in 3 ml of growth medium at a final concentration of 7×10^7 cells/ml. Their respiration (with KOH in the central well) was measured in a Warburg apparatus at 28°C in

When the dark-grown *Chlamydomonas* is returned to light, the cells neither divide nor enlarge appreciably in volume during a 6–8 hour period while chlorophyll is synthesized and the lamellar membranes are restored to their typical light-grown appearance (*84*). Thus the greening process appears to involve inhibition of the normal events leading to cell division during this time, since cells would normally double in 6–8 hours in this medium. Some biochemical changes accompanying the greening process are shown in Fig. 8.12.

The new lamellar membrane appeared to develop from the vesicular remnants scattered through the plastid in the dark-grown cells. The authors found no evidence that the inner chloroplast membrane was the primordium from which new lamellae developed. Hill reaction activity and the photooxidation of cytochrome f developed during the greening process at the same rate as did the synthesis of chlorophyll itself. The rate of chlorophyll synthesis was not increased upon reexposure to light after an interval in the dark. This result is particularly important because it strengthens the evidence that chlorophyll itself or photochlorophyll acts as the regulating molecule in controlling chlorophyll synthesis.

In the same study, cells were exposed to tritiated acetate during the greening process and observed in the electron microscope (*84*). Most of the grains were located over the chloroplast, indicating that most of the macromolecules produced during this 6–8 hour period were synthesized in, or became part of, the chloroplast.

Hudock and Levine followed the greening process with a similar or identical yellow mutant, *y-2* (*56*). They found that the photosynthetic activities of CO_2 fixation, O_2 evolution, and NADP photoreduction correlated directly with the chlorophyll content during the greening process, but that RuDP-carboxylase and ferredoxin did not decrease very much during dark growth, and quickly returned to their previous level when exposed to light. They also found that RuDP-carboxylase activity in wild-type cells grown at various rates in the light and dark showed a maximum of twofold variation.

Both studies show that the greening process in *Chlamydomonas* involves a restricted set of induced syntheses. Most of the components of the chloroplast continue to be synthesized in the dark in the wild-type

---

light (500 ft-c) and dark. Results are plotted as a ratio of O_2 uptake in light vs dark as a function of time. Initial respiration rate in the dark was 6.7 μmoles O_2/10 min/10^8 cells. Dark-grown cells (right) (six generations in semicontinuous culture) were incubated in the light at a final concentration of 10^7 cells/ml; samples of 10^8 cells in 4 ml were incubated in the light and dark in Warburg vessels in a mixture of 0.07 *M* $KHCO_3$ and 0.13 *M* $NaHCO_3$ (pH 8.4), and oxygen evolution measured at 28°C. From Ohad *et al.* (*84*).

and in the yellow mutant. The mutant, having lost chlorophyll-forming ability, has also apparently lost the signal for lamellar formation, but it has not stopped synthesizing the photosynthetic enzymes and any other chloroplast constituents; thus, when the light is turned on, the system is ready for photosynthesis as soon as chlorophyll and membranes appear. This situation offers good experimental opportunities for studying the assembly of lamellar membranes, especially since other macromolecular syntheses appear to be blocked at this time. Building upon the detailed description of the greening process and standardized conditions established by Ohad *et al.* (*83, 84*) with the *y-1* mutant, these investigators and their colleagues (*28a, 33, 36a, 51, 52, 53, 108*) have continued to examine the process of lamellar membrane formation, utilizing the normal greening process as well as blocks imposed by inhibitors of protein synthesis.

Using whole cells, Goldberg and Ohad (*36a*) followed the synthesis of lipids during the greening process. They found that the characteristic galactolipids, sulfolipid, and phospholipids of chloroplasts were the major lipids synthesized during greening, in parallel with the photosynthetic pigments. In a radioautographic study of membrane growth using ^{14}C-acetate as a label, they were able to gauge the relative growth rates of different regions of the chloroplast as inferred from radioisotope incorporation rates. They found that the tubular system of the pyrenoid developed faster than the lamellar discs and that the unstacked discs developed faster than did those seen in pairs of larger stacks. They also inferred that the lamellae grew by incorporation of new material into preexisting structures, namely the remnants of previous lamellae seen as disorganized vesicles in the dark-grown cell (cf. Plates IV, XII, XIII, and XIV).

In a complementary study of membrane composition during greening, de Petrocellis *et al.* (*28a*) found that the ratios of chlorophyll to cytochrome f (i.e., cytochrome 553) and of chlorophyll to carotenoids changed dramatically during the greening process. The reason for this change is that most of the carotenoids and cytochrome f are already present at nearly full cellular concentration in the residual lamellae of dark-grown cells. Thus, considerable movement of molecules must occur in the membranes during their development, for example, to bring the preexisting molecules of cytochrome f into correct alignment with the newly synthesized molecules of chlorophyll and of various proteins in the growing membranes.

Synchronous growth in alternating periods of 12 hours of light and 12 hours of dark provides an opportunity to examine the specific times in the diurnal cycle when particular cell components are synthesized.

Chiang and Sueoka (p. 34) used this regime to examine the times of DNA replication and found replication of chloroplast DNA in the light period and of nuclear DNA in the dark. Schor *et al.* (*108*) used the same regime to examine the cyclical changes in chlorophyll concentration and photosynthetic activity. They found an increase in the ratios of cytochromes 559 + 553 to chlorophyll and in photosystem II activity during the first 3 hours of the light period, during which time no net increase in chlorophyll occurred. They also found a decrease in this ratio and in photosystem II activity during the dark period. The changes in photosystems I and II and in chlorophyll content per cell are shown in Fig. 8.13.

The authors conclude that these changes reflect the movement of cytochrome molecules in and out of existing lamellar membranes, as well as the incorporation of newly synthesized chlorophyll molecules into preexisting membranes. Of course lamellar membrane proteins and lipids are also being synthesized during the light period, so many other components in addition to chlorophyll are being incorporated into growing membranes at this time. Synchronous cultures provide

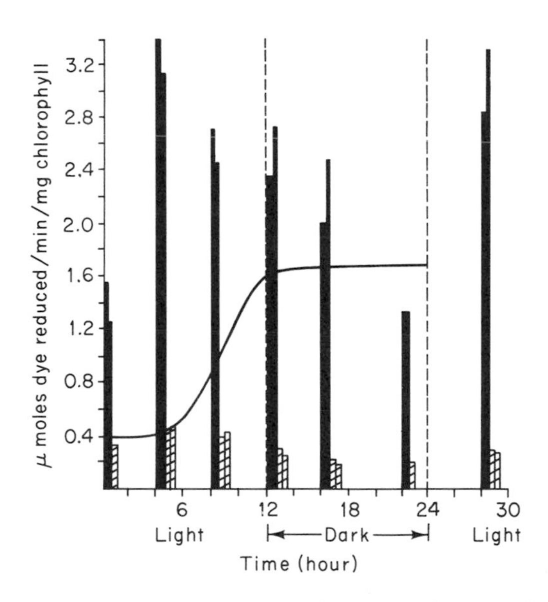

Fig. 8.13. Changes in photoreductive activity during synchronous development. Photosystem I specific activity is measured by the photoreduction of methyl red (striped bars), and Photosystem II specific activity by the photoreduction of dichloroindophenol (black bars). Each bar represents activity determinations made with a different disrupted cell suspension. Chlorophyll (μg/ml) ——. From Schor *et al.* (*108*).

admirable material for the further study of membrane biogenesis under conditions far less drastic than the greening process.

The lamellar membranes themselves show remarkable changes in arrangement during the cell cycle. During the 12 hour dark period, the lamellar discs are tightly stacked into grana, as shown in Plate XIV. At the beginning of the following light period, after cell division, the grana are dissociated and the lamellar discs are found mainly in pairs, as shown in Plate XIII. Numerous regions can be seen in which the paired discs show local separations at the time when cytochromes and chlorophyll are being introduced into the membranes, that is in the period of 3–6 hours in the light. After 6 hours, the discs begin to stack up, and the number of discs per granum increases progressively until the next light period when the number is again reduced. It is important to note that no destruction of membranes occurs during the cycle, but only redistribution and growth. These findings are compatible with the hypothesis that new components are being introduced throughout the preexisting lamellae and that the rearrangement process serves to open up and make available the entire extent of membrane surface for insertion of newly synthesized molecules.

It is difficult to proceed further along these lines either conceptually or experimentally without considering the proteins that constitute some 50% of the membranes. The stumbling block here has been the absence of suitable methods for membrane fractionation and characterization of the individual proteins, a difficulty already discussed with reference to mitochondrial membranes in Chapter 7. Recently, the first attempts to characterize lamellar membrane proteins by gel electrophoresis have been reported and will now be considered.

In these studies, lamellar membranes have been extracted from whole cells rather than from isolated chloroplasts; and purification has depended upon the unique density and banding properties of lamellar membrane fragments in continuous and discontinuous gradients. These procedures, while adequate for examination of gross differences in composition during greening, will not be satisfactory in the future as more precise questions are posed and minor components are investigated. Thus, the need for improved methods of isolating intact chloroplasts remains an important technical problem in studies with this otherwise superior system.

The method of polyacrylamide gel electrophoresis separates proteins according to molecular weight. The special problem with membrane proteins is to solubilize them so that they will migrate in the gel. Solubilization is achieved by detergent action, usually with sodium lauryl sulfate (SLS) [sodium dodecyl sulfate (SDS)], and urea is used to dissociate polymeric proteins into monomers. The detergent treatment denatures the proteins so that they can no longer be identified enzymatically. A second approach to identification of the proteins in the gels involves gentler methods of fractionation and identification of components before solubilizing them for gel electrophoresis. In a third approach, gel patterns from two differently treated samples, e.g., dark-grown vs light-grown, antibiotic-treated vs control, mutant vs wild type are compared by labeling each with a different isotope, e.g., ^{3}H and ^{14}C. The pair of samples are then solubilized together. This latter method has provided some initial success with chloroplast preparations.

Gel patterns of lamellar proteins from *Chlamydomonas* have been described by Hoober (*51*) and Eytan and Ohad (*33*) following somewhat different methods of preparation. Both studies revealed some 15–20 peaks, with two or three major components positioned about 60% of the total distance from the origin. Eytan and Ohad referred to these peaks as "L." They followed the changes in gel pattern during the greening process, and found that L represented the major component which increased markedly during greening, in proportion to the increase in chlorophyll concentration. Incorporation of ^{3}H-acetate into membrane proteins was followed during greening and compared with incorporation during the same time period by cells growing in the dark. Another peak (peak 7) was noted in which the incorporation rates were about the same in light and in dark. Peak 7 was used as a baseline for evaluating the rate of light-induced incorporation into the L region. These results are shown in Fig. 8.14.

The influence of inhibitors of protein synthesis upon the labeling of the L peak was then examined. As previously reported (*53*), both chloramphenicol and cycloheximide inhibit synthesis of chloroplast proteins in *Chlamydomonas* to some extent. Eytan and Ohad found a high rate of incorporation of label into the L region in the presence of chloramphenicol (100 μg/ml), whereas other peaks were poorly labeled under these conditions. When the cells which had been greening in the presence of chloramphenicol were subsequently transferred to cycloheximide (0.2 μg/ml), very little further incorporation into the L peak occurred, but considerable counts were found in other peaks. These results are shown in Fig. 8.15.

Somewhat similar results were obtained by Hoober (*51*) who used preparative methods which gave the sharp peaks and excellent resolution shown in Fig. 8.16. He also used the greening process as the experimental condition to provide maximum synthesis of lamellar membranes and minimal synthesis of other cellular components. He used

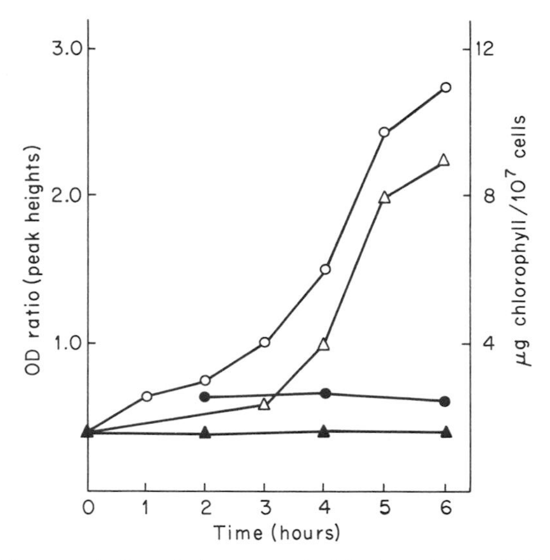

Fig. 8.14. Parallel increase in the relative amount of the L protein and chlorophyll synthesis during greening. Dark-grown cells were incubated in fresh growth medium either in the dark or light. Cells were fractionated on gradients and the membrane proteins from Fraction 5 processed for electrophoresis. Open circles, optical density ratio of Peaks L to 7 of samples incubated in the light; solid circles, optical density ratio of Peaks L to 7 of samples incubated in the dark. Open triangles, chlorophyll content of light-incubated samples; solid triangles, chlorophyll content of samples incubated in the dark. From Eytan and Ohad (33).

radioactive arginine to label the proteins and a double-labeling procedure to provide maximal precision in comparing the effects of chloramphenicol and cycloheximide on lamellar protein synthesis.

As shown in Fig. 8.16, dark-grown cells were allowed to green in a medium containing chloramphenicol (25 μg/ml), and after 4 hours ^{14}C-arginine was added to label the proteins being synthesized in the presence of the drug. After a further $4\frac{1}{2}$ hours, cells were washed free of drug and label and reincubated for 2 hours with cycloheximide (20 μg/ml) and ^{3}H-arginine. Then the cells were harvested, the membranes extracted and subjected to gel electrophoresis. The two large peaks seen in Fig. 8.16 containing ^{14}C but not ^{3}H probably correspond to the L region of Eytan and Ohad. Hoober found a few additional peaks that were labeled in the presence of chloramphenicol and a few labeled only in the presence of cycloheximide, as well as several that were labeled in both regimes. Using molecular weight markers in the gel, he estimated the large peaks labeled in the presence of chloramphenicol at approximately 22,000 daltons. The peaks labeled in the presence of

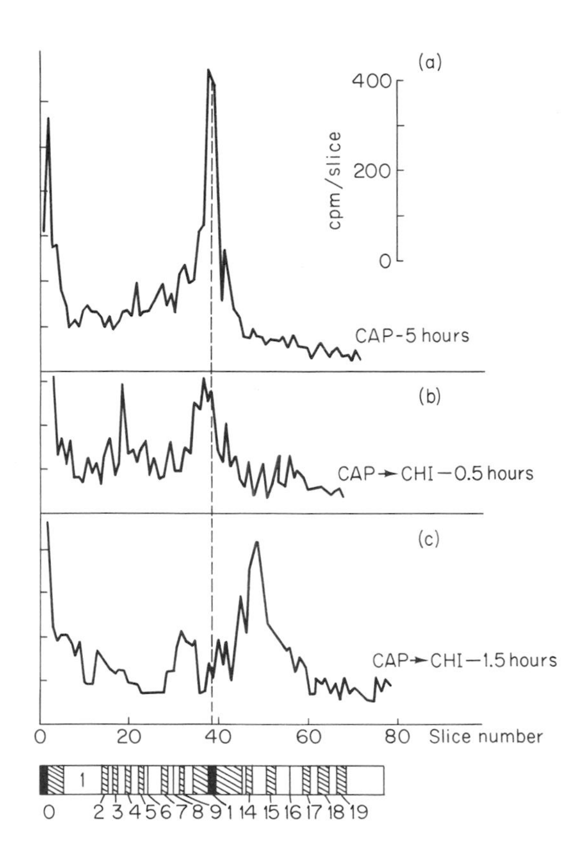

Fig. 8.15. Pulse labeling of membrane proteins with ^{3}H-acetate in cells incubated first in the presence of chloramphenicol (CAP), then transferred to cycloheximide (CHI). Dark-grown cells were incubated in the light in fresh growth medium containing 100 μg/ml of chloramphenicol for 5 hours. The cells were washed by centrifugation and then transferred to fresh medium containing 0.2 μg/ml of cycloheximide and further incubated in the light. Samples of cells were pulse-labeled for 20 minutes with ^{3}H-acetate (20 μCi/μmole, 5×10^{-4} *M*) before transfer to cycloheximide and $\frac{1}{2}$ hour and $1\frac{1}{2}$ hours after transfer. a, b, and c, labeling before the transfer and after $\frac{1}{2}$ and $1\frac{1}{2}$ hours after transfer of cells from chloramphenicol to cycloheximide, respectively. The chlorophyll content at the time of transfer from chloramphenicol to cycloheximide was 5.3 μg/10^7 cells and represented a 50% inhibition as compared to a control system incubated in the absence of chloramphenicol. Notice the progressive reduction of incorporation in the L peak (b and c) after transfer to cycloheximide as compared with the high incorporation in this peak in the presence of chloramphenicol (a). Immediately after transfer to cycloheximide, a high incorporation is found in the region containing Peaks 3, 4, 10, and 11 (b) and later in the region of Peaks 8, 9, 10, and especially 13 to 15 (c). From Eytan and Ohad (33).

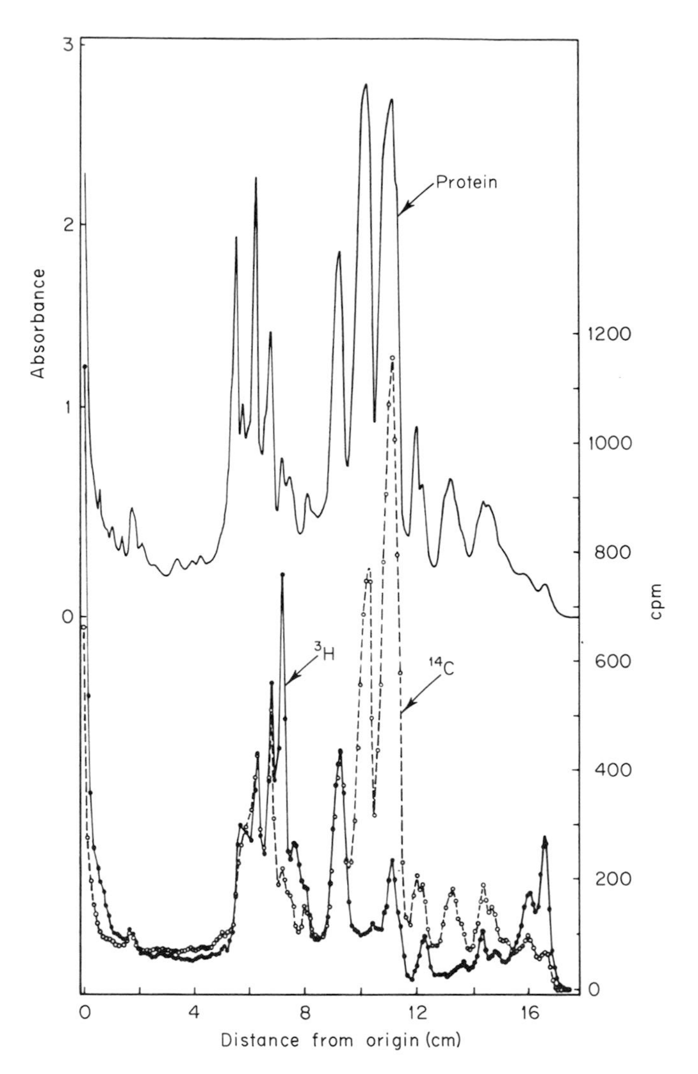

Fig. 8.16. Electrophoretic analysis of the incorporation into disc membrane polypeptides of ^{14}C-arginine in the presence of chloramphenicol and ^{3}H-arginine in the presence of cycloheximide. Etiolated cells of *C. reinhardi* were suspended to 6×10^6 cells/ml in medium containing chloramphenicol (25 μg/ml) and exposed to light. At 4 hours, ^{14}C-arginine (uniformly labeled, 237 mCi/mmole) was added to the medium to a final concentration of 0.04 μCi/ml. After $8\frac{1}{2}$ hours of greening, cells were removed from medium containing chloramphenicol and ^{14}C-arginine, and suspended in medium containing cycloheximide (20 μg/ml). ^{3}H-Arginine (7.32 Ci/mmole) was added to a final concentration of 2.2 μCi/ml. The cells were returned to light for 2 hours, and the membranes were then prepared. About 250 μg of the membrane protein were subjected to electrophoresis at 1.8 V/cm for 30 minutes and 6 V/cm for 7 hours, After staining, the gel was scanned at 540 nm, sliced into 1 mm sections and counted. The scintillation counter settings were optimized for double-label counting, and the results were corrected for spillover and background. ——, protein stain; ○- - -○, ^{14}C; and ●——●, ^{3}H. From Hoober (*51*).

cycloheximide included one of approximately 45,000 daltons and two very light peaks of less than 12,000 molecular weight.

The authors of these two papers (*33, 51*) conclude that some chloroplast lamellar membrane proteins are synthesized on cytoplasmic ribosomes, i.e., those labeled in the presence of chloramphenicol, and that some are synthesized on chloroplast ribosomes, i.e., those labeled in the presence of cycloheximide. It is evident from both studies that the labeling patterns in the presence of the two inhibitors are different and that the major components seen in both studies incorporated label preferentially in the presence of chloramphenicol. However, it is not shown in either study that the concentrations of chloramphenicol used [25 μg/ml (*51*) and 100 μg/ml (*33*)] fully abolished protein synthesis on chloroplast ribosomes. Indeed, more peaks showed incorporation in the presence of 25 μg/ml (*51*) than of 100 μg/ml (*33*).

Evidence that *some* proteins were blocked in synthesis by chloramphenicol at 100 μg/ml was provided by an assay of proton pump activity (*33*), an indicator of intactness of the electron transfer system as well as of energy coupling to it. Lamellae synthesized in the presence of chloramphenicol had almost no activity, whereas a subsequent incubation in the presence of cycloheximide restored 88% of control activity. This evidence does not establish whether 100 μg/ml is a high enough concentration of chloramphenicol to block all synthesis on chloroplast ribosomes.

One of the principal difficulties with *in vivo* inhibitor studies is that internal concentrations are unknown and may not be the same at all sites, especially in a cell full of internal membrane barriers. If inhibitor concentrations are too low, synthesis of some proteins may be blocked preferentially; if concentrations are too high, secondary effects of the drug may complicate the picture. In this regard, chloramphenicol is particularly difficult to gauge, since higher concentrations have been found necessary to block synthesis of some proteins than of others in bacteria (*146*).

The results with cycloheximide are somewhat easier to interpret than those with chloramphenicol. Cycloheximide is a very potent inhibitor of cytoplasmic ribosomes, and the concentration of 15 μg/ml used by Hoober is probably very effective in blocking protein synthesis on cytoplasmic ribosomes. Eytan and Ohad (*33*), using only 0.2 μg/ml cycloheximide, found incorporation into more peaks after shifting from chloramphenicol to cycloheximide than did Hoober (*51*) using 15 μg/ml of the drug.

In conclusion, these studies are important in pioneering the use of gel electrophoresis to examine chloroplast lamellar membrane formation.

Both studies report different patterns of incorporation of radioisotope label in the presence of chloramphenicol and cycloheximide. The results suggest, but do not establish, the synthesis of some lamellar proteins on cytoplasmic ribosomes and of others on chloroplast ribosomes.

The similarities in gel patterns seen in the two investigations are of particular interest since the methods of preparation are quite different. More recently, similar gel patterns were reported (*91*) in a study of lamellar proteins from three higher plants: wheat, barley, and spinach. Only ten peaks were clearly resolved in this study, but the overall pattern shown in Fig. 8.17 resembles those of Fig. 8.15 and 8.16. In an attempt to identify some of the peaks, lamellae were treated to give a fraction enriched in chlorophyll a, and presumably in photosystem I activity, and a second fraction enriched in chlorophyll b, and presumably in photosystem II activity. The results are shown in Fig. 8.18.

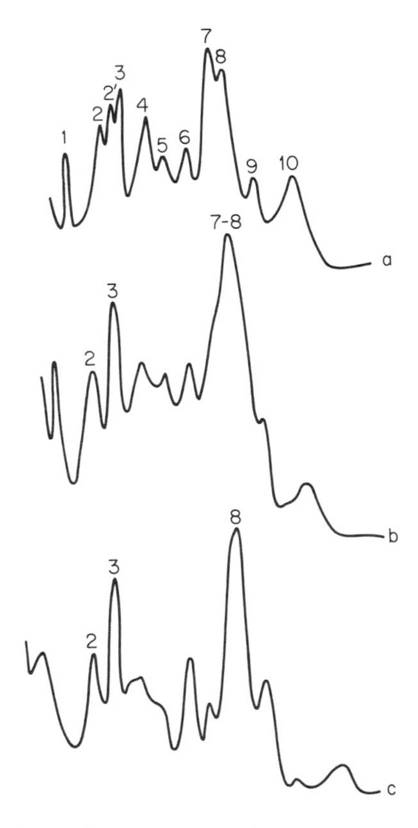

Fig. 8.17. Polyacrylamide gel electrophoresis of lamellar proteins. Densitometric tracings at 600 nm. Migration was from the origin (left) to the anode (right). (a) Spinach chloroplasts, (b) barley chloroplast, (c) wheat chloroplasts. From Remy (*91*).

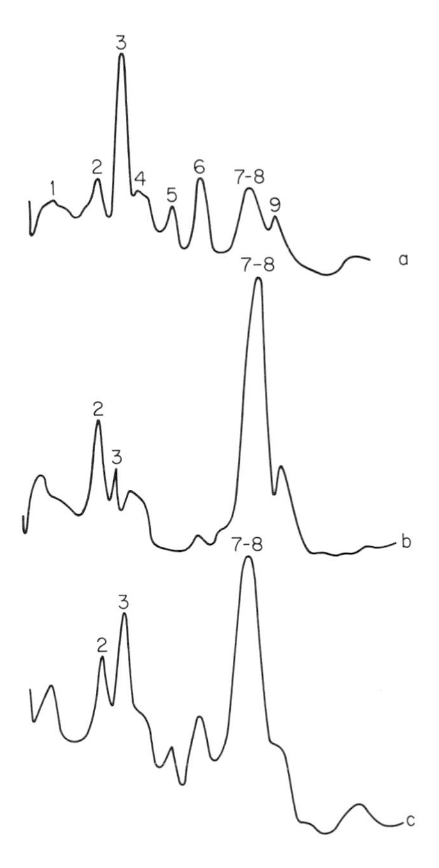

Fig. 8.18. Densitometric tracings of gel electrophoresis of: (a) light fraction (enriched in chlorophyll a), (b) heavy fraction (enriched in chlorophyll b), (c) mixture of light and heavy fraction. See text for further discussion. From Remy (*91*).

Levine and his collaborators have recently used antibiotics, as well as nuclear gene mutations, in *Chlamydomonas* in an effort to distinguish the genetic origins and sites of synthesis of chloroplast proteins (*70, 134*). In previous studies recently reviewed (*70*), a series of nonphotosynthetic acetate-requiring mutant strains were utilized in studies of the electron transfer pathway of photosynthesis. Some of the mutants were shown to be blocked at particular steps in the pathway, but in no instance was the primary effect of the mutation identified. The pleiotropic effects observed are reminiscent of many of the nuclear genes in yeast which influence mitochondrial development and function, as discussed in Chapter 7.

Pleiotropic effects of one nuclear gene mutation, *ac-20*, are of particular interest in relation to chloroplast development. Levine and Togasaki

(*70a*) originally reported that the *ac-20* mutation caused loss of RuDP-carboxylase activity. This was a surprising conclusion to draw since the mutation is "leaky," i.e., permits slow growth in the absence of an added carbon source. Subsequently it was found that photosynthetically grown cells carrying the *ac-20* mutation contained at least 25% of the wild-type activity of this enzyme (*134*).

The most striking difference between mutant cells grown with and without acetate was then shown to be the content of chloroplast ribosomes: acetate-grown mutant cells have only about 8% of the wild-type amount, whereas photosynthetically grown mutant cells contain 25% or more compared with wild-type controls (*37*). The authors propose that all of the pleiotropic effects of the *ac-20* mutation noted in Table 8.10 may be viewed as consequences of the virtual absence of chloroplast ribosomes. Consistent with this conclusion is the fact that the photosynthetically grown mutant is either intermediate in phenotype or normal with respect to all the affected properties listed in Table 8.10.

The primary effect of the *ac-20* mutation is unknown, but it appears to involve nuclear regulation of chloroplast ribosome formation via a

TABLE 8.10

Chloroplast Components Affected, Partially Affected, and Not Affected in Heterotrophic ac-20[a]

Affected	Partially affected	Not affected
RuDP-carboxylase	Chlorophyll (reduced at most by half)	Phosphoribulokinase
Cytochrome 559	Carotenoid (reduced by half)	phosphoriboisomerase
Cytochrome Q	Membrane formation (reduced by half)	3-PGA kinase
Chloroplast membrane organization		G-3-P dehydrogenase (NAD)
Pyrenoid formation		G-3-P dehydrodenase (NADH)
		Triosephosphate isomerase
		FDP aldolase
		Total quinone
		Plastocyanin
		Cytochrome 553
		Cytochrome 564
		P700
		Ferredoxin
		Ferredoxin-NADP reductase
		Eyespot formation
		Starch synthesis

[a] From Goodenough and Levine (*37*).

pathway influenced by feedback from acetate metabolism. The mutant is especially valuable in studies of chloroplast biogenesis because of the possibility it affords to examine the functions of chloroplast ribosomes. The results so far reported lead to the conclusion that remarkably few of the known photosynthetic enzymes are dependent on chloroplast ribosomes for their synthesis. Table 8.10 lists only RuDP-carboxylase, cytochrome 559, cytochrome Q, and unknown components required for stacking of lamellar membranes and for pyrenoid formation.

The evidence from *ac-20* that the carboxylase is synthesized on chloroplast ribosomes is in line with the results of inhibitor studies with *Chlamydomonas* and *Euglena* (see below), as well as with the identification of Fraction I protein in the *in vitro* experiments of Spencer *et al.* (*125*) (cf. p. 311). Since the carboxylase enzyme is present in such a huge amount in normal chloroplasts, it is not surprising that any reduction in ribosome content would be reflected in decreased enzyme synthesis. On the other hand, proteins present in very small amounts might be synthesized on the residual 8% of ribosomes still formed in the heterotrophically grown mutant. Thus, it is not excluded that some of the components listed as "not affected" in Table 8.10 may in fact be synthesized in the chloroplast. It will be particularly interesting to compare the membrane proteins of heterotrophic *ac-20* with the wild type by the method of gel electrophoresis, particularly as the method becomes increasingly sensitive and precise.

Attempts to compare the effects of the *ac-20* mutation with the results of growth of wild-type cells in the presence of various antibiotics (*134*) have led to some confusing results. Because the reports are preliminary and few details have yet been published, these studies will only be briefly noted here.

In short-term experiments of a few hours' duration with synchronous cultures, spectinomycin was found to inhibit the normal doubling of RuDP-carboxylase activity and of photosystem II activity but not of photosystem I activity in the light phase of the diurnal cycle. In long-term experiments, spectinomycin was found not to affect chloroplast ribosome formation at concentrations that did block development of photosystem II activity. Whether or not chloroplast protein synthesis was fully blocked by spectinomycin under these experimental conditions was not established.

The division of labor between chloroplast and mitochondrial genomes remains totally puzzling. *Chlamydomonas* would seem to provide the best experimental system for distinguishing mutations arising in either genome, as well as for examining their phenotypic effects. To date, all non-Mendelian mutations to drug resistance that have been mapped (cf. Chapter 3) are located within the same cytoplasmic linkage group. Why do mutations in the DNA of one organelle lead to phenotypic consequences in both?

Surzycki and Gillham (*133*) have developed a novel procedure, using rifamycin, in an attempt to distinguish between mutations in chloroplast and mitochondrial DNA. This drug reportedly inhibits initiation of transcription of chloroplast but not of mitochondrial DNA. They assumed that the presence of rifamycin would decrease resistance to other antibiotics if that resistance depended on transcription of chloroplast DNA, but not if it depended on transcription of mitochondrial (or nuclear) DNA. Their results are unfortunately not interpretable in any clear fashion, indicating that complex inhibitors cannot be used in this way. In our hands (*101*) different cytoplasmic mutations to drug resistance showing close genetic linkage did not respond uniformly in rifamycin experiments. It seems likely that unraveling the interactions of organelle genomes will require much more direct studies of mutations and their phenotypic consequences than have yet been reported.

The Origin of Chloroplast Proteins in Euglena. The biology of greening in *Euglena* differs sharply from that in *Chlamydomonas* in that wild-type *Euglena* does not form chlorophyll or well-developed plastids when grown in the dark. Dark-grown cells contain small proplastids about 1μ in diameter which contain protochlorophyll and DNA but no chlorophyll, few lamellae, and few ribosomes. In contrast to *Chlamydomonas,* the dark-grown *Euglena* cells have no detectable cytochrome 552 (f) activity, nor do most of the carbon cycle enzymes show any activity. An exception is RuDP-carboxylase, which is synthesized in the dark at a rate that maintains its activity at about 15% that of the light-grown cell. As previously discussed, the lipids of the dark-grown *Euglena* are predominantly those characteristic of mitochondrial membranes (cf. p. 293). When *Euglena* is transferred from dark to light, its lipid composition changes rapidly with the appearance of characteristic chloroplast components.

The only controlling element in the entire greening process which is clearly identified in *Euglena* and *Chlamydomonas* is light itself. What does light do? In all algae and higher plants which have been examined, the action spectrum for chlorophyll formation corresponds to the absorption spectrum of protochlorophyll. Thus, light converts protochlorophyll to chlorophyll (cf., p. 290), and apparently this conversion triggers the subsequent sets of events.

In a normal growth medium, the greening process in *Euglena* is accompanied by unrestricted cell growth and cell division. In *Chlamydomonas,* it may be recalled, the light-induced appearance of chlorophyll not only triggers the formation of lamellar membranes but also represses other cellular activities, such as synthesis of cytoplasmic macromolecules until an advanced stage of chloroplast organization has been achieved.

To study chloroplast development in *Euglena*, therefore, it was desirable to find a condition under which other aspects of growth were inhibited. One such condition was developed by Brawerman and Chargaff (*17*) who showed that *Euglena* could carry out the full transformation from proplastid to chloroplast in a "resting medium" with no exogenous carbon or nitrogen source. Under these conditions, the cells do not divide, and chloroplast formation occurs with no net synthesis of either protein or RNA but with a massive amount of turnover. These conditions have subsequently been used by most investigators of chloroplast development in *Euglena*.

Several investigators have used the antibiotics, cycloheximide and chloramphenicol, during the greening process in an attempt to assess the sites of synthesis of various chloroplast proteins. Smillie *et al.* (*120*) found that chloramphenicol (1 mg/ml) inhibited the formation of RuDP-carboxylase and glyceraldehyde-3-phosphate dehydrogenase, as shown in Table 8.11, but that cycloheximide (15 μg/ml) had no effect on the dehydrogenase synthesis and actually doubled the amount of carboxylase synthesized during 72 hours in the light. Smillie *et al.* (*120*) showed further that cycloheximide at only 5 μg/ml was effective in blocking synthesis of cytoplasmic proteins, as well as blocking cell division and uptake of glucose.

TABLE 8.11

Action of Chloramphenicol and Cycloheximide on the Synthesis In Vivo of Enzymes of the Calvin Cycle[a]

Conditions	Increase during illumination for 72 hours: Ribulose-1,5-diphosphate carboxylase (μmoles substrate/min/10^9 cells)	Increase during illumination for 72 hours: NADP$^+$-glyceraldehyde-3-phosphate dehydrogenase (μmoles substrate/min/10^9 cells)
Control	0.62	13.0
Chloramphenicol	0.07	−0.6
Cycloheximide	1.34	15.3
% Inhibition with chloramphenicol	89	100
Cycloheximide treatment as % of control	216	118

[a] From Smillie *et al.* (*119*).

It was also shown that chloramphenicol inhibited the formation of certain photosynthetic electron transfer proteins in *Euglena*: cytochrome 552 (f), a b-type cytochrome, and ferredoxin-NADP$^+$-reductase.

The difficulty in interpreting results of inhibitor studies was clearly demonstrated by Smillie *et al.* (*119*), with the following experiment. They found that cycloheximide inhibited chlorophyll synthesis during greening, not at the low concentrations of 2–5 μg/ml effective for blocking synthesis of many cytoplasmic proteins, but at 15 μg/ml. However, this high concentration was ineffective in blocking chlorophyll synthesis if greening cells were first exposed to chloramphenicol and subsequently transferred to cycloheximide, in experiments similar to those of Hoober (*51*) and Eytan and Ohad (*33*) with *Chlamydomonas* (p. 321).

A comprehensive study of chloroplast development in *Euglena* carried out by Schiff and Epstein and their students has recently been summarized (*106*). This work has included investigations of structural and biochemical changes accompanying greening in normal strains of *Euglena*, and a detailed examination of a special phenomenon: the loss of chloroplast forming ability by some strains of *Euglena* after exposure to low and nonlethal amounts of UV light. These pioneering studies have provided a wealth of information about *Euglena*, especially with respect to the greening process. Unfortunately, the lack of sexuality in this organism has severely limited the kinds of experiments and interpretations of particular relevance to genetic control of chloroplast development. The application of target theory to the UV irradiation data indicated the presence of about thirty to forty genetic sites involved in chloroplast development. It now seems likely that these sites, described by indirect radiological evidence, correspond to chloroplast DNA.

Origin of Chloroplast Proteins in Higher Plants

Evidence from Greening Studies. Chloroplast development in higher plants differs from that in the green algae in its regulation not only by protochlorophyll but also by the red–far red phytochrome system (*12*, *143*). Since few membrane proteins have yet been identified in the lamellae of either group of organisms, little can be said about their similarities from direct evidence. The basic similarities in photosynthetic pathways (*87*) and in lamellar structure, as seen in electron micrographs (*38*), suggest that the fundamental molecular organization is similar in algae and higher plants, and the similar gel patterns seen in Figs. 8.15–8.18 further support this view.

Most studies of the greening process in higher plants have been carried out with dark-grown etiolated seedlings. Investigators have taken advantage of the ability of seeds to germinate and seedlings to develop for considerable time in the dark utilizing nutrients stored in the seeds. Under these circumstances, the proplastids of the meristematic regions differentiate into structures which are a quarter to one half the size of green chloroplasts, and which are called etioplasts.

By the time seedlings have grown a week or more in the dark, they generally have long slender yellow leaves and are well stocked with etioplasts. For some purposes they provide admirable material for research. However, they are abnormal structures resulting from unbalanced growth, and conclusions based upon studies with etioplasts must be evaluated critically before assuming applicability to the normal condition.

A prime example is the prolamellar body (*41*) shown in Plate XV, a paracrystalline array or lattice seen in etioplasts, in chloroplasts which develop in plants grown at low light intensities, and in certain mutants blocked in normal plastid development. It contains the holochrome protein, and electron microscope studies indicate that it is used in the construction of lamellar membranes. Stages in the development of a normal green chloroplast from an etioplast are shown in Plate XV. Whether the composition of membranes formed from the prolamellar body is the same as that of membranes formed in the normal greening process is not known.

The greening process in higher plants has been examined by many investigators with respect to structural changes, development of photosynthetic ability, appearance of pigments and other lipids, enzymes, etc. (*11, 12, 68*). Recently, much of our knowledge about the normal greening process has been confirmed and extended by studies of nuclear gene mutations affecting chloroplast development. In particular, studies of mutations at some 86 different gene loci in barley, each influencing chloroplast formation, have been collected and studied by von Wettstein and his colleagues, and discussed in a recent review (*143*).

Our present understanding of the overall greening process can most usefully be considered with reference to these barley mutations. The position of the mutant block in the greening of some of the mutant strains is shown diagrammatically in Fig. 8.19, and the changes in membrane organization of five nuclear gene mutants are shown in Plate XVI. In this study, electron microscopic observation of structural changes were correlated with spectrophotometric measurements of pigment absorption and fluorescence, and with biochemical studies of changes in lipid composition of the developing plastids. Virtually no specific information is yet available about lamellar membrane proteins synthesized during greening in higher plants, but with the availability

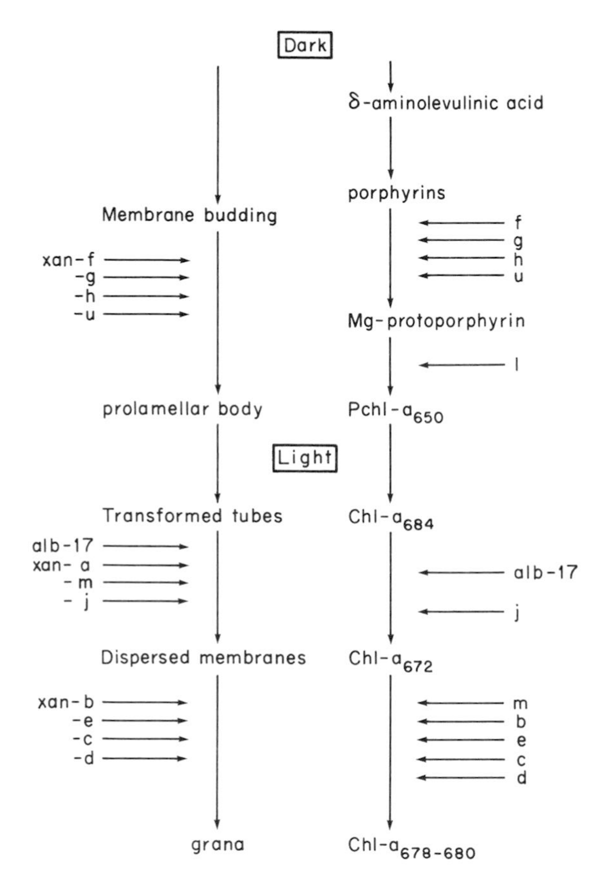

Fig. 8.19. Diagram of the lesions in chloroplast biogenesis (left) and chlorophyll biosynthesis (right) caused by mutations in twelve *xantha* genes and one *albina* gene. From von Wettstein *et al.* (*143*).

of new methods such studies should become increasingly popular.

The following brief account of the greening process is based upon numerous studies, many of them recently reviewed (*11*).

Phase One: The light-induced conversion of preformed protochlorophyllide to chlorophyllide, and tube transformation of the prolamellar body. The conversion of protochlorophyllide to chlorophyllide, studied primarily with etiolated seedlings, has now been shown also to occur in normal development, not only in the initial stage, but also during the rapid subsequent synthesis of chlorophyll in the light (*11a*). Protochlorophyllide itself is the photoreceptor in this process. Chlorophyllide is converted to chlorophyll by the addition of the phytyl side chain in a dark reaction soon after its light-induced origin from protochlorophyllide (*11a*).

No mutant blocked in the photoconversion of protochlorophyllide to chlorophyllide has yet been found (*143*). It may be that the photoconversion step is not under separate enzymatic control, but is a property of the holochrome protein itself (cf. p. 290). If so, the possibility that the holochrome protein is coded by chloroplast rather than nuclear DNA seems worth considering.

Illumination of etioplasts also leads to the rapid dissociation of the prolamellar body. If plants are returned to dark after a short exposure to light, the lattice is reaggregated.

Phase Two: Dispersion of prolamellar body. Following the initial conversion of preformed protochlorophyllide to chlorophyll, there is a lag in further chlorophyll synthesis whether seedings are in the light or in the dark. During the early part of this lag phase, the prolamellar body is dispersed into perforated sheets, which arrange themselves into primary layers (Plate XV). At this time, the first spectral shift in the absorption maximum of chlorophyll a from 683 nm to 672 nm occurs, as discussed earlier (p. 291). This shift may be associated with a transfer of chlorophyll from the holochrome to a different lamellar protein. Several nuclear genes have been associated with this phase of development, as shown in Figs. 8.19 and 8.20. The lag phase in chlorophyll synthesis is regulated by the phytochrome system.

Phase Three: Grana formation. The dispersed perforated sheets are converted into the lamellar discs of the developing grana. The process is influenced by blue light with a maximum at 450 nm, but the receptor pigment has not been identified. During this process, chlorophyll a undergoes a further spectral shift to 678–680 nm, and rapid synthesis of chlorophyll occurs. Numerous *xantha* genes have been shown to block steps in this phase of development.

It seems unlikely that seedlings of higher plants will ever provide as satisfactory material as synchronous cultures of sexual green algae for the detailed analysis of chloroplast biogenesis at the macromolecular level. However, higher plants are of great interest in their own right, both as our primary source of food and as complex organisms with more regulatory systems (such as phytochrome and various plant hormones) than the algae. Furthermore, the plants, such as *Oenothera,* used for studies of chloroplast inheritance since the 1930's, provide a stockpile of mutant material for investigation at the biochemical and macromolecular level.

Evidence from Cytoplasmic Mutants. The genus *Oenothera* has been investigated by geneticists interested in cytoplasmic heredity in higher

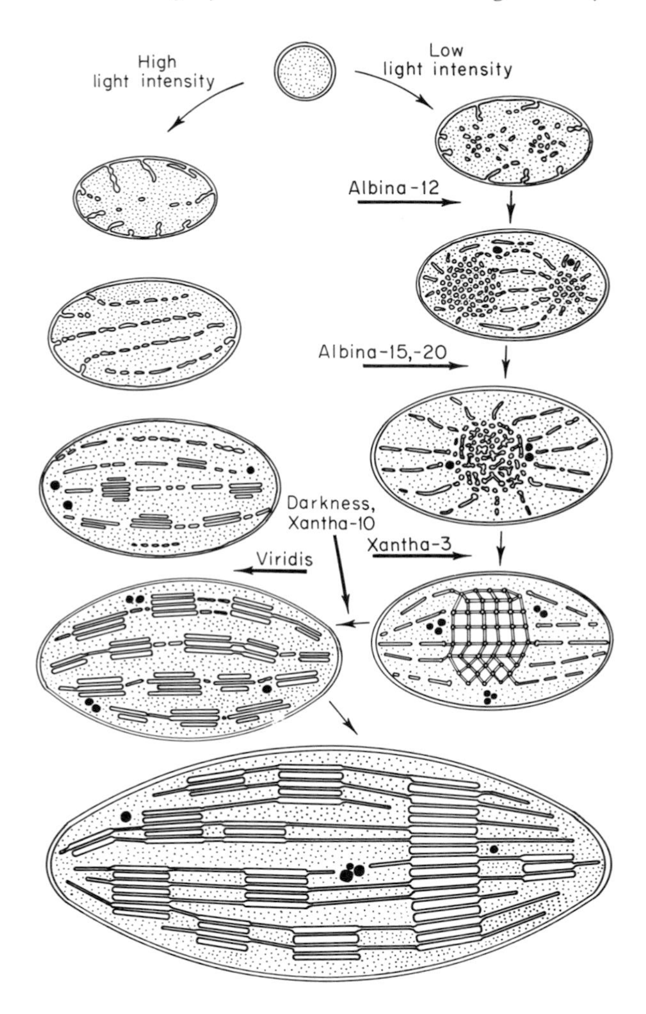

Fig. 8.20. Diagram of the effects of light and of several mutant genes on chloroplast development in barley. Positions of arrows indicate stage at which development is blocked by particular mutants. See text and Plates XV and XVI for further details. From von Wettstein, unpublished.

plants, beginning with the pioneering studies of Renner (Chapter 6). In recent years, cytoplasmic mutant strains of *Oenothera* have been collected and characterized particularly by W. Stubbe (Chapter 6). This material is now being utilized in studies of the molecular basis of the mutant phenotypes by Hallier in collaboration with Stubbe, Heber, and others (*28b, 33a, 42, 43*). They have characterized four mutant strains of identical nuclear genotypes but with different plastid genomes. Two of the mutants, Iγ and Iδ, contain mutant plastids from the *hookeri* parent, and the other two, IIα and IIγ, contain mutant plastids from a related strain, *suaveolens*. The genetic alterations in the mutant plastids interfere with photosynthetic function, and therefore the mutations are lethal; seedlings die when they have used up their stored reserves.

Hallier *et al.* (*42, 43*) compared the four mutant strains with the normal *Oenothera hookeri* control in an attempt to pinpoint the position of the mutant block or defect leading to lethality in each of the mutants. In an electron microscope study, they found little if any structural difference between the mutants and the normal plants examined as very young seedlings just after germination. The photosynthetic capacities of normal and mutant strains were compared in a series of tests. Oxygen evolution, taken as an overall index of photosynthetic activity, was found to be grossly deficient in all four mutants, as shown in Fig. 8.21. In the normal plant, photosynthetic oxygen evolution in the light was countered by oxygen uptake due to respiration in the dark. Two of the mutants, Iγ and IIγ, took up oxygen at the same rate in light and dark, showing no photosynthetic response at all, while the other two, Iδ and IIα, showed a small response to light, but not enough for photosynthetic oxygen evolution to balance respiration. Thus the four mutants are clearly deficient in photosynthetic capacity. Further studies were then undertaken to examine particular steps in the process.

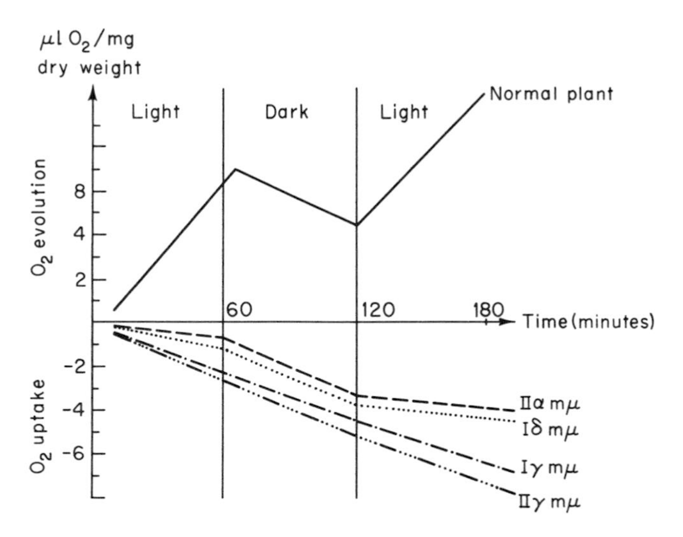

Fig. 8.21. Oxygen exchange reactions of normal and four cytoplasmic mutants of *Oenothera* under light and dark conditions. The four mutant strains are further described in the text. From Hallier (*42*).

They examined distribution of ^{14}C from photosynthetically assimilated $^{14}CO_2$ to survey the pathway of carbon during its assimilation into various sugars. In this study, the four mutants gave similar results: a greatly reduced overall rate of incorporation 100–200 times less than that of normal plants, but no accumulation of intermediates. Thus, the mutants seem to possess all of the enzymes of the carbon cycle, especially since all of them accumulate sucrose, an end product of the pathway. Further detailed studies of several individual enzymes showed no differences from the normal in *in vitro* tests of activity.

The electron transport pathways were then examined. The rates of ATP and NADPH production were found to be similar in normal and mutant plants, but the mutants were found to be very light sensitive, dying faster at high light intensities than at low ones. It was concluded that the position of the mutant block in all four strains was probably in photosystem II, and that each mutant might be blocked at a different point in photosystem II. More recently, a fifth mutant with a block in photosystem I has been described (*28b, 33a*). Further studies of these mutants will be of great interest. Examination of lamellar membrane proteins by one of the new techniques may reveal protein differences between mutant and normal strains.

CONCLUDING REMARKS

The chloroplast is a complex organelle consisting of lamellar discs, the site of photosynthesis and photophosphorylation, sometimes arranged in stacks or grana and sometimes free in the matrix or stroma, surrounded by a double envelope, the outer membrane of which may be similar to the endoplasmic reticulum of the cytoplasm. DNA is present in the matrix as well as are chloroplast ribosomes often found as polysomes, and thereby indicating the presence of messenger RNA's. Starch is stored in chloroplasts, and in the algae starch synthesis is often associated with a special structure, the pyrenoid. Some algal chloroplasts also contain a light-receptor region, the eyespot, which regulates phototactic responses of the organism.

Known chloroplast functions include not only photophosphorylation and energy transfer associated with photosynthesis, but also CO_2 reduction, starch formation, transcription of chloroplast DNA, protein synthesis, and the various other enzymatic activities listed in Table 8.1. The list is far from complete. We may anticipate the identification of many more enzymes and other components involved in biosynthetic and transport functions, as well as new components in the processes of photosynthesis and photophosphorylation.

Our understanding of the organization of molecular constituents in the lamellar membranes is still meager, but strides are now being made, particularly with the aid of integrated biochemical–electron microscope studies. Evidence is mounting that the lamellae are asymmetrical, with components of photosystem II oriented toward the interior of the disc, and those of photosystem I, as well as the coupling factor CF_1 for phosphorylation, oriented outward toward the matrix. Chloroplast lamellae contain unique lipids found nowhere else, but neither their structural nor functional significance is understood.

The evidence available suggests that chloroplast membranes, like those of mitochondria, do not have a fixed stoichiometry, but are multicomponent lipoprotein structures in which new components can be inserted into preexisting membranes.

Chloroplast DNA, with a genomic size (Chapter 2) of $1–2 \times 10^8$ daltons, contains much more informational capacity than does any mitochondrial DNA yet known. Allowing for ribosomal and transfer RNA cistrons, this DNA could code for at least 200 different proteins. Thus, it is evident that we have hardly a clue to the coding functions of chloroplast DNA.

The origin of chloroplast proteins, studied with the use of inhibitors in wild-type cells, appears to lie principally in the cytoplasm. The only proteins which seem to be synthesized within the chloroplast, judging from studies with *Chlamydomonas, Euglena,* and higher plants, are RuDP-carboxylase, one of the cytochromes, some components of photosystem II, and proteins involved in the stacking of lamellar discs and in pyrenoid formation (in *Chlamydomonas*). The huge discrepancy between these meager findings and the presence of coding potential for some 200 proteins emphasizes the need for a more powerful experimental attack upon this problem.

It is likely that chloroplast genes are directly involved in the production of key lamellar proteins, in particular membrane proteins requiring immediate incorporation into a preexisting structure. This possibility will be further discussed in Chapter 9. Here it may be noted that the few cytoplasmic mutants that have been examined thus far show alterations either in photosynthetic electron transport (*Oenothera*) or in ribosome structure, as suggested by altered sensitivity to inhibitors of chloroplast protein synthesis (*Chlamydomonas*). The rapid advances now being made in fractionation and identification of membrane proteins will provide new tools for the examination of the consequences of mutations in chloroplast DNA, and the advances in our understanding of chloroplast functions should provide new tests of functional abnormalities in these mutants.

Thus in the words of an unknown sage: "The future lies ahead." Both the availability of new chloroplast mutations and of new methods for their biochemical and biophysical investigation should lead to great strides in our understanding of chloroplast biogenesis, structure, and function in the next few years.

Suggested Review Articles

Boardman, N. K., Linnane, A. W., and Smillie, R. M., eds. (1971). "Autonomy and Biogenesis of Mitochondria and Chloroplasts," *Aust. Acad. Sci. Symp.* North-Holland Publ., Amsterdam.

Goodwin, T. W., ed. (1967). "Biochemistry of Chloroplasts," Vols. I and II. Academic Press, New York.

Kirk, J. T. O. (1971). Chloroplast structure and biogenesis. *Annu. Rev. Biochem.* **21,** 11.

Kirk, J. T. O., and Tilney-Bassett, R. A. E., eds. (1966). "The Plastids," pts. I, III, and IV. W. H. Freeman, London.

Miller, P. L., ed. (1970). "Control of Organelle Development," *Soc. Exp. Biol. Symp.*, Vol. 24. Cambridge Univ. Press, London.

References

1. Anderson, J. M., and Boardman, N. K. (1966). Fractionation of the photochemical systems of photosynthesis. I. Chlorophyll contents and photochemical activities of particles isolated from spinach chloroplasts. *Biochim. Biophys. Acta* **112,** 403.

1a. App, A. A., and Jagendorf, A. T. (1964). ^{14}C-amino acid incorporation by spinach chloroplast preparation. *Plant Physiol.* **39,** 772.

2. Arntzen, C. J., Dilley, R. A., and Crane, F. L. (1969). A comparison of chloroplast membrane surface visualized by freeze-etch and negative staining techniques; and ultrastructural characterization of membrane fractions obtained from digiton in-treated spinach chloroplasts. *J. Cell Biol.* **43,** 16.

3. Bennun, A., and Avron, M. (1964). Light-dependent and light-triggered adenosine triphosphatase in chloroplasts. *Biochim. Biophys. Acta* **79,** 646.

4. Benson, A. A. (1964). Plant membrane lipids. *Annu. Rev. Plant Physiol.* **15,** 1.

5. Benson, A. A., Gee, R. W., Ji, T-H., and Bowes, G. W. (1971). Lipid–protein interactions in chloroplast lamellar membranes as bases for reconstruction and biosynthesis. *In* "Autonomy and Biogenesis of Mitochondria, and Chloroplasts," *Aust. Acad. Sci. Symp.* (N. K. Boardman, A. W. Linnane, and R. M. Smillie, eds.). pp. 18–26. North-Holland Publ. Amsterdam.

6. Bird, I. F., Porter, H. K., and Stocking, C. R. (1965). Intracellular localization of enzymes associated with sucrose synthesis in leaves. *Biochim. Biophys. Acta* **100,** 366.

7. Bloch, K., Constantopoulos, G., Kenyon, C., and Nagai, J. (1967). Lipid metabolism of algae in the light and in the dark. *In* "Biochemistry of Chloroplasts" (T. W. Goodwin, ed.), Vol. II, p. 197. Academic Press, New York.

7a. Boardman, N. K. (1967). Chloroplast structure and development. *In* "Harvesting the Sun—Photosynthesis in Plant Life" (A. San Pietro, F. A. Greer, and T. J. Army, eds.), pp. 211–230. Academic Press, New York.

8. Boardman, N. K., and Anderson, J. M. (1967). Fractionation of the photochemical systems of photosynthesis. II. Cytochrome and carotenoid contents of particles isolated from spinach chloroplasts. *Biochim. Biophys. Acta* **143,** 187.

9. Boardman, N. K., Francki, R. I. B., and Wildman, S. G. (1965). Protein synthesis by cell-free extracts from tobacco leaves. II. Association of activity with chloroplasts ribosomes. *Biochemistry* **4,** 872.
10. Boardman, N. K., Francki, R. I. B., and Wildman, S. G. (1966). Protein synthesis by cell-free extracts of tobacco leaves. III. Comparison of the physical properties and protein synthesizing activities of 70 S chloroplast and 80 S cytoplasmic ribosomes. *J. Mol. Biol.* **17,** 470.
11. Boardman, N. K., Linnane, A. W., and Smillie, R. M. eds. (1971). "Autonomy and Biogenesis of Mitochondria and Chloroplasts," *Aust. Acad. Sci. Symp.* North-Holland Publ., Amsterdam.
11a. Boardman, N. K., Anderson, J. M., Kahn, A., Thorne, S. W., and Treffry, T. E. (1971). Formation of photosynthetic membranes during chloroplast development. *In* "Autonomy and Biogenesis of Mitochondria and Chloroplasts," *Aust. Acad. Sci. Symp.* (N. K. Boardman, A. W. Linnane, and R. M. Smillie, eds.), p. 70. North-Holland Publ., Amsterdam.
12. Bogorad, L. (1968). Control mechanisms in plastid development. *Develop. Biol., Suppl.* **1,** 1.
13. Bove, J., and Raacke, I. D. (1959). Amino acid-activating enzymes in isolated chloroplasts from spinach leaves. *Arch. Biochem.* **85,** 521.
14. Boynton, J. E., Gillham, N. W., and Burkholder, B. (1970). Mutations altering chloroplast ribosome phenotype in *Chlamydomonas.* II. A new Mendelian mutation. *Proc. Nat. Acad. Sci., U. S.* **67,** 1505.
15. Branton, D. (1969). Membrane structure. *Annu. Rev. Plant Physiol.* **20,** 209.
16. Branton, D., and Park, R. B. (1967). Subunits in chloroplast lamellae. *J. Ultrastruct. Res.* **19,** 283.
17. Brawerman, G., and Chargaff, E. (1959). Changes in protein and ribonucleic acid during the formation of chloroplasts in *Euglena gracilis. Biochim. Biophys. Acta* **31,** 164.
18. Brawerman, G., and Eisenstadt, J. M. (1967). The nucleic acids associated with chloroplasts of *Euglena gracilis* and their role in protein synthesis. *In* "Organizational Biosynthesis" (H. J. Vogel, J. O. Lampen, and V. Ryson, eds.), p. 419. Academic Press, New York.
18a. Buchanan, B. B., Kalberer, P. P., and Arnon, D. I. (1967). Ferredoxin-activated fructose diphosphatase in isolated chloroplasts. *Biochem. Biophys. Res. Commun.* **29,** 74.
19. Butler, W. L., and Briggs, W. R. (1966). The relation between structure and pigments during the first stages of proplastid greening. *Biochim. Biophys. Acta* **112,** 45.
20. Carell, E. F., and Kahn, J. S. (1964). Synthesis of porphyrins by isolated chloroplasts of *Euglena. Arch. Biochem. Biophys.* **108,** 1.
21. Chance, B., and Sager, R. (1957). Oxygen and light induced oxidations of cytochrome, flavoprotein, and pyridine nucleotide in a *Chlamydomonas* mutant. *Plant Physiol.* **32,** 548.
22. Chen, J. L., and Wildman, S. G. (1967). Functional chloroplast polyribosomes from tobacco leaves. *Science* **155,** 1271.
23. Clark, M. F. (1964). Polyribosomes from chloroplasts. *Biochim. Biophys. Acta* **91,** 671.
24. Clark, M. F., Matthews, R. E. F., and Ralph, R. K. (1964). Ribosomes and polyribosomes in *Brassica pekinenesis. Biochim. Biophys. Acta* **91,** 289.
25. Crane, F. L., Arntzen, C. J., Hall, J. D., Ruzicka, F. J., and Dilley, R. A. (1971). Binary membranes in mitochondria and chloroplasts. *In* "Autonomy and Biogenesis of Mitochondria and Chloroplasts," *Aust. Acad. Sci. Symp.* (N. K. Boardman, A. W. Linnane, and R. M. Smillie, eds.), pp. 53–69. North-Holland Publ., Amsterdam.

26. Criddle, R. S. (1966). Protein and lipoprotein organization in the chloroplast. *In* "Biochemistry of Chloroplasts" (T. W. Goodwin, ed.), Vol. I, p. 203. Academic Press, New York.
27. Davenport, H. E., and Hill, R. (1952). The preparation and some properties of cytochrome f. *Proc. Roy. Soc. Ser. B* **139,** 327.
28. Davies, W. H., Mercer, E. I., and Goodwin, T. W. (1966). Some observations on the biosynthesis of the plant sulpholipid by *Euglena gracilis. Biochem. J.* **98,** 369.
28a. De Petrocellis, B., Siekevitz, P., and Palade, G. E. (1970). Changes in chemical composition of thylakoid membranes during greening of the *y-1* mutant of *Chlamydomonas reinhardi. J. Cell Biol.* **44,** 618.
28b. Dolzmann, P. (1968). Photosynthese-Reaktionen einiger Plastom-Mutanten von *Oenothera. Z. Pflanzenphysiol.* **58,** 300.
29. Dure, L. S., III, and Merrick, W. C. (1971). Synthesis of chloroplast tRNA species during plant seed embryogenesis and germination. *In* "Autonomy and Biogenesis of Mitochondria and Chloroplasts," *Aust. Acad. Sci. Symp.* (N. K. Boarman, A. W. Linnane, and R. M. Smillie, eds.), pp. 413–421. North-Holland Publ., Amsterdam.
30. Dyer, T. A., and Leech, R. M. (1968). Chloroplast and cytoplasmic low-molecular-weight ribonucleic acid components of the leaf of *Vicia faba* L. *Biochem. J.* **106,** 689.
31. Erwin, J., and Bloch, K. (1962). The α-linolenic acid content of some photosynthetic microorganisms. *Biochem. Biophys. Res. Commun.* **9,** 103.
32. Everson, R. G., and Slack, C. R. (1968). Distribution of carbonic anhydrase in relation the C_4 pathway of photosynthesis. *Phytochemistry* **7,** 581.
33. Eytan, G., and Ohad, I. (1970). Biogenesis of chloroplast membranes VI. Cooperation between cytoplasmic and chloroplast ribosomes in the synthesis of photosynthetic lamellar proteins during the greening process in a mutant of *Chlamydomonas reinhardi, y-1. J. Biol. Chem.* **245,** 4297.
33a. Fork, D. C., and Heber U. W. (1968). Studies of electron transport reactions of photosynthesis in plastome mutants of *Oenothera. Plant Physiol.* **43,** 606.
34. Francki, R. I. B., Boardman, N. K., and Wildman, S. G. (1965). Protein synthesis by cell-free extracts from tobacco leaves. I. Amino acid incorporating activity of chloroplasts in relation to their structure. *Biochemistry* **4,** 865.
35. Friend, J., and Meyer, A. M. (1960). The enzymic destruction of carotenoids by isolated chloroplasts. *Biochim. Biophys. Acta* **41,** 422.
36. Gillham, N. W., Boynton, J. E., and Burkholder, B. (1970). Mutations altering chloroplasts ribosome phenotype in *Chlamydomonas.* I. Non-Mendelian. *Proc. Nat. Acad. Sci. U. S.* **67,** 1026.
36a. Goldberg, I., and Ohad, I. (1970). Biogenesis of chloroplast membranes. IV. Lipid and pigment changes during synthesis of chloroplast membranes in a mutant of *Chlamydomonas reinhardi y-1:* V. A radioautographic study of membrane growth in a mutant of *Chlamydomonas reinhardi y-1. J. Cell Biol.* **44,** 563 and 572.
37. Goodenough, U. W., and Levine, R. P. (1970). Chloroplast structure and function in *ac-20,* a mutant strain of *Chlamydomonas reinhardi.* III. Chloroplast ribosomes and membrane organization. *J. Cell Biol.* **44,** 547.
37a. Goodenough, U. W., Armstrong, J. J., and Levine, R. P. (1969). Photosynthetic properties of *ac-31,* a mutant strain of *Chlamydomonas reinhardi* devoid of chloroplast membrane stacking. *Plant Physiol.* **44,** 1001.
38. Goodwin, T. W., ed. (1966–1967). "Biochemistry of Chloroplasts," Vol. I. (1966) and Vol. II. (1967). Academic Press, New York.
39. Granick, S. (1963). The plastids: their morphological and chemical differentiation. *In* "Cytodifferentiation and Macromolecular Synthesis" (M. Locke, ed.), pp. 144–174. Academic Press, New York.

40. Gualerzi, C., and Cammarano, T. (1969). Comparative electrophoretic studies on the protein chloroplast and cytoplasmic ribosomes of spinach leaves. *Biochim. Biophys. Acta* **190,** 170.

40a. Gualerzi, C., and Cammarano, T. (1970). Species specificity of ribosomal proteins from chloroplast and cytoplasmic ribosomes of higher plants: *Biochim. Biophys. Acta* **199,** 203.

41. Gunning, B. E. S., and Jagoe, M. P. (1966). The prolamellar body. *In* "Biochemistry of Chloroplasts" (T. W. Goodwin, ed.), Vol. II. Academic Press, New York.

42. Hallier, U. W. (1967). On the use of ^{14}C and ^{32}P in locating genetically caused defects in photosynthesis of some plastid mutants. *In* "Isotopes in Plant Nutrition and Physiology" Int. Atomic Energy Agency, Vienna.

43. Hallier, U. W., Heber, V. and Stubbe, W. (1967). Photosynthese-Reaktionen einiger Plastom-Mutanten von *Oenothera.* I. Der reduktive pentose phosphatzyklus. II. Die Bildung von ATP und NADPH. *Z. Pflanzenphysiol.* **58,** 222 and 289.

44. Hatch, M. D., and Slack, C. R. (1966). Photosynthesis by sugar cane leaves. A new carboxylation reaction and the pathway of sugar formation. *Biochem. J.* **101,** 103.

45. Heber, U. (1960). Vergleichende Untersuchungen an Chloroplasten, die durch Isolierungsoperationen in nicht-wäβrigem und in wäβrigem Milieu erhalten wurden. II. Kritik der Reinheit und Fermentlokalisationen in Chloroplasten. *Z. Naturforsch. B* **15,** 100.

46. Heber, U. (1963). Ribonucleinsäuren in den Chloroplasten der Blattzelle. *Planta* **59,** 600.

47. Henninger, M. D., Gelardi, C., and Crane, F. L. (1966). Rubimedin particles on chloroplast lamellae *Exp. Cell Res.* **44,** 655.

48. Henshall, J. D. and Goodwin, T. W. (1964) Amino acid activating enzymes in germinating pea seedlings. *Phytochemistry* **3,** 677.

49. Hill, R. (1954). The cytochrome b component of chloroplasts. *Nature (London)* **174,** 501.

50. Hill, R., and Bendall, F. (1960). Function of the two cytochrome components in chloroplasts: a working hypothesis. *Nature (London)* **186,** 137.

50a. Homann, P. H., and Schmid, G. H. (1967). Photosynthetic reactions of chloroplasts with unusual structures. *Plant Physiol.* **42,** 1619.

51. Hoober, J. K. (1970). Sites of synthesis of chloroplast membrane polypeptides in *Chlamydomonas reinhardi y-1. J. Biol. Chem.* **245,** 4327.

52. Hoober, J. K., and Blobel, G. (1969). Characterization of the chloroplastic and cytoplasmic ribosomes of *Chlamydomonas reinhardi. J. Mol. Biol.* **41,** 121.

53. Hoober, J. K., Siekevitz, P., and Palade, G. E. (1969). Formation of chloroplast membranes in *Chlamydomonas reinhardi y-1.* Effects of inhibitors of protein synthesis. *J. Biol. Chem.* **244,** 2621.

54. Howell, S. H., Moudrianakis, E. N. (1967). Function of the "Quantasome" in photosynthesis: structure and properties of membrane-bound particles active in the dark reactions of photophosphorylation. *Proc. Nat. Acad. Sci. U. S.* **58,** 1261.

55. Howell, S. H., and Moudrianakis, E. N. (1967). Hill reaction site in chloroplast membranes: non-participation to the quantasome particle in photoreduction. *J. Mol. Biol.* **27,** 323.

56. Hudock, G. A., and Levine, R. P. (1964). Regulation of photosynthesis in *Chlamydomonas reinhardi. Plant Physiol.* **39,** 889.

57. Ingle, J., Possingham, J. V., Wells, R., Leaver, C. J., and Loening, V. E. (1970). Properties of chloroplast ribosomal-RNA. *In* "Control of Organelle Development," *Symp. Soc. Exp. Biol.* (P. L. Miller, ed.), Vol. 24, pp. 303–326. Cambridge Univ. Press, London.

57a. Ingle, J., Wells, R., Possingham, J. V., and Leaver, C. J. (1971). The origins of chloroplast ribosomal-RNA. *In* "Autonomy and Biogenesis of Mitochondria and Chloroplast," *Aust. Acad. Sci. Symp.* (N. K. Boardman, A. W. Linnane, and R. M. Smillie, eds.), pp. 393–401. North-Holland Publ., Amsterdam.
57b. Izawa, S., and Good, N. E. (1966). Effects of salts and electron transport on the conformation of isolated chloroplasts. II. Electron microscopy. *Plant Physiol.* **41,** 544.
57c. Ji, T. H. (1968). The structure of chloroplast lamellar membranes. Ph.D. Thesis, University of California at San Diego.
58. Jones. O. T. G. (1967). Haem synthesis by isolated chloroplasts. *Biochem. Biophys. Res. Commun.* **28,** 671.
59. Joussaume, M., and Bourdu, R. (1966). Conversion of carbon-14-labelled orotic acid into pyrimidine nucleotides by chloroplasts. *Nature (London)* **210,** 1363.
59a. Kaltschmidt, E., and Wittmann, H. G. (1970). Ribosomal proteins, XII. Number of proteins in small and large subunits of *Escherichia coli* as determined by two-dimensional gel electrophoresis. *Proc. Nat. Acad. Sci. U. S.* **67,** 1276.
60. Katoh, S., Shiratori, I., and Takamiya, A. (1961). Distribution of plastocyanin in plants, with special reference to its localization in chloroplasts. *Arch. Biochem.* **94,** 136.
61. Kauzman, W. (1959). Some factors in the interpretation of protein denaturation. *Advan. Protein Chem.* **14,** 1.
62. Kawashima, N., and Wildman, S. G. (1970). Fraction I protein. *Annu. Rev. Plant Physiol.* **21,** 325.
63. Keister, D. L., San Pietro, A., and Stolzenbach, F. E. (1960). Pyridine nucleotide transhydrogenase from spinach. I. Purification and properties. *J. Biol. Chem.* **235,** 2989.
64. Kirk, J. T. O. (1964). DNA-dependent RNA synthesis in chloroplast preparations. *Biochem. Biophys. Res. Commun.* **14,** 393.
65. Kirk, J. T. O. (1964). Studies on RNA synthesis in chloroplast preparations. *Biochem. Biophys. Res. Commun.* **16,** 233.
66. Kirk, J. T. O. (1970). Biochemical aspects of chloroplast development. *Annu. Rev. Plant Physiol.* **21,** 1.
67. Kirk, J. T. O. (1971). Chloroplast structure and biogenesis. *Annu. Rev. Biochem.* **21,** 11.
68. Kirk, J. T. O., and Tilney-Bassett, R. A. E., eds. (1966). "The Plastids." W. H. Freeman, London.
69. Kreutz, W. (1966). The structure of the lamellar system of chloroplasts. *In* "Biochemistry of Chloroplasts" (T. W. Goodwin, ed.), pp. 83–90. Academic Press, New York.
70. Levine, R. P. (1969). The analysis of photosynthesis using mutant strains of algae and higher plants. *Annu. Rev. Plant Physiol.* **20,** 523.
70a. Levine, R. P., and Togasaki, R. K. (1965). A mutant strain of *Chlamydomonas reinhardi* lacking ribulose diphosphate carboxylase activity. *Proc. Nat. Acad. Sci. U. S.* **53,** 587.
71. Loening, V. E., and Ingle, J. (1967). Diversity of RNA components in green plant tissues. *Nature (London)* **215,** 363.
72. Lundegardh, H. (1962). Quantitative relations between chlorophyll and cytochromes in chloroplasts. *Physiol. Plant.* **15,** 390.
73. Lyttleton, J. W. (1962). Isolation of ribosomes from spinach chloroplasts. *Exp. Cell Res.* **26,** 312.
74. McCarthy, R. E., and Racker, E. (1966). Effect of a coupling factor and its antiserum on photophosphorylation and hydrogen ion transport. *Brookhaven Symp. Biol.* **19,** 202.
75. Manton, I. (1966). Some possibly significant structural relations between chloroplasts

and other cell components. *In* "Biochemistry of Chloroplasts" (T. W. Goodwin, ed.), Vol. I, pp. 23–48. Academic Press, New York.

76. Margulies, M. M., and Parenti, F. (1968). *In vitro* protein synthesis by plastids of *Phaseolus vulgaris*. III. Formation of lamellar and soluble chloroplast protein. *Plant Physiol.* **43,** 504.
77. Menke, W. (1966). The structure of the chloroplasts. *In* "Biochemistry of Chloroplasts" (T. W. Goodwin, ed.), Vol. I, pp. 3–18. Academic Press, New York.
78. Mühlethaler, K. (1966). The ultrastructure of the plastid lamellae. *In* "Biochemistry of Chloroplasts" (T. W. Goodwin, ed.), Vol. I, pp. 49–64. Academic Press, New York.
79. Mühlethaler, K., Moor, H., and Szarkowski, J. W. (1965). The ultrastructure of chloroplast lamellae. *Planta* **67,** 305.
80. Nagahisa, M., and Hattori, A. (1964). Studies on oxalic acid oxidase in green leaves. *Plant Cell Physiol.* **5,** 205.
81. Nagai, J., and Bloch, K. (1966). Enzymatic desaturation of stearyl acyl carrier protein. *J. Biol. Chem.* **241,** 1925.
82. Ogawa, T., Kanai, F., and Shibata, K. (1966). Two pigment proteins in spinach chloroplasts. *Biochim. Biophys. Acta* **112,** 223.
83. Ohad, I., Siekevitz, P., and Palade, G. E. (1967). Biogenesis of chloroplast membranes. I. Plastid dedifferentiation in a dark-grown algal mutant (*Chlamydomonas reinhardi*). *J. Cell. Biol.* **35,** 521.
84. Ohad, I., Siekevitz, P., and Palade, G. E. (1967). Biogenesis of chloroplast membranes. II. Plastid differentiation during greening of a dark-grown algal mutant (*Chlamydomonas reinhardi*). *J. Cell Biol.* **35,** 553.
85. Park, R. B. (1966). Chloroplast structure. *In* "The Chlorophylls" (L. P. Vernon and B. Seely, eds.). Academic Press, New York.
86. Peterkofsky, A., and Racker, E. (1961). The reductive pentose phosphate cycle. III. Enzyme activities in cell-free extracts of photosynthetic organisms. *Plant Physiol.* **36,** 409.

86a. Pinto da Silva, P. and Branton, D. (1970). Membrane splitting in freeze-etching. *J. Cell Biol.* **45,** 598.

87. Racker, E. (1969). Function and structure of the inner membrane of mitochondria and chloroplasts. *In* "Membranes of Mitochondria and Chloroplasts" (E. Racker, ed.), pp. 127–171. Van Nostrand-Reinhold, New York.
88. Ragetli, H. W., Weintraub, J. M., and Rink, U. M. (1966). Latent acid phosphatases in chloroplasts. *Can. J. Bot.* **44,** 1723.
89. Ranalletti, M., Gnanam, A., and Jagendorf, A. T. (1969). Amino acid incorporation by isolated chloroplasts. *Biochim. Biophys. Acta* **186,** 192.
90. Rawson, J. R., and Stutz, E. (1969). Isolation and characterization of *Euglena gracilis* cytoplasmic and chloroplast ribosomes and their RNA components. *Biochim. Biophys. Acta* **190,** 368.
91. Remy, R. (1971). Resolution of chloroplast lamellar proteins by electrophoresis in polyacrylamide gels. Different patterns obtained with fractions enriched in either chlorophyll a or chlorophyll b. *FEBS Lett.* **13,** 313.
92. Ritenour, G. L., Joy, K. W., Bunning, J., and Hageman, R. H. (1967). Intracellular localization of nitrate reductase, nitrite reductase, and glutamic acid dehydrogenase in green leaf tissue. *Plant Physiol.* **42,** 233.
93. Rogers, L. J., Shah, S. P. J., and Goodwin, T. W. (1967). The intracellular localization of mevalonate activating enzymes: Its importance in the regulation of terpenoid biosynthesis. *In* "Biochemistry of Chloroplasts" (T. W. Goodwin, ed.), Vol. II, p. 283. Academic Press, New York.

94. Sager, R. (1959). The architecture of the chloroplast in relation to its photosynthetic activities. *Brookhaven Symp. Biol.* **11,** 101.
95. Sager, R. (1961). Photosynthetic pigments in mutant strains of *Chlamydomonas. Carnegie Inst. Washington Yearb.* **60,** 374.
96. Sager, R., and Hamilton, M. G. (1967). Cytoplasmic and chloroplast ribosomes of *Chlamydomonas:* ultracentrifugal characterization. *Science* **157,** 709.
97. Sager, R., and Palade, G. E. (1954). Chloroplast structure in green and yellow strains of *Chlamydomonas. Exp. Cell Res.* **7,** 584.
98. Sager, R., and Palade, G. E. (1957). Structure and development of the chloroplast in *Chlamydomonas.* I. The normal green cell. *J. Biophys. Biochem. Cytol.* **3,** 463.
99. Sager, R., and Ramanis, Z. (1970). A genetic map on non-Mendelian genes in *Chlamydomonas. Proc. Nat. Acad. Sci. U. S.* **65,** 593.
100. Sager, R., and Ramanis, Z. (1972). Genetic circularity of chloroplast DNA in *Chlamydomonas. Proc. Nat. Acad. Sci. U. S.* (in press).
101. Sager, R., and Ramanis, Z (unpublished data).
102. Sager, R., and Zalokar, M. (1958). Pigments and photosynthesis in a carotenoid-deficient mutant of *Chlamydomonas. Nature (London)* **182,** 98.
103. San Pietro, A., and Lang, H. M. (1958). Photosynthetic pyridine nucleotide reductase. I. Partial purification and properties of the enzyme from spinach. *J. Biol. Chem.* **231,** 211.
104. Sane, P. V., Goodchild, D. J., and Park, R. B. (1970). Characterization of chloroplast photosystems 1 and 2 separated by a non-detergent method. *Biochim. Biophys. Acta* **216,** 162.
105. Sastry, P. S., and Kates, M. (1964). Hydrolysis of monogalactosyl and digalactosyl diglycerides by specific enzymes in runner-bean leaves. *Biochemistry* **3,** 1280.
106. Schiff, J. A., and Epstein, H. T. (1965). The continuity of the chloroplast in *Euglena. Symp. Soc. Develop. Biol.* **24,** 131.
107. Schopfer, P., and Siegelman, H. W. (1968). Purification of photochlorophyllide holochrome. *Plant Physiol.* **43,** 990.
108. Schor, S., Siekevitz, P., and Palade, G. E. (1970). Cyclic changes in thylakoid membranes of synchronized *Chlamydomonas reinhardi. Proc. Nat. Acad. Sci. U. S.* **66,** 174.
109. Schweiger, H. G. (1969). Cell biology of *Acetabularia. Curr. Top. Microbiol. Immunol.* **50,** 1.
110. Schweiger, H. G., (1970). Synthesis of RNA in *Acetabularia. In* "Control of Organelle Development," *Symp. Soc. Exp. Biol.* (P. L. Miller, ed.), Vol. 24, pp. 327–344. Cambridge Univ. Press, London.
110a. Schweiger, H. G., Dillar, W. L., Gibor, A., and Berger, S. (1967). RNA synthesis in *Acetabularia.* I. RNA synthesis in enucleated cells. *Protoplasma* **64,** 1.
110b. Scott, N. S., and Smillie, R. M. (1967). Evidence for the direction of chloroplast ribosomal RNA synthesis by chloroplast DNA. *Biochem. Biophys. Res. Commun.* **28,** 598.
111. Scott, N. S., Shah, V. C., and Smillie, R. M. (1968). Synthesis of chloroplast DNA in isolated chloroplasts. *J. Cell Biol.* **38,** 151.
112. Semal, J., Spencer, D., Kim, Y. T., and Wildman, S. G. (1964). Properties of a ribonucleic acid-synthesizing system in cell-free extracts of tobacco leaves. *Biochim. Biophys. Acta* **91,** 205.
113. Shah, V. C., and Lyman, H. (1966). DNA-dependent RNA synthesis in chloroplasts of *Euglena gracilis. J. Cell Biol.* **29,** 174.
114. Shepard, D. (1965). Chloroplast multiplication and growth in the unicellular alga *Acetabularia mediterranea. Exp. Cell Res.* **37,** 93.

115. Shepard, D. C. (1965). An autoradiographic comparison of the effects of enucleation and actinomycin D on the incorporation of nucleic acid and protein precursors by *Acetabularia* chloroplasts. *Biochim. Biophys. Acta* **108,** 635.
116. Shibata, K. (1957). Spectroscopic studies on chlorophyll formation in intact leaves. *J. Biochem.* **44,** 147.
117. Slack, C. R., and Hatch, M. D. (1967). Comparative studies on the activity of carboxylases and other enzymes in relation to the new pathway of photosyntetic carbon dioxide fixation in tropical grasses. *Biochem. J.* **103,** 660.
118. Smillie, R. M. (1963). Formation and function of soluble proteins in chloroplasts. *Can. J. Bot.* **41,** 123.
119. Smillie, R. M., Scott, N. S. (1969). "Organelle Biosynthesis: The Chloroplast" (B. W. Agranoffee *et al.,* eds.), *Progr. Mol. Subcellular Biol.,* Vol. 1, pp. 136–202. Springer-Verlag, Heidelberg and New York.
119a. Smillie, R. M., Evans, W. R., and Lyman, H. (1963). Metabolic events during the formation of a photosynthetic from a nonphotosynthetic cell. *Brookhaven Symp. Biol.* **16,** 89.
120. Smillie, R. M., Scott, N. S., and Graham, D. (1968). Biogenesis of chloroplasts: Roles of chloroplast DNA and chloroplast ribosomes. *In* "Comparative Biochemistry and Biophysics of Photosynthesis" (K. Shibata, A. Takamiya, A T. Jagendorf, and R. F. Fuller, eds.), p. 332. Univ. Tokyo Press, Tokyo.
120a. Smillie, R. M., Anderson, K. S., and Bishop, D. G. (1971). Plastocyanin-dependent photoreduction of NADP by agranal chloroplasts from maize. *FEBS Lett.* **13,** 318.
121. Smith, J. H. C. (1960). Protochlorophyll transformations. *In* "Comparative Biochemistry of Photoreactive Systems" (M. B. Allen, ed.), pp. 257–278. Academic Press, New York.
122. Spencer, D. (1965). Protein synthesis by isolated spinach chloroplasts. *Arch. Biochem. Biophys.* **111,** 381.
123. Spencer, D., and Wildman, S. G. (1964). The incorporation of amino acids into protein by cell-free extracts from tobacco leaves. *Biochemistry* **3,** 954.
124. Spencer, D., and Whitfeld, P. R. (1967). Ribonucleic acid synthesizing activity of spinach chloroplasts and nuclei. *Arch. Biochem. Biophys.* **121,** 336.
125. Spencer, D., Whitfeld, P. R., Bottomley, W., and Wheeler, A. M. (1971). The nature of the proteins and nucleic acids synthesized by isolated chloroplasts. *In* "Autonomy and Biogenesis of Mitochondria and Chloroplasts," *Aust. Acad. Sci. Symp.* (N. K. Boardman, A. W. Linnane, and R. M. Smillie, eds.), pp. 372–382. North-Holland Publ., Amsterdam.
126. Stocking, C. R. (1959). Chloroplast isolation in nonaqueous media. *Plant Physiol.* **34,** 56.
127. Stumpf, P. K., Brooks, J., Galliard, T., Hawke, J. C., and Simoni, R. (1967). Biosynthesis of fatty acids by photosynthetic tissues of higher plants. *In* "Biochemistry of Chloroplasts" (T. W. Goodwin, ed.), Vol. II, pp. 213. Academic Press, New York.
128. Stutz, E. (1971). Characterization of *Euglena gracilis* chloroplast single strand DNA. *In* "Autonomy and Biogenesis of Mitochondria and Chloroplasts," *Aust. Acad. Sci. Symp.* (N. K. Boardman, A. W. Linnane, and R. M. Smillie, eds.), pp. 277–281. North-Holland Publ., Amsterdam.
129. Stutz, E., and Noll, H. (1967). Characterization of cytoplasmic and chloroplast polysomes in plants. Evidence for three classes of ribosomal RNA in nature. *Proc. Nat. Acad. Sci. U. S.* **57,** 774.
130. Stutz, E., and Rawson, J. R. (1970). Separation and characterization of *Euglena gracilis* chloroplast single-strand DNA. *Biochim. Biophys. Acta* **209,** 16.

131. Sudyina, E. G. (1963). Chlorophyllase reaction in the last stage of biosynthesis of chlorophyll. *Photochem. Photobiol.* **2,** 181.
132. Surzycki, S. J. (1969). Genetic functions of the chloroplast of *Chlamydomonas reinhardi:* Effect of rifampin on chloroplast DNA-dependent RNA polymerase. *Proc. Nat. Acad. Sci. U. S.* **63,** 327.
133. Surzycki, S. J., and Gillham, N. W. (1971). Organelle mutations and their expression in *Chlamydomonas reinhardi. Proc. Nat. Acad. Sci. U. S.* **68,** 1301
134. Surzycki, S. J., Goodenough, U. W., Levine, R. P., and Armstrong, J. J. (1970). Nuclear and chloroplast control of chloroplast structure and function in *Chlamydomonas reinhardi. In* "Control of Organelle Development," *Symp. Soc. Exp. Biol.* (P. L. Miller, ed.), Vol. 24, pp. 13–35. Cambridge Univ. Press, London.
135. Surzycki, S. J., and Hastings, P. J. (1968). Control of chloroplast RNA synthesis in *Chlamydomonas reinhardi. Nature* (*London*) **200,** 786.
135a. Szarkowski, J. W., and Ombach, M. (1962). Die Ribonuklease aktivitat Plastiden. *Naturwissenschaften* **49,** 135.
136. Tewari, K. K., and Wildman, S. G. (1970). Information content in the chloroplast DNA. *In* "Control of Organelle Development," *Symp. Soc. Exp. Biol.* (P. L. Miller, ed.), Vol. 24, pp. 147–180. Cambridge Univ. Press, London.
137. Thornber, J. P. (1969). Comparison of a chlorophyll a–protein complex isolated from a blue-green alga with chlorophyll-protein complexes obtained from green bacteria and higher plants. *Biochim. Biophys. Acta* **172,** 230.
138. Thornber, J. P., Stewart, J. C., Hatton, M. W. C., and Bailey, J. L. (1967). Studies on the nature of chloroplast lamellae. II. Chemical composition and further physical properties of two chlorophyll protein complexes. *Biochemistry* **6,** 2006.
139. Tillack, T. W., and Marchesi, V. T. (1970). Demonstration on the outer surface of freeze-etched red blood cell membrane. *J. Cell Biol.* **45,** 649.
140. Vernon, L. P., Ke, B., Katoh, S., San Pietro, A. and Shaw, E. R. (1966). Properties of subchloroplast particles prepared by the action of digitonin, Triton X-100 and sonication. *Brookhaven Symp. Biol.* **19,** 102.
141. von Wettstein, D. (1959). The formation of plastid structures. *Brookhaven Symp. Biol.* **11,** 138.
142. von Wettstein, D. (1961). Nuclear and cytoplasmic factors in development of chloroplast structure and function. *Can. J. Bot.* **39,** 1537.
143. von Wettstein, D., Henningsen, K. W., Boynton, J. E., Kannangara, G. C., and Nielsen, O. F. (1971). The genic control of chloroplast development in barley. *In* "Autonomy and Biogenesis of Mitochondria and Chloroplasts," *Aust. Acad. Sci. Symp.* (N. K. Boardman, A. W. Linnane, and R. M. Smillie, eds.), pp. 205–223. North-Holland Publ., Amsterdam.
144. Weier, T. E., and Benson, A. A. (1966). The molecular nature of chloroplast membranes. *In* "Biochemistry of Chloroplasts" (T. W. Goodwin, ed.), Vol. I, pp. 91–113. Academic Press, New York.
145. Weier, T. E., Stocking, C. R., Thomson, W. W., and Drever, H. (1963). The grana as structural limits in chloroplasts of mesophyll of *Nicotiana rustica* and *Phaseolus vulgaris. J. Ultrastruct. Res.* **8,** 122.
146. Weisblum, B., and Davies, J. (1968). Antibiotic inhibitors of the bacterial ribosome. *Bacteriol. Rev.* **32,** 493.
147. Wildman, S. G. (1971). An approach towards ascertaining the function of chloroplast DNA in tobacco plants. *In* "Autonomy and Biogenesis of Mitochondria and Chloroplasts," *Aust. Acad. Sci. Symp.* (N. K. Boardman, A. W. Linnane, and R. M. Smillie, eds.), pp. 402–412. North-Holland Publ. Amsterdam.

147a. Wildman, W. C., and Bonner, W. D. (1947). Proteins of green leaves. I. Isolation, enzymatic properties and auxin contents of spinach cytoplasmic proteins. *Arch. Biochem.* **14,** 381.

148. Zelitch, I. (1971). "Photosynthesis, Photorespiration, and Plant Productivity." Academic Press, New York.

Concluding Remarks

The preceding two chapters have discussed the problem posed at the outset: to define the functions of organelle DNA's and of organelle protein-synthesizing systems. The preliminary results that we have reviewed, based on the recent burst of research interest and activity in this field, have not provided definite answers to our questions. Indeed, some investigators have gone so far as to suggest that organelle DNA's do little more than act as templates for ribosomal and transfer RNA's and that organelle protein synthesis is of very minor importance, at least in a quantitative sense. This minimal view is not in accord with the available evidence, and is far out of line if one considers the implications of what we already know.

The extensive evidence summarized in Chapters 2–6 has made it abundantly clear that organelle DNA's play an essential role in cell organization as shown by their unique nucleotide sequences and composition, semiconservative replication, systematic transmission at cell division, and well-integrated mechanisms for their segregation and recombination. Furthermore, a wide spectrum of mutations are now being isolated and identified with these DNA's. Mutations to drug resistance similar to those in bacteria suggest mutationally induced alterations in ribosomal proteins; and mutations to temperature-sensitivity (both hot and cold) suggest mutational alterations in other organelle proteins and in some assembly processes.

Similarly, the essential role of organelle protein-synthesizing systems is attested by the presence in organelles of functional ribosomes with striking differences from those of the cell sap. The differences in ribosome size and composition and in the initiating triplet, as well as differences in many of the tRNA's and cofactors for protein synthesis, make it abundantly clear that the two systems are both of vital importance in cell growth. Indeed, it is likely that the differences between organelle and cell-sap protein-synthesizing systems presage the existence of regulatory mechanisms at the translation level. On this assumption, it might be well worthwhile to search explicitly for the cell's equivalents of cycloheximide and chloramphenicol.

Despite our assurance that both the DNA's and the protein-synthesizing systems of organelles have essential roles to play, preliminary efforts to pinpoint these functions have not met with much success. The one unambiguous finding in all organisms examined is that organelle DNA's are transcribed to produce the ribosomal and some if not all transfer RNA's of the corresponding organelles. Evidence for the existence of heterogeneous RNA's in the size range of messenger RNA's has also been reported for both mitochondria and chloroplasts, but incorporation experiments with amino acids have indicated that most organelle proteins are synthesized in the cell sap and imported into the organelles.

Indeed, attempts to identify which proteins are synthesized within the organelles have been surprisingly unsuccessful. In chloroplasts, it is fairly well established that RuDP-carboxylase, which itself makes up one-third to one-half of the total chloroplast protein, is synthesized within the organelle. This is an exceptional instance. Otherwise a very small amount of protein synthesis has been detected in chloroplasts or mitochondria. Most of this synthesis in chloroplasts and all of it in mitochondria have been associated with the membrane fraction.

Why have the results thus far been so sparse? Essentially, I think, because the techniques have not been adequate to cope with the complexities of the problem. First, as most investigators have stressed in discussing their results, macromolecular syntheses in organelles are so tightly coupled to syntheses in the rest of the cell that efforts to uncouple them by the use of inhibitors have given uninterpretable or misleading results. Second, some of the proteins of interest may be present in very small amounts, representing a minute quantity of the total protein and therefore very difficult to detect with methods such as acrylamide gel separation. In addition, some or most of the proteins synthesized in the organelles are membrane-bound, and techniques to isolate and identify this class of proteins are still inadequate. Finally, one must consider the possibility that organelle gene products, either RNA's or proteins, may be exported to the cell sap or even to the nucleus to function as feedback regulators. The tight coupling between the genomes is most likely a two-way process.

These considerations relate primarily to detecting the products of protein synthesis in organelles. Turning now to the RNA transcripts of organelle DNA's, it is possible that messenger RNA's of nuclear origin are translated in organelles. If so, the fact that a protein is synthesized within the organelle does not constitute proof that it was coded there. Thus, identifying the functions of organelle DNA's depends upon the analysis of biochemical changes resulting from mutations in the corresponding organelle genes. Such mutations provide a powerful analytical tool to use in consort with biochemical and biophysical methods, just as have mutations in bacterial and viral systems.

Looking ahead, what seem to be the most promising lines for future research?

1. *Physiological studies.* Comparisons of organelle structure and composition, especially protein changes, following physiological shifts, e.g., from dark to light or from anaerobic to aerobic, have already provided very significant leads toward the identification of particular proteins present in functional but not in nonfunctional organelles. Studies of chemical, enzymological, and structural changes over the cell cycle are also providing a powerful physiological approach to the *in situ* assembly problem, especially in chloroplasts.

2. *Studies of isolated organelles.* Ideally, the way to study organelle formation is with isolated organelles. So far the results have been disappointing because of very low incorporation rates of labeled substrates. Nonetheless, this approach is potentially very powerful. Perhaps methods involving the adding back of cell-sap components or even of other cell organelles will be necessary to restore high synthetic activities.

3. *Membrane assembly.* The rapid developments now taking place in the technology of separation and reassembly of membrane components will surely play a central role in future investigations of organelle biogenesis. Leading examples include the experimental study of mitochondrial membrane assembly and double-labeling experiments to compare membrane proteins of mutant vs wild type or of treated vs control cells.

4. *Cytoplasmic gene mutations.* Above all, I wish to stress the value of using mutant strains carrying specifically defined cytoplasmic gene mutations as fine analytical dissection tools. In screening for mutations, one must to some extent anticipate what kinds of phenotypes will

appear. From present knowledge, mutations affecting three classes of proteins can be expected. (1) *Ribosomal proteins and cofactors in protein synthesis.* The occurrence of organelle mutations to drug-resistance, analogous to bacterial mutations which result from altered ribosomal proteins, strongly implicates organelle genes in coding for at least some proteins involved in the translation apparatus. (2) *Certain membrane proteins.* The presence of recognizable mitochondrial membranes in *petite* yeast containing little or no mitochondrial DNA is the strongest evidence that this organelle DNA does not uniquely code for mitochondrial membrane proteins. Nonetheless, mutations in mitochondrial DNA lead to loss (at least) of electron transport activity, indicating an essential role of this DNA in mitochondrial function. Evidence which is indirect but substantial supports a similar view of chloroplast DNA: that it codes for proteins with essential though not yet identified functions in photosynthesis and photophosphorylation. (3) *Regulatory proteins.* Since integration of organelle activity is undoubtedly a two-way process, some organelle genes may code for proteins exported to the cell sap or even to the nucleus as regulators. If genes of this type exist, then mutations provide a powerful and probably the only known method to find them.

In summary, we have seen that the old views of genetic autonomy of organelles has been replaced by the more modern concept of regulatory interactions between two (or three) separate genomes, the nuclear one containing the bulk of genetic information and the organelle genomes retaining certain essential but as yet undefined genetic functions which probably include feedback loops to the cell sap and to the nucleus. We may anticipate rapid research strides in the near future based on the availability of cytoplasmic gene mutations and of new techniques to study membrane-bound proteins and organelle functions. With this integrated approach we may expect considerable progress toward the understanding of organelle biogenesis and its role in the control of cell growth.

9

Cytoplasmic Genes and Cell Heredity

In contrast to the preceding chapters which have concentrated on fact, this last chapter will be speculative. We have learned a good deal about cytoplasmic genetics in this book, but we have not found the answer to the key question: Why do cytoplasmic genomes exist?

The aim of this chapter is threefold: (1) to evaluate possible advantages to the cell in having cytoplasmic genes; (2) to consider aspects of cytoplasmic heredity not discussed in previous chapters, especially in relation to DNA's of exogenous origin (viruses) and to the genetic control of structure and pattern; and (3) to note some applications of cytoplasmic genetics in medical science and in agriculture. The chapter will end with a list of fundamental biological problems for future research which cytoplasmic genetics may help elucidate.

SOME POSTULATED ADVANTAGES TO CELLS OF CYTOPLASMIC GENOMES

The cytoplasmic genomes we have chiefly considered, those of mitochondria and of chloroplasts, stand in a special relation to the nucleus. The organelles are not genetically autonomous, but the organelle DNA's are unique. We know from the study of evolution that nonessential genes are quickly lost; thus we start with the premise that cytoplasmic genomes play an essential role in eukaryotic cells.

The essential role of these genomes is further emphasized by examining those exceptional cells which can survive in the apparent absence of their organelle DNA's. Three such instances have been documented: loss of chloroplast DNA in the facultative heterotroph *Euglena* (*19*), loss of mitochondrial DNA in the facultative anaerobe *Saccharyomyces cerevisiae* (*26*), and loss of kinetoplast DNA in certain parasitic *Trypanosomes* (*60*). In each instance the loss of organelle DNA is paralleled by loss of organelle function, and survival of the organism results from the availability of an alternative energy supply.

With these rare exceptions, the presence of mitochondrial DNA is universal in all eukaryotic organisms, in parallel with the essential role of the organelle itself in the life of the organism. Similarly, in plants the presence of chloroplast DNA is universally associated with plastid functions, primarily photosynthesis and photophosphorylation.

The essential role of organelles in energy metabolism does not itself explain the essential role of organelle *genes.* What advantages do these genomes confer upon cells and organisms? To assess this question, let us first consider the activities of cytoplasmic DNA's themselves: their replication, mutation, segregation, and recombination processes. We will then consider the question from the viewpoint of transcriptional control and protein synthesis.

DNA Synthesis

Studies with synchronous cultures discussed in Chapters 2 and 3 have shown that nuclear and organelle DNA's are replicated at different but specified times in the cell cycle. The replicating systems appear tightly coupled in the sense that the amounts of organelle DNA relative to nuclear DNA do not vary nearly as much as does organelle size (especially membrane surface area) under conditions making widely different demands on the organelle's energy-generating capacity. An extreme example is the transition from anaerobic to aerobic or from glucose-repressed to glucose-derepressed yeast; membrane surface area, especially the inner mitochondrial membrane, increases manyfold, but the amount of mitochondrial DNA remains unchanged (*22*).

Nonetheless, the mitochondrial and nuclear DNA replicating systems in yeast can be uncoupled by cycloheximide (*28*). Similarly, in *Chlamydomonas,* the replication of nuclear and chloroplast DNA is tightly coupled in vegetative growth, but can be uncoupled by streptomycin (*20*). It seems, therefore, that nuclear and organelle DNA's are independently tied to the cell cycle and may, under special circumstances, be replicated independently.

In the sexual cycle, independent replication of organelle DNA's appears to play a highly important role, as shown by the preferential transmission of one of the two parental genomes. Here, a mechanism like the restriction system, which operates in bacteria to destroy foreign DNA's, is apparently involved in regulating the normal pattern of inheritance of organelle DNA's. We have proposed that one advantage to the organism of uniparental or preferential inheritance of organelle DNA's in the sexual cycle lies in limiting the possibility of genetic recombination (*53*).

Mutation

Specificity of mutagenic response has been well documented for both chloroplast and mitochondrial DNA's. Mitochondrial DNA's are especially susceptible to acridine dyes, such as acriflavine and ethidium bromide in yeast (*55*) and trypaflavine in *Trypanosomes* (*60*), under conditions that are reportedly not mutagenic for the nuclear complement. In algae, streptomycin is a specific mutagen for chloroplast DNA's, with no reported effect on nuclear genes (cf. Chapter 3). In the corn plant, mutations affecting either chloroplast function or pollen sterility (possibly a mitochondrial function) can be induced by an intracellular mutagen produced by cells of a particular nuclear genotype (*iojap*) (*51*). There is no evidence that this mutagen affects nuclear genes.

With respect to mutagenicity, the most remarkable fact is that mutations of organelle DNA's are found at all. If organelle DNA's behaved like those of polytene or polyploid nuclei, most mutations would never be detected, since only one of many copies would have been altered. Nonetheless, both spontaneous and induced mutations are found in mitochondrial and chloroplast DNA's with frequencies not unlike those of nuclear or bacterial genes. The feasibility of selecting for mitochondrial mutations, even in large and complex organisms, is attested by the recent report of an erythromycin-resistant strain of *Paramecium* (*8*) in which the resistance showed a cytoplasmic pattern of transmission which was attributed to a mutation in mitochondrial DNA (*1*).

Recombination

Genetic studies with *Chlamydomonas* (Chapter 3) and yeast (Chapter 4) have demonstrated that recombination of organelle genes can occur in cells which are cytoplasmic heterozygotes (i.e., cytohets). However, opportunities for genetic recombination are rare because of the operation of two unique mechanisms that strikingly distinguish the behavior of organelle DNA's from nuclear DNA's. One mechanism is the exclusive or preferential transmission of organelle DNA's from one parent to all progeny in the sexual cycle seen as "maternal inheritance" in plants and "mitochondrial sex" in yeast. The second mechanism is somatic segregation, a process that liquidates any cytohets that may have formed, restoring the growing clones to homozygosis. Each of these mechanisms, which limit recombination, depends upon an elaborate sequence of gene-controlled events, the existence of which bespeaks their evolutionary value to the organism.

A speculation on the advantage to the organism of limiting recombination of organelle genes is the following. In the nucleus, recombination is severely curtailed and regulated by the presence of nucleoproteins, by the spindle mechanism, and by a special structure, the synaptinemal complex. In bacteria, on the contrary, no mechanisms regulating recombination have been detected. The multitude of studies of transformation, transduction, and conjugation have all shown that bacterial DNA's are constantly available for recombination with homologous DNA's of exogenous origin. Present evidence strongly indicates that organelle DNA's are naked like bacterial DNA's except for a postulated attachment site that regulates replication and distribution. The genetic evidence that recombination of organelle DNA's occurs at any time (i.e., whenever cytohets are present so that the process can be detected) is consistent with this postulated similarity between bacterial and organelle DNA's. These considerations lead us to postulate that organelle DNA's undergo unrestricted exchange events during vegetative growth. If organelle DNA's were heterozygous, unrestricted exchanges would lead to extensive clonal variation. We propose therefore, that homozygosis of organelle DNA's is favored as a means of restricting recombination and maintaining clonal uniformity.

Transcription

Very little is yet known about the regulation of transcription of organelle DNA's per se. Nonetheless, abundant evidence for the transcription of ribosomal and transfer RNA's and preliminary evidence for the transcription of messenger RNA's from organelle DNA's have been reported and are discussed in the preceding chapters. The mitochondrial RNA polymerase of *Neurospora* is a single polypeptide chain (*34*), in contrast to most RNA polymerases which are more complex, with several different subunits. The great differences in composition and organization between nucleoprotein chromosomes of the nucleus and naked DNA's of the organelles are undoubtedly matched by different modes of transcriptional control. One indication is the inhibition of organelle RNA polymerases by rifamycin, a drug without effect on nuclear transcription.

The regulation of transcription lies at the heart of the interaction between nuclear and organelle genomes. Several investigators have speculated that messenger RNA's from the nucleus may enter the organelle for translation on organelle ribosomes. This proposal has the merit of providing a function for the organelle protein-synthesizing system beyond the translation of organelle genes. But if some messenger RNA's can be imported, why not all? In that event, the organelle genomes would be shorn of all coding functions within the organelle. The size and genomic complexity of organelle DNA's and the known classes of organelle mutations argue very strongly against this reductionist view. The possibility that some messenger RNA's of nuclear origin do enter the organelles is open to experimental verification and deserves to be tested.

Another possibility that deserves serious consideration is that many organelle gene products are exported out of the organelle for regulatory purposes. If messenger RNA's can be transported into organelles, then they should also be transportable outward. The same is true for proteins. What tighter coupling between two genomes than that each regulates the transcription of relevant genes on the other genome? Evidence is growing that much of the nuclear DNA is a great superstructure controlling the transcription of relatively few structural genes (*17, 48*). Bacterial genomes appear to have a minimum of regulation of this kind. Organelle genomes may represent an intermediate condition containing both structural genes and regulatory genes, some effective on organelle DNA and others upon nuclear DNA's.

Some experimental support for this view comes from the presence of fast renaturing regions in chloroplast DNA (*63*) and evidence of short runs of high AT content in yeast mitochondrial DNA (*10*). Another kind of indirect support comes from considering the size difference between animal and microbial mitochondrial DNA's; both are highly evolved. If anything, microbial systems have been subjected to more stringent selection pressures and for a longer evolutionary time than have animal cells. Perhaps the microbial eukaryotes solved the problem of regulatory control between nucleus and organelle in a different manner than did the animals, the difference lying in the amount of feedback from the organelle. On this hypothesis, one would predict that the fivefold excess of genomic size in microbial mitochondrial genomes is primarily devoted to regulatory rather than to structural genes.

Evolution

The important role of organelle genomes per se in evolution has been well documented in certain genera of higher plants, especially *Oenothera* (*61*) and *Solanum* (*29*), as discussed in Chapter 6. For example, the existence of plastids with different multiplication rates, vis-a-vis the nuclear-determined cell phenotypes in *Oenothera*, provides substantial confirmation of this role. Similarly, in the genus *Solanum*, cytoplasmic genes have been shown to have played a role in speciation by controlling pollen sterility and some forms of sexual incompatibility. Cytoplasmically inherited sterility has also been cited as a factor in evolution of the mosquito, *Culex pipiens* (*12*).

In summary, we have reviewed some of the evident advantages to the cell resulting from the existence of unique organelle DNA's. The presence of these genomes provides opportunities for the independent operation of mechanisms of replication, mutation, recombination, transcription, and evolution in parallel with, but separate from, those of the nucleus. Ultimately, of course, the selective advantage and therefore the persistence of organelle genomes have depended upon their value in optimizing organelle function. What have we learned about the special functions of the organelle genomes that would illuminate their selective advantage to the organism?

Functions of Organelle Genomes

Chloroplasts and mitochondria function as the main suppliers of cellular energy, so we may be sure that evolutionary selection has operated with great force upon the organelle genomes. We assume that the direction of this selective pressure has been toward insuring optimal response by the organelle to the changing food and energy requirements of the organism. We must conclude, therefore, that the presence of organelle genomes facilitates and improves organelle function.

We have seen that organelle DNA's are implicated in organelle biogenesis as judged from the consequences of organelle mutations. Mitochondrial *petites* cannot respire, surviving solely by fermentation. Similarly, many chloroplast mutations lead to loss of photosynthetic and photophosphorylating abilities. Whether these phenotypes are the result of directly or indirectly induced alterations in organelle proteins, the consequences are the same at the cellular level: loss of an essential function as the result of an organelle mutation.

Somewhat more puzzling are the organelle mutations conferring resistance to antibiotics such as streptomycin, spectinomycin, chloramphenicol, erythromycin, and many related compounds. In bacteria, one mechanism of antibiotic resistance has been shown to be mutationally induced changes in individual ribosomal proteins, decreasing the binding capacity of ribosomes for particular drugs. On this basis, it has been widely predicted that organelle mutations to drug resistance would also be found to result from alterations in ribosomal proteins induced by mutations of organelle genes coding for those proteins.

However, this prediction is at variance with the results of *in vivo* radioisotope-labeling experiments which have shown that organelle ribosomal proteins are synthesized in the presence of chloramphenicol but not in the presence of cycloheximide. As discussed in Chapter 7, this result has been interpreted to mean that these proteins are synthesized in the cytoplasm and imported into the organelles. Whether this interpretation is correct is not yet clear. What *is* clear is the chemical difference between ribosomes of organelles and of the cell sap within each organism and the consequent differences in response of the two (or more) protein-synthesizing systems within the cell to poisons of exogenous origin and, presumably, to intracellular signals. It seems most likely, though not yet rigorously established, that the organelle genomes contribute some specificity to the organelle protein-synthesizing apparatus.

Finally, we need to consider the organelle genomes as repositories of regulatory information. Suppose a cell suddenly needs a greatly increased amount of electron transport activity to oxidize a high concentration of substrate. We assume that intramitochondrial concentrations of certain key metabolites will be more sensitive to these changing energy requirements than the concentration of the same compounds located elsewhere in the cell. Either the fast reaction kinetics or the sequestering of key regulatory metabolites within the organelle could explain the differences in concentration of these substances across the organelle membranes. Thus, the recognition of a shortage of oxidative capacity would become apparent within the organelle by the accumulation or disappearance of key substances. This change in concentration could then trigger the transcription of regulatory genes concerned with biogenesis within the organelle as well as genes that regulate macromolecular synthesis elsewhere in the cell.

Thus, the available evidence about organelle genome function, in addition to the known genes transcribed as ribosomal and tRNA's has led us to propose three classes of organelle genes: structural genes coding for particular membrane components; structural genes coding for particular ribosomal proteins; and regulatory genes coding for molecules which mediate gene expression both within the organelle and elsewhere in the cell.

ASPECTS OF CELL HEREDITY NOT PREVIOUSLY CONSIDERED

This book has mainly considered classes of cytoplasmic genes and cytoplasmic DNA's whose functions, while not understood in detail, have been clarified enough so that molecular mechanisms can be investigated. These cytoplasmic genes are carried in DNA, and their mode of action appears to follow the rules of molecular genetics: DNA coding for other macromolecules which in turn provide the basis for growth and specificity. Now we may ask: Do the known genomes, nuclear and cytoplasmic, constitute the total hereditary apparatus of the eukaryotic cell?

In the past, three kinds of phenomena have been discussed under the heading of non-Mendelian cytoplasmic heredity: (1) the inheritance of organelles, structure, and patterns; (2) the inheritance of intracellular parasites, symbionts, and viral genes; and (3) clonal transmission of established phenotypes, such as mating types and surface antigens in *Paramecium* (*6, 7, 24, 25, 49, 50, 52, 57, 58, 59*). The problem of clonal transmission refers to phenomena such as the development of different phenotypes in subclones of identical genotype and the switching during clonal multiplication from one phenotype to another. Where analyzed, phenomena of this kind appear not to involve cytoplasmic heredity per se, but rather to reflect changes in nuclear gene expression regulated at the DNA level (*7, 15, 24*). These phenomena are surely of great importance in differentiation, but in the absence of evidence of any direct involvement with cytoplasmic genetics, they will not be considered here.

We will consider the other two phenomena: the inheritance of structure and the inheritance of DNA's of exogenous origin. A serious summary of the voluminous literature on this subject is, however, far beyond the scope of this book. Fortunately, much of it, especially that dealing with microbial systems, has recently been reviewed (*13, 24, 49, 50*).

Cytoplasmic Genes and the Inheritance of Structure

In gazing at good electron micrographs of cell sections, one is quickly aware of the complexity and abundance of subcellular structures. What is known and unknown about the genetic control of their formation? In the past, genetics lent itself beautifully to the investigation of temporal biosynthetic pathways. Now cell biologists are concerned with spatial considerations: the biogenesis of structures. Beyond DNA, is there hereditary continuity in structures? Can mutations be utilized here too to dissect the processes of biogenesis? Is structural continuity itself a kind of non-DNA heredity?

In order to discuss this problem, it is necessary to digress briefly to consider what is known about the origin of biological structures. The first level in the development of cell structure may be taken as the folding of polypeptide chains to form unique three-dimensional configurations. It is now well established that the amino acid sequence itself (in the right environment) determines the folded configuration or tertiary structure of the proteins. Although complete details are not known, the overall process can be shown to proceed *in vitro* under defined conditions (*3, 4*). DNA dictating the amino acid sequence, together with internal noncovalent molecular forces, produce a novel structure by spontaneous arrangement. The geneticists' question has been answered: no additional genetic information is required to determine tertiary structure of proteins beyond that present in the linear sequence of the polypeptide chain.

At the next level of complexity, folded protein subunits may aggregate to form specific quaternary structures, which are endowed with new properties, in particular the ability to undergo allosteric conformational changes, as well as to bind particular substrates or cofactors. This aggregation, too, is a spontaneous process: no additional genetic information is needed to determine the correct and specific alignment of quaternary structures (*16, 32b*).

The importance of the allosteric properties gained by quaternary association of protein subunits (i.e., protomers) can hardly be overstated. The oxygen-carrying capacity of hemoglobin, the antigenic specificity of antibiotics, and the regulatory role of enzymes in feedback inhibition are examples of biological functions dependent upon quaternary structure, yet genetically determined by DNA and inherent though not expressed in amino acid sequence.

An example of further complexity in spontaneous assembly of a novel structure is seen in RNA plant viruses composed of a single RNA molecule and many identical protein subunits. The steric fit between the subunits is determined by their individual configurations, and these in turn are established by their amino acid sequence. The assembly is spontaneous because no covalent bonds, requiring high energy of activation, are involved: only very many low energy noncovalent bonds (*21*). The polyhedral viruses containing several different coat proteins have been shown to assemble spontaneously, their overall structure being determined by the so-called architectonic properties of the component molecules (*11, 32c*).

A far greater complexity is encountered in the assembly of bacterial ribosomes (*46*) and phage T4 (*32a, 64*). Here the essential first step in understanding their assembly, namely, identifying all of the component parts, itself a long and complicated work, is just being completed. In both systems, mutants have played a central role in the investigation, and here too the total genetic message appears to lie in DNA.

Both systems have in common the fact that each particle is assembled *de novo*. In the phage system (*64*), at least three principles of assembly have been established: (1) control of protein precursors by "clipping," i.e., the conformational change of particular proteins by splitting off a part of the molecule at the time of assembly; (2) stepwise spontaneous assembly of subassemblies (e.g., head, tail fibers); and (3) joining of different finished subassemblies. Although mainly a spontaneous process, at least one catalyzed step has been identified. The study of ribosome assembly, which is also stepwise, has so far not revealed any additional mechanisms, but an important class of mutants, called *sad* (subassembly defective), have not yet been fully analyzed (*46*). Some surprises may still be in store for us.

The systems discussed so far have provided evidence for two principles of morphogenesis: DNA-determined specificity of components, and stepwise spontaneous or enzyme-catalyzed assembly. Are these two principles sufficient to explain all known phenomena of morphogenesis? To approach this question, let us consider certain more complicated subcellular systems: membranes, centrioles, and cortical inheritance in *Paramecium*.

Let us begin with membrane biogenesis. Here the situation is very different from the systems described above. First of all, in no membrane system do we yet have a complete compositional inventory, and that deficiency alone limits our understanding. The principal classes of membrane components are proteins and lipids, and the lipids play an important role in assembly. In sharp contrast to the structures discussed above, membranes do not have a fixed stoichiometry, but may vary in composition within rather wide limits (*33*). Despite their variable composition, membranes exhibit specificity, as shown by the localization of various enzymes and other proteins that give the membranes their identity and by the presence of particular lipids in some membranes and not others (i.e., galactolipids in chloroplast membranes, cardiolipin in mitochondrial inner membranes).

Membrane specificity is itself a kind of heredity. How is this kind of heredity determined? Membranes apparently do not arise *de novo*. While this proposition is difficult if not impossible to disprove, there is powerful indirect evidence in its support. First, cytoplasmic continuity, and with it membrane continuity, is unbroken at any stage in the life cycle of any organism. Second, strong evidence of membrane continuity comes from recent studies of membrane growth: membranes grow by the insertion of newly synthesized molecules into preexisting structures (*33, 37, 38, 54*). The detailed mechanism of insertion is not well understood as yet but is evidently related to the lipoprotein composition of the membrane. As discussed in Chapter 7, membranes are somewhat loosely constructed, with space available into which new components with the correct charge distribution and shape can be inserted.

Thus, membranes can be thought of as hereditary structures on two counts: their specificity and their continuity. Is there any evidence that membrane specificity is directly DNA based? DNA's have been reported in association with various membrane fractions of eukaryotic cells (e.g., *30*), but as yet no rigorous evidence has identified a particular DNA with a particular membrane fraction or structure other than chloroplasts and mitochondria. This lack of evidence may primarily result from the technical difficulties involved in identifying small DNA's of cytoplasmic origin.

Indeed, it is a reasonable hypothesis that additional DNA's may yet be found in the cytoplasm associated with particular membrane fractions such as the plasma membrane, Golgi apparatus, and endoplasmic reticulum, just as DNA's have been associated with chloroplasts and mitochondria. Such DNA's could play a part in maintaining the specificity of various membranes of the eukaryotic cell by coding for special proteins, and these in turn could permit or block the insertion of other components from the surrounding cell sap during membrane growth.

Thus, a relatively small amount of DNA could have a decisive effect on membrane specificity. Such DNA's could be visualized as bound to the membrane, with messenger RNA's peeling off like a pinwheel (cf. Plate XI), each attached in turn to ribosomes which spin off nascent lipophilic proteins that have affinity for the membrane (*46a*). This octopus-like structure might be detectable with the newest techniques of electron microscope autoradiography.

The octopus hypothesis, which is addressed to the molecular basis of membrane *specificity*, does not fully cope with the genetic significance of membrane *continuity* (*58*). If membranes do not arise *de novo*, but only from preexisting membranes, must we then consider these structures themselves as hereditary elements?

This question forces us to reconsider the geneticists' dichotomy between genotype and phenotype. Perhaps the single most fundamental property of genetic information is its exclusion from the phenotype of the organism. The study of genetic mechanisms and of evolution has taught us that selection does not operate directly on DNA;

the genome is shielded from selection and varies only as the result of random events. This dichotomy represents such a profound principle of biological organization that it seems unlikely that cells contain structures which are both genotype and phenotype. (An exception is DNA polymerase itself which provides a feedback loop through which the environment may alter DNA.)

This abstract formulation is supported by a number of new experimental observations. Two structures which have often been cited as examples of genetic continuity have now been shown rather conclusively to be capable of arising *de novo,* that is in the absence of a preexisting structure of the same kind: bacterial cell walls and eukaryotic centrioles or basal bodies.

The bacterial cell wall can be removed by growing bacteria in the presence of penicillin or lysozyme, leading to the formation of protoplasts which can grow as wall-less structures called L-forms. Unstable L-forms revert because they contain remnants of cell wall, which appear to act as primers for further wall synthesis. Stable L-forms have been found not to revert, presumably because no remnants of cell wall remain. Recently, however, it has been shown (*14*) that stable L-forms can be induced to revert to normal bacteria by placing them in a suitable physical environment, such as solid gelatin, hard agar, or even a bacterial filter. These substances do not act as chemical primers, but do induce wall synthesis. The experiments show clearly that preexisting wall fragments are not essential and therefore not to be classed as hereditary elements, although, of course, they do greatly *facilitate* synthesis when present. Facilitation is not heredity.

The formation of centrioles or basal bodies (now considered identical structures) is a very complex problem, and this discussion will be limited to one facet: the question of *de novo* origin. Until recently, the classic view of cytologists has been that centrioles arise only in association with preexisting centrioles. The mechanism by which the preexisting structure aided in morphogenesis was not discussed, and, indeed, could hardly be considered on the basis of light microscope observations. With the rise of interest in morphogenesis and the availability of new electron microscope techniques, cytologists have reexamined the origin of basal bodies in a number of particularly suitable materials.

The results are uniform in showing that basal bodies can arise *de novo,* in the sense that no structural precursors resembling centrioles or procentrioles have been detected. Favored materials have included *Naegleria gruberi* (*23*) during transformation from its centriole-free amoeboid stage to its flagellate form and the ciliated epithelium of the oviduct, either during basal body and cilia formation in newborn mice (*18*) or in estrogen-driven reformation of basal bodies after ovariectomy in *Rhesus* monkeys (*2*).

Although the molecular analysis of centriole formation is just beginning, it now seems likely that the two principles: DNA determination of protein specificity and stepwise self-assembly of components will suffice as the conceptual basis for understanding the detailed process. The existence of centriolar DNA remains, as discussed previously, an open question.

In all the studies of centrioles and basal bodies it has been clear that their location is not random, but rather that some forces in the cell, exterior to the organelle itself, dictate their location, orientation, and number. Can these forces in any sense be considered hereditary? This question, more generally stated as the genetic control of the patterned organization of structures, has been the subject of an extraordinarily beautiful study by Sonneborn and his associates (*9, 59*). The availability of a recent review of this work (*59*) relieves me of the necessity of providing the reader with more than a skeleton of the facts, and allows me to proceed to a brief discussion of some implications of the results.

The cortical structures of the ciliated protozoan, *Paramecium aurelia*, are arranged in a precise repeating pattern which can be resolved in the light microscope after silver impregnation. These structures, which include ciliary basal bodies, trichocysts, various fibers and microtubules, and a parasomal sac, are arranged in unit territories so precise that, in Sonneborn's words: "examination of any territory *in situ* reveals the cell's right, left, anterior, and posterior" (*59*). In a grafting experiment (*9*), a small patch of unit territories was rotated 180°. Subsequently "the inverted patch grew during cell division until its rows extended full length along the body surface. Thereafter, the progeny inherited the inverted row or rows" (*59*). The only progeny cells containing the inverted patch were those derived by fission from preexisting cells with the inverted patch, and neither the micronucleus, the macronucleus, nor the free-flowing cytoplasm had any influence on the transmission of this trait.

This experiment is one of many which have led Sonneborn to the view that "there is no evident escape from the conclusion that essential aspects of development in *Paramecium* are encoded partly in cortical geography, not solely in DNA" (*59*). Is there any alternative to Sonneborn's conclusion?

In my view, the problem as posed is primarily a semantic one. Most steps in self-assembly are carried out in relation to the structure formed in the preceding step. In ribosome and phage assembly for example, regulatory controls can be exerted by the cell (stopping assembly at intermediate stages) by withholding components needed for later steps. Thus, a partially assembled particle could be viewed as a bit of preformed structure essential for further assembly. A known example is the procentriole, which may be present for some time as such; it develops further into a fully formed centriole or basal body at a later time in the scenario of development.

It may be argued that the location of organelles and other structures in precise patterns such as the unit territory in *Paramecium* involves some molecular mechanisms beyond self-assembly. However, increasingly complex multicomponent systems are being found capable of self-assembly (*11, 32a, 32c*). Perhaps the molecular processes behind localization and patterning are different in kind from those we now recognize, but I think it is far more likely that they only *appear* to be different because they so extensively utilize preexisting structures.

In this connection, we need to recognize the importance of facilitated morphogenesis. In bacterial cell wall synthesis, we saw that a fragment of preformed wall, while not essential, greatly facilitated further synthesis. A more general example is the role of primers in the polymerization of macromolecules. The distinction between a *primer,* utilized for end-on addition of new monomers, and a *template,* which determines specificity, needs to be recalled. The only templates we know about in biological systems are nucleic acids.

In summary, the origin of structures and patterned arrangements within cells can probably be understood as the consequence of two fundamental processes: genetics, operating by DNA-based determination of protein specificity; and patterned growth, operating by intramolecular conformational changes and stepwise self-assembly.

Viruses and Cytoplasmic Genes

The literature of cytoplasmic heredity contains several instances of non-Mendelian inheritance which have proved to be the consequence of infection by a virus or a bacterial endosymbiont. Although the distinction between viruses and cytoplasmic genes is often obvious, there are borderline examples in which the distinction is difficult to make. Three criteria have been useful.

1. The tight coupling of the replication of organelle and nuclear DNA's is not usually seen in viral DNA's, which replicate independent of the cell cycle.

2. The functions of known cytoplasmic genes are essential and closely tied to cellular metabolism by regulatory mechanisms, whereas these properties are not characteristic of viral genes.

3. The infectivity of viruses is not characteristic of cytoplasmic genes, although, of course, cytoplasmic genes may be transferred experimentally, e.g., by microinjection. The literature on intracellular symbionts showing some properties of cytoplasmic genes has recently been reviewed by Preer (*50*), and so the following discussion will make no attempt at full coverage.

The operational difficulty of distinguishing an endosymbiont from a cytoplasmic gene was first shown dramatically in the studies of *kappa* by Sonneborn and his students (*57*). Indeed, for many years, *kappa* represented the model cytoplasmic factor, and its investigation provided a virtual handbook of methodology to be applied to other organisms. Then Preer (*49*) found that *kappa* did not always multiply at the same rate as did the cell and could be diluted out. Mathematical analysis of the dilution kinetics provided estimates of the number of *kappa* particles per cell, and x-ray inactivation data provided an estimate of particle size. Preer (*49*) then found that *kappa* could be seen in the light microscope by staining with Feulgen reagent, which is a relatively specific stain for DNA.

It has now become well established that *kappa* is a bacterial symbiont and that various strains of *Paramecium* contain, in addition to *kappa*, many (perhaps hundreds) other symbiotic particles. Some of them have been studied in detail, such as *mu, lambda, sigma, gamma, delta, alpha,* and *tau,* all of which confer new phenotypic properties on the host (*57*). None of these particles have been shown to be essential to the survival of the host, although some host genes are involved in their maintenance. Their principal known value to the host is that they confer immunity to the killing action of particle-bearing *Paramecium* toward particle-free cells. Recently, a system which is formally similar to *kappa* has been found in yeast (*56*); the particulate basis, whether viral or endogeneous, has not yet been established.

In the past, many geneticists felt that they would only be convinced of the reality of cytoplasmic genes if they were found in *Drosophila,* the organism whose genetics was the most extensively investigated. Consequently, the discovery of cytoplasmically transmitted CO_2 sensitivity (*39a*) was greeted with great interest, as was another cytoplasmically transmitted trait, "sex-ratio" (*41*), the production of largely or exclusively female progeny. After many years of investigation, the property of CO_2 sensitivity was shown to be the consequence of infection by a virus called *sigma* (*40*); and most instances of "sex-ratio" were shown to be the result of infection by spirochaetes or by a DNA-containing virus which they carry (*50*). Other examples of apparent cytoplasmic inheritance in *Drosophila* in which viral infection may be involved include the induction of tumors in *Drosophila* (*5*), the induction of sterility in some strains but not others by the cytoplasmic factor *delta* (*43*), male sterility in *Drosophila paulistorum* (*50*), a factor responsible for chromosome breakage (*39*), and the cytological discovery of viruslike particles in *Drosophila* tissues (*31*).

The presence of many virus-associated traits in *Drosophila* raises the possibility that in the insects, as in *Paramecium,* endosymbionts play a significant role in survival and evolution within the genus (*40*). In *Drosophila,* however, the systematic investigation of cytoplasmic genetics per se by mutagenesis and selection has not yet been carried out. Now that *Drosophila* is becoming a favorite organism for the analysis of differentiation, it is a matter of some concern that nothing is yet known of its cytoplasmic genome. The presence of viruses in *Drosophila* need not interfere with the analysis of cytoplasmic genetics any more than endosymbionts in *Paramecium* interfere with the analysis of mitochondrial erythromycin resistance (*1, 8*).

In *Culex,* the mosquito, extensive investigations by Laven (*35, 36*) have revealed a complex system of nuclear and cytoplasmic gene interactions regulating viability of progeny. In some crosses no progeny are produced at all. Laven's findings, in addition to their basic biological interest, have had an unexpected and important application: the use of cytoplasmic sterility in the eradication of disease-bearing strains of the mosquito in Southeast Asia. The molecular identity of this class of cytoplasmically inherited determinants has not yet been established, but whether or not they are of viral origin is secondary to their functional significance.

Thus, in the insects as well as in microbial systems, phenotypes of great biological interest are determined by genes found in the border region between endosymbionts and true cytoplasmic genes. Endosymbionts are of particular interest because they provide a means for increasing the gene pool, bringing nucleic acids of exogenous origin into cells where they may become permanently lodged.

Among the most dramatic examples of this process are the oncogenic animal viruses, which transform the cells they infect from a normal to a neoplastic state (*27*). Indeed a wide and diverse array of animal viruses are now known, from those that lyse cells to those that enter and disappear from view. Within this array, some become integrated into the cell's own DNA.

A recent report (*41a*) describes infecting human fibroblasts from a patient lacking the enzyme α-D-galactose-1-phosphate uridyl transferase with a transducing bacteriophage carrying the bacterial gene that codes for this enzyme in *E. coli.* The fibroblasts, growing in tissue culture, were shown to acquire transferase activity after exposure to the virus and to retain this activity during repeated subcultures. The enzyme production was blocked by cycloheximide, an antibiotic which blocks protein synthesis in mammalian cells but not in bacteria.

The authors suggest that the phage DNA may be preserved by integration into the host genome by plasmid-like survival in the cytoplasm or by interaction with mitochondria. In our view, if the bacterial gene has become integrated at all, mitochondrial DNA would provide a far more accessible site than would nuclear DNA. Indeed, the hypothesis that cytoplasmic DNA's serve as ports of entry of viral genes into animal genomes may be directly testable with this system.

We conclude that the distinction between viral and endogenous cytoplasmic genes may sometimes be difficult to establish; in one organism or another stages have been seen in a continuum from lytic virus to integrated gene. Rather than attempting to achieve hard and fast definitions and distinctions, it would seem more rewarding to recognize this continuity and to examine each instance on its merits. In particular, the methods of investigation and techniques of analysis employed with animal viruses should be utilized in the study of organelles and vice versa.

I should like to comment on the endosymbiotic origin of chloroplasts and mitochondria, which has become a popular subject of debate. In my view, chloroplasts and mitochondria may well have arisen from free-living prokaryotic cells which participated in the origin of eukaryotes. After all, chloroplasts and mitochondria have all the essential equipment for life: DNA, transcription enzymes for making RNA, and a full protein-synthesizing apparatus. The central issue in the evolution of the eukaryotic cell is the origin of the nucleus, a question which has hardly yet been touched upon. The nucleus, totally lacking the machinery for protein synthesis, may well be the most recently evolved organelle of all. The tight integration we have seen between the nucleus and the organlles may then be the consequence of evolutionary shifts of some DNA from the cytoplasm to a new organelle, the nucleus, principally concerned with centralized genetic regulation.

SOME RECENT APPLICATIONS OF CYTOPLASMIC GENETICS

Hybrid Seed Production

Mutations of cytoplasmic genes resulting in the formation of sterile pollen are widespread in nature, found in most families of higher plants as discussed in Chapter 6. Cytoplasmic pollen sterility has been successfully incorporated into commercial inbred lines to insure cross pollination in production of hybrid seed in several crop plants, the most important being the corn plant, *Zea mays*. As discussed in Chapter 6, the use of cytoplasmic pollen sterility has resulted in simplified seed production methods. Inbred lines used throughout the world for corn seed production now contain selected cytoplasmic genes to insure the success of the method.

The use of this method is now being imperiled by a pathogenic fungus, *Helminthosporium maydis,* which produces a toxin highly destructive to the corn plant, causing a disease known as southern leaf blight (*32*). Both nuclear and cytoplasmic genes strongly influence susceptibility to the fungus. In particular, the cytoplasmic gene known as (*T*), most widely used in corn seed production, is implicated in susceptibility to a new race of *H. maydis,* called race T, which in 1970 destroyed over 50% of the corn crop in southern United States and as much as 25% in some parts of the midwest. Whereas the toxin produced by this strain has been shown to destroy the inner membrane of mitochondria from plants carrying the (*T*) cytoplasmic gene for pollen sterility, mitochondria from other inbred lines are resistant (*42*). So far, susceptibility and pollen sterility have shown complete association, but it is not known whether the apparent linkage is genetic or whether one gene is responsible for both phenotypic traits.

The seriousness of the problem has led to a retreat in production methods to hand-detasseling, as inbred lines carrying (*T*) cytoplasm have been withdrawn.

It may be hoped that the discovery of linkage between cytoplasmic pollen sterility and fungal susceptibility will lead to renewed interest and support for research investigation into the cytoplasmic genetics of corn. Already other effects of (*T*) cytoplasm upon growth of corn plants are beginning to be noticed, and it seems likely that other mutations of cytoplasmic genes may be of great value to the corn breeder.

Eradication of *Culex pipiens fatigans*

This mosquito is the chief vector of filariasis in southeast Asia. As mentioned above, Laven (*35*) intensively investigated the cytoplasmically controlled incompatibility between different populations of this species. He subsequently established the effectiveness of introducing incompatible males into natural populations. Females do not discriminate between normal and incompatible males, and the total population rapidly declines. In a field experiment in Burma (*36*), an entire mosquito population of 4000 to 20,000 mosquitos was eradicated by releasing 5000 incompatible males per day in the region over a 2-month period. This work, supported by the World Health Organization, is a model of the use of biological control, rather than pesticides, in the elimination of disease-carrying strains of insects.

Developmental Anomalies in Mouse and Man

The consequences of cytoplasmic gene mutations in embryological development are almost totally unknown. Recently, a few suggestive reports have been published which merit consideration. Nance examined the inheritance in twins of anencephaly and spina bifida, which are congenital malformations leading to neurological defects (*44, 45*). Both of these traits are seen more frequently in the mothers' relatives than in the fathers' relatives of individuals with the defect, suggesting a maternal genetic or developmental influence. In twins, however, Nance found that of 125 selected pairs of twins, in only one pair did both twins have the defect. In the remaining 124 pairs only one of the twins had the defect. This surprising result, which is statistically highly significant, has been interpreted by Nance as evidence of a cytoplasmic genetic effect, resulting from unequal partition of some cytoplasmic genetic material.

Cytoplasmic inheritance has also been implicated in a developmental abnormality in the mouse (*47, 62*). In studies of the effects of treatment with the teratogenic compound 6-aminonicotinamide (6-AN), some strain differences were noted in the susceptibility of different inbred lines of mice to this teratogen. This difference in susceptibility was transmitted by susceptible females to their F_1 progeny, but not by males. When F_1 females from the reciprocal crosses were backcrossed to resistant males, only the females whose mothers were susceptible transmitted the susceptibility to the F_2 generation. In the next backcross generation, the susceptibility was lost, but that is not surprising, since nuclear genes from the male parent may be expected to interact with the cytoplasmic genome. Two generations of maternal transmission suffice to demonstrate the presence of a maternal inherited *gene* rather than a nongenetic maternal *"influence."*

Serious consideration of the cytoplasmic genome as influential in development has hardly begun, so it is not surprising that little information is yet available. As recognition of the existence of cytogenes grows, it may be anticipated that evidence of their role in the development of higher organisms will also grow.

Control of Disease-Causing Hemoflagellates

The hemoflagellates or trypanosomids are a group of parasitic protozoa responsible for serious and widespread diseases of man, including African sleeping sickness, leishmaniasis (Kala Azar), and Chaga's disease. The trypanosomes characteristically contain a specialized organelle, the kinetoplast, which is mitochondria-like in structure and function, and contains its own kinetoplast DNA (cf. Chapter 2).

Treatment of trypanosomes with acridine dyes such as trypaflavine and acriflavine, which differentially block replication of kinetoplast DNA, leads to loss of respiratory function and death. A few species, including those that cause African sleeping sickness, can survive in the mammalian bloodstream with an auxiliary respiratory system. However, even those species cannot survive in the other stages of their complex life cycles. Thus, our knowledge of the specificity of response of kinetoplast DNA should permit the development of drugs with minimal toxicity to man for use in the eradication of these parasites.

Cytoplasmic Genes and the Etiology of Cancer

Since the transformation of cells from normal to neoplastic is hereditary in clones, cancer is a genetic disease at least at the cellular level. Which genes are involved? In the past, only nuclear genes were taken into consideration in the planning and evaluation of cancer research studies. With the development of our knowledge about cytoplasmic genetics, it would be useful to reconsider past studies and to design new investigations aimed specifically at examining the possible role of cytoplasmic genes in the neoplastic transformation.

At present, two lines of experimental evidence are particularly intriguing. As discussed in Chapter 2, Vinograd and associates found high frequencies of dimeric forms of mitochondrial DNA in the leukocytes of patients with certain forms of leukemia. The mechanism of origin of these forms, as well as their functional relation, if any, to neoplasia, deserves serious consideration. Second, there is evidence dating back to Warburg, and repeatedly confirmed since then, showing an energy shift from mitochondrial respiration to glycolysis in tumor tissues. This change is permanent in clones of tumor cells; it is genetically determined. The possibility that mitochondria of tumor cells are functionally impaired leads directly to the question of whether any changes in mitochondrial DNA are associated with the etiology of cancer.

These observations are merely suggestive of a link between mitochondria and cancer, but they may provide a starting point for future investigations. The possibility that oncogenic viruses may use cytoplasmic DNA's as initial foci for integration suggests another line of attack upon the cancer problem.

CONCLUDING REMARKS

The outlook throughout this book has been toward the future, attempting to formulate present-day understandings of cytoplasmic genetic systems as they relate to the overall problems of biology. It is fitting therefore to conclude with a list of questions, representing what appear today as outstanding research problems for the near future. These questions are grouped into two sets, those concerning DNA and genetic analysis per se and those concerning the functions of cytoplasmic genomes.

A. Cytoplasmic DNA's and genetic analysis.

1. What is the molecular basis of maternal inheritance and preferential transmission of cytogenes?
2. Are there classes of cytoplasmic DNA's not yet identified, for example, membrane-associated DNA's conferring membrane specificity?
3. What is the basis of the apparent discrepancy between the number of genetic copies and physical copies of mitochondrial DNA in yeast and of chloroplast DNA in *Chlamydomonas?*
4. Is there intracellular competition between different (e.g., mutated and unmutated) organelle DNA molecules at the level of replication?
5. Can a systematic cytoplasmic genetics, including mutagenesis, mapping, and recombination analysis, be developed for higher plants and animals?

B. Functions of cytoplasmic DNA's.

1. What functions of cytoplasmic DNA's necessitate their sequestering in organelles or dispersal in the cytoplasm?
2. Do some cytoplasmic genes function as structural genes for particular proteins?
3. Do some cytoplasmic genes function as regulatory genes controlling the expression of other organelle or nuclear genes?
4. What are the regulatory mechanisms that couple the growth of organelles to the rest of the cell?
5. With respect to plants, are there special mechanisms of interaction between chloroplast and mitochondrial genomes?
6. What is the functional significance of miniribosomes?

Beyond the solutions to these specific questions lies their significance for our expanding understandings of biological mechanisms, and these in turn contribute scientific potential toward the promotion of health and human welfare. Genetics is the core science of biology, and we have long known that each new fragment of understanding in genetics may have enormous ramifications in medical science and in agriculture. So too with cytoplasmic genetics, which is a recently rediscovered, but primordial, part of our genome. As problems in cytoplasmic genetics are solved and as the awareness of the existence of cytogenes becomes widespread we can anticipate fundamental contributions to the analysis of metabolism and development, to the solution of medical problems, including the etiology of cancer and the molecular basis of aging, to the biological control of pathogens, and to increased productivity in agriculture.

The future welfare of man depends not only upon the availability of knowledge, but also upon its usage. In a democratic society, the uses of science can be regulated by the people. Constructive regulation requires knowledge as well as wisdom. This book has been addressed to students of all ages who have found excitement and challenge in the struggle for new knowledge. Perhaps we lucky ones who have discovered the profound pleasures in research and in learning can try harder than ever to communicate these pleasures, and the values on which they are based, to the growing population of the world, upon whose knowledge and wisdom the future of man depends.

References

1. Adoutte, A., and Beisson, J. (1970). Cytoplasmic inheritance of erythromycin resistant mutations in *Paramecium aurelia. Mol. Gen. Genet.* **108,** 70.
2. Anderson, R. G., and Brenner, R. M. (1971). The formation of basal bodies (centrioles) in the Rhesus monkey oviduct. *J. Cell Biol.* **50,** 10.
3. Anfinsen, C. B. (1967). The formation of the tertiary structure of proteins. *Harvey Lect.* **61,** 95.
4. Anfinsen, C. B. (1968). I. Self-assembly of macromolecular structures. Spontaneous formation of the three-dimensional structure of proteins. *Symp. Soc. Develop. Biol. Suppl. 2,* **27,** 1.
5. Barigozzi, C. (1963). Relationship between cytoplasm and chromosome in the transmission of melanotic tumors in *Drosophila. In* "Biological Organization at the Cellular and Supercellular Level" (R. J. C. Harris, ed.), p. 261. Academic Press, New York.
6. Beale, G. H. (1954). "The Genetics of *Paramecium aurelia,*" Cambridge Univ. Press, London.
7. Beale, G. H. (1957). The antigen system of *Paramecium aurelia. Int. Rev. Cytol.* **6,** 1.
8. Beale, G. H. (1969). A note on the inheritance of erythromycin-resistance in *Paramecium aurelia. Genet. Res.* **14,** 341.
9. Beisson, J., and Sonneborn, T. M. (1965). Cytoplasmic inheritance of the organization of the cell cortex in *Paramecium aurelia. Proc. Nat. Acad. Sci. U. S.* **53,** 275.
10. Bernardi, G., Faures, M., Piperno, G., and Slonimski, P. P. (1970). Mitochondrial DNA's from respiratory-sufficient and cytoplasmic respiratory-deficient mutant yeast. *J. Mol. Biol.* **48,** 23.
11. Casper, D. L. D. (1966). Design principles in organized biological structures. *In* "Principles of Biomolecular Organization" (G. E. W. Wolstenholme and M. O'Connor, eds.), *Ciba Found. Symp.* pp. 7–35, Little, Brown, Boston, Massachusetts.
12. Caspari, E., and Watson, G. S. (1959). On the evolutionary importance of cytoplasmic sterility in mosquitoes. *Evolution* **13,** 568.
13. Charles, H. P., and Knight, B. J., eds., (1970). "Organization and Control in Prokaryotic and Eukaryotic Cells," *Soc. Gen. Microsc. Symp.,* Vol. 20. Cambridge Univ. Press, London.

14. Clive, D., and Landman, O. E. (1970) Reversion of *Bacillus subtilis* protoplasts to the bacillary form induced by exogenous cell wall, bacteria and by growth in membrane filters. *J. Gen. Microbiol.* **61,** 233.
15. Cohn, M., and Horibata, J. (1959). Inhibition by glucose of the induced synthesis of β-galactoside enzyme system of *Escherchia coli. J. Bacteriol.* **78,** 601.
16. Cold Spring Harbor Symposia on Quantitative Biology (1963). "Synthesis and Structure of Macromolecules," Vol. 28, Cold Spring Harbor Laboratory of Quantitative Biology, Cold Spring Harbor, New York.
17. Crick, F. (1971). General model for the chromosomes of higher organisms. *Nature (London)* **234,** 25.
18. Dirksen, E. R. (1971). Centriole morphogenesis in developing ciliated epithelium of the mouse oviduct. *J. Cell Biol.* **51,** 286.
19. Edelman, M., Schiff, J. A., and Epstein, H. T. (1965). Studies of chloroplast development in *Euglena.* XII. Two types of satellite DNA. *J. Mol. Biol.* **11,** 769.
20. Flechtner, V., and Sager, R. (1972). Effects of streptomycin and other antibiotics on DNA replication in *Chlamydmonas. Fed. Proc. Fed. Amer. Soc. Exp. Biol.* (in press).
21. Fraenkel-Conrat, J., and William, R. C. (1955). Reconstitution of active tobacco mosaic virus from its inactive protein and nucleic acid components. *Proc. Nat. Acad. Sci. U. S.* **41,** 690.
22. Fukahara, J. (1969). Relative proportions of mitochondrial and nuclear DNA in yeast under various conditions of growth. *Eur. J. Biochem.* **11,** 135.
23. Fulton, C., and Dingle, A. D. (1971). Basal bodies, but not centrioles in *Naegleria. J. Cell Biol.* **51,** 826.
24. Gibson, I. (1969). Interacting genetic systems in *Paramecium. Advan. Morphog.* **8,** 159.
25. Gibson, I. (1970). The genetics of protozoan organelles. *Symp. Soc. Exp. Biol.* **24,** 379.
26. Goldring, E. S., Grossman, L. I., Krupnick, D., Cryer, D. R., and Marmur, J. (1970). The petite mutation in yeast: loss of mitochondrial deoxyribonucleic acid during induction of petites with ethidium bromide. *J. Mol. Biol.* **52,** 323.
27. Green, M. (1970). Oncogenic viruses. *Annu. Rev. Biochem.* **39,** 701.
28. Grossman, L., Goldring, E. S., and Marmur, J. (1969). Preferential synthesis of yeast mitochondrial DNA in the absence of protein synthesis. *J. Mol. Biol.* **46,** 367.
29. Grun, P., and Aubertin, M. (1965). Evolutionary pathways of cytoplasmic male sterility in *Solanum. Genetics* **51,** 399.
30. Guerineau, M., Grandchamp, C., Paoletti, C., and Slonimski, P. (1971). Characterization of a new class of circular DNA molecules in yeast. *Biochem. Biophys. Res. Commun.* **42,** 550.
31. Hiromu, A., Galeff, E., Davis, L., and Schneiderman, H. A. (1967). Virus-like particles in normal and tumerous tissues of *Drosophila. Science* **157,** 810.
32. Hooker, A. L., Smith, D. R., Lim, S. M., and Beckett, J. B. (1970). Reaction of corn seedlings with male-sterile cytoplasm to *Helminthosporium maydis. Plant Dis. Rep.* **54,** 708.
32a. Kellenberger, E. (1966). Control mechanisms in bacteriophage morphopoiesis. *In* "Principles of Biomolecular Organization" (G. E. W. Wolstenholme and M. O'Connor, eds.), *Ciba Found. Symp.*, pp. 192–226. Little, Brown, Boston, Massachusetts.
32b. Klotz, I. M., Langerman, N. R., and Darnall, D. W. (1970). Quaternary structure of proteins. *Annu. Rev. Biochem.* **39,** 25.
32c. Klug, A. (1967). Design of self-assembling systems of equal units. *In* "Formation and Fate of Cell Organelles" (K. B. Warren, ed.), *Symp. Int. Soc. Cell Biol.*, Vol. 6, pp. 1–18. Academic Press, New York.
33. Korn, E. D. (1968). Cell membranes: Structure and synthesis. *Annu. Rev. Biochem.* **38,** 263.

34. Küntzel, H., and Schafer, K. P. (1971). Mitochondrial RNA polymerase from *Neurospora crassa. Nature New Biol.* **231,** 265.
35. Laven, H. (1957). Verebung durch kerngene und das Problem der ausserkaryotischen Vererbung bei *Culex pipiens.* II. Ausserkaryotische Vererbung. *Z. Indukt. Abstamm. Vererbungsl.* **88,** 478.
36. Laven, H. (1967). Eradication of *Culex pipiens fatigans* through cytoplasmic incompatibility. *Nature (London)* **216,** 383.
37. Leskes, A., Siekevitz, P., and Palade, G. E. (1971). Differentiation of endoplasmic reticulum in hepatocytes. I. Glucose-6-phosphatase distribution *in situ. J. Cell Biol.* **49,** 264.
38. Leskes, A., Siekevitz, P., and Palade, G. E. (1971). Differentiation of endoplasmic reticulum in hapatocytes. II. Glucose-6-phosphatase in rough microsomes. *J. Cell Biol.* **49,** 288.
39. Levitan, M., and Williamson, D. L. (1965). Evidence for the cytoplasmic and possibly episomal nature of a chromosome breaker. *Genetics* **52,** 456.
39a. L'Héritier, P. (1958). The hereditary virus of *Drosophila. Advan. Virus Res.* **5,** 195.
40. L'Héritier, P. (1970). *Drosophila* viruses and their role as evolutionary factors. *In* "Evolutionary Biology" (T. Dobzhansky, ed.), Vol. 4, pp. 185–209, Appleton, New York.
41. Malogolowkin, C. (1958). Maternally inherited "sex-ratio" conditions in *Drosophila willistoni* and *Drosophila paulistroum. Genetics* **43,** 274.
41a. Merril, C. R., Grier, M. R., and Petricciani, J. C. (1971). Bacterial virus gene expression in human cells. *Nature (London)* **233,** 398.
42. Miller, R. J., and Koeppe, D. E. (1971). Southern corn leaf blight: Susceptible and resistant mitochondria. *Science* **173,** 67.
43. Minamori, S., Fujika, N., Ito, K., and Ikebuchi, M. (1970) Extrachromosomal element *delta* in *Drosophila melanogaster.* IV. Variation and persistence of *delta*-associating second chromosomes in a natural population. *Evolution* **24,** 735.
44. Nance, W. E. (1969). Anencephaly and spina bifida: a possible example of cytoplasmic inheritance in man. *Nature (London)* **224,** 373.
45. Nance, W. E., Davies, J., and Chazen, E. (1969). Symmelia: A possible example of cytoplasmic inheritance in man. *Clin. Res.* **17,** 316.
46. Nomura, M. (1970). Bacterial ribosome. *Bacteriol. Rev.* **34,** 228.
46a. Pardee, A. B., personal communication.
47. Pollard, D. R., and Fraser, F. C. (1968). Further studies on a cytoplasmically transmitted difference in response to the teratogen 6-aminonicotinamide. *Teratology* **1,** 335.
48. Pontecorvo, G. (1963). Microbiol Genetics: retrospect and prospect. *Proc. Roy. Soc. London Ser. B.* **158,** 1.
49. Preer, J. R., Jr. (1968). Genetics of the protozoa. *In* "Research in Protozology" (T. T. Chen, ed.), pp. 139–278, Pergamon, New York.
50. Preer, J. R., Jr. (1971). Extrachromosomal inheritance: hereditary symbionts, mitochondria, and chloroplasts. *Annu. Rev. Genet.* **5,** 626.
51. Rhoades, M. M. (1950). Gene induced mutation of a heritable cytoplasmic factor producing male sterility in maize. *Proc. Nat. Acad. Sci. U. S.* **36,** 634.
52. Sager, R. (1965). On non-chromosomal heredity in microorganisms. *In* "15th Symposium of the Society for General Microbiology" (M. R. Pollock and M. H. Richmond, eds.), pp. 324–342. Cambridge Univ. Press, London.
53. Sager, R. (1971). Evolution of preferential transmission mechanisms in cytoplasmic genetic systems. *Brookhaven Symp.* (Submitted for publication).

54. Schor, S., Siekevitz, P., and Palade, G. E. (1970). Cyclic changes in thylakoid membranes of synchronized *Chlamydomonas reinhardi*. *Proc. Nat. Acad. Sci. U. S.* **66,** 174.
55. Slonimski, P. P., Perrodin, G., and Croft, J. H. (1968). Ethidium bromide induce mutation of yeast mitochondria: complete transformation of cells into respiratory deficient non-chromosomal *"petites." Biochem. Biophys. Res. Commun.* **30,** 232.
56. Somers, J. M., and Bevan, E. A. (1969). The inheritance of the killer character in yeast. *Genet. Res.* **13,** 7.
57. Sonneborn, T. M. (1959). *Kappa* and related particles in *Paramecium*. *Advan. Virus Res.* **6,** 229.
58. Sonneborn, T. M. (1963). Does preformed cell structure play an essential role in cell heredity? *In* "The Nature of Biological Diversity" (J. M. Allen, ed.), pp. 165–221, McGraw-Hill, New York.
59. Sonneborn, T. M. (1970). Gene action in development. *Proc. Roy. Soc. London Ser. B* **176,** 347.
60. Steinert, M., and Van Assel, S. (1967). The loss of kinetoplastic DNA in two species of *Trypanosomatidae* treated with acriflavine. *J. Cell Biol.* **34,** 489.
61. Stubbe, W. (1964). The role of the plastome in evolution of the genus *Oenothera*. *Genetica* **35,** 28.
62. Verrusio, A. C., Pollard, D. R., and Fraser, F. C. (1968). A cytoplasmically transmitted, diet-dependent difference in response to the terotogenic effects of 6-aminonicotinamide. *Science* **160,** 206.
63. Wells, R., and Sager, R. (1971). Denaturation and renaturation kinetics of chloroplast DNA from *Chlamydomonas reinhardi*. *J. Mol. Biol.* **58,** 611.
64. Wood, W. B., Edgar, R. S., King, J., Liehausis, I., and Henninger, M. (1968). Bacteriophage assembly. *Fed. Proc. Fed. Amer. Soc. Exp. Biol.* **27,** 1160.

Appendix

CESIUM CHLORIDE DENSITY GRADIENT CENTRIFUGATION

A general method for equilibrium ultracentrifugation of macromolecules in density gradients was developed by Meselson *et al.* (6). This method has provided a powerful means of characterizing individual DNA's and of identifying and separating DNA's of different average base composition from the same preparation. Some results of this method are shown in Figs. 2.2, 2.3, 2.9, 3.8–3.11, and 5.9.

In this procedure, concentrated solutions of cesium chloride are used, under specified ultracentrifuge conditions, to establish linear density gradients of continuously increasing density along the direction of centrifugal force. Cesium chloride was chosen because at high molarities it has a range of bouyant densities which covers the range of most DNA's.

Macromolecules such as DNA or virus particles become concentrated at equilibrium in Gaussian bands centered at the position corresponding to their bouyant densities within the CsCl density gradient. Bandwidth is a function of molecular weight and can be used to estimate molecular weight for homogeneous material, but cannot readily be used for samples heterogeneous in either composition or molecular weight.

If the compositional heterogeneity is sufficient, more than a single band may form. Since the bouyant densities of samples across the gradient can be experimentally determined, the actual bouyant density of individual macromolecular species can be established, as can the density differences of two or more components present in the same gradient. The relationship between bouyant density and nucleotide composition is given by the empirical equation (*7*)

$$\rho = 0.098(\mathrm{GC}) + 1.660 \text{ gm/cm}^3$$

where (GC) = mole fraction guanine plus cytosine. Nucleotide composition can also be estimated from the *Tm* [°C temp. at midpoint of the rise in OD_{260} resulting from temperature-induced breakage of hydrogen bonds (hyperchromicity) (*3a*)].

$$Tm = 69.3 + 0.41(\mathrm{GC}) \quad \text{in solutions containing } 0.2\ M\ \mathrm{Na}^+$$

DENSITY TRANSFER METHOD TO EXAMINE THE MOLECULAR PATTERN OF DNA REPLICATION

The density transfer method was developed by Meselson and Stahl (*5*) as a means to determine whether the pattern of DNA replication in *E. coli* was semiconservative. The method has since been widely applied to the study of DNA replication and repair in many systems, including organelle DNA's (cf. Fig. 2.9).

DNA from cells grown for many doublings with an ^{15}N source differs from ^{14}N-DNA by 0.015 gm/cm^3 bouyant density in cesium chloride, a difference easily detected in cesium chloride gradients. This density difference can be exploited to distinguish an "old" or presynthesized strand of DNA from a "new" strand synthesized after transfer of cells to a fresh medium. When cells grown in either ^{14}N or ^{15}N until fully labeled are transferred to a medium containing the isotope not originally used, the next round of replication after transfer (providing that "old" precursor pools are negligible) gives rise to DNA molecules of density intermediate between ^{14}N and ^{15}N duplex densities.

Meselson and Stahl showed that the intermediate density was a consequence of hybridization between one "old" and one "new" complementary single-stranded DNA molecule. The cells were allowed to continue growth and undergo a second round of DNA replication. Two density bands with equal amounts of DNA were then found, one corresponding to the hybrid density and the other to the correct position for duplex DNA containing both "new" strands. This result was interpreted as evidence of semiconservative replication: that the initially present "old" duplex molecules had been strand-separated during replication, each serving as a template for one "new" strand. Thus, after one round of replication only hybrid molecules were present, whereas after two rounds half of the molecules were hybrid and half were entirely "new." This interpretation was confirmed by denaturing the hybrid molecules to demonstrate that indeed one was of the ^{15}N denatured density and the other of the ^{14}N denatured density. (Since the hybrid position is only 0.006–0.007 gm/cm^3 from either homogeneous duplex, it is difficult to use ^{15}N alone to study repair synthesis in which much smaller density shifts are involved. For such purposes, bromuracil, which greatly increases the bouyant density of DNA, has been used as a substitute for thymine.)

NEAREST-NEIGHBOR FREQUENCY ANALYSIS

This method was developed by Josse *et al.* (*2*) as a means to examine the relationship between a template DNA and the new DNA product synthesized under the influence of DNA polymerase. In the absence of a way to determine nucleotide sequences in DNA, this highly ingenious method enables one to examine the frequency of each of the sixteen possible pairs of dinucleotides (doublets) as an approach to comparing similarities and differences in sequence.

The *in vitro* synthesizing system requires all four nucleoside triphosphates as well as the enzyme and various cofactors. In the nearest-neighbor frequency analysis procedure, diagrammed in Fig. A.1, the synthesis is run four times, each time with a different one of the four nucleoside triphosphates labeled with ^{32}P in the phosphate closest to the nitrogen base. Unlabeled pyrophosphate is released in the reaction,

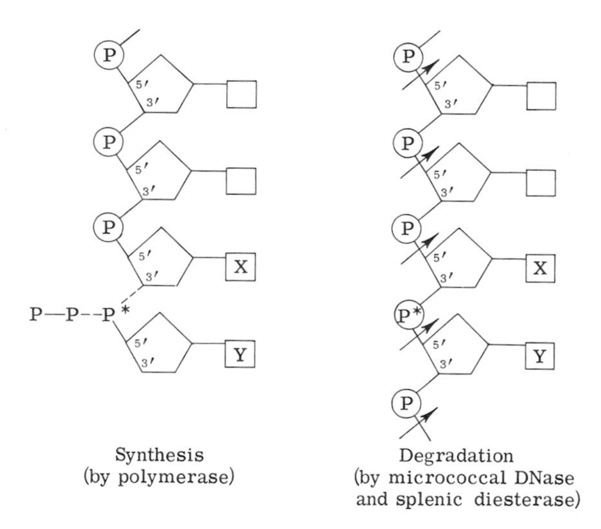

Fig. A.1. Synthesis of a ^{32}P-labeled DNA chain and its subsequent enzymatic degradation to 3′-deoxyribonucleotides. The arrows indicate the linkage cleaved by micrococcal deoxyribonuclease and calf spleen phosphodiesterase, yielding a digest composed entirely of 3′-deoxyribonucleotides (*3*).

while the labeled phosphate is incorporated into the growing molecule in the 5′ position. After synthesis, the product DNA is degraded with enzymes which cleave the phosphodiester bonds between the phosphate and the sugar to which it was initially attached. Consequently, after degradation the label is associated with the next nucleotide up the chain. In a particular experiment, if the labeled precursor were, for instance, cytosine triphosphate, the relative radioactivity associated with each of the four bases would indicate the relative frequencies of the four dinucleotides: CpC, GpC, ApC, and TpC. (By convention, GpC stands for G–3′–p–5′–C, etc.) Similarly, the other three runs would provide data for the remaining twelve dinucleotide pairs. The values from the four experiments are combined by multiplying each observed doublet frequency by the mole fraction of the total DNA contributed by the nucleotide used as label. In this way the sixteen frequencies add up to 1. An example of results obtained with this method is shown in Fig. 2.4.

Historically, this method was not only important in demonstrating the relationship between template and product, but also in providing an independent demonstration that the complementary DNA strands are of opposite polarity with respect to the 3′–5′ direction of the chain, as shown in Fig. A.2. The values for key doublets involved in this test, given in parenthesis, are from an experiment using DNA from *Mycobacterium phlei* as template (3)

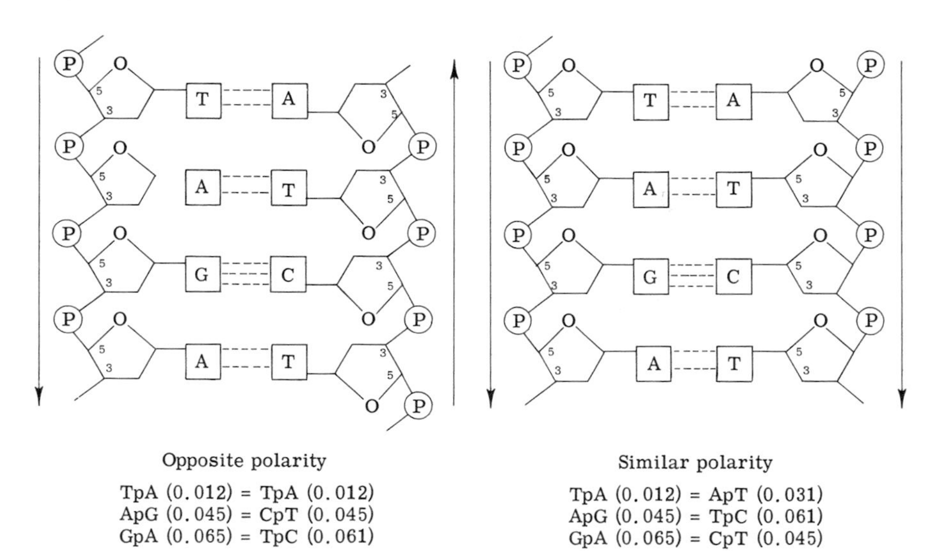

Fig. A.2. Contrast between a Watson and Crick DNA model with strands of opposite polarity and a model of strands of similar polarity. The predicted matching nearest-neighbor sequence frequencies are different. Values in parentheses are sequence frequencies from an experiment with *Mycobacterium phlei* DNA (3).

KINETICS OF RENATURATION OF DNA

Duplex DNA can be fully denatured by heat, as well as by other means, dissociating the two complementary strands. The strands can then be renatured by placing the DNA under conditions favorable to realignment of the molecules and reformation of the hydrogen bonds. The process of renaturation, or reannealing, first discovered by Marmur and Lane (*4*), has been developed by Wetmur and Davidson (*8*) and by Britten and Kohne (*1*) for investigation of the genomic size and complexity of native DNA's.

Under conditions of constant temperature and salt concentration, the rate of renaturation of a homogeneous DNA is dependent upon the concentration of the dissociated single-stranded molecules, and thus follows second order kinetics. By measuring the ratio of reassociated to dissociated strands during reannealing, one can determine the reannealing rate, expressed as the experimental second order rate constant k_2. A typical second order rate plot is shown in Fig. A.3.

The genomic size of the sample DNA, sheared to the same molecular weight as T4 DNA and with a similar base composition, can be determined directly or by comparison with T4 DNA. The renaturation rates of the two DNA's determined under the same conditions are inversely proportional to their genomic sizes.

Reannealing data can also be expressed in the form of a C_0t plot as shown in Fig. A.4. This method of plotting the data is particularly useful for comparing DNA's of different genomic complexities (*1*).

The percent of reassociated DNA can be determined experimentally by either of two procedures: (1) separation of single- from double-

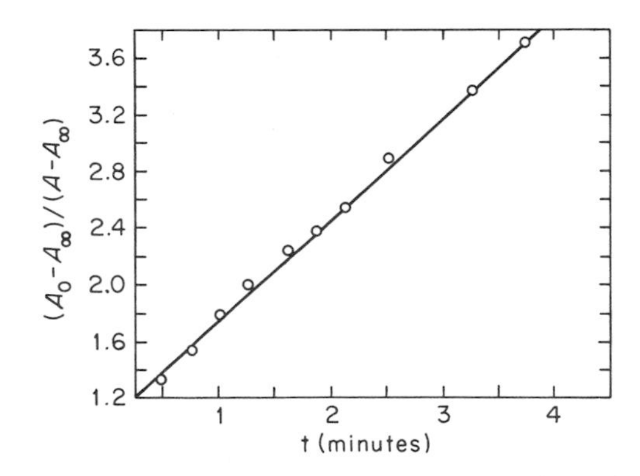

Fig. A.3. A second order rate plot. A_0 is the absorbance of the fully denatured sample; A_∞ is the absorbance of the double-stranded or native sample; A is the absorbance measured at each time point during the reaction.

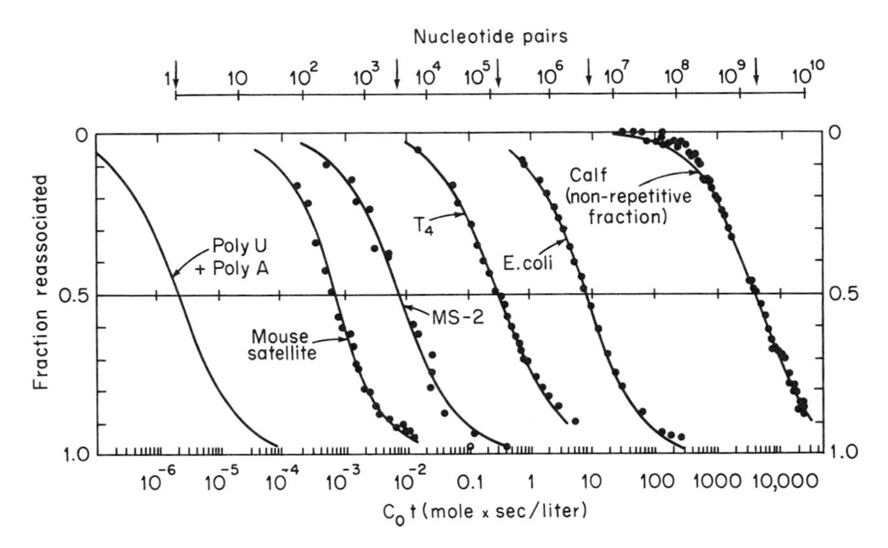

Fig. A.4. Reassociation of double-stranded nucleic acids from various sources. The genome size is indicated by the arrows near the upper nomographic scale. Over a factor of 10^9, this value is proportional to the C_0t required for half reaction. C_0t refers to the concentration of DNA in moles of nucleotides per liter times the number of seconds of reaction time.

stranded molecules on hydroxyapatite columns, or (2) measurement of optical density at 260 nm. The OD_{260} is approximately 35% higher for single-stranded than for double-stranded DNA of the same concentration (i.e., hyperchromicity). Thus, the process of reannealing can be followed by monitoring the decrease in OD_{260} in a spectrophotometer.

References

1. Britten, R. J., and Kohne, D. E. (1966). Nucleotide sequence repetition in DNA. *Carnegie Inst. Wash. Yearb.* **65,** 78–106.
2. Josse, J., Kaiser, A. D., and Kornberg, A. (1961). Enzymatic synthesis of DNA. *J. Biol. Chem.* **236,** 864.
3. Kornberg, A. (1961). "Enzymatic Synthesis of DNA." CIBA Lectures in Microbial Biochemistry. Wiley, New York.

3a. Marmur, J. and Doty, P. (1962). Determination of the base composition of deoxyribonucleic acid from its thermal denaturation temperature. *J. Mol. Biol.* **5,** 109.

4. Marmur, J., and Lane, D. (1960). Strand separation and specific recombination in deoxyribonucleic acids. *Biol. Studies Proc. Nat. Acad. Sci. U.S.* **46,** 453.
5. Meselson, M., and Stahl, F. W. (1958). The replication of DNA in *Escherichia coli. Proc. Nat. Acad. Sci. U. S.* **44,** 671.
6. Meselson, M., Stahl, F. W., and Vinograd, J. (1957). Equilibrium sedimentation of macromolecules in density gradients. *Proc. Nat. Acad. Sci. U. S.* **43,** 581.
7. Schildkraut, C. L., Marmur, J., and Doty, P. (1962). Determination of the base composition of deoxyribonucleic acid from its bouyant density in CsCl. *J. Mol. Biol.* **4,** 430.
8. Wetmur, J. G., and Davidson, N. (1968). Kinetics of renaturation of DNA. *J. Mol. Biol.* **31,** 349–370.

Glossary

actinomycin D a drug that inhibits transcription of DNA by DNA-dependent RNA polymerase

aerobic living or active only in the presence of oxygen

allele one of two or more alternate forms of a gene

allotopic altered sensitivity of an enzyme depending on its association with a membrane

α-amanitin a drug that inhibits transcription of DNA by binding to some DNA-dependent RNA polymerases

amino acyl acceptor a tRNA which accepts acylated (activated) amino acids in an initial step in protein synthesis

$$(\text{AA} \sim \text{AMP} + \text{tRNA} \longrightarrow \text{AA–tRNA} + \text{AMP})$$

aminoglycosides a class of antibiotics, some of which (e.g., streptomycin) bind to the 30 S subunit of bacterial ribosomes, leading to miscoding and inhibition of protein synthesis

amphipathic molecules with spatially separated hydrophilic and hydrophobic regions

anaerobic living or active in the absence of oxygen

aneuploidy having one or more chromosomes less than or in excess of the usual number

apoenzyme protein component of a complex enzyme which also contains a nonprotein prosthetic group

ascospore *see* **ascus**

ascus (asci) a spore sac in which meiosis occurs, giving rise to haploid nuclei around which haploid cells called ascospores are formed (yeast and fungi)

attachment point postulated point of attachment of chloroplast DNA to membrane in *Chlamydomonas* identified genetically by mapping procedure

auxotrophic cells dependent on exogenous growth factors not required by prototrophic strains

basal body *see* **blepharoplast**

bimolecular leaflet *see* **unit membrane**

biogenesis the development of organisms, cells, or subcellular structures from similar preexisting forms

biosynthetic pathway a series of enzyme-catalyzed reactions leading to synthesis of a particular metabolite

blepharoplast (basal body, kinetosome) the deeply staining granule or basal body at the base of flagella, homologous to the kinetosome found at the base of cilia, and structurally identical to the centriole

bouyant density density of a solute expressed in grams per cubic centimeter (gm/cm^3)

catabolite repression a mechanism of regulation at the DNA level by which more efficient energy sources such as glucose prevent the formation of enzymes needed to catabolize poorer energy sources (operates via cyclic AMP at least in some systems)

catonated forms (of DNA) circular DNA molecules which are topologically interlinked

cell sap (cytosol) cell cytoplasm excluding organelles but including ribosomes

centromere the region on a chromosome to which the spindle fibers attach during mitosis and meiosis

chloramphenicol antibiotic which inhibits protein synthesis in bacteria, chloroplasts, and mitochondria

chondriome collectively the mitochondria of the cell

chondriosome former term for a mitochondrion

cistron a segment of the genome (DNA or RNA) which codes for a specific gene product, either a protein or an RNA

clone a population of cells derived from a single cell

cristae folds of the inner mitochondrial membrane to which ATPase knobs are attached

cycloheximide an antibiotic that inhibits protein synthesis on 80 S cell-sap ribosomes of eukaryotes

cytochromes electron-transport enzymes containing heme or a related iron-binding prosthetic group

cytohet cell containing two cytoplasmic genomes which differ in one or more genes

cytosol *see* **cell sap**

dalton unit of molecular weight equal to the weight of a hydrogen atom

density gradient centrifugation *see* Appendix

density shift method *see* Appendix

derepression the release of repression of enzyme production

digitonin a detergent which dissolves or substantially alters biological membranes

dikaryon a binucleate cell, spore, or mycelium containing two nuclei, identical or different

endoplasmic reticulum (ER) a system of cytoplasmic membranes and vesicles (rough ER has associated ribosomes, whereas smooth ER does not)

erythromycin an antibiotic that blocks protein synthesis in bacteria, chloroplasts, and mitochondria

ethidium bromide an acridine dye which binds preferentially to circular DNA; used to separate circular from linear DNA in CsCl gradient; mutagen for yeast mitochondrial DNA

etiolated pale or colorless as a result of growth in the absence of light

etioplast plastid of etiolated leaves

dominance, phenotypic the expression of one allele, but not the other, when both are present

dominance, replicative the replication or transmission of one DNA molecule or linkage group, but not its homolog, when both are present (*see also* **suppressiveness**)

episome a dispensable DNA (and perhaps RNA) element which is additional to the wild type genome and may exist either free or integrated into host DNA

eukaryotic cells or organisms having a visibly evident nucleus, i.e., nucleoprotein chromosomes and nucleolus surrounded by nuclear envelope

f-met-tRNA (*N*-formylmethionyl tRNA) the species of methionyl tRNA used as initiating amino acylated tRNA in protein synthesis on bacterial, mitochondrial, and chloroplast ribosomes, but not on 80 S ribosomes of the cell sap

facultative anaerobe an organism that can live in either aerobic or anaerobic environment

facultative phototroph organism that can grow either photosynthetically or heterotrophically

Feulgen reaction staining of DNA, based on the ability of its partial hydrolysis products (aldehydes) to restore a magenta color to basic fuchsin which has been made colorless by reaction with sulfurous acid

fluctuation analysis developed by Luria and Delbruck, distinguishes between the adaptive and mutational origin of a new trait by a statistical test of the variance in the number of mutant cells arising in samples from many independent cultures

freeze-etch method preparation for electron microscopy by rapid freezing and sectioning to induce fracture formation; the exposed fracture faces are immediately layered with a thin platimum–carbon film, and the resulting replica is observed in the electron microscope

frets unstacked lamellar discs which connect separate grana (also called stroma lamellae)

gel electrophoresis a method of separating a mixture of molecular species according to charge, size, and shape; depends on migration through a gel under the influence of an electric field

gene unit of inheritance defined biochemically as a specific sequence of nucleotides that codes for a single gene product either an RNA (e.g., tRNA, ribosomal RNA) or a polypeptide

gene conversion process leading to aberrant allelic ratios, seen in heterozygous diploids at meiosis or mitosis

gamete a mature reproductive cell, normally haploid, capable of fusing with a gamete of opposite sex to form a diploid zygote

genome the full complement of genes of an organism

genotype the genetic constitution of an organism

grana structures within the chloroplast consisting of stacks of flattened lamellar discs

haploid cells or individuals in which each chromosome (or linkage group) is present in one copy

heterogamous different size gametes of the two sexes or mating types (opposite of isogamous)

heterocyton (heterocytosome) a cell or mycelium containing a mixture of different cytoplasmic genomes with a uniform nuclear genome (*see also* **cytohet**)

heterokaryon a cell or mycelium containing genetically different nuclei in a common cytoplasm (opposite of homokaryon)

heterothallic a mating system in which sex or mating type of the gametes is determined genetically

heterotrophic organisms incapable of manufacturing some organic molecules essential for their growth

Hill reaction release of molecular O_2 from H_2O by illuminated chloroplasts in the presence of a suitable electron acceptor, first described by R. Hill.

holochrome protein a protein to which protochlorophyllide is bound during photoreduction to chlorophyllide

homothallic a mating system in which one haploid cell gives rise to gametes which mate, form zygotes, and undergo meiosis

hybridization, molecular formation of a double-stranded structure, either DNA–DNA or DNA–RNA, by hydrogen-bonding of complementary single-stranded molecules

hydrophilic molecules or parts of molecules that readily associate with water; usually containing polar groups that form hydrogen bonds in water

hydrophobic molecules or parts of molecules that do not readily associate with water; usually nonpolar, poorly soluble or insoluble in water

hyperchromic shift of DNA an increase in the molecular absorption at 260 nm by breakage of hydrogen bonds to give single-stranded structures; a further increase results from hydrolysis of the single-stranded molecules to individual nucleotides

hyphae the individual filaments of filamentous fungi; a mass of hyphae is termed a mycelium

intralamellar space *see* **lumen**

isogamous equal size gametes of the two sexes or mating types

isogenic cells or organisms with the same genotype

kinetoplast specialized giant mitochondrion containing DNA; unique to parasitic flagellates

kinetosome *see* **blepharoplast**

Krebs cycle (citric acid cycle, tricarboxylic acid cycle) most common pathway for oxidative metabolism of pyruvate, the usual end product of glucose fermentation

lamellar membrane a specialized membrane of chloroplasts containing chlorophyll arranged in closed discs or thylakoids

leucoplast a plastid containing no chlorophyll

lumen the intralamellar space contained within the lamellar discs of chloroplasts

Mendelian gene a gene showing segregation pattern meiosis based on chromosomal location, i.e., obeying Mendel's Laws

meristematic region a region of rapidly dividing cells in plants

microsomal fraction rough-surfaced membranes of endoplasmic reticulum with attached ribosomes

modification, host-induced a modification in phage DNA (e.g., by methylation) that protects the modified form from a nuclease (i.e., restriction enzyme) which attacks the unmodified DNA (*see also* **restriction, host-induced**)

monosome a single ribosome; also a chromosome lacking a homologous partner

morphogenesis the formation of biological structures and differentiation of cells and tissues

mutagenesis the induction of mutations by physical or chemical agents

mycelia *see* **hyphae**

nearest-neighbor frequency analysis *see* Appendix

negative interference higher frequency of recombination between closely linked mutational sites than expected from recombination frequencies involving outside markers

non-Mendelian gene unit of inheritance that does not follow Mendel's Laws of segregation and reassortment

oligomycin an antibiotic that uncouples oxidative phosphorylation from electron transport

organelle a subcellular structure with a specific function, i.e., organ within a cell

oxidative phosphorylation production of ATP coupled to electron flow from substrates to O_2 (e.g., Krebs cycle intermediates)

Pasteur effect a control mechanism by which oxidative metabolism inhibits the fermentation of glucose

periclinal chimeras chimeras (usually in plants) composed of tissues of two or more genotypes with the genetically different tissues arranged in concentric layers

petite mutation a class of mutations in yeast leading to loss of respiratory ability

phenotype the visible traits of an organism resulting from the interaction of the genotype and the environment

photoautotroph an organism able to grow with light as a source of energy

photophosphorylation production of ATP from ADP coupled with photosynthetic electron transport

photoreactivation the reversal of lethal or mutagenic effects of ultraviolet irradiation by exposure of cells to visible wavelengths of light; mechanism involves dissociation by longer wavelengths of pyrimidine dimers formed during ultraviolet irradiation

photosystem I a photochemical reaction center in photosynthesis which transfers electrons from photosystem II to NADP

photosystem II a photochemical reaction center in photosynthesis which splits water producing oxygen and weak reductant; coupled to photosystem I

phototactic movement of an organism in response to stimulation by light

phototrophic turning response of an organism to stimulation by light

phytochrome system pigmented protein which acts as a hormone in higher plants, controlling germination and flowering in response to day length

plasmon the total cytoplasmic genome

plastom the total genome of a plastid

polysome aggregates of ribosomes held together by messenger RNA

prokaryotic cells or organisms (such as bacteria) these *lack* a nuclear envelope, nucleoprotein chromosomes, and nucleolus

prolamellar body a paracrystalline lattice seen mainly in proplastids of etiolated plants (see Plate XV)

protoperithecium a perithecial initial which develops into a perithecium after fertilization has occurred

pyrenoid a special organelle of most green algae that functions in the production of starch or other carbohydrate storage materials

reannealing kinetics *see* Appendix

recombination the appearance of new combinations of genes in progeny of parents differing in genetic constitution

recombination, linked an exchange of genetic material between two homologous chromosomes

recombination, unlinked new combinations of genes resulting from reassortment of parental genomes

replication, semiconservative the usual mode of DNA replication in which each newly replicated doubled-stranded molecule consists of one parental strand and one complementary newly synthesized strand

restriction, host-induced destruction of foreign DNA by a specific nuclease which does not attack host (modified) DNA (*see also* **modification, host-induced**)

rifampicin an antibiotic which inhibits initiation of transcription by binding to the DNA-dependent RNA polymerase

sedimentation coefficient (Svedberg, S) a quantitative measure of the rate of sedimentation of a given substance in a centrifugal field

somatic segregation (mitotic segregation) the origin of new genotypes at mitotic cell division by redistribution of parental genome

stroma the proteinaceous matrix of the chloroplast which bathes the grana and stroma lamellae

stroma lamellae *see* **frets**

suppressiveness a property of some mitochondrial *petite* mutations of yeast, defined as the percent *petite* colonies formed by zygotes from crosses of *petite* and wild type; considered by some as replicative dominance

tetrad the four chromatids of the paired homologous chromosomes during the first meiotic division; or the four progeny cells of a single meiotic event

thylakoids *see* **lamellar discs**

transcription (of DNA) a process by which the base sequence of DNA is used to produce the complementary base sequence in a single-stranded RNA

translation (of genetic message) a process by which the base sequence of messenger RNA directs the sequence of amino acids assembled into polypeptides

unit membrane a model proposed to account for the structure of biological membranes; the model consists of two bimolecular leaflets of phospholipids with nonpolar components facing inward and polar components facing outward toward the proteins which line both interfaces

zoospores progeny cells produced from the zygote following germination in *Chlamydomonas*

zygote the diploid product of fusion of two haploid cells (gametes) in sexual reproduction

Subject Index

D

R

Z